# Methods in Enzymology

Volume 64
ENZYME KINETICS AND MECHANISM
Part B
Isotopic Probes and Complex Enzyme Systems

# METHODS IN ENZYMOLOGY

EDITORS-IN-CHIEF

Sidney P. Colowick     Nathan O. Kaplan

# Methods in Enzymology

## Volume 64

# Enzyme Kinetics and Mechanism

### Part B
### Isotopic Probes and Complex Enzyme Systems

EDITED BY

**Daniel L. Purich**
DEPARTMENT OF CHEMISTRY
UNIVERSITY OF CALIFORNIA
SANTA BARBARA, CALIFORNIA

1980

## ACADEMIC PRESS
*A Subsidiary of Harcourt Brace Jovanovich, Publishers*

New York    London    Toronto    Sydney    San Francisco

COPYRIGHT © 1980, BY ACADEMIC PRESS, INC.
ALL RIGHTS RESERVED.
NO PART OF THIS PUBLICATION MAY BE REPRODUCED OR
TRANSMITTED IN ANY FORM OR BY ANY MEANS, ELECTRONIC
OR MECHANICAL, INCLUDING PHOTOCOPY, RECORDING, OR ANY
INFORMATION STORAGE AND RETRIEVAL SYSTEM, WITHOUT
PERMISSION IN WRITING FROM THE PUBLISHER.

ACADEMIC PRESS, INC.
111 Fifth Avenue, New York, New York 10003

*United Kingdom Edition published by*
ACADEMIC PRESS, INC. (LONDON) LTD.
24/28 Oval Road, London NW1 7DX

Library of Congress Cataloging in Publication Data
Main entry under title:

Enzyme kinetics and mechanism.

  (Methods in enzymology; v. 64, pt. B)
  Includes bibliographical references and index.
  1. Enzymes.  2. Chemical reaction, Rate of.
I. Purich, Daniel L. II. Series. [DNLM: 1. Enzymes
--Metabolism. 2. Kinetics. W1 ME9615K v. 63 pt. A /
QU135.3 E61]
QP601.M49 vol. 63 pt. A-B  574.1'925'08s  [574.1'925]
ISBN 0-12-181964-7 v. 64                  79-18746

PRINTED IN THE UNITED STATES OF AMERICA

80 81 82    9 8 7 6 5 4 3 2 1

# Table of Contents

CONTRIBUTORS TO VOLUME 64 . . . . . . . . . . . . . . . . vii

PREFACE . . . . . . . . . . . . . . . . . . . . . . . . . . ix

VOLUMES IN SERIES . . . . . . . . . . . . . . . . . . . . xi

## Section I. Isotope Probes of Mechanism

1. Isotope Exchange Methods for Elucidating Enzymic Catalysis — DANIEL L. PURICH AND R. DONALD ALLISON — 3

2. The Isotope Trapping Method: Desorption Rates of Productive E·S Complexes — IRWIN A. ROSE — 47

3. Oxygen-18 Probes of Enzymic Reactions of Phosphate Compounds — DAVID D. HACKNEY, KERSTIN E. STEMPEL, AND PAUL D. BOYER — 60

4. Determination of Heavy-Atom Isotope Effects on Enzyme-Catalyzed Reactions — MARION H. O'LEARY — 83

5. Measurement of Isotope Effects by the Equilibrium Perturbation Technique — W. WALLACE CLELAND — 104

6. Use of Secondary $\alpha$-Hydrogen Isotope Effects in Reactions of Pyrimidine Nucleoside/Nucleotide Metabolism: Thymidylate Synthetase — THOMAS W. BRUICE, CHARLES GARRETT, YUSUKE WATAYA, AND DANIEL V. SANTI — 125

## Section II. Complex Enzyme Systems

7. Cooperativity in Enzyme Function: Equilibrium and Kinetic Aspects — KENNETH E. NEET — 139

8. Hysteretic Enzymes — KENNETH E. NEET AND G. ROBERT AINSLIE, JR. — 192

9. The Kinetics of Immobilized Enzyme Systems — KEITH J. LAIDLER AND PETER S. BUNTING — 227

10. Subsite Mapping of Enzymes: Application to Polysaccharide Depolymerases — JIMMY D. ALLEN — 248

11. The Kinetics and Processivity of Nucleic Acid Polymerases — WILLIAM R. MCCLURE AND YVONNE CHOW — 277

12. Covalently Interconvertible Enzyme Cascade Systems — P. B. CHOCK AND E. R. STADTMAN — 297

13. The Use of Alternative Substrates to Study Enzyme-Catalyzed Chemical Modification — DONALD J. GRAVES AND TODD M. MARTENSEN  325

14. Enzyme Kinetics of Lipolysis — R. VERGER  340

AUTHOR INDEX . . . . . . . . . . . . . . . . . . . . . . 393

SUBJECT INDEX . . . . . . . . . . . . . . . . . . . . . 404

# Contributors to Volume 64

Article numbers are in parentheses following the names of contributors.
Affiliations listed are current.

G. ROBERT AINSLIE, JR. (8), *Department of Pediatrics, Cleveland Metropolitan General Hospital, Cleveland, Ohio 44109*

JIMMY D. ALLEN (10), *International Diagnostic Technology, Santa Clara, California 95050*

R. DONALD ALLISON (1), *Department of Chemistry, University of California, Santa Barbara, California 93106*

PAUL D. BOYER (3), *Department of Chemistry and Molecular Biology Institute, University of California, Los Angeles, California 90024*

THOMAS W. BRUICE (6), *Department of Biochemistry and Biophysics, University of California, San Francisco, California 94143*

PETER S. BUNTING (9), *Department of Biochemistry, Sunnybrook Medical Center, Toronto, Ontario M4N 3M5, Canada*

P. B. CHOCK (12), *Laboratory of Biochemistry, National Heart, Lung, and Blood Institute, National Institutes of Health, Bethesda, Maryland 20205*

YVONNE CHOW (11), *Department of Biochemistry and Molecular Biology, Harvard University, Cambridge, Massachusetts 02138*

W. WALLACE CLELAND (5), *Department of Biochemistry, University of Wisconsin, Madison, Wisconsin 53706*

CHARLES GARRETT (6), *Department of Biochemistry and Biophysics, University of California, San Francisco, California 94143*

DONALD J. GRAVES (13), *Department of Biochemistry and Biophysics, Iowa State University, Ames, Iowa 50011*

DAVID D. HACKNEY (3), *Department of Biological Sciences, Carnegie-Mellon University, Pittsburgh, Pennsylvania 15213*

KEITH J. LAIDLER (9), *Department of Chemistry, University of Ottawa, Ottawa, Ontario K1N 9B4, Canada*

WILLIAM R. MCCLURE (11), *Department of Biochemistry and Molecular Biology, Harvard University, Cambridge, Massachusetts 02138*

TODD M. MARTENSEN (13), *Laboratory of Cell Biology, National Institutes of Health, Bethesda, Maryland 20014*

KENNETH E. NEET (7, 8), *Department of Biochemistry, Case Western Reserve University, Cleveland, Ohio 44106*

MARION H. O'LEARY (4), *Department of Chemistry, University of Wisconsin, Madison, Wisconsin 53706*

DANIEL L. PURICH (1), *Department of Chemistry, University of California, Santa Barbara, California 93106*

IRWIN A. ROSE (2), *Institute for Cancer Research, Philadelphia, Pennsylvania 19111*

DANIEL V. SANTI (6), *Department of Biochemistry and Biophysics, and Department of Pharmaceutical Chemistry, University of California, San Francisco, California 94143*

E. R. STADTMAN (12), *Laboratory of Biochemistry, National Heart, Lung, and Blood Institute, National Institutes of Health, Bethesda, Maryland 20205*

KERSTIN E. STEMPEL (3), *Department of Chemistry and Molecular Biology Institute, University of California, Los Angeles, California 90024*

R. VERGER (14), *Centre de Biochimie et de Biologie Moléculaire du Centre National de la Recherche Scientifique, Marseille Cedex 2, France*

YUSUKE WATAYA (6), *Faculty of Pharmaceutical Sciences, Okayama University, Tsushima, Okayama 700, Japan*

# Preface

In the early years of chemistry the identification of new reactions preceded serious consideration of reaction kinetics, and it was not until Berthelot derived the bimolecular rate equation in 1861 that chemical kinetics offered any real value to the practicing chemist. Fortunately, biochemistry, which had its roots in the late nineteenth century, experienced the benefit of developments in kinetic theory. In fact, kinetic arguments have played a major role in defining the metabolic pathways, the mechanistic action of enzymes, and even the processing of genetic material. Nevertheless, it is amusing to witness the disdain of many investigators toward mechanistic conclusions drawn from kinetic data. After all, kinetic arguments are frequently tediously detailed with algebra and calculus, and so many refuse to believe that such abstract constructs truly apply to real systems. For those of us who derive much fascination, excitement, and satisfaction from the combination of chemical and kinetic probes of enzyme mechanism and regulation, the statement that "kinetics never proves anything" is especially amusing. When one views the definition of the word "proof" as an operation designed to test the validity of a fact or truth, the preceding statement serves only to demonstrate that we have failed to communicate the power and scope of kinetic arguments. The purpose of this volume is to initiate those who are interested in an advanced treatment of enzyme kinetic theory and practice. Indeed, this area of biochemistry is rich in information and experimental diversity, and it is the only means to examine the most fundamental characteristic of enzymes—catalytic rate enhancement.

Part A (Volume 63) and Part B (Volume 64) are the first of a series of volumes to treat enzyme kinetics and mechanism, and the chapters presented have been written to provide practical as well as theoretical considerations. However, there has been no attempt on my part to impose a uniform format of symbols, rate constants, and notation. Certainly, uniformity may aid the novice, but I believe that it would also present a burden to those wishing to examine the literature. There, the diversity of notation is enormous, and with good reason, because the textural meaning of particular terms must be considered. In this respect, the practice of utilizing a variety of notations should encourage the student to develop some flexibility and thereby ease the entry into the chemical literature of enzyme dynamics and mechanism. Each of the contributors is an expert in the literature, and I have been especially pleased by the constant reference to key sources of experimental detail.

I wish to acknowledge with pleasure and gratitude the cooperation and ideas of these contributors, and I am indebted in particular to Professors

Fromm and Cleland for many suggestions during the initial stages of developing the scope of this presentation. My students, certainly R. Donald Allison, also deserve much praise for surveying the literature and convincing me that a balanced view of the field could be presented in the confines of this series. The staff of Academic Press has also provided great encouragement and guidance, and to them I am deeply indebted. Finally, I wish to acknowledge the wisdom and friendship offered to me by Sidney Colowick and Nathan Kaplan.

DANIEL L. PURICH

# METHODS IN ENZYMOLOGY

### EDITED BY

### Sidney P. Colowick and Nathan O. Kaplan

VANDERBILT UNIVERSITY
SCHOOL OF MEDICINE
NASHVILLE, TENNESSEE

DEPARTMENT OF CHEMISTRY
UNIVERSITY OF CALIFORNIA
AT SAN DIEGO
LA JOLLA, CALIFORNIA

I. Preparation and Assay of Enzymes
II. Preparation and Assay of Enzymes
III. Preparation and Assay of Substrates
IV. Special Techniques for the Enzymologist
V. Preparation and Assay of Enzymes
VI. Preparation and Assay of Enzymes (*Continued*)
    Preparation and Assay of Substrates
    Special Techniques
VII. Cumulative Subject Index

# METHODS IN ENZYMOLOGY

## EDITORS-IN-CHIEF

### Sidney P. Colowick    Nathan O. Kaplan

VOLUME VIII. Complex Carbohydrates
*Edited by* ELIZABETH F. NEUFELD AND VICTOR GINSBURG

VOLUME IX. Carbohydrate Metabolism
*Edited by* WILLIS A. WOOD

VOLUME X. Oxidation and Phosphorylation
*Edited by* RONALD W. ESTABROOK AND MAYNARD E. PULLMAN

VOLUME XI. Enzyme Structure
*Edited by* C. H. W. HIRS

VOLUME XII. Nucleic Acids (Parts A and B)
*Edited by* LAWRENCE GROSSMAN AND KIVIE MOLDAVE

VOLUME XIII. Citric Acid Cycle
*Edited by* J. M. LOWENSTEIN

VOLUME XIV. Lipids
*Edited by* J. M. LOWENSTEIN

VOLUME XV. Steroids and Terpenoids
*Edited by* RAYMOND B. CLAYTON

VOLUME XVI. Fast Reactions
*Edited by* KENNETH KUSTIN

VOLUME XVII. Metabolism of Amino Acids and Amines (Parts A and B)
*Edited by* HERBERT TABOR AND CELIA WHITE TABOR

VOLUME XVIII. Vitamins and Coenzymes (Parts A, B, and C)
*Edited by* DONALD B. MCCORMICK AND LEMUEL D. WRIGHT

VOLUME XIX. Proteolytic Enzymes
*Edited by* GERTRUDE E. PERLMANN AND LASZLO LORAND

VOLUME XX. Nucleic Acids and Protein Synthesis (Part C)
*Edited by* KIVIE MOLDAVE AND LAWRENCE GROSSMAN

VOLUME XXI. Nucleic Acids (Part D)
*Edited by* LAWRENCE GROSSMAN AND KIVIE MOLDAVE

VOLUME XXII. Enzyme Purification and Related Techniques
*Edited by* WILLIAM B. JAKOBY

VOLUME XXIII. Photosynthesis (Part A)
*Edited by* ANTHONY SAN PIETRO

VOLUME XXIV. Photosynthesis and Nitrogen Fixation (Part B)
*Edited by* ANTHONY SAN PIETRO

VOLUME XXV. Enzyme Structure (Part B)
*Edited by* C. H. W. HIRS AND SERGE N. TIMASHEFF

VOLUME XXVI. Enzyme Structure (Part C)
*Edited by* C. H. W. HIRS AND SERGE N. TIMASHEFF

VOLUME XXVII. Enzyme Structure (Part D)
*Edited by* C. H. W. HIRS AND SERGE N. TIMASHEFF

VOLUME XXVIII. Complex Carbohydrates (Part B)
*Edited by* VICTOR GINSBURG

VOLUME XXIX. Nucleic Acids and Protein Synthesis (Part E)
*Edited by* LAWRENCE GROSSMAN AND KIVIE MOLDAVE

VOLUME XXX. Nucleic Acids and Protein Synthesis (Part F)
*Edited by* KIVIE MOLDAVE AND LAWRENCE GROSSMAN

VOLUME XXXI. Biomembranes (Part A)
*Edited by* SIDNEY FLEISCHER AND LESTER PACKER

VOLUME XXXII. Biomembranes (Part B)
*Edited by* SIDNEY FLEISCHER AND LESTER PACKER

VOLUME XXXIII. Cumulative Subject Index Volumes I-XXX
*Edited by* MARTHA G. DENNIS AND EDWARD A. DENNIS

VOLUME XXXIV. Affinity Techniques (Enzyme Purification: Part B)
*Edited by* WILLIAM B. JAKOBY AND MEIR WILCHEK

VOLUME XXXV. Lipids (Part B)
*Edited by* JOHN M. LOWENSTEIN

VOLUME XXXVI. Hormone Action (Part A: Steroid Hormones)
*Edited by* BERT W. O'MALLEY AND JOEL G. HARDMAN

VOLUME XXXVII. Hormone Action (Part B: Peptide Hormones)
*Edited by* BERT W. O'MALLEY AND JOEL G. HARDMAN

VOLUME XXXVIII. Hormone Action (Part C: Cyclic Nucleotides)
*Edited by* JOEL G. HARDMAN AND BERT W. O'MALLEY

VOLUME XXXIX. Hormone Action (Part D: Isolated Cells, Tissues, and Organ Systems)
*Edited by* JOEL G. HARDMAN AND BERT W. O'MALLEY

VOLUME XL. Hormone Action (Part E: Nuclear Structure and Function)
*Edited by* BERT W. O'MALLEY AND JOEL G. HARDMAN

VOLUME XLI. Carbohydrate Metabolism (Part B)
*Edited by* W. A. WOOD

VOLUME XLII. Carbohydrate Metabolism (Part C)
*Edited by* W. A. WOOD

VOLUME XLIII. Antibiotics
*Edited by* JOHN H. HASH

VOLUME XLIV. Immobilized Enzymes
*Edited by* KLAUS MOSBACH

VOLUME XLV. Proteolytic Enzymes (Part B)
*Edited by* LASZLO LORAND

VOLUME XLVI. Affinity Labeling
*Edited by* WILLIAM B. JAKOBY AND MEIR WILCHEK

VOLUME XLVII. Enzyme Structure (Part E)
*Edited by* C. H. W. HIRS AND SERGE N. TIMASHEFF

VOLUME XLVIII. Enzyme Structure (Part F)
*Edited by* C. H. W. HIRS AND SERGE N. TIMASHEFF

VOLUME XLIX. Enzyme Structure (Part G)
*Edited by* C. H. W. HIRS AND SERGE N. TIMASHEFF

VOLUME L. Complex Carbohydrates (Part C)
*Edited by* VICTOR GINSBURG

VOLUME LI. Purine and Pyrimidine Nucleotide Metabolism
*Edited by* PATRICIA A. HOFFEE AND MARY ELLEN JONES

VOLUME LII. Biomembranes (Part C: Biological Oxidations)
*Edited by* SIDNEY FLEISCHER AND LESTER PACKER

VOLUME LIII. Biomembranes (Part D: Biological Oxidations)
*Edited by* SIDNEY FLEISCHER AND LESTER PACKER

VOLUME LIV. Biomembranes (Part E: Biological Oxidations)
*Edited by* SIDNEY FLEISCHER AND LESTER PACKER

VOLUME LV. Biomembranes (Part F: Bioenergetics)
*Edited by* SIDNEY FLEISCHER AND LESTER PACKER

VOLUME LVI. Biomembranes (Part G: Bioenergetics)
*Edited by* SIDNEY FLEISCHER AND LESTER PACKER

VOLUME LVII. Bioluminescence and Chemiluminescence
*Edited by* MARLENE A. DELUCA

VOLUME LVIII. Cell Culture
*Edited by* WILLIAM B. JAKOBY AND IRA H. PASTAN

VOLUME LIX. Nucleic Acids and Protein Synthesis (Part G)
*Edited by* KIVIE MOLDAVE AND LAWRENCE GROSSMAN

VOLUME LX. Nucleic Acids and Protein Synthesis (Part H)
*Edited by* KIVIE MOLDAVE AND LAWRENCE GROSSMAN

VOLUME 61. Enzyme Structure (Part H)
*Edited by* C. H. W. HIRS AND SERGE N. TIMASHEFF

VOLUME 62. Vitamins and Coenzymes (Part D)
*Edited by* DONALD B. MCCORMICK AND LEMUEL D. WRIGHT

VOLUME 63. Enzyme Kinetics and Mechanism (Part A: Initial Rate and Inhibitor Methods)
*Edited by* DANIEL L. PURICH

VOLUME 64. Enzyme Kinetics and Mechanism (Part B: Isotopic Probes and Complex Enzyme Systems)
*Edited by* DANIEL L. PURICH

VOLUME 65. Nucleic Acids (Part I)
*Edited by* LAWRENCE GROSSMAN AND KIVIE MOLDAVE

VOLUME 66. Vitamins and Coenzymes (Part E)
*Edited by* DONALD B. MCCORMICK AND LEMEUL D. WRIGHT

VOLUME 67. Vitamins and Coenzymes (Part F)
*Edited by* DONALD B. MCCORMICK AND LEMUEL D. WRIGHT

VOLUME 68. Recombinant DNA
*Edited by* RAY WU

VOLUME 69. Photosynthesis and Nitrogen Fixation (Part C) (in preparation)
*Edited by* ANTHONY SAN PIETRO

# Section I

# Isotope Probes of Mechanism

## [1] Isotope Exchange Methods for Elucidating Enzymic Catalysis

By DANIEL L. PURICH and R. DONALD ALLISON

Isotope exchange methods constitute an entire domain of enzyme kinetics that until 20 years ago was largely concerned with net velocity measurements. The availability of isotopes and appropriate sensing devices (e.g., mass spectroscopy and scintillation counting) has encouraged the development of new approaches to understand enzyme catalysis and regulation. Among the early investigators concerned with exchange studies,[1-3] Boyer was probably the first to recognize their power and scope,[4,5] and his recent review outlines many of the major findings.[6] At one time, equilibrium exchange studies were employed almost exclusively as an adjunct to initial-rate studies of the ordering of substrate binding and product release.[7-10] Now, the approach has been considerably extended to involve rate measurement of loss, or exchange, of essentially all possible substrate atoms or functional groups of atoms, the determination of kinetic isotope effects, the definition of hitherto hidden stereochemical processes, and the examination of the interference of regulatory activators, inhibitors, and interconverting enzymes with the detailed chemical steps in the catalytic process. It is also interesting to note that oxidative phosphorylation and photophosphorylation, while outside the scope of this chapter, have also yielded important information through exchange studies.

To thoroughly examine each of the above aspects of exchange studies would require excessive space, and the reader is referred to several additional sources.[6,11,12] Generally, we will deal with representative applica-

[1] P. D. Boyer, R. C. Mills, and H. J. Fromm, *Arch. Biochem. Biophys.* **81,** 249 (1959).
[2] I. A. Rose, *Proc. Natl. Acad. Sci. U. S. A.* **44,** 10 (1958).
[3] L. F. Hass and W. L. Byrne, *J. Am. Chem. Soc.* **82,** 947 (1960).
[4] P. D. Boyer, *Arch. Biochem. Biophys.* **82,** 387 (1959).
[5] P. D. Boyer and E. Silverstein, *Acta Chem. Scand.* **17,** Suppl. 1, 195 (1963).
[6] P. D. Boyer, *Acc. Chem. Res.* **11,** 218 (1978).
[7] H. J. Fromm, E. Silverstein, and P. D. Boyer, *J. Biol. Chem.* **239,** 3645 (1964).
[8] E. Silverstein and P. D. Boyer, *J. Biol. Chem.* **239,** 3901 (1964).
[9] E. Silverstein and P. D. Boyer, *J. Biol. Chem.* **239,** 3908 (1964).
[10] J. F. Morrison and W. W. Cleland, *J. Biol. Chem.* **241,** 673 (1966).
[11] W. W. Cleland, *in* "The Enzymes" (P. Boyer, ed.), 3rd ed., Vol. 2, p. 1. Academic Press, New York, 1970.
[12] H. J. Fromm, "Initial Rate Enzyme Kinetics." Springer-Verlag, Berlin and New York, 1975.

tions of isotopic exchange at, or near, equilibrium. Here, one may find information on the binding and release of substrates, the rapidity of certain exchanges relative to each other and to the rate-limiting or rate-determining steps, the occurrence of abortive complexes, the likelihood of substrate synergism and possibly cooperativity, and the validity of the rapid-equilibrium assumption. We shall also examine exchange processes away from equilibrium, a condition of obvious importance when one studies irreversible or essentially irreversible processes. Again, valuable information frequently obtained in such cases includes the order of substrate binding and product release, the participation of covalent enzyme–substrate intermediates, and the nature of the irreversible step.

## Systems at Equilibrium

Even at equilibrium enzymes relentlessly shuttle substrates and products forth and back, and the flux in each direction is essentially constant and balanced over the time period used in measurements. Inasmuch as the rate of enzymic catalysis depends upon the concentration of enzyme–substrate(s) and enzyme–product(s) complexes, the enzyme's behavior at equilibrium is a complex composite of the net reaction rates in the forward and reverse directions observed away from equilibrium. It is also true that the equilibrium flux in each direction depends upon enzyme concentration, and one may adjust the enzyme level to suit the limitations on the exchange experiment.

As noted earlier, introduction of a labeled substrate or product may be utilized to trace the course of the reaction quantitatively. Depending upon the position of the isotopic atom(s) in the labeled substrate or product, various exchanges may be examined. The types of exchanges subject to measurement can be illustrated by considering a bisubstrate reaction in the following form:

$$A\text{—}X + B\text{—}Y \rightleftharpoons A\text{—}Y + B\text{—}X \qquad (1)$$

In all, one might expect that there are a number of exchange reactions to be examined. However, not all hypothetical exchanges actually involve the transfer of atoms (or functional groupings of atoms) between the various substrate–product partners; thus, only three exchanges occur. For example, in the hexokinase reaction it is possible to observe exchange reactions between glucose and glucose 6-phosphate, ADP and ATP, ATP and glucose 6-phosphate, but not glucose and ADP. Likewise, NAD-dependent dehydrogenases will never undergo exchange between the oxidized substrate and the oxidized coenzyme. Nonetheless, more complicated enzyme systems may have a number of exchange processes associated with the catalytic process, and the glutamine synthetase reaction provides an excellent example. Here, the following exchanges are possible.[13,14]

$$[^{14}C]ATP \leftrightarrow ADP$$
$$[^{14}C]Glutamate \leftrightarrow glutamine$$
$$[^{32}P]ATP \leftrightarrow P_i$$
$$[^{15}N]NH_3 \leftrightarrow glutamine$$
$$[^{18}O]ATP \leftrightarrow P_i$$
$$[^{18}O]Glutamate \leftrightarrow P_i$$
$$[^{18}O]Glutamine \leftrightarrow P_i$$
$$[^{18}O]ATP \leftrightarrow glutamate$$
$$[^{18}O]ATP \leftrightarrow glutamine$$
$$[^{18}O]Glutamate \leftrightarrow glutamine$$

The oxygen fluxes can provide additional probes of the dynamics of the reaction including stereochemical behavior of the enzyme. A detailed description of heavy-oxygen techniques is presented in this volume [3].

To maintain equilibrium, attention must be given to the mass-action ratio, which is the product of each reaction product concentration divided by the product of each substrate.

$$K_{eq} = \frac{[P][Q]\ldots}{[A][B]\ldots} \tag{2}$$

Typically, the system is treated in terms of substrate–product pairs (e.g., $\alpha = [P]/[A]$ and $\beta = [Q]/[B]$). One substrate–product pair is held constant with respect to the absolute concentrations of each component; the other substrate–product pair may be adjusted to a variety of absolute levels. Nonetheless, the experimental conditions must allow the product of $\alpha$ and $\beta$ to be $K_{eq}$. The nonvaried substrate–product pair may be prepared with buffer, salts, and modifiers (if desired), and enzyme is added to the complete reaction system to equilibrate the system. Since the experimenter has a good estimate of $K_{eq}$, the composition of the system will not change much from the preset conditions. A small aliquot of labeled reactant (substrate or product) is added, and the progress is followed by periodic sampling. Although the progress curve for the exchange is always first order (see below), the magnitude of the rate constant depends upon the absolute concentrations of all reactants, level of enzyme, and the kinetic mechanism.

The most common application of such exchange experiments is the determination of the kinetic mechanism with respect to binding preference or order. Let us examine the Ordered Bi Bi mechanism with ternary complexes.

---

[13] F. C. Wedler and P. D. Boyer, *J. Biol. Chem.* **247**, 984 (1972).
[14] B. O. Stokes and P. D. Boyer, *J. Biol. Chem.* **251**, 5558 (1976).

$$(E) \xrightleftharpoons[k_2]{k_1[A]} (EA) \xrightleftharpoons[k_4]{k_3[B]} (EAB) \xrightleftharpoons[k_6]{k_5} (EPQ) \xrightleftharpoons[k_8[P]]{k_7} (EQ) \xrightleftharpoons[k_{10}[Q]]{k_9} (E) \quad (3)$$

The A ↔ Q exchange velocity as a function of the absolute levels of the [B]/[P] pair is the diagnostic test for compulsory ordered mechanisms wherein one substrate, A, must precede the other's binding and one product, P, must precede the other product's release. As the [B]/[P] pair is raised, the central complexes become the favored enzyme species, and the uncomplexed enzyme form becomes scarce. Exchange between A and Q requires that there be a form of the enzyme which may adsorb the labeled species. Since the uncomplexed enzyme form is required for A* to be converted to Q*, and vice versa, the A ↔ Q exchange will increase as we go from low to moderate levels of the [B]/[P] pair, but the A ↔ Q exchange will be depressed at high [B]/[P] absolute levels. The A ↔ Q exchange will not be depressed at high [A]/[Q] absolute levels since increasing A and Q concentrations favor the binding of A* or Q*. Interestingly, the B ↔ P exchange requires that (EA) or (EQ) be present for combination with B* or P*, respectively. Thus, raising the [A]/[Q] or [B]/[P] pair will result in increased B ↔ P exchange in a hyperbolic fashion consistent with rate saturation kinetics. In summary, the depression of the A ↔ Q exchange at high [B]/[P] serves to identify the ordered mechanism. However, in the random (i.e., noncompulsory) mechanism [Eq. (4)] we have alternative routes for exchange processes, and no exchanges will be inhibited at high levels of any substrate–product pair. This is true since A* may bind to (E) and (EB), and

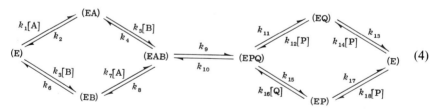 (4)

raising the [B]/[P] pair will not inhibit the combination of A* to the (EB) species. Because the mechanism is completely symmetrical, a similar statement is true for all exchanges.

These qualitative conclusions will be justified later in terms of exchange-rate expressions, but it is significant to note that exchange kinetic studies may segregate substrate order into ordered and random processes. Likewise, it is possible to draw other conclusions based upon the relative rapidity of exchanges. For example, if ternary complex interconversion limits the rate of a random mechanism, then the maximal exchange rate of each exchange will be identical. One such case is provided by rabbit muscle creatine kinase at pH 8.0.[10] For yeast hexokinase, however, the nucleotide exchange is approximately twice the rate of the glucose ↔ glucose-6-P exchange,[7,15] and one must consider release of glucose or glucose-6-P as rate contributing. In the case of ovine brain or *Escherichia coli* glutamine synthetase,[13,16,17]

$$R_{\max,\text{Glu}\leftrightarrow\text{Gln}} > R_{\max,\text{NH}_3\leftrightarrow\text{Gln}} > R_{\max,\text{P}_i\leftrightarrow\text{ATP}} \simeq R_{\max,\text{ADP}\leftrightarrow\text{ATP}} \tag{5}$$

suggesting that nucleotide release contributes more to the overall rate than either interconversion or the association and dissociation of other substrates and products. On the other hand, with pea seed glutamine synthetase the $NH_3$ ↔ glutamine exchange is slowest.[18]

In summary, the power of exchange studies is quite considerable, as illustrated in Table I. Aside from the large body of information about the kinetic mechanisms of many enzymes, the references included in Table I contain a wealth of data on separating and measuring particular metabolites.

*Rate Equations for Isotopic Exchange.* There are a number of useful approaches to deriving equilibrium exchange rate expressions. Boyer's method[4,5] is reasonably straightforward, and Fromm[12] has a good step-by-step outline to obtain such rate laws. Morales *et al.*[19] criticized this method and presented a more formal and generalized isotope distribution equation; however, Darvey[20] has shown that Boyer's approach is sound. Numerous alternative routes for obtaining rate equations have been developed,[4,5,21-24] and there are two basic types of equations, depending upon whether the system is at, or significantly displaced from, equilibrium.

The derivation of exchange-rate laws is based upon the following assumptions:

1. The total enzyme concentration is much less than the concentration of any of its substrates or modifiers.
2. The concentration of the isotopically labeled compound is much

---

[15] D. L. Purich and H. J. Fromm, *Biochem. Biophys. Res. Commun.* **47**, 916 (1972).
[16] F. C. Wedler, *J. Biol. Chem.* **249**, 5080 (1974).
[17] R. D. Allison, J. A. Todhunter, and D. L. Purich, *J. Biol. Chem.* **252**, 6046 (1977).
[18] F. C. Wedler, *J. Biol. Chem.* **249**, 7715 (1974).
[19] M. F. Morales, M. Horovitz, and J. Botts, *Arch. Biochem. Biophys.* **99**, 258 (1962).
[20] I. G. Darvey, *J. Theor. Biol.* **42**, 55 (1973).
[21] G. Yagil and H. D. Hoberman, *Biochemistry* **8**, 352 (1969).
[22] H. G. Britton, *Arch. Biochem. Biophys.* **117**, 167 (1966).
[23] W. W. Cleland, *Annu. Rev. Biochem.* **36**, 77 (1967).
[24] W. W. Cleland, *Biochemistry* **14**, 3220 (1975).

TABLE I
SUMMARY OF ENZYME MECHANISMS DEDUCED BY ISOTOPE EXCHANGE DATA

| Enzyme class | Specific reaction | Proposed kinetic mechanism | Diagnostic exchange(s) | References[a] |
|---|---|---|---|---|
| Dehydrogenases | Lactate dehydrogenase (bovine heart and rabbit muscle) | Ordered (pH 7.9), partially ordered (pH 9.7) | $NAD^+ \leftrightarrow NADH$, pyruvate $\leftrightarrow$ lactate, $NADH \leftrightarrow$ lactate | (1, 2) |
| | Alcohol dehydrogenase (yeast and equine liver) | Partially ordered | $NAD^+ \leftrightarrow NADH$, acetaldehyde $\leftrightarrow$ ethanol | (3, 4) |
| | Shikimate dehydrogenase (pea seedlings) | Ordered | $NADP^+ \leftrightarrow NADPH$, shikimate $\leftrightarrow$ dehydroshikimate | (5) |
| | Malate dehydrogenase (porcine heart) | Ordered (pH 8), partially ordered (pH 9) | $NAD^+ \leftrightarrow NADH$, oxaloacetate $\leftrightarrow$ malate | (6) |
| | Glutamate dehydrogenase (bovine liver) | Partially ordered (pH 8) | $NAD(P)^+ \leftrightarrow NAD(P)H$, glutamate $\leftrightarrow$ $\alpha$-ketoglutarate | (7) |
| | Glutamate dehydrogenase, alanine deamination (bovine liver) | Random (pH 8) | $NAD^+ \leftrightarrow NADH$, alanine $\leftrightarrow$ pyruvate | (8) |
| | $NADP^+$-isocitrate dehydrogenase (porcine heart) | Random | $NADP^+ \leftrightarrow NADPH$, $CO_2 \leftrightarrow$ isocitrate | (9) |
| | 20$\beta$-Hydroxysteroid dehydrogenase (*Streptomyces hydrogenans*) | Partially ordered | $NAD^+ \leftrightarrow NADH$ | (10) |
| | 17$\beta$-Estradiol dehydrogenase (human placenta) | Random | $NAD(P)^+ \leftrightarrow NAD(P)H$, steroid ketone $\leftrightarrow$ steroid alcohol | (11) |
| Kinases | Hexokinase (yeast) | Random | $ATP \leftrightarrow ADP$, glucose $\leftrightarrow$ glucose-6-P | (12, 13) |
| | Nucleoside diphosphate kinase (yeast) | Ping Pong | $ATP \leftrightarrow ADP$ and $UTP \leftrightarrow UDP$ | (14) |
| | Adenylate kinase (myokinase) (yeast and rabbit muscle) | Random | $ADP \leftrightarrow ATP$ and $ADP \leftrightarrow AMP$ | (15, 16) |
| | Arginine kinase (sea crayfish) | Random (minor Ping Pong pathway observed) | Arginine $\leftrightarrow$ phosphoarginine and $ATP \leftrightarrow ADP$ | (17) |
| | Galactokinase (*Escherichia coli*) | Random | Galactose $\leftrightarrow$ galactose-1-P and $MgATP \leftrightarrow MgADP$ | (18) |
| | Creatine kinase (rabbit muscle) | Random (Rapid Equilibrium) (pH 8) | Creatine $\leftrightarrow$ creatine phosphate and $ATP \leftrightarrow ADP$ | (19) |

| | Enzyme | Mechanism | Reaction | Ref. |
|---|---|---|---|---|
| | Acetate kinase (E. coli) | Random ("activated Ping Pong") | Acetate ⇌ acetyl phosphate and ATP ⇌ ADP | (20) |
| | Aspartate kinase, lysine sensitive (E. coli) | Random on, ordered off | ADP ⇌ ATP and ATP ⇌ β-aspartyl phosphate | (21) |
| | Fructokinase (bovine liver) | Partially ordered | Fructose ⇌ fructose-1-P and ATP ⇌ ADP | (22) |
| | Phosphofructokinase, nonallosteric (Lactobacillus plantarum) | Ordered | Fructose-6-P ⇌ fructose-1,6-diP and ATP ⇌ ADP | (23) |
| Synthetases | Succinyl-CoA synthetase (E. coli) | Partially ordered (with other minor pathways) | ATP ⇌ ADP and succinate ⇌ succinyl-CoA | (24, 25) |
| | Asparagine synthetase (E. coli) | Ping Pong (random addition of aspartate and ATP) | $PP_i$ ⇌ ATP | (26) |
| | Short-chain fatty acyl-CoA synthetase (pine seeds) | Ping Pong (ordered addition of ATP and acid) | $PP_i$ ⇌ ATP and $PP_i$ ⇌ dATP | (27) |
| | γ-Glutamylcysteine synthetase (toad and rat liver) | Ping Pong (toad), sequential (rat) | Cysteine ⇌ γ-glutamylcysteine and glutamate ⇌ γ-glutamylcysteine | (28) |
| | Arginyl tRNA synthetase (E. coli) | Random | $PP_i$ ⇌ ATP, AMP ⇌ ATP, arginine ⇌ arg-tRNA, tRNA ⇌ arg-tRNA | (29) |
| | Isoleucine tRNA synthetase (E. coli) | Ping Pong (random release) | $PP_i$ ⇌ ATP | (30, 31) |
| | Glutamine synthetase (pea, E. coli, ovine brain) | Random | $P_i$ ⇌ ATP, glutamate ⇌ glutamine, ATP ⇌ ADP | (32–36) |
| | Glutamine synthetase (ovine brain) | Partially ordered | ATP ⇌ ADP, ATP ⇌ $P_i$, glutamate ⇌ glutamine | (35) |
| Phosphorylases | Sucrose phosphorylase (Pseudomonas saccharophila) | Ping Pong | Glucose-1-P ⇌ $P_i$ | (37) |
| | Maltodextrin phosphorylase (E. coli) | Random | Glucose-1-P ⇌ $P_i$ and glucose-1-P ⇌ maltoheptaose | (38) |
| | Glycogen phosphorylase a (rabbit muscle) | Random (rapid equilibrium) | Glucose-1-P ⇌ $P_i$ and glucose-1-P ⇌ glycogen | (39, 40) |
| | Glycogen phosphorylase b (rabbit muscle) | Random (rapid equilibrium) | Glucose-1-P ⇌ $P_i$ and glucose-1-P ⇌ glycogen | (41) |
| Hydrolases | Glucose-6-phosphatase (rat liver) | Ordered off | Glucose ⇌ glucose-6-P | (42) |
| | Phosphoserine phosphatase (rat and chicken liver) | Ordered off | Serine ⇌ phosphoserine | (43, 44) |

(continued)

TABLE I (continued)

| Enzyme class | Specific reaction | Proposed kinetic mechanism | Diagnostic exchange(s) | References[a] |
|---|---|---|---|---|
| | Pepsin | Ordered off (pH 4.7), Random (rapid equilibrium, pH 3.3) | $N$-Acetyl-L-phenylalanine-L-tyrosine ethyl ester ⇌ L-tyrosine ethyl ester | (45) |
| Others | Aspartate transcarbamylase ($E.\ coli$) | Ordered | Aspartate ⇌ carbamyl aspartate carbamyl phosphate ⇌ $P_i$ | (46) |
| | Citrate cleavage enzyme (rat liver) | Mixed | Citrate ⇌ acetyl-CoA, citrate–oxaloacetate, $P_i$ ⇌ ATP | (47) |
| | Phosphoglucomutase (rabbit muscle) | Random (Mg vs. sugar–phosphate) | Glucose-1-P ⇌ glucose-6-P | (48) |
| | Galactose-1-P uridylyltransferase ($E.\ coli$ and human erythrocyte) | Ping Pong | UDP-glucose ⇌ glucose-1-P and galactose-1-P ⇌ UDP-galactose | (49, 50) |
| | Aldolase (yeast and rabbit muscle) | Ordered | Dihydroxyacetone-phosphate ⇌ water | (51) |
| | CoA Transferase (porcine heart) | Ping Pong | Succinate ⇌ succinyl-CoA and acetoacetate ⇌ acetoacetyl-CoA | (52) |

[a] Key to references:
1. E. Silverstein and P. D. Boyer, *J. Biol. Chem.* **239**, 3901 (1964).
2. G. Yagil and H. D. Hoberman, *Biochemistry* **8**, 352 (1969).
3. E. Silverstein and P. D. Boyer, *J. Biol. Chem.* **239**, 3908 (1964).
4. G. R. Ainslie, Jr., and W. W. Cleland, *J. Biol. Chem.* **247**, 946 (1972).
5. D. Balinsky, A. W. Dennis, and W. W. Cleland, *Biochemistry* **10**, 1947 (1971).
6. E. Silverstein and G. Sulebele, *Biochemistry* **8**, 2543 (1969).
7. E. Silverstein and G. Sulebele, *Biochemistry* **12**, 2164 (1973).
8. E. Silverstein and G. Sulebele, *Biochemistry* **13**, 1815 (1974).
9. M. L. Uhr, V. W. Thompson, and W. W. Cleland, *J. Biol. Chem.* **249**, 2920 (1974).
10. G. Betz and P. Taylor, *Arch. Biochem. Biophys.* **137**, 109 (1970).
11. G. Betz, *J. Biol. Chem.* **246**, 2063 (1971).
12. H. J. Fromm, E. Silverstein, and P. D. Boyer, *J. Biol. Chem.* **239**, 3645 (1964).
13. D. L. Purich and H. J. Fromm, *Biochem. Biophys. Res. Commun.* **47**, 916 (1972).
14. E. Garces and W. W. Cleland, *Biochemistry* **8**, 633 (1969).
15. S. Su and P. J. Russell, Jr., *J. Biol. Chem.* **243**, 3826 (1968).
16. D. G. Rhoads and J. M. Lowenstein, *J. Biol. Chem.* **243**, 3963 (1968).

17. E. Smith and J. F. Morrison, *J. Biol. Chem.* **244**, 4224 (1969).
18. J. S. Gulbinsky and W. W. Cleland, *Biochemistry* **7**, 566 (1968).
19. J. F. Morrison and W. W. Cleland, *J. Biol. Chem.* **241**, 673 (1966).
20. M. T. Skarstedt and E. Silverstein, *J. Biol. Chem.* **251**, 6775 (1976).
21. J.-F. Shaw and W. G. Smith, *J. Biol. Chem.* **252**, 5304 (1977).
22. F. M. Raushel and W. W. Cleland, *Biochemistry* **16**, 2176 (1977).
23. W. A. Simon and H. W. Hofer, *Eur. J. Biochem.* **88**, 175 (1978).
24. W. A. Bridger, W. A. Millen, and P. D. Boyer, *Biochemistry* **7**, 3608 (1968).
25. F. J. Moffet and W. A. Bridger, *Can. J. Biochem.* **51**, 44 (1973).
26. H. Cedar and J. H. Schartz, *J. Biol. Chem.* **244**, 4122 (1969).
27. O. A. Young and J. W. Anderson, *Biochem. J.* **137**, 435 (1974).
28. J. S. Davis, J. B. Balinsky, J. S. Harington, and J. B. Shepherd, *Biochem. J.* **133**, 667 (1973).
29. T. S. Papas and A. Peterkofsky, *Biochemistry* **11**, 4602 (1972).
30. F. X. Cole and P. R. Schimmel, *Biochemistry* **9**, 480 (1970).
31. F. X. Cole and P. R. Schimmel, *Biochemistry* **9**, 3143 (1970).
32. P. D. Boyer, R. C. Mills, and H. J. Fromm, *Arch. Biochem. Biophys.* **81**, 249 (1959).
33. D. J. Graves and P. D. Boyer, *Biochemistry* **1**, 739 (1962).
34. F. C. Wedler and P. D. Boyer, *J. Biol. Chem.* **247**, 984 (1972).
35. F. C. Wedler, *J. Biol. Chem.* **249**, 5080 (1974).
36. R. D. Allison, J. A. Todhunter, and D. L. Purich, *J. Biol. Chem.* **252**, 6046 (1977).
37. M. Doudoroff, H. A. Baker, and W. Z. Hassid, *J. Biol. Chem.* **168**, 725 (1947).
38. J. Chao, G F. Johnson, and D. J. Graves, *Biochemistry* **8**, 1459 (1969).
39. A. M. Gold, R. M. Johnson, and J. K. Tseng, *J. Biol. Chem.* **245**, 2564 (1970).
40. H. D. Engers, W. A. Bridger, and N. B. Madsen, *Can. J. Biochem.* **48**, 755 (1970).
41. H. D. Engers, W. A. Bridger, and N. B. Madsen, *J. Biol. Chem.* **244**, 5936 (1969).
42. L. F. Hass and W. L. Byrne, *J. Am. Chem. Soc.* **82**, 947 (1960).
43. F. C. Neuhaus and W. L. Byrne, *J. Biol. Chem.* **234**, 113 (1959).
44. L. F. Borkenhagen and E. P. Kennedy, *J. Biol. Chem.* **234**, 849 (1959).
45. L. M. Ginodman and N. G. Lutsenko, *Biokhimya* **37**, 81 (1972).
46. F. C. Wedler and F. J. Gasser, *Arch. Biochem. Biophys.* **163**, 57 (1974).
47. Y. J. K. Farrar and K. M. Plowman, *J. Biol. Chem.* **246**, 3783 (1971).
48. W. J. Ray, Jr., G. A. Roscelli, and D. S. Kirkpatrick, *J. Biol. Chem.* **241**, 2603 (1966).
49. L.-J. Wong and P. A. Frey, *Biochemistry* **13**, 3889 (1974).
50. H. B. Marcus, J. W. Wu, F. S. Boches, T. A. Tedesco, W. J. Mellman, and R. G. Kallen, *J. Biol. Chem.* **252**, 5363 (1977).
51. I. A. Rose, E. L. O'Connell, and A. H. Mehler, *J. Biol. Chem.* **240**, 1758 (1965).
52. L. B. Hersh and W. P. Jencks, *J. Biol. Chem.* **242**, 3468 (1967).

less than that of the unlabeled compound. Thus, there is no significant change in the concentration of the unlabeled species.
3. All unlabeled species are present at their equilibrium concentrations prior to the addition of the labeled species (the enzyme system being reversible).
4. The pathway for the exchange of the labeled species is the same as that of the corresponding unlabeled species and this exchange is via the reaction path being studied. If there is a second enzyme present that catalyzes the same exchange, the value for $R$, the rate of exchange, will be a superimposition of two or more exchanges.
5. Kinetic isotope effects, if any, are negligible.
6. The labeled enzyme intermediates are at steady state levels (the following results are not valid for the presteady state).[20]

For a detailed description about the derivation of isotope rate expressions, the reader is referred to Chapter [4] by Huang in this series, Vol. 63. To conserve space in the presentation of rate laws, we have adopted the format used by Fromm,[12] and those who are interested in rewriting the equations in terms of individual rate constants should consult the appendix to his text. In Table II, we list the exchange-rate expressions for many typical enzyme kinetic mechanisms. These equations are relatively long and complex by comparison to initial-rate expressions, and it is generally cumbersome, if not completely impractical, to obtain kinetic constants for each step from isotope exchange data. Instead, one deals with the equations to obtain qualitative patterns of rate behavior, and the most frequently used method deals with the effect of raising particular substrate–product pairs. By far the most reliable way to do this uses L'Hospital's rule to determine the limits.[17] Treatment of the ordered mechanism provides a useful example. Let $[Q]/[A]$ be $\alpha$ and $[P]/[B]$ be $\beta$ (or $K_{eq}/\alpha$). From Table II, the rate of the A ⇌ Q exchange becomes

$$R_{A \leftrightarrow Q} = V_1[A][B] \bigg/ \left\{ K_{ia}K_b + K_a[B] + \frac{K_{eq}K_{ia}K_bK_q[B]}{\alpha K_{iq}K_p} \left[ 1 + \frac{[A]}{K_{ia}} \left(1 + \frac{[B]}{K_{eb}}\right) + \frac{\alpha[A]}{K_{iq}} \right] \right\} \quad (6)$$

Likewise,

$$R_{B \leftrightarrow P} = V_1[B] \bigg/ K_b \left[ 1 + \frac{K_{ia}}{[A]} + \frac{[B]}{K_{eb}} + \frac{K_{ia}\alpha}{K_{iq}} \right] \quad (7)$$

Now, we may establish the limiting value of $R_{A \leftrightarrow Q}$ as the absolute concentration of the B-P pair becomes very large, i.e., $[B] \to \infty$ in Eq. (6). In this case, $R$ can be treated as a quotient of two functions: $f(B) = k[B]$ and $g(B) = k' + k''[B] + k'''[B]^2$; thus

$$R_{A \leftrightharpoons Q \atop \text{limiting B-P}} = \lim_{B \to \infty} \frac{f(B)}{g(B)} = \lim_{B \to \infty} \frac{f'(B)}{g'(B)} = \lim_{B \to \infty} \frac{k}{k'' + 2k'''[B]} = 0 \quad (8)$$

On the other hand, $R_{A \leftrightharpoons Q}$ versus the A-Q pair will have its limit given by

$$R_{A \leftrightharpoons Q \atop \text{maximum}} = V_1[B] \bigg/ \frac{K_{eq} K_{ia} K_b K_q[B]}{\alpha K_{iq} K_p} \left[ \frac{1}{K_{ia}} \left( 1 + \frac{[B]}{K_{eb}} \right) + \frac{\alpha}{K_{iq}} \right] \quad (9)$$

Thus, $R_{A \leftrightharpoons Q}$ falls to zero as the B-P pair is increased, but $R_{A \leftrightharpoons Q}$ achieves a finite plateau value as the A-Q pair is increased.

This approach is quite useful when a single substrate–product pair is increased; however, Wedler and Boyer[13] have introduced a protocol in which all substrates and products are increased simultaneously. Their qualitative conclusions about the exchange behavior under such conditions may be appreciated mathematically by the following treatment.[17] Consider again Eq. (6) for the $R_{A \leftrightharpoons Q}$ exchange in the Ordered Bi Bi mechanism. Let $[B] = \gamma[A]$, then,

$$R_{A \leftrightharpoons Q} = \gamma V_1[A]^2 \bigg/ \left\{ K_{ia} K_b + K_a \gamma[A] + \frac{K_{eq} K_{ia} K_b K_q \gamma[A]}{\alpha K_{iq} K_p} \left[ 1 + \frac{[A]}{K_{ia}} \left( 1 + \frac{\gamma[A]}{K_{eb}} \right) + \frac{\alpha[A]}{K_{iq}} \right] \right\} \quad (10)$$

Now, we may again express $R_{A \leftrightharpoons Q}$ as the quotient of two functions: $f(A) = k[A]^2$ and $g(A) = k' + k''[A] + k'''[A]^2 + k''''[A]^3$. Applying L'Hospital's rule once again, we find that $R_{A \leftrightharpoons Q}$ reaches a limiting value of zero as $[A]$, $[B]$, $[P]$, and $[Q]$ are increased indefinitely. Thus, mechanisms with noncompetitive interactions will lead to depression in exchanges, but competitive interactions (even the formation of abortive complexes) will be balanced as $[A]$, $[B]$, $[P]$, and $[Q]$ are increased. The only major limitation with the Wedler and Boyer protocol is that it cannot be applied to reactions involving a different number of substrates and products. For example, a Uni Bi reaction has an equilibrium constant of $[P][Q]/[A]$ and serial dilution will move the system away from equilibrium.

Examination of the equations presented in Table II indicates that there are a number of types of exchange-rate profiles. In terms of plots of $R$ vs. a substrate–product pair, we may observe hyperbolic (H), or sigmoidal (S) plots with no, partial, or complete depression of the exchange. These profiles are illustrated in Fig. 1, and it should be noted that variation of a single substrate–product pair should yield hyperbolic plots. The presence

TABLE II
Isotope Exchange-Rate Expressions for Selected Uni and Bi Substrate Systems[a]

1. Uni Uni

$$R_{A \rightleftharpoons P} = \frac{V_1[A]}{K_a[1 + [A]/K_{ia} + [P]/K_{ip}]}$$

2. Ordered Bi Uni

$$R_{A \rightleftharpoons P} = \frac{V_1[A][B]}{(K_{ia}K_b + K_a[B])[1 + [A]/K_{ia} + [P]/K_{ip}]}$$

$$R_{B \rightleftharpoons P} = \frac{V_1[B]}{K_b[1 + K_{ia}/[A] + K_{ia}[P]/K_{ip}[A]]}$$

3. Random Bi Uni (Rapid Equilibrium)

$$R_{all} = \frac{V_1}{1 + K_a/[A] + K_b/[B] + (K_{ia}K_b/[A][B])(1 + [P]/K_{ip})}$$

4. Random Bi Uni

$$R_{A \rightleftharpoons P} = \frac{[k_1k_3k_9[A][B] + k_7k_9[A](k_2 + k_3[B])[B]/K_{ib}]E_0}{[k_2(k_4 + k_8 + k_9) + k_3[B](k_8 + k_9)][1 + [A]/K_{ia} + [B]/K_{ib} + [A][B]/K_{ia}K_b]}$$

$$R_{B \rightleftharpoons P} = \frac{[k_5k_7k_9[A][B] + k_3k_9[B](k_6 + k_7[A])([A]/K_{ia})]E_0}{[k_6(k_4 + k_8 + k_9) + k_7[A](k_4 + k_9)][1 + [A]/K_{ia} + [B]/K_{ib} + [A][B]/K_{ia}K_b]}$$

5. Ordered Bi Bi

$$R_{A \rightleftharpoons Q} = \frac{V_1[A][B]}{[K_{ia}K_b + K_a[B] + K_{ia}K_bK_q[P]/K_{iq}K_p][1 + ([A]/K_{ia})(1 + [B]/K_{eb}) + [Q]/K_{iq}]}$$

$$R_{A \rightleftharpoons P} = \frac{V_1[A][B]}{[K_{ia}K_b + K_a[B]][1 + ([A]/K_{ia})(1 + [B]/K_{eb}) + [Q]/K_{iq}]}$$

$$R_{B \rightleftharpoons Q} = \frac{V_1[B]}{[K_b + K_bK_q[P]/K_{iq}K_p][1 + K_{ia}/[A] + [B]/K_{eb} + K_{ia}[Q]/K_{iq}[A]]}$$

$$R_{B \rightleftharpoons P} = \frac{V_1[B]}{K_b[1 + K_{ia}/[A] + [B]/K_{eb} + K_{ia}[Q]/K_{iq}[A]]}$$

[1]    ISOTOPE EXCHANGE METHODS    15

6. Theorell–Chance Bi Bi

$$R_{A \rightleftharpoons Q} = \frac{V_1[A][B]}{[K_a[B] + K_{ia}K_b + K_{iq}[P]/K_{eq}][1 + [A]/K_{ia} + [Q]/K_{iq}]}$$

$$R_{A \rightleftharpoons P} = \frac{V_1[A][B]}{[K_{ia}K_b + K_a[B]][1 + [A]/K_{ia} + [Q]/K_{iq}]}$$

$$R_{B \rightleftharpoons Q} = \frac{V_1[B]}{[K_b + K_{iq}[P]/K_{ia}K_{eq}][1 + K_{ia}/[A] + K_{ia}[Q]/K_{iq}[A]]}$$

$$R_{B \rightleftharpoons P} = \frac{V_1[B]}{K_b[1 + K_{ia}/[A] + K_{ia}[Q]/K_{iq}[A]]}$$

7. Random Bi Bi (Rapid Equilibrium)

$$R_{all} = \frac{V_1}{1 + K_a/[A] + K_b/[B] + K_{ia}K_b/[A][B][1 + [P]/K_{ip} + [Q]/K_{ic} + [P][Q]/K_{ip}K_q]}$$

8. Random Bi Bi

$$R_{A \rightleftharpoons Q} = \frac{[k_9k_{11} + k_{13}(k_{10}[P] + k_{11})](k_1k_3[A][B] + k_7([A][B]/K_{ib})(k_2 + k_3[B])E_0}{(Z)\{k_2[k_{10}[P](k_4 + k_8 + k_{13}) + k_{11}(k_4 + k_8 + k_9 + k_{13})] + k_3[B][k_{10}[P](k_8 + k_{13}) + k_{11}(k_8 + k_9 + k_{13}) + k_3[B]]\}}$$

$$R_{A \rightleftharpoons P} = \frac{[k_{13}k_{15} + k_9(k_{14}[Q] + k_{15})](k_1k_3[A][B] + k_7([A][B]/K_{ib})(k_2 + k_3[B]))E_0}{(Z)\{k_2[k_{14}[Q](k_4 + k_8 + k_9) + k_{15}(k_4 + k_8 + k_9 + k_{13})] + k_3[B][k_{14}[Q](k_8 + k_9) + k_{15}(k_8 + k_9 + k_{13})]\}}$$

$$R_{B \rightleftharpoons Q} = \frac{[k_9k_{11} + k_{13}(k_{10}[P] + k_{11})](k_5k_7[A][B] + k_3([A][B]/K_{ia})(k_6 + k_7[A]))E_0}{(Z)\{k_6[k_{10}[P](k_4 + k_8 + k_{13}) + k_{11}(k_4 + k_8 + k_9 + k_{13})] + k_7[A][k_{10}[P](k_4 + k_{13}) + k_{11}(k_4 + k_9 + k_{13})]\}}$$

$$R_{B \rightleftharpoons P} = \frac{[k_{13}k_{15} + k_9(k_{14}[Q] + k_{15})](k_5k_7[A][B] + k_3([A][B]/K_{ia})(k_6 + k_7[A]))E_0}{(Z)\{k_6[k_{14}[Q](k_4 + k_8 + k_9) + k_{15}(k_4 + k_8 + k_9 + k_{13})] + k_7[A][k_{14}[Q](k_4 + k_9) + k_{15}(k_4 + k_9 + k_{13})]\}}$$

where $Z = \{1 + [A]/K_{ia}(1 + [B]/K_b) + [B]/K_{ib} + [P]/K_{ip} + [Q]/K_{iq}\}$

(*continued*)

TABLE II (*continued*)

9. Random On–Ordered Off Bi Bi

$$R_{A \to Q} = \frac{k_9 k_{11}[k_1 k_3[A][B] + k_7([A][B]/K_{1b})(k_2 + k_3[B])] E_0}{\{k_2[k_{10}[P](k_4 + k_8) + k_{11}(k_4 + k_8 + k_9) + k_3[B][k_8 k_{10}[P] + k_{11}(k_8 + k_9)]\} (Y)}$$

$$R_{A \to P} = \frac{k_9[k_1 k_3[A][B] + k_7([A][B]/K_{1b})(k_2 + k_3[B])] E_0}{\{k_3[B](k_8 + k_9) + k_2(k_4 + k_8 + k_9)\} (Y)}$$

$$R_{B \to Q} = \frac{k_9 k_{11}[k_5 k_7[A][B] + k_3([A][B]/K_{ia})(k_6 + k_7[A])] E_0}{\{k_6[k_{10}[P](k_4 + k_8) + k_{11}(k_4 + k_8 + k_9) + k_7[A][k_4 k_{10}[P] + k_{11}(k_4 + k_9)]\} (Y)}$$

$$R_{B \to P} = \frac{k_9[k_5 k_7[A][B] + k_3([A][B]/K_{ia})(k_6 + k_7[A])] E_0}{\{k_7[A](k_4 + k_9) + k_6(k_4 + k_8 + k_9)\} (Y)}$$

where $Y = \{1 + ([A]/K_{ia})(1 + [B]/K_b) + [B]/K_{ib} + [Q]/K_{iq}\}$

10. Ping Pong Bi Bi

$$R_{A \to P} = \frac{V_1[A]}{K_a[1 + [A]/K_{ia} + K_{ip}[A]/K_{ia}[P] + [Q]/K_{iq}]}$$

$$R_{B \to Q} = \frac{V_1[B]}{K_b[1 + [B]/K_{ib} + [P]/K_{ip} + K_{iq}[B]/K_{ib}[Q]]}$$

$$R_{A \to Q} = \frac{V_1[A][B]}{[K_a[B] + (K_{ia} K_b[P]/K_{ip})[1 + [A]/K_{ia} + [Q]/K_{iq} + K_{ip}[A]/K_{ia}[P]]}$$

[a] The expressions are derived for systems at chemical equilibrium and in the absence of abortive complexes. All expressions, except for the Uni Uni and Rapid Equilibrium cases, were derived assuming only one central catalytic complex.

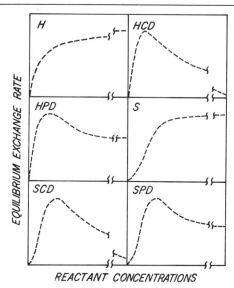

FIG. 1. Types of exchange-rate profiles for enzyme-catalyzed reactions. Hyperbolic (H), hyperbolic with complete depression (HCD), hyperbolic with partial depression (HPD), sigmoidal (S), sigmoidal with complete depression (SCD), sigmoidal with partial depression (SPD), and linear (L).

of sigmoidal plots may provide preliminary evidence for cooperativity; however, the Wedler and Boyer protocol,[13] in which all substrates and products are varied, should yield sigmoidal plots irrespective of the occurrence of cooperativity (of course, hyperbolic profiles in the Wedler–Boyer protocol may arise fortuitously, depending upon the magnitude of various kinetic parameters). In any case, the expected exchange-rate profiles for the systems defined in Table II are explicitly described in Table III. These do not take into account abortive complex effects, which are described later.

It may be valuable to note that the rapid-equilibrium ordered and rapid-equilibrium random mechanisms are virtually indistinguishable by isotope exchange data alone. Since the central complex interconversion rate is limiting in both cases, hyperbolic plots of $R$ vs. any substrate–product pair are to be expected. With the Wedler–Boyer method, $R_{max}$ will be $V_1V_2/(V_1 + V_2)$ where $V_1$ and $V_2$ are the forward and reverse $V_{max}$ values obtained in initial-rate experiments. Under favorable conditions, a distinction might be possible by use of slope replots of $(1/R)$ vs. the A-Q pair at various constant values of the B-P pair. This method would distinguish a rapid-equilibrium ordered mechanism by the intersection at the origin and the random at a minimal nonzero value of $K_a/V_1$. Of course,

## TABLE III
### EXCHANGE PROFILES FOR ENZYME KINETIC MECHANISMS[a]

| Mechanism | Exchange | Varied substrate–product pair(s) | | | | |
|---|---|---|---|---|---|---|
| | | A-P | B-P | A-Q | B-Q | A-B-P-Q[b] |
| 1. Uni Uni | A ↔ P | H | | | | |
| 2. Ordered Bi Uni | A ↔ B | H | HCD | | | |
| | B ↔ P | H | H | | | |
| 3. Random Bi Uni (Rapid Equilibrium) | Any | H | H | | | |
| 4. Random Bi Uni | A ↔ P | H | H | | | |
| | B ↔ P | H | H | | | |
| 5. Ordered Bi Bi | A ↔ P | H | HCD | H | HCD | SCD |
| | B ↔ P | H | H | H | H | S |
| | A ↔ Q | HCD | HCD | H | HCD | SCD |
| | B ↔ Q | HCD | HCD | H | H | SCD |
| 6. Theorell–Chance Bi Bi | A ↔ P | H | H | H | HCD | S |
| | B ↔ P | H | L | H | H | SL |
| | A ↔ Q | HCD | H | H | HCD | S |
| | B ↔ Q | HCD | H | H | H | S |
| 7. Random Bi Bi (Rapid Equilibrium) | Any | H | H | H | H | S |
| 8. Random Bi Bi | Any | H | H | H | H | S |
| 9. Random On–Ordered Off | A ↔ P | H | H | H | H | S |
| | B ↔ P | H | H | H | H | S |
| | A ↔ Q | HCD | HCD | H | H | SCD |
| | B ↔ Q | HCD | HCD | H | H | SCD |

[a] See Table II for a parallel listing of rate equations. See Fig. 1 for types of exchange profiles.

[b] A-B-P-Q refers to the Wedler–Boyer protocol of varying all substrates and products in a constant ratio.

initial-rate studies may provide a clear indication of the mechanism if certain abortives do not form (see below). Likewise, the steady-state mechanisms referred to as Ordered Bi Bi and Theorell–Chance Bi Bi may be distinguished by observation of the linear dependence of the B ↔ P exchange vs. the absolute concentration of the B–P pair. This is characteristic of the Theorell–Chance pathway because EA + B → EQ + P is a simple, bimolecular, process without ternary complex formation.

Finally, one may infer from the complexity of the equations in Table II that quantitative estimates of kinetic parameters, especially Michaelis and dissociation constants, are difficult to obtain from the exchange data. The Ping Pong mechanisms are notable exceptions, and replots may provide valuable data.

## Experimental Determination of Equilibrium Exchange Rates

The estimation of exchange rates may be achieved by adding an isotopically labeled substrate (or product) to a fully equilibrated reaction system containing enzyme and by subsequently determining the extent of interconversion to labeled product (or substrate) or periodic sampling. The basic equation for relating the rate of isotopic exchange ($R$) to the concentrations of two reactants, X and Y, undergoing exchange is given by

$$R = \frac{[X][Y][\ln(1 - F)]}{([X] + [Y])t} \quad (11)$$

where $t$ is the period elapsed between initiating and sampling the exchange process, and $F$ is the fractional attainment of isotopic equilibrium.[25] Labeled species may be designated $X^*$ and $Y^*$, and the label will be so distributed at isotopic equilibrium.

$$[X^*]/[Y^*] = [X]_e/[Y]_e \quad (12)$$

Using the subscripts t and e to indicate [X] and [Y] at their time-dependent and equilibrium values, respectively, $F$ may be expressed as the dimensionless number as follows

$$F = [X^*]_t/[X^*]_e = [Y^*]_t/[Y^*]_e \quad (13)$$

In practice, $[X^*]_t$ and $[X^*]_e$ may be expressed in any consistent units that are proportional to their concentration or abundance.[26] For radioactive determinations, either total activity or specific radioactivity may be used; and for stable isotopes, mole fraction or gram atom excess may be employed.

The expression given above for $R$ indicates that the exchange process is first order, and it will remain so provided tracer quantities are employed. It is also possible to roughly measure $R$ by single-point determinations, but it is preferable to use several sampling times and to confirm the first-order rate law as shown in Fig. 2. In this case, the labeled NADH was added to a fully equilibrated lactate dehydrogenase reaction system,[21] and $F$ was measured at various intervals. It will be noted that the concentrations of cosubstrate (pyruvate) and coproduct (NAD$^+$) do not enter into Eq. (11), but this does not mean that $R$ is independent of the concentration of cosubstrate and coproduct. Indeed, $R$ depends upon the fraction of enzyme capable of reacting with the labeled substrate (or product),

---

[25] A. A. Frost and R. G. Pearson, "Kinetics and Mechanism," 2nd ed. Wiley, New York, 1961.

[26] T. H. Norris, *J. Phys. Colloid Chem.* **54**, 777 (1950).

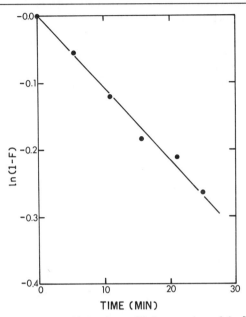

FIG. 2. Plot of ln(1−F), extent of isotopic equilibrium vs. time of the [4A-³H]NADH lactate exchange of bovine heart lactate dehydrogenase. The conditions of the assay were as follows: pH 8.03; $I = 0.2~M$ Tris-KCl; $T = 28.5°$.

and cursory examination of the enzyme exchange-rate expressions presented earlier in Table II indicates that all substrates and products affect $R$. This is true because the concentrations of substrate(s) and product(s) control the concentration of all Michaelis complexes as well as any "mixed" substrate–product–enzyme complexes resulting from abortive complex formation. In some cases, Eq. (11) may be further simplified if the level of one exchange partner is considerably greater than the other. For example, consider the tritium exchange from labeled dihydroxyacetone phosphate to water in the aldolase reaction[27] or the exchange from labeled malate to water in the fumarase reaction.[28] In such cases, the rate of exchange will become

$$R = -\frac{[\ln(1-F)][X][H_2O]}{([X]+[H_2O])t} = -\frac{[\ln(1-F)][X]}{t} \quad (14)$$

To establish valid conditions for estimating an equilibrium isotope exchange rate, one must have an accurate value for the apparent reaction

[27] I. A. Rose, E. L. O'Connell, and A. H. Mehler, *J. Biol. Chem.* **240**, 1758 (1965).
[28] J. N. Hansen, E. C. Dinovo, and P. D. Boyer, *J. Biol. Chem* **244**, 6270 (1969).

equilibrium constant. This is necessary not only for estimating $[X^*]_e$ and $[Y^*]_e$, but also for manipulating the experimental conditions. One should note that the state of equilibrium is independent of the path by which it is achieved, and rarely does the practicing kineticist merely mix the substrate(s) and enzyme to attain equilibrium. Instead, it is more valuable to preset the values of substrate/product ratios as noted earlier, and this requires a value for $K_{app}$. If an incorrect value is chosen and an insufficient time is allowed for enzyme-catalyzed equilibration, the observed equilibrium rate may be incorrect because it will be the composite of the equilibrium rate plus or minus the net rate of reaction required to restore equilibrium. Such observed rate values nullify the validity of conclusions drawn from the experiment. One practical check on such problems may be accomplished by including the radiolabeled substrate or product at the time the enzyme is first added; the isotopic distribution at the time, which one would have otherwise added the isotope to initiate the exchange, should be consistent with the equilibrium value. A more strict condition is the demonstration of equal rates using $X^*$ or $Y^*$ to trace the exchange phenomena. Application of such controls may be especially helpful at the highest concentrations of reactants, where there may be a problem for the enzyme to adequately equilibrate the system. Another excellent procedure to monitor the equilibrium condition also permits the redetermination of $K_{app}$.[29,30] All products and substrates are present at known concentrations such that $([P][Q] \ldots /[A][B] \ldots )$ is near the anticipated value of the equilibrium constant. Following the addition of enzyme, the change in product or substrate concentration is measured after quenching the reaction and determining the analytical concentrations. If the amount of product has increased, then the initial value of $K_{app}$ was less than the apparent equilibrium constant. If the amount of product decreased, the initial ratio was greater than $K_{app}$. Plotting $\Delta P$ (or $\Delta A$) as a function of various initial mass-action ratios, the value of $K_{app}$ becomes the point where $\Delta P = \Delta A = 0$. This is shown in Fig. 3 for a hypothetical case. In practice, the above protocol should be applied at several absolute concentrations of substrates and products because it is insufficient to assume that $K_{app}$ will be completely independent of reactant concentrations. At very high levels of substrates, deviations from ideal behavior may be large. Likewise, metal ion and proton concentration changes during equilibration may shift the mass-action ratio. Incidentally, some buffers may actually shift the equilibrium constant by reacting with substrates, products, or cofactors. Tris buffer, for example, reacts with some aldehydes, such

---

[29] K. M. Plowman, "Enzyme Kinetics." McGraw-Hill, New York, 1972.
[30] I. H. Segel, "Enzyme Kinetics." Wiley, New York, 1975.

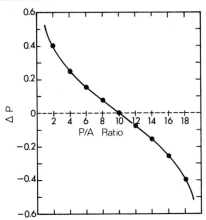

FIG. 3. Estimation of equilibrium constant by approximating the substrate(s)/product(s) ratio and measuring the direction of shift upon enzyme addition. Several initial ratios are prepared, and the direction of shift is determined. The intersection of $\Delta P$ or $\Delta A$ at zero establishes the equilibrium position.

as glyceraldehyde 3-phosphate and acetaldehyde, and pyrophosphate is a powerful metal ion-chelating agent. In this respect, $K_{app}$ must be redetermined if the experimenter changes the pH, buffer components, metal ion concentrations, or reaction temperature. The experimenter may even elect to use the substrates and products as the pH buffer instead of adding additional ions.[17]

Before leaving equilibrium constants completely, it may be valuable again to note that the time required to attain equilibrium in the presence of enzyme may be deceptively long. Depending upon the enzyme levels, this may require several hours, but many investigators fail to allow adequate equilibration or to demonstrate the sufficiency of their protocol.

The range of substrate and product levels required adequately to distinguish ordered and random substrate addition must be experimentally determined. Ideally, the absolute levels of substrate and products should be varied to allow the observed value of $R$ to fall between 0.5 and 20 times $R_{0.5}$, the concentrations of A, B, P, Q, etc., required to yield one-half the observed maximal exchange. Computer simulation studies indicate that the range of substrate(s) and product(s) concentration necessary for this condition may have little relation to the Michaelis constants obtained in initial-rate experiments. To observe inhibition of exchange at high "inner-pair" substrate–product levels, the upper range of concentrations should be extended to its practical limits. For rapid-equilibrium ordered cases, the depression of the A ⇌ Q exchange as [P]/[B] is increased will never be observed. Thus, systems approaching this condition will fail to demonstrate inhibition even though the system does not involve random enzyme–substrate binding pathways.

As Fromm[12] clearly indicates, the proper [P]/[A], [Q]/[A], [P]/[B], and [Q]/[B] ratios are also very important. These values determine the ultimate isotopic distribution of label at isotopic equilibrium, and it is frequently advantageous to use the isotopic label that corresponds to the least abundant reactant in a substrate–product pair. This permits the greatest redistribution of label and the most accurate estimate of $F$. (One exception is the use of [$^{18}$O]H$_2$O in H$_2$O ↔ substrate exchanges. If one uses $^{18}$O-labeled substrate, the atom excess in the water after exchange may not be appreciably above the natural abundance of the heavy-oxygen isotope.) Ideally, the ratio of substrate to product should fall between 0.05 and 20. For values outside this range, one may encounter difficulty estimating accurately the rate of isotopic exchange. Of course, this limitation deals only with the substrate–product pair undergoing exchange (i.e., the substrate and product that becomes labeled at isotopic equilibrium). Thus, one may compensate for choosing a favorable ratio of this substrate–product pair by altering the ratio of the other substrate(s) or product(s). If the value of $K_{app}$ is very large or very small, the experimenter may wish to alter the value of $K_{app}$ by working at a different pH. For example, the glutamine synthetase equilibrium constant at pH 8 is approximately 10,000–12,000, but at pH 6.5 it is around 444. If one wanted a [glutamine]/[glutamate] ratio of 10 at pH 8, the product of the ratios of [ADP]/[ATP] and [P$_i$]/[NH$_3$] must equal 1000–1200. Under these conditions, even a slight redistribution of the system in the presence of the enzyme may alter the ratios of [ADP]/[ATP] and [P$_i$]/[NH$_3$] from their preset values. Thus, one would experience difficulties in obtaining meaningful plots of $R$ vs. the absolute levels of a particular substrate–product pair. In this respect, a change of pH can be useful to obtain more favorable experimental conditions, but it may also lead to a change in mechanism. Creatine kinase, for example, appears to be rapid-equilibrium random at pH 8 but equilibrium ordered at pH 7.[10,31] Another complication that results from lowering the pH is that stability constants for metal–substrate complexes may be appreciably altered. This may lead to a redistribution of substrate between active and inactive forms.

Another factor that is frequently overlooked in exchange studies is the occurrence of time-dependent changes in a chemical system supposedly at equilibrium. If an enzyme is not particularly stable, the fraction of active catalyst may be affected by the protective effects of ligands toward thermal denaturation. In this case, the exchange-rate profiles may yield

[31] M. I. Schimerlik and W. W. Cleland, *J. Biol. Chem.* **248**, 8418 (1973).

completely ambiguous results. In the event of substrate instability, the equilibrium concentration of each component may slowly change, and the experimental conditions will be difficult to control. One example is provided by the need to examine the malate dehydrogenase reaction at 1° to avoid problems arising from decomposition of oxaloacetate.[32] Such changes may be even more deceptive with polymerases and depolymerases that also catalyze minor transferase reactions. For example, the presence of debranching enzyme activity in the glycogen phosphorylase system may alter the distribution of available phosphorylase binding sites on the glycogen. Thus, the value of $R$ and its dependence upon substrate/product ratios may be time-dependent. Finally, there are even cases where a supposed allosteric modifier of a particular reaction turns out to be an alternative substrate for the reaction itself or some contaminating enzymic activity. If the first occurs, then the equilibrium distribution will change as the "modifier" concentration is altered. If the second case takes place, the observed exchange-rate behavior may be a complex composite of several effects. These situations may be surprisingly common for nucleotide-dependent enzymes where the presence of contaminating adenylate kinase, nucleoside diphosphate kinase, etc., may compromise the value of the experiment.

Upon chosing the ratios of products and substrates, the reaction samples may be prepared. Rather than laboriously adding each component to each reaction mix, it is advisable to prepare three stock solutions: one containing buffer, salts, and cofactors; the second containing one substrate–product pair, say [Q]/[A], at the highest desired concentrations for the experiment with allowance for dilution upon combination with the other stock solutions; and the third containing [P]/[B] in a similar fashion. Now, various dilutions may be made with the second and third stock solutions to obtain the desired absolute levels of each component, and this may be done without altering the ratios. Upon mixing the first stock solution with suitable dilutions of the others, the enzyme is added, and the reaction system is permitted to equilibrate. The isotope is then added to initiate the measurements of the exchange process. Another protocol, sometimes useful, involves the mixing of enzyme, substrates, and products at the desired level of the nonvaried substrate–product pair. After equilibration, appropriate aliquots of a stock solution containing the varied substrate–product pair are added at a ratio satisfying the equilibrium distribution. Then, isotope is added to begin the rate measurement. This protocol is fine provided the experimenter rigorously demonstrates that the system is at equilibrium prior to isotope addition. Examples of the

---

[32] E. Silverstein and G. Sulebele, *Biochemistry* **8**, 2543 (1969).

two protocols are given for the hexokinase[7] and aspartokinase[33] reactions, respectively.

It is advisable to give adequate thought to the method used for quenching the reaction. Additions of acid, base, phenol, ethanol, EDTA, ion exchange resin, sulfhydryl reagents, or rapid boiling or freezing have been utilized to stop the reaction with minimal uncertainty about the reaction period. Several methods should be used before one adopts a standard protocol, and the best method as regards convenience and accuracy should be adequately examined. By far, the best check is to stop the reaction after various periods and to demonstrate that the exchange obeys Eq. (11). The time required for quenching the process is an experimental parameter, and only trial and error will uncover the best protocol.

In the absence of subunit dissociation and association events, the exchange rate should be directly proportional to the enzyme concentration. This is an important point that should be verified. A typical example is provided by the work of Silverstein and Sulebele[32] on the porcine heart malate dehydrogenase reaction (Fig. 4). Here, the oxaloacetate ↔ malate and $NAD^+$ ↔ NADH exchange rates are linearly dependent upon enzyme concentration. It is also important to note that the levels of enzyme used in exchange reactions are frequently higher than in initial-rate studies; thus, contaminating enzyme activities may become a problem. This might lead to a nonlinear dependence of $R$ upon enzyme level. A nonlinear relation may also be realized if the enzyme level is not vanishingly small by comparison to the levels of substrates and products. Quadratic terms compensating for the depletion of free substrate(s) or product(s) level by the bound forms will lead to this nonlinearity, and the experimenter will be confronted by the inadequacy of the rate laws in Table II to describe the observed behavior.

In the case of Ping Pong mechanisms one may deal with three equilibrium conditions: A, B, P, and Q present; A and P only; and B and Q only. The last two cases are possible because the amount of enzyme linking two components of a half-reaction is very small compared to the substrate and product levels.

$$E + A \rightleftharpoons EX \rightleftharpoons E' + P \tag{15}$$

At any level of [A] and [P] the system will equilibrate by shifting the [E]/[E'] ratio. One experimental advantage of the Ping Pong case is that more quantitative information is available. One experimental difficulty of measuring partial exchanges, however, is that the presence of a small contaminant amount of the second substrate or product can make a sequen-

---

[33] J.-F. Shaw and W. G. Smith, *J. Biol. Chem.* **252**, 5304 (1977).

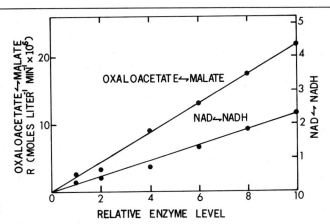

FIG. 4. Plot of the equilibrium reaction rate ($R$) vs. enzyme concentration. The reaction mixture contained 0.33–3.3 μg/ml of pig heart mitochondrial malate dehydrogenase, 4.95 m$M$ NAD$^+$, 48 μ$M$ NADH, 23.3 m$M$ malate, 233 μ$M$ oxaloacetate, and 70 m$M$ Tris-NO$_3$ at pH 8.0 and 1°.

tial mechanism appear to be catalyzing a partial reaction. For example, Switzer proposed the intermediary participation of a pyrophosphoryl enzyme in the PRPP synthetase reaction.[34] However, later experiments discounted this idea.[35] There are many such cases, and those with apparently slow exchanges are discussed later. One source of the problem is substrate impurity, and the availability of high pressure liquid chromatography may mitigate this problem.

It may also be helpful to make some practical comments on the Wedler and Boyer[13] method of raising all reaction components in an equilibrium ratio. This may be achieved by making a single stock solution at the highest analytical concentrations and then diluting this into buffer containing metal ions and cofactors whose free concentration is to remain constant. With metal–ligand complexes acting as substrates and products, it is necessary to estimate the amount of bound metal and to allow for this in the substrate–product stock solution. Likewise, the buffer should contain the estimated amount of uncomplexed metal ion with an allowance for dilution upon preparation of the complete reaction mix. The dilution of the substrate–product stock mix can then be easily accomplished with water, and upon combination with the buffer mix, the free uncomplexed metal ion level will be identical for each absolute level of substrates and products. One experimental limitation to the Wedler and

[34] R. L. Switzer, *J. Biol. Chem.* **245**, 483 (1970).
[35] R. L. Switzer and P. D. Simcox, *J. Biol. Chem.* **249**, 5304 (1974).

Boyer technique is that large changes in ionic strength can affect stability constants and enzyme-rate behavior. The sensitivity of an enzyme to such changes may be evaluated by initial-rate experiments.

Finally, there are several practical considerations regarding isotope addition and exchange-rate assays. Since the addition of labeled reactant is used to initiate the exchange measurement, it is desirable that the label be of sufficient specific radioactivity to prevent a shift in the equilibrium distribution of reactants. In some cases, this may be difficult to achieve, especially if the labeled species cannot be prepared in high specific activity. In such situations, one may estimate the amount of perturbation on the basis of the specific activity. If this labeled reactant is of a low specific activity, one may mix it with nonlabeled product such that the [substrate*]/[product] ratio is close to the expected equilibrium distribution. When this mixture is added to the system at equilibrium, there should be little disturbance of the relative reactant concentrations. Likewise, all isotope additions should be made in a manner that obviates substantial dilution of the equilibrium samples. This is especially important in reaction systems having a different number of substrates and products (e.g., phosphatases, aldolases, and esterases). Such systems have equilibrium constants with units of molarity, and they will shift the distribution upon dilution. After the isotope is added, one may withdraw samples at various intervals for analysis. When it becomes impractical to withdraw identical aliquots, then one may use a ratio counting method. For example, if we were measuring an ADP* ↔ ATP exchange, the initial aliquots could be large, such that the number of ATP* counts was maximized. After separation on DEAE-ion exchange paper, the radioactivity of the ADP and ATP spots could be measured and normalized as (cpm ATP*)/[(cpm ATP*) + (cpm ADP*)], the percentage of radioactivity as ATP*. The same may be done after five to six half-lives of exchange to obtain the ratio at isotopic equilibrium. The quotient of these yields a value for $F$. This approach has the added advantage that no corrections are required for decreases in radioactivity, which are frequently a problem with $^{32}P$.

*Examples of Equilibrium Exchange-Rate Studies*

Table I summarizes some of the systems that have been characterized by isotope exchange studies, but it is of value to consider several examples in more detail. The emphasis of this section will be on qualitative conclusions that may be drawn from exchange-rate profiles and the relative exchange velocities.

*Lactate Dehydrogenase.* Silverstein and Boyer[8] were the first to examine the rates of exchange between lactate and pyruvate as well as

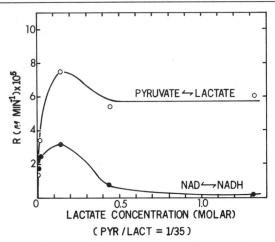

FIG. 5. The effect of lactate and pyruvate concentrations on equilibrium reaction rates with rabbit muscle lactate dehydrogenase at pH 7.9. Reaction mixtures contained 1.68–1.70 m$M$ NAD$^+$, 30.2–45.5 μ$M$ NADH, and pyruvate and lactate as shown, in 145 m$M$ Tris-NO$_3$ at pH 7.9 and 25°.

between NAD$^+$ and NADH in the lactate dehydrogenase system. Convenient ratios of [NADH]/[NAD$^+$] and [pyruvate]/[lactate] were chosen to satisfy the apparent equilibrium constant. They established that each exchange rate was proportional to time of reaction and proportional to enzyme concentration. They also demonstrated the equality of the pyruvate ↔ lactate exchange rates with [1-$^{14}$C]pyruvate or [1-$^{14}$C]lactate to within 10%. To examine the substrate binding order, the effect of lactate and pyruvate concentrations on equilibrium reaction rates of the pyruvate ↔ lactate and NAD$^+$ ↔ NADH exchanges were examined. As shown in Fig. 5, the pyruvate ↔ lactate exchange rises to an essentially stable plateau, but the coenzyme exchange is markedly depressed. This observation excludes random substrate addition (see Table III, line 8), and indicates that the coenzymes form the binary Michaelis constants (i.e., NAD$^+$ is the first to bind and NADH is the last product to be released in each catalytic cycle).

*Alcohol Dehydrogenase.* Liver alcohol dehydrogenase is among the most characterized bisubstrate enzyme systems. Sund and Theorell[36] suggested that the liver enzyme may form all four binary complexes with substrates or products, and Silverstein and Boyer[9] demonstrated that the

---

[36] H. Sund and H. Theorell, *in* "The Enzymes" (P. D. Boyer, H. Lardy, and K. Myrbäck, eds.), 2nd ed., Vol. 7, p. 25. Academic Press, New York, 1963.

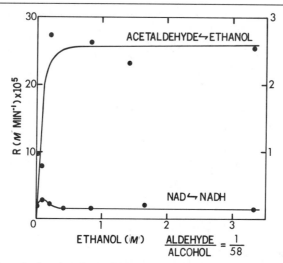

Fig. 6. Effect of ethanol and acetaldehyde concentrations on the equilibrium reaction rates with liver alcohol dehydrogenase. The scale for the NAD$^+$ ↔ NADH rate, given on the right ordinate, is 10 times that for the acetaldehyde ethanol exchange (at the left). Reaction mixtures contained, at 25°, 21 m$M$ diethylbarbiturate buffer, pH 7.9, 18.2 m$M$ NaNO$_3$, ethanol, and acetaldehyde as given in the figure, 1.55–1.49 m$M$ NAD$^+$, and 51–115 $\mu M$ NADH. In this and other experiments, it was not possible to hold NAD$^+$ and NADH ratios constant with a constant acetaldehyde to ethanol ratio because of shift in the equilibrium at the high concentrations of ethanol and acetaldehyde. The measured concentrations of NAD$^+$ for the various points given in the figure as ethanol was increased were 1.55, 1.56, 1.56, 1.56, 1.55, 1.54, 1.53, and 1.49 m$M$; the corresponding points for NADH were 56, 51, 52, 54, 60, 66, 80, and 115 $\mu M$. Thus only above 1 $M$ ethanol was the equilibrium shift appreciable.

reaction does not proceed with a strictly compulsory substrate binding order. As shown in Fig. 6, the acetaldehyde ↔ ethanol and NAD$^+$ ↔ NADH exchanges have an exchange-rate profile very different from that in the analogous experiments with lactate dehydrogenase. The acetaldehyde ↔ ethanol exchange rises to a plateau value of about $2.6 \times 10^{-4}$ $M$ min$^{-1}$, but the NAD$^+$ ↔ NADH exchange rises to a plateau value about 100–160 times less rapid. This finding indicates that appreciable dissociation of enzyme · NADH · acetaldehyde to enzyme · ethanol + NAD$^+$ must occur, but the principal pathway of net reaction must exhibit coenzyme release as the slowest step. These investigators also recognized that the wide disparity of the two exchanges in the absence of a compulsory pathway eliminates the possibility that ternary complex interconversion is a slow step in the catalysis by the liver enzyme.

*Yeast Hexokinase.* Isotope exchange studies played an important role in defining the kinetic mechanism of yeast hexokinase.[7] The exchange

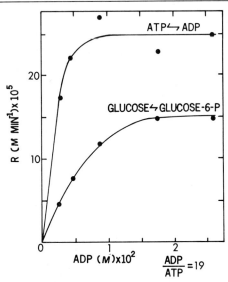

FIG. 7. (A) Effect of ATP and ADP concentrations on equilibrium reaction rates catalyzed by yeast hexokinase. The reaction mixtures contained, at 25°, 57.8 m$M$ imidazole-NO$_3$, pH 6.5; 13 m$M$ Mg(NO$_3$)$_2$; 2.5 m$M$ glucose; 38.5 m$M$ glucose 6-phosphate; 16.8 µg (29 Kunitz–McDonald units) of yeast hexokinase per milliliter; ATP and ADP as shown in the figure; and 0.34 mg of bovine serum albumin per milliliter.

work was carried out at pH 6.5 ($K_{app}$ = 490) to obviate the experimental difficulties with a large $K_{app}$ at higher pH. As shown in Figs. 7A, and 7B the ADP ↔ ATP exchange exceeds the glucose ↔ glucose 6-phosphate exchange rates by approximately 1.8-fold. Nonetheless, there is no evidence that the exchanges become depressed over the wide range of concentrations employed, and one must conclude that the mechanism is random but certainly not rapid-equilibrium random. This is to say, the interconversion of ternary complexes is relatively slow but not rate determining. This conclusion has been disputed by Noat and Ricard,[37] who favor an ordered mechanism with sugar binding as the obligatory first step, but Fromm[38] has shown by computation that any depression in the sugar ↔ sugar phosphate exchange would have been detected under the experimental regime used in the exchange studies. At present, the bulk of evidence accords with a random mechanism.[39,40] This is illustrated in Fig. 8.

[37] G. Noat and J. Ricard, *Eur. J. Biochem.* **5,** 71 (1968).
[38] H. J. Fromm, *Eur. J. Biochem.* **7,** 385 (1969).
[39] D. L. Purich, H. J. Fromm, and F. B. Rudolph, *Adv. Enzymol.* **39,** 249 (1974).
[40] W. W. Cleland, *Adv. Enzymol.* **45,** 273 (1977).

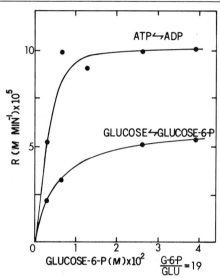

FIG. 7. (B) Effect of glucose and glucose 6-phosphate concentrations on equilibrium reaction rates catalyzed by yeast hexokinase. The reaction mixtures contained, at 25°, 57.8 m$M$ imidazole-NO$_3$, pH 6.5; 13 m$M$ Mg(NO$_3$)$_2$; 0.99–2.2 m$M$ ATP; 25.6 m$M$ ADP; glucose and glucose 6-phosphate as in the figure; 7.83 μg (13.5 Kunitz–McDonald units) of yeast hexokinase per milliliter; and 0.624 mg of bovine serum albumin per milliliter.

It may be useful to point out several technical features of this report. The equilibrium constant at pH 6.5 is about 490. As noted earlier, this permits one far more latitude than the use of a very large equilibrium constant at higher pH. These investigators chose to set the [ADP]/[ATP] and [glucose-6-P]/[glucose] ratios at 19 and 15.4, respectively, giving an initial mass-action ratio of 293. After the addition of enzyme, the reaction mixtures were maintained at 25° for 1 hr to achieve equilibration prior to isotope additions. If the system completely equilibrated, the [ADP]/[ATP] ratio at 10 and 25 m$M$ ADP in Fig. 7A would have been 29 and 66, respectively. Thus, there is no doubt that the ratio of [ADP]/[ATP] varies considerably from the assumed constant value of 19. This flaw in experimental design is fairly common, and it illustrates the importance that must be attributed to satisfying the value of $K_{app}$ in the initial design. Another experimental feature of this report is the great care taken to establish the radiopurity of the [$^{14}$C]glucose and [$^{14}$C]ATP. The calculation of exchange rates must always take radioimpurity into account, and it is best to rely upon enzymic assays because of their high specificity. Finally, these investigators obtained some information about the minimum values of dissociation constants from plots of $R^{-1}$ vs. $S^{-1}$. These approximate values

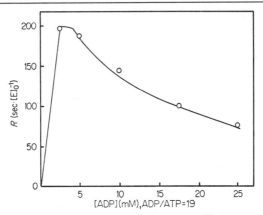

FIG. 8. The effect of ATP and ADP concentrations on equilibrium reaction rates (R) catalyzed by the enzyme yeast hexokinase. The points on the graph were calculated using the equation for an ordered mechanism, the rate constants provided by G. Noat and J. Ricard [*Eur. J. Biochem.* **5,** 71 (1968)], and the experimental conditions from Fig. 7A for the glucose ↔ glucose 6-phosphate exchange.

were found to agree rather closely with $K_d$ values obtained from initial-rate studies.

*Creatine Kinase.* Important mechanistic data about the rabbit muscle creatine kinase reaction was provided by Morrison and Cleland.[10] These investigators found that the initial rates of the ADP ↔ ATP and creatine-P ↔ creatine exchanges at equilibrium are approximately equal, indicative of a rapid-equilibrium random mechanism. While the initial ADP ↔ ATP exchange rate increased hyperbolically to a maximum value as the creatine-P–creatine pair was raised in concentration, higher concentrations of the MgATP–MgADP pair caused an inhibition of the exchange. This was shown to result from the inhibitory effect of NaCl that is obtained from magnesium chloride and sodium salts of the nucleotides. These investigators also provided excellent evidence for an enzyme·creatine·ADP abortive complex by observing the ADP ↔ ATP exchange rate as a function of increases in the creatine–MgADP pair (see Fig. 9). The depression in the exchange is clearly evident, and this agrees with initial-rate studies by Morrison and James.[41] One technical advance presented in this exchange study was the use of initial rates of isotope exchange. These were obtained by the ratio (counts per minute of product per minute of reaction per microgram of enzyme)/(counts per minute of substrate per micromole), yielding exchange velocity ($v^*$) as micromoles

---

[41] J. F. Morrison and E. James, *Biochem. J.* **97,** 37 (1965).

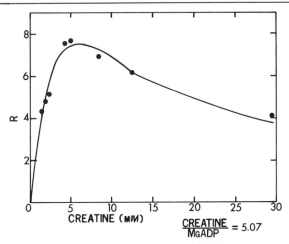

FIG. 9. Effect of increasing concentrations of the MgADP–creatine pair on the initial velocity of the ATP ⇌ ADP exchange. Basic reaction mixtures contained, in 0.5 ml; 0.1 $M$ triethanolamine–HCl buffer (pH 8.0), 0.01 m$M$ EDTA, 3.85 m$M$ ATP, 0.323 m$M$ ADP, 5.52 m$M$ MgCl$_2$, 1.39 m$M$ creatine, 0.756 m$M$ phosphocreatine, and 1 µg of creatine kinase. The concentrations of MgADP$^-$ and creatine were increased as indicated. The exchange reaction was started by the addition of 40 µl (0.4 µCi); temperature, 30°. The exchange rate ($R$) is expressed as millimicromoles per minute per microgram of enzyme.

per minute per microgram of enzyme. Here, a higher radiospecific activity is required to follow the course of the isotope exchange, but one does not need to use Eq. (11). Thus, the requirement for full attainment of equilibrium implicit in Eq. (11) is no longer necessary. It is also true that one need not rely upon accurate determination of the final distribution of label between substrate and product. Finally, Morrison and Cleland[10] introduced the use of calculated theoretical curves to demonstrate the correspondence of theoretical and observed rate behavior.

*Glutamine Synthetase.* Isotope exchange studies of this enzyme are rather extensive, and only the Wedler and Boyer[13] method is described here. This method is especially valuable for three-substrate systems in that the number of exchanges that may be dealt with by varying different substrate–product pairs is large. For distinguishing ordered and random addition pathways, the reader may already have noted that the former involve noncompetitive interactions but the latter cannot. By varying the levels of all substrates and products in a constant ratio, noncompetitive interactions in ordered systems will lead to a depression in the exchange. Competitive effects remain balanced under these conditions, and the relative concentrations of each enzyme species remain effectively constant. Thus, there is no tendency to change the availability of a particular enzyme form required for exchange. Figure 10 presents the results of varying all substrate levels simultaneously in a constant ratio corresponding to the equilibrium constant, as probed by the glutamate ↔ glutamine and $P_i$ ↔ ATP exchanges. The maximum levels of substrates were raised considerably beyond the reported Michaelis constants for the substrates. These findings rule out compulsory binding orders or noncompetitive effects as being responsible for the previous inhibitions.[13]

It is interesting to note that Wedler and Boyer[13] did not obtain any evidence for partial exchange reactions between ADP and ATP in the presence of glutamate when ammonium ion was scrupulously omitted. This finding indicates that the following partial reaction is not kinetically significant in glutamine synthetase catalysis.

$$\text{Glutamate + ATP} \xrightleftharpoons{\text{enzyme}} \text{glutamyl-P + ADP} \qquad (16)$$

Contrary to their conclusion that such evidence indicates that no enzyme-bound phosphorylated, enzyme-bound adenosine diphosphoryl, or γ-glutamyl-P participates in the reaction, there is sufficient chemical[42,43] and stereochemical[44] evidence for the latter. One alternative conclusion that may be drawn from the failure to observe a partial exchange reaction is that ADP release from the enzyme·ADP·γ-glutamyl-P complex is extremely slow, if it occurs at all.

*Nucleoside Diphosphokinase.* Of the kinase-type phosphotransferases, nucleoside diphosphokinase has drawn interest in terms of its stable phosphoenzyme intermediate. Initial velocity and product inhibition data are in full agreement with a Ping Pong Bi Bi mechanism.[45–47] Garces and Cleland[48] presented additional evidence for such a mechanism by examination of the MgADP ↔ MgATP and MgUDP ↔ MgUTP exchange processes. Initial velocities of each exchange were measured at various concentrations of the exchange partners in the absence of the other nucleoside diphosphate and triphosphate components. The MgADP ↔ MgATP exchange is linear, but there is evidence for competitive substrate inhibition by MgADP. A similar effect was observed with MgUTP inhibition of the MgUDP ↔ MgUTP exchange. To confirm the

---

[42] A. Meister, in "The Enzymes" (P. Boyer, ed.), 3rd ed., Vol. 10, p. 699. Academic Press, New York, 1974.
[43] J. A. Todhunter and D. L. Purich, *J. Biol. Chem.* **250**, 3505 (1975).
[44] C. F. Midelfort and I. A. Rose, *J. Biol. Chem.* **251**, 5881 (1976).
[45] N. Mourad and R. E. Parks, Jr., *J. Biol. Chem.* **241**, 271 (1966).
[46] N. Mourad and R. E. Parks, Jr., *J. Biol. Chem.* **241**, 3838 (1966).
[47] A. W. Norman, R. T. Wedding, and M. K. Black, *Biochem. Biophys. Res. Commun.* **20**, 703 (1965).

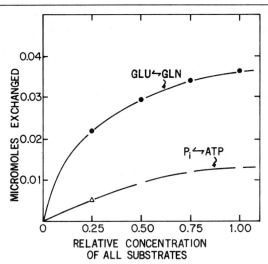

FIG. 10. The effects on equilibrium exchange rates of varying all substrate concentrations simultaneously and in constant ratio. The 1.0-ml reaction at pH 6.50, 37° with 1.00 relative concentration contained 2 m$M$ NH$_3$, 2 m$M$ glutamate, 1 m$M$ ATP, 20 m$M$ glutamine, 20 m$M$ P$_i$, and 4 m$M$ ADP. The reaction mixtures also contained 200 m$M$ KCl, 50 m$M$ $\beta,\beta$-dimethylglutarate buffer, 1 m$M$ MnCl$_2$, MgCl$_2$ equal to nucleotide, plus 0.4 mg of *Escherichia coli* W glutamine synthetase, $E_{10}$.

basic parallel nature of these exchange patterns by allowing for competitive substrate effects, Garces and Cleland[48] found that variation in the absolute levels of the two reactants in a constant ratio (tri-/diphosphate = 14) gave completely linear reciprocal plots. Thus, the basic kinetic mechanism may be represented as follows:

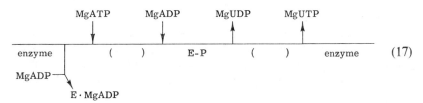

(17)

The viability of a phosphoryl enzyme intermediate was verified also with the bovine liver enzyme when the rates of phosphorylation and dephosphorylation were studied with a rapid mixing technique.[49] Finally, Cleland has presented cogent analyses of complex Ping Pong mechanisms,[40,50] and

---

[48] E. Garces and W. W. Cleland, *Biochemistry* **8**, 633 (1969).
[49] O. Wålinder, Ö. Zetterqvist, and L. Engström, *J. Biol. Chem.* **244**, 1060 (1969).
[50] D. V. Santi, R. W. Webster, Jr., and W. W. Cleland, this series, Vol. 29 [49].

these reports are highly worthwhile reading, especially for those interested in three-substrate synthetase-type reactions.

Enzyme Systems Away from Equilibrium

Although most exchange studies have been desired for processes at equilibrium, the back exchange of labeled product while the reaction is proceeding in the forward direction can provide valuable information about enzymic catalysis. Interestingly, some of the first attempts to gain mechanistic insight about enzyme action were of this sort. Under favorable conditions, investigators may utilize such isotope exchange data to learn about the order of product release and the presence of covalent enzyme–substrate compounds. One of the first systems to be characterized in this way was glucose-6-phosphatase,[3] which has the following basic kinetic mechanism.

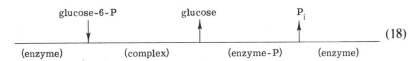

(18)

During the course of glucose 6-phosphate hydrolysis, radiolabeled glucose was added and the back exchange to form labeled glucose-6-P was examined. (Likewise, the possibility of incorporation of labeled $P_i$ into glucose-6-P was examined under identical conditions, but none was observed.) As shown in Table IV, the rate of the glucose ⇌ glucose-6-P exchange correlated rather well with the amount of glucose inhibition of the phosphatase. These findings led Hass and Byrne[3] to conclude that glucose is the first product to leave and that the negative free energy of hydrolysis is preserved as a phosphoryl–enzyme compound. It is also interesting to note that Zatman, Kaplan, and Colowick[51] used such exchange phenomena away from equilibrium to synthesize [$^{14}$C]NAD$^+$ from [$^{14}$C]nicotinamide and NAD$^+$ with bovine spleen NADase. They also observed that labeled adenosine diphosphoribosylpyrophosphate (ADPR) failed to exchange in a similar fashion, and the following reaction sequence was favored.

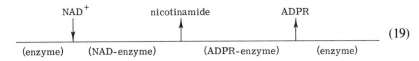

(19)

Such mechanisms involve the ordered release of products, and the

---

[51] L. J. Zatman, N. O. Kaplan, and S. P. Colowick, *J. Biol. Chem.* **200**, 197 (1953).

TABLE IV
EXAMINATION OF THE GLUCOSE ⇌ GLUCOSE 6-PHOSPHATE EXCHANGE DURING
ENZYME-CATALYZED HYDROLYSIS OF GLUCOSE 6-PHOSPHATE[a]

| Enzyme | Hydrolase activity ($v$) | Hydrolase activity with glucose ($v_i$) | Difference ($v - v_i$) | Exchange activity ($v^*$) |
|---|---|---|---|---|
| Normal rat liver enzyme | 7.17 | 5.05 | 2.12 | 2.12 |
| Diabetic rat liver enzyme | 7.40 | 5.25 | 2.15 | 1.95 |

[a] Reaction samples at 37° (pH 6.0) contained 8.5 m$M$ glucose 6-phosphate, and hydrolytic activity is presented as micromoles of $P_i$ liberated per grams of liver per minute. Glucose, when present, was 72 m$M$. In the exchange measurements [$^{14}$C]glucose was added to a final specific activity of 4700 cpm/$\mu$mole, and the exchange rate is in the same units as above.

scheme is formally of the Ping Pong Bi Bi type with $H_2O$ presumably entering the catalytic process after release of the first product. Labeled P can exchange back to A in the absence of Q only when there is a significant level of EQ in the steady state and P* is sufficiently high (i.e., equal or greater than the respective $K_i$ for P). An exchange from Q* back to A may only occur when there is a significant level of P present. If Q* does exchange without P present, one must conclude that the order is noncompulsory (random) with adequate EP complex in the steady state to support significant exchange. In this respect, observation of a P* ⇌ A exchange is not strict proof of ordered release, and one may argue that a random mechanism with slow EQ breakdown to enzyme and Q is operative.

In the case of multisubstrate enzyme systems, one may gain evidence about the order of product release by adapting the experimental procedure to withdraw one product from participation in exchange. This approach was developed by Kosow and Rose[52] to examine product release order in the hexokinase system. Glucose-6-P was rapidly depleted by excess NADP$^+$ and glucose-6-P dehydrogenase; labeled ADP was added to measure the ADP ⇌ ATP exchange. Without reference to the mode of substrate binding, we may write the following scheme:

$$(\quad) \rightleftharpoons E \cdot Glu \cdot MgATP \rightleftharpoons E \cdot MgADP \cdot Glu\text{-}P \begin{array}{c} \nearrow \text{MgADP*} \quad \nearrow E \cdot Glu\text{-}P \searrow\!\!\!\times \text{Glu-P} \\ \\ \searrow \text{Glu-P} \searrow\!\!\!\times \quad \nearrow E \cdot MgADP \nearrow E \end{array} \qquad (20)$$

[52] D. P. Kosow and I. A. Rose, *J. Biol. Chem.* **245**, 198 (1970).

If the enzyme·glucose-6-P complex is significant in the steady-state phosphorylation of glucose, then ADP* will combine with it and re-form E·MgADP·glucose-6-P which will interconvert to form E·ATP*·glucose and lead to ATP* synthesis. Kosow and Rose[52] found the rate of exchange to be 22% the rate of the forward reaction, where the forward reaction is inhibited 70% by ADP. The significance of this observed exchange may only be judged in qualitative terms (i.e., as evidence for enzyme·glucose-6-P complex). Unless the partition coefficient for E·MgADP·glucose-6-P interconversion to E·MgATP·glucose and E·MgADP is known, it becomes difficult to make any quantitative interpretations.

Enzyme Interactions Affecting Exchange Behavior

*Substrate Synergism*

The cardinal feature of Ping Pong mechanisms is the ability of the ensyme to catalyze partial exchange reactions as a result of the independence of the substrate's interactions with the enzyme. This is reflected in the fact that the second substrate is obliged to await the dissociation of the first product before it may bind to the enzyme. Multisubstrate enzymes frequently mediate such partial reactions which may be related to important steps in catalysis. One enzyme proposed to be of this sort is succinyl-CoA synthetase, but the partial reactions are relatively slow, and the participation of such reactions in catalysis becomes difficult to assess. Indeed, slow partial exchanges have been interpreted as proof of contamination of a particular enzyme with another enzyme or a small amount of the second substrate, an indication that the mechanism is not Ping Pong, or that the presence of other substrates may markedly increase the rates of elementary reactions giving rise to the partial exchange. Bridger *et al.*[53] proposed that the latter phenomenon be termed substrate synergism, and they examined this enzyme–substrate interaction by deriving appropriate raw laws for various exchanges. Their conclusion is that the rate of a partial exchange reaction must exceed the rate of the same exchange reaction in the presence of all the other substrates if the same catalytic steps and efficiencies are involved. If the opposite relation is observed, one must consider the possibility that synergism exists.

Lueck and Fromm[54] have also examined the significance of partial exchange rate comparisons, and they focused on often misleading compari-

---

[53] W. A. Bridger, W. A. Millen, and P. D. Boyer, *Biochemistry* **7**, 3608 (1968).
[54] J. D. Lueck and H. J. Fromm, *FEBS Lett.* **32**, 184 (1973).

sons of exchange rates made with respect to initial velocity data. For the Ping Pong Bi Bi mechanism, it can be shown that the following relationship pertains.

$$1/R_{max,A \leftrightharpoons P} + 1/R_{max,B \leftrightharpoons Q} = 1/V_1 + 1/V_2 \tag{21}$$

where $R_{max,A \leftrightharpoons P}$ and $R_{max,B \leftrightharpoons Q}$ refer to the maximal rates of the partial exchange reactions between the specified substrates, and $V_1$ and $V_2$ refer to the maximal velocity in the forward and reverse reactions, respectively. Thus, one may only evaluate the possibility of substrate synergism after these parameters are quantitated. For example, we may rewrite Eq. (21) to define a new parameter, $Q_{syn}$, the synergism quotient.

$$Q_{syn} = \frac{(R_{max,A \leftrightharpoons P})^{-1} + (R_{max,B \leftrightharpoons Q})^{-1}}{V_1^{-1} + V_2^{-1}} \tag{22}$$

Only when $Q_{syn}$ is substantially greater than one, may the experimenter conclude that there is a possibility of substrate synergism or that the mechanism is not Ping Pong. With yeast nucleoside diphosphokinase, $Q_{syn}$ is about 1.3 based upon our replots of the data of Garces and Cleland.[48] Since the enzyme's specific activity may vary somewhat, a value of 1.3 is indicative of a Ping Pong mechanism. On the other hand, the acetyl $\leftrightharpoons$ acetyl-P exchange in the *E. coli* acetate kinase reaction is quite feeble, and $Q_{syn}$ is about 32.[55] Thus, a Ping Pong mechanism is not likely for acetate kinase but substrate synergism may be involved. Indeed, the so-called "activated Ping Pong" mechanism of Skarstedt and Silverstein[56] is a form of substrate synergism. Lueck and Fromm[54] also presented another criterion for comparing partial exchange rates in the absence and in the presence of the substrate–product pair not involved in the isotope exchange. For the Ping Pong Bi Bi mechanism, one may compare the rates of various exchanges using Eq. (23)

$$1/R_{A \leftrightharpoons Q} = 1/R_{A \leftrightharpoons P} + 1/R_{B \leftrightharpoons Q} \tag{23}$$

This expression indicates that the partial exchange rates must equal or exceed the overall exchange rate.

*Abortive Complex Formation*

As noted for creatine kinase, a variety of enzymes may form abortive complexes that are nonproductive forms of the enzyme. Such complexes form as a result of the adsorption of ligands under conditions where the

---

[55] C. A. Janson and W. W. Cleland, *J. Biol. Chem.* **249**, 2567 (1974).
[56] M. T. Skarstedt and E. Silverstein, *J. Biol. Chem.* **251**, 6775 (1976).

enzyme may not carry out its usual chemistry. For example, the binding of ribulose and NAD$^+$ to ribitol dehydrogenase leads to the formation of an enzyme·NAD$^+$·ribulose complex, which cannot allow for hydrogen transfer because both ligands are already in their oxidized states. The variety of possible binary and ternary abortive complexes (especially in three-substrate systems) presents a problem in analyzing exchange data in some cases. With regard to bisubstrate reactions and their exchange rate behavior, Wong and Hanes[57] were among the early investigators to describe several abortive effects that may limit the application of isotope exchange as an unambiguous tool for segregating ordered and random catalytic pathways. One case that attracted their attention involves the following ordered pathway with the release of A and Q from the central complexes to form EB and EP abortives.

$$E \rightleftharpoons^{A} EA \rightleftharpoons^{B} EAB \rightleftharpoons EPQ \rightleftharpoons^{P} EQ \rightleftharpoons^{Q} E$$

$$A + EB \qquad EP + Q \qquad (24)$$

They reasoned that such abortive complex formation provides a second route of exchange between A and Q that may be active even when high levels of the B − P pair reduce the uncomplexed enzyme form (E). Thus, $R_{A \leftrightarrow Q}$ will vary in a hyperbolic fashion with respect to the B − P pair even though the kinetic mechanism is ordered.

Rudolph and Fromm[58] considered yet another mechanism involving the formation of binary abortives of the sort described above. To illustrate their mechanism it is helpful to discuss it in terms of a hypothetical dehydrogenase reaction

$$S_{red} + NAD^+ \rightleftharpoons P_{ox} + NADH \qquad (25)$$

where $S_{red}$ and $P_{ox}$ represent the reduced and oxidized reactants. (This is done only for illustrative purposes, and the scheme may be applied to any type of enzyme reaction.) The detailed scheme, based on the Theorell–Chance mechanism, may be represented as follows:

---

[57] J. T.-F. Wong and C. S. Hanes, *Nature (London)* **203**, 492 (1964).
[58] F. B. Rudolph and H. J. Fromm, *J. Biol. Chem.* **246**, 6611 (1971).

$$
\begin{array}{c}
\phantom{XXXXXXXXXXXX} S_{red} \\
E \rightleftharpoons E \cdot NAD \searrow \phantom{XX} \nearrow E \cdot NADH \rightleftharpoons E \\
\updownarrow \phantom{XXXXX} P_{ox} \phantom{XXX} \updownarrow \\
E \cdot NAD \cdot P_{ox} \phantom{XXX} E \cdot NADH \cdot S_{red} \\
\updownarrow \phantom{XXXXXXXXXXX} \updownarrow \\
NAD + E \cdot P_{ox} \phantom{XX} E \cdot S_{red} + NADH
\end{array}
\qquad (26)
$$

Here, the binding of $NAD^+$ (or NADH) facilitates the binding of $P_{ox}$ (or $S_{red}$) to form the oxidized (or reduced) abortive ternary complexes, which may dissociate to give the oxidized (or reduced) abortive binary complexes $E \cdot P_{ox}$ and $E \cdot S_{red}$. The complicated equilibrium exchange law for the $NAD^+ \leftrightarrow NADH$ exchange predicts that this exchange is depressed at high $S_{red}$–$P_{ox}$. Thus, this sort of abortive complex scheme does not behave differently from the ordered schemes considered in Table II.

Boyer[6] has stated that these mechanisms are somewhat unlikely but that such possibilities should be considered. Certainly, the catalytic efficiency of enzymes attests to their tendency to avoid unnecessary detours, but for systems at equilibrium (the condition under which the exchange is measured) these complexes will be present if these schemes apply. In this respect, the reader is well advised to remember that isotope exchange at equilibrium is only one of many kinetic tools and that there are limitations in the overall scope of each.

To detect the presence of abortive ternary complexes, the kineticist may raise the concentration of dissimilar substrate–product pairs (e.g., glucose-ADP and ATP-glucose-6-P in the hexokinase reaction). This will lead to the inhibition of all exchanges irrespective of the kinetic mechanism. Nonetheless, product inhibition is still unrivaled as the means for detecting abortive complex formation. From the generality of abortive complex formation, it becomes obvious that enzyme sites are fairly flexible. The physiologic conditions of the cell are such that abortive complexes probably have more mechanistic significance than regulatory value. Indeed, rather high concentrations of substrates and products are generally required for abortive complexation, but there has been some speculation on physiologic roles, especially for the dehydrogenases.[59,60] It is also interesting to note that EB and EP abortive complexes might easily

---

[59] R. D. Cahn, N. O. Kaplan, L. Levine, and E. Zwilling, *Science* **136**, 962 (1962).
[60] D. L. Purich and H. J. Fromm, *Curr. Top. Cell. Regul.* **6**, 131 (1972).

form in certain ordered schemes. One attractive example is postulated in the ligand exclusion model wherein A binds in a cleft and B binds over A to form the productive EAB complex.[61]

Frieden[62] recently identified limitations on using initial velocity data to distinguish certain ordered and random mechanisms. He found that the rapid-equilibrium ordered mechanism with EB and EP abortive complexes cannot be distinguished from the random-addition case except by equilibrium exchange measurements. Frieden considered the following mechanism, in which A is the leading substrate in the sequential formation of the productive ternary complex.

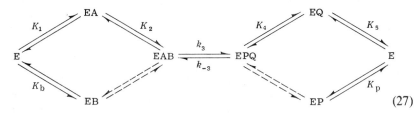
(27)

In this scheme, $K_1$, $K_2$, $K_4$, $K_5$, $K_b$, and $K_p$ are dissociation constants, and the rate constant of the slow step is either $k_3$ or $k_{-3}$, depending upon the reaction direction considered. The dashed arrows represent the additional pathways that do not occur in this mechanism but are common to the random addition pathway. Frieden[62] has shown that the rate expressions in the absence [Eq. (28)] and in the presence [Eq. (29)] of EB abortive formation are

$$V_0 = V_{mf} \bigg/ \left\{ 1 + \frac{K_2}{[B]} + \frac{K_1 K_2}{[A][B]} \right\} \tag{28}$$

$$V_0 = V_{mf} \bigg/ \left\{ 1 + \frac{K_2}{[B]} + \frac{K_1 K_2}{K_b [A]} + \frac{K_1 K_2}{[A][B]} \right\} \tag{29}$$

One cannot uniquely distinguish Eq. (29) from the random pathway by initial rates, Haldane relationships, Dalziel $\phi$ relationships, or the battery of inhibition techniques (including product inhibition, if EP forms).[62]

When the rapid-equilibrium assumption is also applied in the derivation of isotope exchange rate equations, Purich *et al.*[63] find that this method also is incapable of providing a rigorous distinction. The rate law

---

[61] D. G. Cross and H. F. Fisher, *J. Biol. Chem.* **245**, 2612 (1970).
[62] C. Frieden, *Biochem. Biophys. Res. Commun.* **68**, 914 (1976).
[63] D. L. Purich, R. D. Allison, and J. A. Todhunter, *Biochem. Biophys. Res. Commun.* **77**, 753 (1977).

obtained for the A ⇌ Q exchange under the rapid-equilibrium assumption requires that

$$\lim_{B \to \infty} R_{A \rightleftharpoons Q} = k_3 E_0 \bigg/ \left\{ \frac{1}{[A]} \left[ \frac{\beta}{K_p} + \frac{1}{K_b} \right] + \frac{1}{K_1 K_2} + \frac{K_{eq}}{K_4 K_5} \right\} K_1 K_2 \quad (30)$$

where $\beta = [P]/[B]$. This limit shows that as the level of B (and therefore P) is raised enormously high, the exchange rate reaches a maximum and will not decrease. One may anticipate this, since A* is in rapid equilibrium with the EA and EAB forms and the gross exchange rate depends only upon $k_3[EAB]$.

Conclusions based on these exchange equations can in fact be somewhat misleading because these rapid-equilibrium cases are obtained by eliminating terms from the complete rate expressions. To circumvent this Purich et al.[63] have numerically evaluated the complete equation for the ordered mechanism with EB and EP abortives. The basic idea stems from the fact that the rapid equilibrium condition is valid only for certain values of the rate constants. For this reason, they began with the more general expressions for the rates at steady state [Eq. (31)] and equilibrium [Eq. (32)].

$$\frac{E_0}{v} = \frac{[E] + [EA] + [EAB] + [EPQ] + [EQ] + [EB] + [EP]}{k_3[EAB]} \quad (31)$$

$$\frac{R}{E_0} = k_1 k_2 k_3 [A][B][E] \left\{ (k_{-1} + k_2[B]) - \frac{k_2 k_{-2}[B]}{k_{-2} + k_3} \right\}^{-1} \bigg/ \quad (32)$$
$$\{[E] + [EA] + [EAB] + [EPQ] + [EQ] + [EB] + [EP]\}$$

The determinants for the various enzyme species were obtained using the ENZ EQ program of Fromm.[12] These expressions can be examined by using values for rate constants to give rapid equilibration of all enzyme species except ternary complex interconversion. Bimolecular rate constants were set at $10^{7.5}$ $M^{-1}$ $sec^{-1}$, which is a reasonable value for enzyme–substrate reactions.[64] With the exception of ternary complex interconversion, the unimolecular rate constants were set at $10^{3.5}$ $sec^{-1}$, and thus all dissociation constants are $10^{-4}$ $M$. The ternary complex interconversion was described by various values. There are many combinations of rate constants that can be considered, but these are quite representative. It was noted that, for $k_3$ values of around 20 $sec^{-1}$ or less, initial rate plots in the absence of EB and EP complex formation were characteristic of equilibrium ordered mechanisms (i.e., there was a characteristic con-

---

[64] G. G. Hammes and P. R. Schimmel, in "The Enzymes" (P. Boyer, ed.), 3rd ed., Vol. 2, p. 67. Academic Press, New York, 1970.

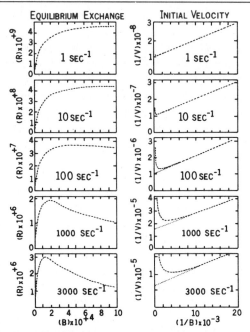

FIG. 11. Comparison of equilibrium exchange rates ($R$) and initial velocity ($v$) measurements for various values of the rate constants for ternary complex interconversion. Total enzyme was held at 20 n$M$; the bimolecular rate constants were $10^{7.5} M^{-1}$ sec$^{-1}$, and the unimolecular constants were $10^{3.5} M^{-1}$ sec$^{-1}$. For the equilibrium rate calculations, A and Q were maintained at 0.1 m$M$ ($K_m$ levels), and the absolute levels B and P were raised in a constant ratio as shown on the graph. For the initial-velocity comparisons, A was maintained at 0.05 m$M$ (0.5 $K_m$), and P and Q were zero.

vergence of all lines at the $1/v$ axis in plots of $1/v$ vs. $1/[B]$). However, by including EB and EP abortives, all plots give convergence to the left of the $1/v$ axis, and this shows that the general model behaves as predicted by the rapid-equilibrium equation.[62] The important finding of this study is shown in Fig. 11, where the A ⇌ Q equilibrium exchange rate and the initial velocity plots are compared at various values of $k_3$ and $k_{-3}$. When $k_3$ is fairly small (i.e., 10 sec$^{-1}$ or less), one must raise B and P to levels corresponding to 10–20 times $K_b$ and $K_p$ to observe any depression in the exchange. As $k_3$ gets greater than 30 or 50 sec$^{-1}$ the steady-state assumption becomes relevant, and the A ⇌ Q exchange is depressed by raising B and P. Significantly, the initial velocity plots show a strong substrate inhibition, characteristic of EB abortive formation, at or above such $k_3$ values.

The applicability of the rapid-equilibrium assumption thus depends on the rate of the ternary complex step. By definition, the rate constant for this step is $k_{cat}$ in either the rapid-equilibrium random or the Frieden mechanism. Since initial-rate or isotope exchange methods can easily distinguish these mechanism for $k_{cat}$ values above 30 or 50 sec$^{-1}$ (see Fig. 11), the problem pointed out by Frieden[62] is not at all general. Indeed, the $k_{cat}$ values of a number of enzyme systems are clearly considerably above this range.[63] Undoubtedly, there are a number of enzymes with lower catalytic constants, and these will not submit to a definitive distinction by these approaches. Nonetheless, knowledge of the $k_{cat}$ value can be used in conjunction with the plots like those in Fig. 11. If the $k_{cat}$ is above 30 or 50 sec$^{-1}$, then the ordered mechanism with EB and EP abortives will be discernible by the characteristic substrate inhibition and depressed equilibrium exchange rates. Both phenomena will be measurable at experimentally accessible values of B and P. The key point is that the rapid-equilibrium assumption must be valid for Frieden's mechanism to be considered, and $k_{cat}$ must be very small. Fromm[65] has suggested three experimental criteria for detecting the EB and EP abortives, and it will be of interest to determine their practical value.

## Enzymes at High Concentrations

Because the intracellular concentration of enzyme may be higher than feasible in most steady-state experiments, the properties of enzymes at high concentrations becomes of theoretical and metabolic interest. Purich and Fromm[15] have noted that equilibrium exchange reactions provide a valuable means to investigate catalysis at high enzyme levels, thereby eliminating the need for coupled enzyme assays or for continuous direct assay (as in the case of the dehydrogenases). The initial-rate phase of most systems away from equilibrium is too short to be examined except by stopped-flow methods. Since equilibrium exchange is first order, the accuracy of such experiments is enhanced, and it is possible to study enzymes at moderately high levels even by manual mixing procedures. This is shown in Fig. 12 for the yeast hexokinase P-II isozyme. The linearity of the enzyme concentration dependence of $R$ indicates that no kinetically important changes occur in the catalytic power of the phosphotransferase. It was also possible to show that a steady-state random mechanism prevails under such conditions.[15] Interestingly, the availability of rapid mixing–quenching devices may allow one to examine enzymes at concentrations in excess of 10 mg/ml with 1-sec reaction periods.[15] Provided the condition that total enzyme concentration is substantially below the least abundant substrate or product level, plots of $R$ vs. enzyme level

---

[65] H. J. Fromm, *Biochem. Biophys. Res. Commun.* **72,** 55 (1976).

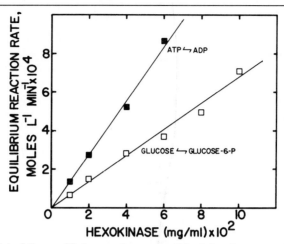

FIG. 12. Plot of the equilibrium exchange rate ($R$) of the glucose ⇌ glucose-6-P and ATP ⇌ ADP exchanges vs. the concentration of yeast hexokinase. Reaction mixtures (final volume, 0.125 ml) contained 58 m$M$ imidazole–NO$_3$ buffer (pH 6.5), 13 m$M$ Mg(NO$_3$)$_2$, 50 m$M$ glucose-6-P, 2.5 m$M$ glucose, 38.0 m$M$ ADP, 2.0 m$M$ ATP, and a variable concentration of the P-II isozyme of yeast hexokinase. Reaction samples were preincubated for 75 min at 28° before the addition of approximately 200,000 cpm of either [$^{14}$C]glucose-6-P (specific radioactivity, 20 mCi/mmole).

may provide evidence about the catalytic potential of self-associating enzymes. A linear dependence is expected in the absence of significant kinetic differences between dissociated and oligomeric systems. If oligomerization increases activity, concave upward plots will be observed.

## Concluding Remarks

The application of isotopic exchange methods to understand biological processes is truly broader than one may infer on the basis of this chapter. Such diverse topics as energy transduction, allosteric regulation, enzyme cooperativity, metabolic transport, and the behavior of supramolecular processes have all been enriched by the clever application of isotope exchange measurements. In this respect, the power and scope of the technique is still undergoing rapid expansion in theory and experiment. The only unfortunate aspect of this growth is the impossible task of adequately describing all the theoretical and technical advances in a single chapter. Obviously, our only choice was to limit the scope of this chapter, but we highly recommend that the reader examine Boyer's synopsis[6] of the present status of isotope exchange methodology.

## [2] The Isotope Trapping Method: Desorption Rates of Productive E·S Complexes[1]

### By Irwin A. Rose

Two of the basic questions asked about enzyme reaction mechanisms are: Are there covalent intermediates, and when do substrates combine and products leave in the catalytic cycle? To establish the presence of a covalent intermediate such as glutamyl-P in glutamine synthetase, Meister's group[2] used a novel isotope chase experiment in which [$^{14}$C]glutamate, ATP, and enzyme first were brought together without the third substrate $NH_3$ or $NH_2OH$. A solution containing $NH_2OH$ with a large amount of unlabeled glutamate, the so-called chase solution, was then added so that most of the radioactivity in glutamine hydroxamate would be derived from any [$^{14}$C]glutamyl-P held to the enzyme and only a small amount would come from turnover of the enzyme in the steady state that follows the completion of the reaction mixture. The latter would make a small contribution of radioactivity because of the low specific activity of free substrate at the time the reaction mixture was completed. Examples of the use of this approach to inquire into the possible role of covalent intermediates in two- and three-substrate reactions have been especially important when isotope exchange studies fail to reveal the presence of an intermediate for reasons that may be connected not to the mechanism but rather to the failure to have significant rates of product release in the partial system. In the isotope exchange mode of study, the order and rates of release of reactants from central complexes are important in determining rates of isotope exchange. The significance of any such exchange rate measured is based on a comparison with normal catalytic rates of the complete system, which may allow additional possible routes for product release. The absence of ADP:ATP exchange by glutamine synthetase when glutamate was present suggested that glutamyl-P was not formed. On the other hand if an enzyme-bound covalent intermediate is formed upon combining an enzyme with some components of the reaction system, any tightly bound products will have many seconds or even minutes to dissociate from the enzyme before the complex is required to react with the reaction-completing substrate in the chase phase of the

---

[1] This work was supported by United States Public Health Service Grants GM-20940, CA-06927, and RR-05539 and by an appropriation from the Commonwealth of Pennsylvania.

[2] P. R. Krishnaswamy, V. Pamiljans, and A. Meister, *J. Biol. Chem.* **237**, 2932 (1962).

experiment. Because the formation of covalent intermediate is pulled by product release, there may be a close equivalence in amount of complex with the number of active sites. Although the complex that reacts in the chase will lack first product and will be an abnormal intermediate in a kinetic sense, the method of Meister constitutes a major chemical device for recognizing tightly bound covalent compounds that can be formed on and transformed by an enzyme. The most general explanation for a negative result where a covalent intermediate is expected is that the missing substrate is required to induce the proper conformation in the enzyme for the intermediate to form. For these reasons failure to trap is not proof that a covalent intermediate is not formed in the net reaction.

A positive result in this test for a covalent intermediate has an interesting alternative interpretation: that a covalent intermediate has not been formed but that the unmodified radioactive substrate in a Michaelis or noncovalent complex is converted to product at a greater rate than it is released from the enzyme. This alternative interpretation requires that the extent of the isotope partition be a function of the second substrate concentration, whereas if a bound covalent intermediate is formed it is probable that it will survive a short period of dilution in unlabeled substrate before the reaction-completing substrate is added. This chapter will deal with noncovalent presumed Michaelis complexes that one would not expect to be successfully chased into product if dilution with carrier substrate intervened before completion of the reaction mixture. Since E·S complexes must have rates of substrate release that are at least as rapid as the net reaction in the reverse direction, rarely slower than 1 $sec^{-1}$, the addition of diluting substrate should clear the enzyme rapidly.

Based on these considerations, an example of trapping from a noncovalent Michaelis complex was sought; it was found in the first example that was tried, yeast hexokinase and glucose.[3] Not only could the glucose of the binary E·glucose complex be trapped when ATP was added, but the ternary complex, E·glucose·ATP was shown to have its glucose completely locked in, relative to the rate of its progress to product.

*What Can Be Learned?* In a properly controlled experiment the amount of substrate (*A) that is trapped compared with the amount known to be bound provides an estimate of the minimum fraction that one can say is *bound properly* in the sense that the enzyme does not have to dissociate *A before it enters into productive reaction. Therefore the enzyme is able to do catalysis by the sequence A + E → E·A → E·A·B → product. This may not be the major sequence of events

[3] I. A. Rose, E. L. O'Connell, S. Litwin, and J. Bar Tana, *J. Biol. Chem.* **249,** 5163 (1974).

under normal steady-state conditions. However, it is the sequence that is approached as [A] → ∞ if indeed the *steady-state* form of E is one to which A can bind. The ability to recognize the occurrence of a proper binary complex or of complexes in rapid equilibrium with a proper complex is a major feature of the isotope partition approach. Improper binding of substrate is not revealed by steady-state kinetic approaches. Likewise the demonstration of binding, as in an equilibrium dialysis study, is no assurance of the formation of a functional complex. Therefore the ability to trap *A from E*A provides information that relates to both kinetic and equilibrium aspects of complex quality.

The failure to trap any or less than all of E*A in the experiment has three interpretations.

1. More than one species of complex are present, not all of which are competent Michaelis complexes.
2. The dissociation rate of EA may be so rapid that even at $>K_m$ concentrations of second substrate the exchange rate with carrier A is too rapid for a practical study with sufficiently large concentrations of B. The upper limit to the off rate of A can be estimated from the equation $K_{dis} = k_{off}/k_{on}$ where $k_{on}$, when it is diffusion limited, is $\sim 10^8 \ M^{-1} \ sec^{-1}$, so that knowing $V_{max}/K_m$ of trapping substrate (B) and $K_{dis}$ of EA one can use the kinetic equation derived in the next section to give the maximum concentration of B that would be necessary to trap a significant amount of *A if E*A is functional. The amount of B necessary to trap half of E*A that is known to be present cannot be greater than $2 \times K_{dis(EA)} \times 10^8 \ M^{-1} \ sec^{-1}/(V_{max}/K_{m(B)})$. [If it is not feasible to study this concentration of B owing to fiscal, solubility, viscosity, or inhibitory factors, no decision can be reached.] Failing to trap *A at this concentration of B therefore implies either the first interpretation above or interpretation 3.
3. *A dissociates from the ternary complex E·*A·B at a rate much greater than $V_{max}$ of the reaction. On the other hand, if one is able to trap some of E*A in the first turnover of the partition, then it must follow that $k_{off}$ of A from E·A·B is not $\gg V_{max}$. A mathematical treatment for a two-substrate problem follows.

Kinetic Expressions

The partition of E*A between free *A and product, *P, will depend on its decay along alternative routes given in the minimum scheme.

$$E \xrightarrow{k_1[*A]} E*A \underset{k_4}{\overset{k_3B}{\rightleftarrows}} E_B^{*A} \xrightarrow{k_5} \xrightarrow{k_9} E + *P$$

$$\downarrow k_2 \qquad\qquad \downarrow k_7$$

$$*A + E \qquad\qquad *A + E_B$$

Steps 2 and 7 are written as irreversible because of the presence of unlabeled A in the reaction mixture. Formation of labeled product may depend on several steps, but trapping depends only on the steps up to and including the first irreversible step in each path toward product formation. Therefore, in the scheme presented the trapping of E*A in the single turnover that is meaningful does not depend on $k_9$. For readily reversible reactions, $k_5$ will include all steps from $E_B^{*A}$ to release of a product. Litwin, in Rose et al.,[3] showed that integration of the differential equations that describe this system provides the same partition of E*A into *P and *A as derived from simple steady-state partition analysis.

*Desorption from the Binary Complex.* Assume that all the E*A present in the pulse, $[E*A]_0$, is functional with the average rate constants shown. In the limit as [B] increases the fraction of bound *A that is trapped

$$\frac{*P_{max}}{[E*A]_0} = \frac{k_5}{(k_5 + k_7)} \quad \text{or} \quad k_7 = k_5 \left[\frac{[E*A]_0}{*P_{max}} - 1\right]$$

Rate constant $k_5$ is at least as great as $V_{max}/E_T$, $V_{max}/E_T = k_5k_9/(k_5 + k_9)$, and $k_9$ is a bulk constant containing all other terms required to generate E. Thus a lower limit to $k_7$ is obtained from

$$k_7 \geq \frac{V_{max}}{E_T} \left[\frac{[EA]_0}{*P_{max}} - 1\right]$$

By a partition analysis approach, Cleland[4] has shown that

$$*P = *P_{max}/(1 + K_B'/[B])$$

where $K_B' = [k_2(k_4 + k_5 + k_7)]/[k_3(k_5 + k_7)]$ is the concentration of B that causes half the maximum amount of trapping of $[E*A]_0$ that is found when [B] increases without limit. Thus $K_B'$ is to an isotope trapping experiment as $K_m$ is to a steady-state experiment.

Although the steady-state parameter $(V_{max}/E_t)/K_B = k_3k_5/(k_4 + k_5)$ does not quite serve to dissect $k_2$ out of $K_B'$ exactly, it can be seen to be very useful.

---

[4] W. W. Cleland. *Biochemistry* **14**, 3220 (1975).

$$\frac{V_{\max}/E_T}{K_B} K_B' = \frac{k_2(k_4 + k_5 + k_7)k_5}{(k_5 + k_7)(k_4 + k_5)}$$

$$= k_2 \bigg/ \left[1 + \frac{k_7}{k_5}\frac{k_4}{k_4 + k_5 + k_7}\right]$$

$$k_2 = \frac{K_B'}{K_B}\frac{V_{\max}}{E_T}\left(1 + \left[\frac{[E*A]_0}{*P_{\max}} - 1\right]\frac{k_4}{k_4 + k_5 + k_7}\right)$$

and $k_2$ will lie between

$$\frac{K_B'}{K_B}\frac{V_{\max}}{E_T} \quad \text{and} \quad \frac{K_B'}{K_B}\frac{V_{\max}}{E_T}\frac{[E*A]_0}{*P_{\max}}$$

The upper limit of $k_2$ is observed if $k_4$ is greater than $k_5 + k_7$, that is, if B is in rapid equilibrium with the ternary substrates complex. If *A does not leave at a significant rate from the ternary complex, $k_7 < k_5$, $k_2$ will correspond to the lower limit. Therefore, for this model, the upper and lower limits of $k_2$ (as a bulk constant for release of functional A) can be determined relative to $V_{\max}/E_T$.

It is expected that $[E*A]_0$ will be composed of more than one species, which are in equilibrium with each other in the pulse solution. B may react properly with some of these, and some of the binary complexes of EA will not be able to release A directly.

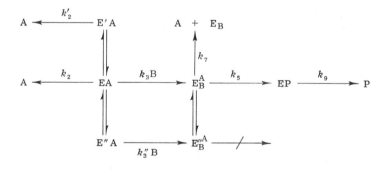

The fraction of $[E*A]_0$ that is trapped into product, $*P_{\max}/[E*A]_0$ as $B \to \infty$ will depend not only on the partition at the ternary complex, $k_5/(k_5 + k_7)$, but on the position of equilibrium among the binary forms and the rates of their interconversion. In a comparison of different A substrates for the same enzyme, it is expected that $K_B'$ will be greater for those that bind weakly. This effect will not be seen in the $K_m$ of B, which is always determined at saturating concentrations of A.

In a careful study of the NAD$^+$-specific malic enzyme of *Ascaris*,

Landsperger et al.[5] found that $*P_{max}$ for NAD trapping was close to $E_T$ (1.33 vs. 1.50 nmol) whereas $*P_{max}$ for malate was only 0.6 nmol or about 40% of theoretical. Although part of the loss of malate might be at the ternary complex or at the binary complex in catalytically functional forms that do not react rapidly enough, as in $E'A \rightarrow EA$ above, it is also possible that ~60% of the [$^{14}C$]malate on the enzyme is bound incorrectly. Thus mesotartrate acts well as a competitive inhibitor, suggesting that both carboxyl sites could accept the C-1 carboxyl of malate. The binary $E*NAD^+$ complex could be fully trapped in spite of ~60% incorrect ternary complex formed if malate leaves rapidly and returns again in the correct manner ~40% of the time, whereas the correct species form product rapidly compared with a slow desorption of $*NAD^+$.

*Desorption from the Ternary Complex.* Failure to trap $*A$ in $[E*A]_0$, i.e., $*P_{max} \ll [E*A]_0$ cannot be ascribed definitely to rapid desorption of A from the ternary substrate complex relative to product formation unless it is certain that the radioactive ternary complex was present at the beginning of the chase period. When some $*A$ is trapped it can be concluded that $k_7$ is not great compared with the interconversion of $E_B^{*A}$ to first product. However, only if $k_9$ is much greater than $k_5$ is it possible to relate the trapping to $V_{max}/E_T$ and therefore evaluate $k_7$. It may be possible to rule out a small $k_9$ by failure to show noncompetitive inhibition by product. In four cases substrate dissociation from the ternary substrate complex has been shown to be negligible or very slow: glucose from hexokinase,[3] $NAD^+$ from *Ascaris* malic enzyme,[5] phosphoenol pyruvate from pyruvate kinase,[6] and fructose from fructose kinase.[7]

Procedures

Three steps are used in the isotope partition analysis of binary complex competence and stability as follows:

1. Formation of the complex or *pulse* step $E + *A \rightleftharpoons E*A$
2. Partition of the complex between product and free substrate or *chase* step

$$E*A + (A + B) \longrightarrow \begin{array}{c} *P \\ *A \\ P \end{array}$$

3. Termination of the reaction and analysis

[5] W. J. Landsperger, D. W. Fodge, and B. G. Harris, *J. Biol. Chem.* **253**, 1868 (1978).
[6] L. G. Dann and H. G. Britton, *Biochem. J.* **169**, 39 (1978).
[7] R. W. Raushel and W. W. Cleland, *Biochemistry* **16**, 2176 (1977).

*Step 1. Formation of the E\*A Complex ($E + {}^*A \rightarrow E{}^*A$).* The enzyme and labeled substrate should be mixed under conditions and at concentrations that give a good proportion of the substrate as a stable complex with the enzyme. Upon addition of acid, alkali, detergent, phenol, etc., radioactive reaction product should be minimal ($<[E{}^*A]_0$) indicating that reaction-completing substrates or cofactors are indeed absent. All elements necessary to form the complex to be tested may be included short of allowing product formation to occur during this period, the pulse period. Therefore, in testing hexokinase, the following mixes could be made: E + glucose, E + ATP, E + ATP + Mg, E + glucose + Mg, and E + ATP + glucose. The last mix is most difficult to study successfully because of the problem of traces of metal ion contamination leading to product formation in the pulse.

The amount of radioactivity that is bound to the enzyme *must* be known accurately for a full use of the method. It is frequently[5,6] not safe to assume that published values for the dissociation constant can be applied to the enzyme at hand at the concentration of enzyme that will be used. The Colowick and Womack[8] rate-of-dialysis method for determining the dissociation constant of E·A is technically sound and rapid; it can be done with a reasonable amount of material in most cases and is nondestructive of enzyme. This method can detect the presence of an isotopic impurity that binds more tightly than the main isotopic species. Such an impurity, whether substrate or not, would influence the interpretation of an isotope partition experiment. The pulse solution may be of any volume, but it is usually necessary that the enzyme and substrate be as concentrated as possible to attain a high fraction of the total substrate in the bound form. Unbound enzyme and substrate contribute to irrelevant radioactive product that decreases the sensitivity of the experiment. If the data obtained from the experiment are to be used quantitatively to give rate constants, the state of the E·A complex should be that of an initial-rate kinetic experiment; e.g., the state of polymerization should not differ. On the other hand, some important statements can be made of a qualitative nature from a demonstration of isotope trapping, and in this case the requirement may be relaxed.

*Step 2. Mixing and Reaction.* The pulse mixture is next brought together with missing components plus unlabeled test substrate, the chase solution, in such a way that reaction will proceed under conditions used for initial-rate kinetic studies. This is necessary because the steady-state parameter ($V_{\max}/K_B$) is used in the calculations for $k_2$. For this reason, the high concentration of substrate used to dilute the isotopic

---

[8] S. P. Colowick and F. C. Womack, *J. Biol. Chem.* **244**, 774 (1969).

substrate should not cause inhibition in the usual assay. (The concentration of second substrate used in the chase solution will determine the rate of formation of the ternary Michaelis complex in a second-order reaction. The faster the dissociation rate of E*A, the more B is necessary to prevent it.) It is recommended that the trapping of bound substrate be given at least five half-lives before *termination* to be assured that at least 90% of whatever original complex was present has an opportunity to be trapped. Waiting this long for termination allows about five turnovers with the diluted substrate. Therefore, assuming that termination could be made by rapid quenching, it might be sufficient to dilute the radioactive substrate with ~50 times as much carrier substrate. On the other hand, if a rapid quench device is not available, manual quenching usually requires 1–2 sec after the mixing is completed. Therefore dilution with unlabeled substrate may have to be much greater so that the amount of radioactivity converted to product in this time will not make the determination of the trapped counts inaccurate. This phase of the experiment may be conducted in very large volumes to allow good dilution without high concentrations of substrate. A large volume helps to avoid the problem of an unfavorable reaction equilibrium in the trapping reaction. It is not desirable to allow the radioactive product to approach chemical equilibrium with unlabeled substrate because simple isotope exchange would lead to loss of label from product. In order to give sufficient sensitivity for the trapping of only 10% of the bound substrate the extent of dilution by unlabeled A contained in the chase solution should be at least 10(× chase time × turnover rate)-fold. Therefore, if an enzyme turns over at 100 sec$^{-1}$ with 2 sec before termination, a 2000-fold dilution is required.

In mixing the pulse volume with a generally much larger chase volume containing a mixture of carrier substrate and reaction-completing species, it would be ideal if full dilution of the isotope occurred at the time when free enzyme came into contact with second substrate. In this way there would be only two conditions of labeled substrate to be considered in evaluating the source of labeled product: [E*A]$_0$ and unbound substrate at the final low specific activity. Such mixing is impossible to achieve in fact. Ray and Long[9] have provided a simple qualitative color test to compare the effectiveness of mixing of a small-volume stream into a large volume, thereby allowing one to optimize geometry and rate of stirring.

The nature of the mixing problem is shown in Fig. 1. At the time of half mixing ($t_{1/2}$) the average substrate specific activity has been brought to two times its final value. During this period, B has reached half of its final concentration, which, in some situations, may be much greater than $K_m$.

[9] W. J. Ray, Jr., and J. W. Long, *Biochemistry* **15,** 3990 (1976).

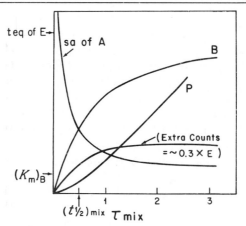

FIG. 1. Graph showing the extra counts in product (P) due to incomplete mixing at the time the reaction begins, $k_{mix} \simeq 3\ k_{cat}$; sa = specific activity.

Meanwhile the $[E^*A]_0$ is partitioning to $^*A + E$ and $^*P + E$, and these molecules of E and any free E present initially will interact with substrates present in the immediate environment. Formation of radioactive product formed from free E will have nothing to do with $k_2$, and the proposed ordered path of reaction and should not be included in $^*P$ of the primary process, if indeed direct trapping is real. As in the case of the trapping of $E^*A$, the trapping of these extra counts will not depend on $k_9$ and therefore may occur in the pre-steady state at a rate greater than $V_{max}/E_t$. However, if $V_{max}$ is very large there could be multiple turnovers of the enzyme with substrate at a specific activity greater than final. Figure 1 shows a plot of extra counts incorporated in the case that $\tau_{mixing}^{-1} \simeq 3\ k_{cat}$, where the extra counts will reach ~0.3 of an equivalent of E. Increasing the concentration of B when it is already greater than $K_m$ will not affect extra counts very much. However, lowering the concentration of B in the chase to the $K_m$ value will obviate the mixing artifact and may not decrease trapping of $E^*A$ if $k_2$ is small. Therefore it is important to cover a wide range of [B] in any experiment. Hand mixing is adequate when $K'_B < K_B$, and excessive concentrations of B can be avoided. This was the case in the yeast hexokinase study[3] with $k_{cat} \sim 300\ sec^{-1}$ and $K'_{ATP} \simeq 0.3\ K_m$ of ATP.

Mixing by machine can provide mixing half-times of the order of a millisecond.[10,11] About half an equivalent of product would be formed in this

---

[10] Q. H. Gibson, this series, Vol. 16, p. 187.
[11] D. P. Ballou and G. A. Palmer, *Anal. Chem.* **46**, 1248 (1974).

time by an enzyme with $V_m/E_T \cong 1000$ sec$^{-1}$ if B was present at $> K_m$ levels in the chase solution. This would result in a significant mixing contribution of label to P not due to direct trapping of $[E^*A]_0$. It is important to note at this point that this mixing artifact will go undetected if the conditions of the experiment are changed to have *A present in the chase solution added to free E.

Clearly, if $^*P/[E^*A]_0$ is $>1$, a mixing artifact is indicated. Proper behavior would be indicated by a linear reciprocal plot: 1/*P vs. 1/B. Any change in technique that increases the mixing/catalysis rate ratio should not influence the result if there is no mixing artifact. Some ways of changing this ratio are to change the flow rate in a mechanical mixer; to alter solution viscosity with a neutral component such as glycerol[11]; or to change temperature to decrease catalysis preferentially.[12] On the other hand, if a mixing contribution to labeled product cannot be avoided, it would be necessary to quantitate it and correct for it. The most exact way to do this is to provide conditions in the pulse in which [E*A] cannot form and yet catalysis begins normally upon mixing. Some approaches to this are as follows: Include a competitive inhibitor of the binding of *A in the pulse solution, the effect of which will be immediately overcome by the high concentration of A in the chase mixture; also, if the pulse is set up at a pH unfavorable for the formation of a Michaelis complex the rapid change of pH in the mixing step will trigger reaction of substrate as it is being diluted. A variation on this approach is to delete $Mg^{2+}$ from a pulse of E + *ADP (or *ATP), if $Mg^{2+}$ is needed for productive binding. The presence of MgADP (or MgATP) in the chase will result in very rapid, post-mixing formation of Mg*ADP (or Mg*ATP) with half-times of $\sim$msec.

*Step 3. Termination and Analysis.* Because reaction time is not a critical matter if a high dilution of the isotope is made in the chase step, it may not be necessary to have stopping times shorter than 1 sec. However, a simple flow-through apparatus can easily be made to give much shorter chase times. When very reproducible reaction times are necessary, a machine that drives the syringe contents through a mixing and quenching cell may be needed. Such an apparatus is described by Gibson[10] and by Ballou[11] and is a component of most modern stopped-flow machines.

When stopping time is not very reproducible, and even when it is thought to be, it is important to separate the substrate and product and determine their specific activities. Knowing the specific activity (SA) of substrate in the pulse, the amount of substrate (nanomoles) trapped is

---

[12] I. A. Rose and E. L. O'Connell, *J. Biol. Chem.* **252**, 479 (1977).

$$\frac{\text{(SA product} - \text{SA final substrate)}}{\text{(SA initial substrate} - \text{SA final substrate)}} \times \text{nanomoles of product}$$

When reaction time is quite reproducible, as it can be by machine drive or manually in 2–5 sec, one can determine the trapped counts by using an approach that does not require determinations of the specific activity of product and finally mixed substrate. In this approach a control experiment is done in which the pulse solution contains the fully diluted substrate: (E + *A + A) followed by B with termination at the same time. In this case, the counts found in the product will be fewer by the counts trapped. Compared with the initial specific activity of *A in the parallel experiment, one obtains the amount of *A trapped. As mentioned earlier, this approach does not measure mixing artifacts and therefore assumes that mixing is adequate. Because time of reaction must be the same in each experiment when this kind of approach is used, it is important that the enzyme be completely inactive when the reaction is terminated with acid. It is also important that the enzyme not be reactivated at any time prior to product isolation. Thus if the reaction is terminated with $HClO_4$ that, after centrifugation, is neutralized with KOH prior to separation by ion exchange, it is desirable to delay the neutralization until immediately before the columning step to avoid spurious reaction due to enzyme that has been reactivated at neutrality.

Working with large amounts of enzyme it is well to be aware that mere acidification and centrifugation are often insufficient to remove all enzyme and that often there is some recovery of enzyme activity upon neutralization. In a study where irreversible quenching of muscle aldolase was necessary prior to neutralization, the only method found that was mild enough to preserve the triose-Ps present was digestion of the aldolase by pepsin at pH $\sim 1$.[12]

## Limitations and Extensions of the Methodology

The isotope partition approach can be applied in almost any situation with substrate of sufficient affinity, or, lacking that, with sufficient amount of enzyme to give a useful amount of label in the enzyme-bound state. The initial-rate kinetics of the enzyme saturated with first substrate in going from pulse to chase conditions should not show hysteretic phenomena in approaching the steady state. A lag in generating product due to some rate-limiting conformational change or dissociation into subunits will give a falsely higher $k_{off}$ for the bound substrate. Although a burst or lag may be too short-lived to be observed in any usual approach to steady state study, it may be possible to observe it by using lower temperatures.

In the end this uncertainty can be dealt with by a rapid-quench experiment over a range of reaction times, leaving out the diluting substrate from the chase mixture. A plot of product formed vs. time would indicate whether a hysteretic phenomenon occurs at the time of mixing. Likewise a hysteretic development of the normal substrate release kinetics from the binary or ternary complexes relative to product formation would result in a false rate constant for substrate release. The latter possibility can be recognized if the calculated rate is slower than $V_{max}$ for the reverse process.

Because the isotope partition method depends on product formation rates that can be very large, it can be used to determine rates of substrate release that are more rapid than those measured by stopped flow, where dead time of the optics limits the method to more than a millisecond. Furthermore, rates that are too fast for the usual quench method can be determined because the limitations imposed by mixing rates and time of stopping may be considerably relaxed for the isotope partition experiment.

Where a slow conformational change may exist between the binary and functional ternary complex, it may be possible to show trapping only in the presence of an abortive inhibitor. For example, suppose glucose of $E \cdot {}^*G$ could not be trapped for this reason, but that $E^{*G}_{ADP}$ would form a complex with the proper conformation during the pulse period. With a rapid exchange of ATP for ADP in the chase, trapping of $^*G$ becomes possible. A similar mode of action of an allosteric activator would be seen when it was present in the pulse mixture, but not in the chase.

Powers and Meister[13] used the isotope trapping method to investigate the possibility of two separate ATP sites on carbamyl-P synthetase.

$$HCO_3 + ATP + NH_3 \longrightarrow H_2NCO_2^- \longrightarrow H_2NC\begin{smallmatrix}O\\OPO_3\end{smallmatrix} + ADP$$
$$\downarrow \qquad\qquad \uparrow$$
$$ADP + P_i \qquad ATP$$

Using [$^{32}$P]ATP and $HOCO_2^-$ in the pulse, the authors report the trapping of $^{32}$P into both carbamyl-P and $P_i$ after dilution with unlabeled ATP + $NH_3$. This result requires that two sites be occupied in the pulse.

Isotope trapping has been used to determine whether an effector molecule that is also a substrate is acting at the substrate site. $^{14}CO_2$, which activates ribulose-1,5-bisphosphate carboxylase, could not be trapped into PGA after dilution in $^{12}CO_2$ and substrate,[14] suggesting that it may activate at a unique site. Results of this kind cannot be given positive interpretation unless the minimal conditions for trapping have been met, $B > K_{dis(EA)} \times K_B/V_m \times 10^8$.

[13] S. G. Powers and A. Meister, *J. Biol. Chem.* **253**, 800 (1978).
[14] G. H. Lorimer, M. R. Badger, and J. J. Andrews, *Biochemistry* **15**, 529 (1976).

Isotope partition analysis may offer unique approaches to the question of whether crystals of enzyme·substrate complexes are closely related to the functional structures found in solution.[15] If the crystal that is examined by X-ray diffraction is grown in radioactive substrate, one can perform the usual isotope partition experiment in which the crystal is dissolved in a large volume of unlabeled substrate and second substrate. Presumably, a crystal form that represents the functional EA complex will react as it dissolves, whereas an improper structure will require rearrangements that will allow little or no trapping by second substrate. Ease of trapping, as judged by the magnitude of $K'_B$, may prove to be greater or less for a crystalline complex of E*A than for the equilibrium solution form of E*A. Therefore it may be possible to arrange different crystalline forms of E*A in the sequence they have in the catalytic path.

## Summary

The method of isotope trapping of E·S complexes provides a useful additional approach to enzyme kinetics and is unique in that it evaluates the catalytic competence and the quality of the range of complexes that may be present under the designation E·S. With substrates of good affinity the method can be used at protein concentrations at the nanomolar level. Incomplete trapping under $V_{max}$ conditions may indicate that an incorrect interaction of substrate occurs among the bound forms. Alternatively, the ternary substrates complex may not be completely trapped; in this case its rate of partition between first product and free substrate gives new information on the rate profile of the whole system. A distinction between these alternative interpretations may be possible in some cases by the direct evaluation of the competence and partition of ternary substrate complexes using the chase methodology (unpublished studies of K. D. Wilkinson and I. A. Rose).

As has been indicated, the isotope trapping method needs to be undertaken with a good appreciation of its limitations and sources of error. To avoid these it is necessary that the dependence of trapping on second substrate concentration in the chase solution be measured and that suitable controls for a mixing artifact be included.

---

[15] L. Stryer, *Annu. Rev. Biochem.* **37**, 25 (1968).

## [3] Oxygen-18 Probes of Enzymic Reactions of Phosphate Compounds[1]

By DAVID D. HACKNEY, KERSTIN E. STEMPEL, and PAUL D. BOYER

Oxygen-18 probes of enzymic catalyses involving formation or cleavage of O—P bonds are commonly of two types. In one the source of oxygen in inorganic phosphate ($P_i$) or other phosphate compounds is determined; in the other the extent of exchange of oxygens between $P_i$ or other phosphate compounds and the source of oxygens is measured. Additional important refinements may concern the position of the oxygen, whether bridge or branch, in phosphate esters or diesters, and the distribution of $^{18}O$-labeled species present based on the number of $^{18}O$ atoms associated with each phosphate molecule.

Most past experimental work has made use of determination of the total $^{18}O$ present in isolated $P_i$ by conversion of the oxygens to $CO_2$ and measurement of the mass 46/44 ratio. This is still a useful procedure for many purposes, but more recently methods based on formation and measurement of volatile phosphate derivatives with direct measurement of their mass in a spectrometer have become available and offer certain advantages. Also, a valuable newer procedure is the use of $^{31}P$ nuclear magnetic resonance (NMR) to measure the position, number, and amount of $^{18}O$ in suitable phosphate samples.

This chapter gives information about the following topics: preparation of $^{18}O$-labeled $P_i$ standards; measurement of $^{18}O$ in water; preparation of $P_i$ and ATP highly labeled with $^{18}O$; isolation of $P_i$ from reaction mixtures; measurement of $^{18}O$ in $P_i$ and other phosphate compounds; some applications of $^{18}O$ measurements.

Procedures for $^{18}O$ measurements based on conversion of $P_i$ oxygens to $CO_2$ and their application to study of oxidative phosphorylation have been described in earlier volumes of this series.[2,3] The present chapter will give some refinements in these procedures, but will emphasize the methodology and application of volatile phosphate procedures as developed and used in this laboratory in the past several years.

---

[1] Preparation of this paper and researches in our laboratory were supported by U. S. Public Health Service Grant GM 11094, NSF Grant PCM 75-18884, and Energy Research and Development Administration Contract EY-76-S-03-0034.

[2] P. D. Boyer and D. M. Bryan, this series, Vol. 10, p. 60.

[3] P. D. Boyer and K. E. Stempel, this series, Vol. 55, p. 245.

## Preparation of $^{18}$O-Labeled $P_i$

Oxygen-18-labeled $P_i$ is not readily available commercially and is expensive. For some purposes it is desirable to have standards of known $^{18}$O content or $P_i$ nearly fully labeled with $^{18}$O. Procedures for preparation of $[^{18}O]P_i$ are thus given here.

### Preparation of Phosphate Standards

*Principle.* Precisely measured amounts of enriched $[^{18}O]P_i$ and nonisotopic HOH are brought to isotopic equilibrium with the aid of exchange catalyzed by inorganic pyrophosphatase. The $^{18}$O content of the water is determined accurately, and from this the total $^{18}$O in the original $[^{18}O]P_i$ is calculated. This procedure eliminates any isotope effect on the equilibrium distribution of $^{18}$O between HOH and $P_i$ as almost all the $^{18}$O is transferred to water.[4] Standards of lower $^{18}$O content are prepared by appropriate dilution of this accurately measured primary high standard.

*Materials*
Water, not enriched with $^{18}$O, but with known $^{18}$O content, measured by the same procedure used for $^{18}$O-enriched water after equilibration with $[^{18}O]P_i$
$[^{18}O]KH_2PO_4$, about 8–20 atom%, recrystallized from 50% ethanol and dried in a vacuum oven
$MgCl_2$ 1 $M$
(17 $N$) NaOH, saturated
Inorganic pyrophosphatase (Sigma Chemical Company), lyophilized powder

*Procedure.* Weighed amounts of water, $[^{18}O]KH_2PO_4$, 1 $M$ $MgCl_2$, and 17 $N$ NaOH are placed in a sealed weighing vial. Convenient amounts are 2.0 g of HOH, 50 mg of $[^{18}O]KH_2PO_4$, 11 mg of 1 $M$ $MgCl_2$ ($\simeq$ 10 mg of HOH), and 15 mg of 17 $N$ NaOH ($\simeq$ 5 mg of HOH) to give a solution at about pH 7.2 containing approximately 0.17 $M$ $P_i$ and 5 m$M$ $MgCl_2$ at room temperature. Pyrophosphatase (4 mg) is added and the solution allowed to stand for 24 hr or longer. The approximate half-time for exchange under these conditions is about 0.7 hr and pyrophosphatase remains active. Thus exchange should be 99.9% complete in 24 hr. The final atom % of the water under the above conditions is about 0.3–0.6, a convenient level for analysis.

After equilibration the water is recovered by freezing, evacuating, and

---

[4] The equilibrium distribution favors $^{18}$O accumulation in $P_i$ by 3–4% (J. F. Kirsch, private communication).

lyophilizing in a Y-tube connected to a vacuum line. The $^{18}$O-content of the water is determined as described below.

The $^{18}$O-content of the $P_i$ can then be calculated as follows:

$$\frac{\text{a.p.e. in equilibrated water} \times (\text{moles water} + \text{moles } P_i \times 4)}{\text{moles } P_i \times 4}$$

$$= \text{a.p.e. in original } P_i.$$

(a.p.e. = Atom % excess above natural isotope abundance.)

Replicate determinations by the above method agreed within range of 0.05%.

More dilute [$^{18}$O]$P_i$ standards prepared as described above and analyzed by the guanidine hydrochloride pyrolysis procedure[1,2] gave $^{18}$O values about 1.5% lower than the true value. This may reflect a slight contribution of glass oxygens to the $CO_2$ pool during pyrolysis. Variation of such contribution could account for the slight variability encountered in $^{18}$O analysis by guanidine hydrochloride pyrolysis. The small error inherent in the guanidine hydrochloride pyrolysis procedure is negligible for many experimental purposes. It can be partially compensated for by inclusion of appropriate [$^{18}$O]$P_i$ internal standards in experimental runs. Inclusion of such standards is also of value to correct for any extraneous oxygens from other sources appearing in the final $P_i$ used for analysis.

It is often useful to dilute highly $^{18}$O-enriched $P_i$ with unenriched $P_i$. This is necessary for preparation of standards and, when sample size must be limited to prevent excessive use of enzymes, so that a pulse of $^{18}$O can be added without disturbing the chemical equilibrium. Analysis of such mixed samples by the guanidine hydrochloride pyrolysis method could pose problems if the $CO_2$ preferentially derived both of its oxygens from one $P_i$ molecule. Control experiments with such heterogeneous samples indicate that this is not the case as the distribution of $CO_2$ species of mass 44, 46, and 48 is that expected from the binomial distribution for the average $^{18}$O enrichment indicating that essentially complete randomization has occurred.

Measurement of $^{18}$O Content of Water

*Principle.* When ample volumes of water are available, precise measurement of the $^{18}$O content is possible through isotopic equilibration with $CO_2$ and measurement of the mass 46/44 ratio. In our experience, with water of 0.2–1.0 atom% $^{18}$O, replicate analyses agree within a standard deviation of 0.1–0.2%, that is, within 1–2 parts per thousand.

With samples of water less than a few tenths of a milliliter, equilibration by shaking becomes difficult. For such samples, methanol is added to give adequate volume for shaking without undue increase in the half-time for equilibration.

*Materials*

Nonisotopic $CO_2$. We maintain a reservoir of 500-ml capacity connected both to an 800-mm leg manometer and, via a small dosing chamber of measured volume (about 0.5 ml), to the vacuum line. The storage bulb is filled with $CO_2$ from a source such as Dry Ice, with prior freezing in $N_2$ and evacuation to remove impurities. From the volume, pressure, and temperature, the micromoles of $CO_2$ admitted to the sample can be satisfactorily estimated assuming that Boyle's law holds.

Equilibration bulbs of about 2.5-ml volume joined through a 10/30 joint to a stopcock with a male 10/30 joint to connect to the vacuum line

*Procedure with Ample Volumes of HOH.* An accurately measured volume of water (0.5 ml is convenient), is placed in an equilibration bulb. The bulb is capped with the vacuum stopcock (using Apiezen H grease), and connected to the vacuum line. If necessary, the stopcock is evacuated. The water is degassed by freezing in Dry Ice–isopropanol, evacuation, and closing, thawing, and refreezing cycles, so that any dissolved $CO_2$ will also be removed. Three such cycles are usually sufficient. The water is then frozen in liquid nitrogen, and the appropriate dose (40–50 $\mu$mol) of $CO_2$ is admitted from the dosing chamber and frozen onto the water. The bulb and stopcock are removed from the line, capped with a small rubber stopper and Parafilm to prevent moisture from getting into the inside of the 10/30 joint, and placed in a 30° shaking water bath for 24–48 hr. $T_{1/2}$ for $CO_2 \rightleftharpoons H_2O$ equilibration is about 0.9 hr.

After equilibration is complete, the $CO_2$ is collected by rapidly freezing the small bulb in Dry Ice–isopropanol, connecting it to the vacuum line, and evacuating above the stopcock. The $CO_2$ is expanded into a much larger evacuated vessel. This vessel is then frozen in liquid $N_2$ and evacuated to remove any air that might have leaked in during equilibration. The vessel is then warmed in Dry Ice–isopropanol to liberate the $CO_2$ with retention of any water that may have transferred. The $CO_2$ is transferred by freezing to a vessel containing $P_2O_5$, which is then warmed to about room temperature and shaken to ensure complete removal of water. A 2 × 10-cm vessel fitted with a 24/40 joint connected through a right-angle tube, stopcock, 10/30 joint, second right-angle tube, and 18/9 ball joint to the line allows convenient evacuation and shaking of the $P_2O_5$ without removal from the vacuum line. The $CO_2$ can then be transferred to a collection bulb as previously described[1] and analyzed on a mass spectrometer.

*Procedure with Smaller HOH Samples.* For small volumes of water of enrichment so low that they cannot be diluted to a more convenient volume (200–500 μl), methanol can be substituted for up to 95% of the volume required for equilibration up to a total volume of 1 ml. The methanol should be dried, for example, over Drierite. Somewhat longer equilibration times are allowed if methanol is present, up to 72 hr if final volume is 0.5 ml. $T_{1/2}$ for $CO_2 \rightleftharpoons H_2O$ equilibration in 95% MeOH is 1–3 hr, depending somewhat on the total liquid volume and shaking conditions.

Degassing of the $H_2O$–methanol mixture is more difficult, as methanol behaves somewhat like $CO_2$ at Dry Ice and liquid nitrogen temperatures. Therefore, the initial degassings, primarily of air, are done from liquid $N_2$ and the final one from Dry Ice–isopropanol, relying on the methanol vapor to sweep out any $CO_2$.

Collection presents similar problems of methanol mimicking $CO_2$. The collection bulb is frozen in Dry Ice–isopropanol, then liquid $N_2$ and any leaked air is evacuated. The bulb is warmed to Dry Ice temperature once more and only the initial burst of $CO_2$ is expanded into the larger vessel, mimimizing transfer of MeOH. As a further precaution, concentrated sulfuric acid, which is more efficient for MeOH removal, can be substituted for $P_2O_5$ for drying the $CO_2$. The concentrated $H_2SO_4$, about 0.5 ml in the bottom of a vessel similar to the $P_2O_5$ setup, is evauated for several hours, preferably overnight, to degas prior to use as a methanol trap.

Calculation of the enrichment of water is made on the basis of the known equilibrium distribution of $^{18}O$,[5,6] and standard procedures for calculation of atom percentage excess $^{18}O$ from the ratio, $R$, of mass 46/44. A typical calculation, where the value of 1.0407 is the correction for the equilibrium isotope distribution between HOH and $CO_2$,[6] is as follows:

$$\text{Atom\% } ^{18}O \text{ in } CO_2 = 100 \, R/(2 + R)$$

$$\text{Atom\% } ^{18}O \text{ in original } H_2O = \frac{\mu g \text{ atoms O as } CO_2}{\mu g \text{ atoms O as } H_2O}$$

$$(\text{final atom\% in } CO_2 - \text{original atom\% in } CO_2) + \frac{\text{final atom\% } CO_2}{1.0407}$$

where microgramatoms of O as $CO_2 = 2 \times$ micromoles of $CO_2$, and microgramatoms of O as $H_2O$ = microliters of $H_2O \times 55.5$.

---

[5] J. R. O'Neil and S. Epstein, *J. Geophys. Res.* **71**, 4955 (1966).
[6] I. Dostrovsky and F. S. Klein, *Anal. Chem.* **24**, 414 (1952).

## Preparation of $P_i$ and ATP Highly Labeled with $^{18}O$

[$^{18}O$]$P_i$ *Preparation*

*Principle.* Enriched $P_i$ has previously been made by equilibration of $KH_2PO_4$ with enriched water at high temperature[6,7] or may be prepared by pyrophosphatase-accelerated equilibration as described above for [$^{18}O$]$P_i$ standards. Such procedures dilute the [$^{18}O$]$H_2O$ water with the unenriched $P_i$ oxygens. Other procedures have utilized hydrolysis of $P_2O_5$[8] and $POCl_3$,[9] which, although more reactive, still require use of $^{18}O$-enriched starting material to prevent dilution of the $^{18}O$-enrichment of the water. Hydrolysis of $PCl_5$ in enriched water[10] overcomes both these problems, as $PCl_5$ is highly reactive and contains no oxygen. The usefulness of the hydrolysis of $PCl_5$ for the formation of highly enriched $P_i$ has been established by Risley and Van Etten.[11] The procedure described here is a modification of their method, which results in essentially no dilution in the $P_i$ of the original $^{18}O$ enrichment of the water.

*Materials*

$PCl_5$, imidazole

Anion exchange resin; AG 1-X4, 100–200 mesh, $Cl^-$, (Bio-Rad Co., Richmond, California)

KOH, 1 $N$ and 5 $N$

HCl, 30 m$M$

Tris, 1 $M$

$H^{18}OH$, up to nearly 100% $^{18}O$ (Norsk Hydro, Oslo, Norway)

*Procedure.* All manipulations of $PCl_5$ and very enriched [$^{18}O$]$H_2O$ should be performed in a glove bag under $N_2$ to prevent contamination by atmospheric water. Imidazole should be recrystallized from ethyl acetate, ground with a mortar and pestle, and dried under vacuum over $P_2O_5$. (All glassware should be scrupulously cleaned to avoid $P_i$ contamination.)

Water of 99% $^{18}O$ enrichment (4 ml, 222 mmol) is added dropwise with shaking to $PCl_5$ (1.79 g, 8.6 mmol) in a 50-ml Erlenmeyer flask. After 15 min, imidazole (5 g, 74 mmol) is added to neutralize the reaction mixture. Most of the unreacted [$^{18}O$]$H_2O$ (3.0 ml) can be recovered by vacuum transfer overnight into a vessel at Dry Ice temperature and the remaining

---

[7] M. Cohn and G. Drysdale, *J. Biol. Chem.* **216**, 831 (1955).
[8] M. Cohn, *J. Biol. Chem.* **180**, 771 (1949).
[9] C. F. Midelfort and I. A. Rose, *J. Biol. Chem.* **251**, 5881 (1976).
[10] R. L. Metzenberg, M. Marshall, P. P. Cohen, and W. G. Miller, *J. Biol. Chem.* **234**, 1534 (1959).
[11] J. M. Risley and R. L. Van Etten, *J. Labelled Compd. Radiopharm.* **15**, 533 (1978).

[$^{18}$O]H$_2$O is recovered at lower enrichment by addition of 10 ml of unenriched water and recollection by vacuum transfer.

The reaction mixture is diluted to 4 liters with deionized water, adjusted to pH 9 with 1 $N$ KOH, and loaded onto a 2 × 19-cm column of AG 1-X4. The column is eluted with 30 m$M$ HCl, and the peak P$_i$ fractions are pooled and lyophilized. The H$_3$PO$_4$ is dissolved in 80 ml of water, the pH is adjusted to 4 with 5 $N$ KOH and relyophilized to give 1.0 g of KH$_2$PO$_4$ (7.7 mmol, 89% overall yield from PCl$_5$). Mass spectral analysis of a typical preparation indicated that the P$^{18}$O$_3$ species was 1.6% of the P$^{18}$O$_4$ species with no species of lower enrichment detectable. This corresponds to an $^{16}$O content of 0.4% and demonstrates that essentially no contamination with $^{16}$O had occurred. The presence of a P$^{18}$O$_3$$^{17}$O$_1$ species indicated an $^{17}$O content of approximately 0.1%.

Pyrophosphate (PP$_i$) may also be eluted from the column with 100 m$M$ HCl. The peak fractions are identified by P$_i$ analysis after acid hydrolysis, pooled, adjusted to pH 9 with Tris, and loaded onto a 0.7 × 7-cm column of AG 1-X4. The column is washed with 30 m$M$ HCl, and the PP$_i$ is eluted with cold 100 m$M$ HCl and immediately neutralized with 1 $M$ Tris. The yield of PP$_i$ was 50 $\mu$mol and it was of approximately the same $^{18}$O enrichment as the P$_i$.

*Comments*. The use of an unenriched base for neutralization of the acidic reaction mixture results in some incorporation of $^{16}$O into P$_i$ during the quench.[11] This problem is avoided in the present procedure by neutralization with solid imidazole, which does not contain a source of $^{16}$O and does not form a stable phosphoamide derivative under acidic conditions. The P$_i$ that is obtained has undergone no detectable isotopic dilution and has the same $^{18}$O content as the starting water. Similar procedures can be used with water of lower $^{18}$O enrichment to yield randomly labeled P$_i$ at any average $^{18}$O enrichment.

The 25:1 molar ratio of H$_2$O:PCl$_5$ described here results in incomplete hydrolysis, a fraction of the phosphate being present as PP$_i$ and higher polyphosphates. These $^{18}$O-enriched polyphosphates are often a desirable by-product, whose yield can be increased by using a lower ratio of water, to PCl$_5$.

The [$^{18}$O]H$_2$O recovered by vacuum transfer contains some imidazole. This does not interfere with reuse of the water for further [$^{18}$O]P$_i$ synthesis, but is undesirable in other uses. It can be removed by addition of solid NaHSO$_4$ and collection of the water by vacuum transfer.

The KH$_2$PO$_4$ obtained is of satisfactory purity for most uses, but it does contain some residual KCl resulting from incomplete removal of HCl. The KCl can be removed by recrystallization from ethanol–water.[2]

[$\gamma$-$^{18}O$]ATP

*Principle.* ATP that is highly $^{18}$O-enriched in its three terminal oxygens can be prepared by the procedure of Penefsky et al.[12] originally designed for synthesis of [$\gamma$-$^{32}$P]ATP.

*Materials*
HClO$_4$, 1 $M$
KOH, 5 $N$
KCl, 4 $M$
HCl, 30 m$M$, 100 m$M$, and 500 m$M$
AG 1-X4, 100–200 mesh, Cl$^-$ form

A reaction mixture at 10° and pH 7.2 in 30 ml consists of the following components:

25 m$M$ HEPES
0.25 m$M$ NAD
15 m$M$ MgCl$_2$
30 m$M$ Pyruvate
20 m$M$ DL-Glyceralde-
  hyde-3-P

10 m$M$ ADP
11 m$M$ 99% [$^{18}$O]KH$_2$PO$_4$
250 U Glyceraldehyde-3-P dehydrogenase
600 U Phosphoglycerate kinase
100 U Lactate dehydrogenase

*Procedure.* The preceding reaction mixture is stopped after about 25 min (70% loss of free P$_i$) by addition of 15 ml of cold 1 $M$ HClO$_4$. The protein is removed by centrifugation, the supernatant is adjusted to pH 8.5 with 5 $N$ KOH, and 2 ml of 4 $M$ KCl are added to complete precipitation of KClO$_4$. After 60 min at 0°C, the KClO$_4$ is removed by centrifugation. The supernatant is diluted to 1 liter and loaded onto a 0.7 × 7-cm column of AG 1-X4. The P$_i$ is eluted with 30 m$M$ HCl, and then ADP is eluted with 35 ml of cold 100 m$M$ HCl. The ATP is eluted with cold 500 m$M$ HCl and immediately neutralized with 5 $N$ KOH. The yield was 210 $\mu$mol of ATP in a typical preparation, with 38 m$M$ in the peak fraction.

*Comments.* Glyceraldehyde-3-P is unstable at neutral pH and releases its P$_i$ with resulting dilution of the [$^{18}$O]P$_i$. This is minimized by use of high enzyme concentrations, low temperature, and only partial conversion of P$_i$ to ATP. Typically 5–10% of the ATP is the ATP$^{18}$O$_0$ species even with these precautions. Little or no exchange occurs and high ratios of ATP$^{18}$O$_3$ : ATP$^{18}$O$_2$ are obtained.

For most applications the presence of KCl in the ATP fractions presents no problem and this fast procedure is satisfactory. When KCl must be avoided, chromatography on DEAE-cellulose or Sephadex should be used with elution by volatile triethylamine bicarbonate.

[12] H. S. Penefsky, M. E. Pullman, A. Datta, and R. Racker, *J. Biol. Chem.* **235,** 3330 (1960).

## Isolation of $P_i$ from Reaction Mixtures

*Isolation of $P_i$ for Conversion of Oxygens to $CO_2$*

*Principle.* The triethylamine–molybdate precipitation procedure described in an earlier volume[2] has been replaced by a more convenient extraction procedure. The $P_i$ in samples is removed by extraction of the phosphomolybdate complex into organic solvent. The $P_i$ is then transferred into aqueous ammonia followed by precipitation as magnesium ammonium phosphate.

*Materials*

Isobutanol–benzene,[13] 1:1, v/v or water-saturated butyl acetate
HCl, 12 $M$
Ammonium molybdate, 60 m$M$, in 10 m$M$ HCl
$NH_4OH$, 5 $M$, containing 1 $M$ $NH_4Cl$

*Procedure.* Samples, such as perchloric acid extracts of reaction mixtures containing 5–15 μmol of $P_i$ at a concentration not greater than 10 m$M$ or considerably more dilute if necessary, are used. Make the sample about 1 $M$ in H$^+$ by addition of 12 $M$ HCl. Extract with about one-third volume (less if $P_i$ is much less than 10 m$M$), but not less than 2 ml, of isobutanol–benzene by vigorous mixing (for example, with the aid of a Vortex mixer). Teflon-lined, screw-cap conical centrifuge tubes are convenient for extraction. After separation, remove the upper layer with a Pasteur pipette and discard.

Make the sample 12 m$M$ in molybdate by addition of the 60 m$M$ molybdate solution. Add 12 $M$ HCl to retain the 1 $M$ H$^+$ concentration, then extract with isobutanol–benzene as before. Transfer the upper layer into a capped centrifuge tube. Repeat the extraction if recovery of nearly all $P_i$ is critical. Wash the isobutanol–benzene layer by mixing with about 0.5 volume of 1 $M$ HCl and discard the aqueous layer. Extract the $P_i$ by shaking with 1–1.5 ml of the $NH_4OH$–$NH_4Cl$ solution. Remove the isobutanol–benzene layer. The sample is then ready for precipitation as magnesium ammonium phosphate and isolation of the $P_i$ for analysis as described previously.[1,2] Lyophilization, however, is more convenient than heating for drying the samples and avoids exchange of $^{16}O$ in glass with $P_i$. Alternatively, the $P_i$ may be removed from the organic layer as described below and separated by AG 1-X4 chromatography followed by neutralization and lyophilization to give dried $KH_2PO_4$ + KCl for pyrolysis.

[13] Benzene may be a carcinogenic hazard. Appropriate precaution should be maintained in its use, or other solvents, such as butyl acetate, should be used. However, separation of layers is much better with isobutanol–benzene mixture than with butyl acetate.

## Isolation of $P_i$ for Preparation of a Volatile Phosphate

*Principle.* Phosphate samples can be purified and prepared for derivatization by ion exchange chromatography using AG 1-X4 anion exchange resin. The $P_i$ is loaded onto the resin at slightly basic pH, washed with 10 m$M$ HCl, and eluted with 30 m$M$ HCl. This simple procedure is adequate for most samples; however, samples containing phosphate esters or other compounds that elute similarly to $P_i$ require further purification by either gas chromatography after derivatization or by the molybdate extraction procedure.

*Materials*
  AG 1-X4, Cl⁻ form, anion exchange resin
  Butyl acetate or isobutanol–benzene (1:1, v/v)[13]
  HCl, 10 m$M$, 30 m$M$, and 12 $M$
  Tris, 0.2 $M$ and 1 $M$; and KCl, 0.1 $M$
  Molybdate, 60 m$M$, in 10 m$M$ HCl

*Procedure.* AG 1-X4 COLUMNS. A 0.5 × 2.0-cm column conveniently prepared in a 5.75-in. Pasteur pipette can be used for $P_i$ samples of ≤20 $\mu$mol. The pipette is scratched with a file and broken off just below the upper constriction. The cut end is lightly fire polished. A glass wool plug is inserted into the drawn-out region of the pipette, and the column is filled with a slurry of resin to the proper height. The column is washed with several column volumes of 1 $N$ HCl and then with deionized water until the pH of the eluent is ≥4.

A large number of columns can be supported by a Lucite rack containing 8-mm circular holes through which the columns are inserted and held in place at the top by a collar of 0.25-in. rubber tubing. For loading moderate volumes an extension tube can be joined to the top by rubber tubing, and for larger volumes a funnel or other reservoir can be used. The flow rate can be increased by connecting a length of ⅛-in. rubber tubing to the bottom of the column. Good retention of $P_i$ can still be obtained with flow rates of 5 ml/min and higher.

SAMPLE PREPARATION. In order to ensure complete retention of $P_i$ by the column, the sample should be diluted with deionized water to an ionic strength of ≤0.01 and the pH should be adjusted to ≈8.5. A 1 $M$ solution of Tris is useful for adjusting the pH of neutral and acidic samples, as this base has a low $P_i$ content is uncharged in the basic form and thus does not add to the ionic strength, and has a p$K_a$ of 8.2, which minimizes the over-titration that can occur with strong bases. Application of the sample at high pH results in greater OH⁻ binding and greatly increases the amount of 10 m$M$ HCl needed to acidify the column.

ELUTION. After the sample has been loaded, the column should be washed with several column volumes of deionized water. The column is then acidified with 10 m$M$ HCl in 0.5-ml aliquots until the eluent is pH ≤ 2.5 by pH paper. With small $P_i$ samples containing no other titratable components, this requires approximately 1 ml of 10 m$M$ HCl. The 10 m$M$ HCl wash is continued for an additional 2 × 0.5 ml after the eluent becomes ≤pH 2.5 and can be continued even further if the sample contains impurities that are removed in this wash. This 10 m$M$ HCl wash, however, will remove a significant amount of $P_i$ with small samples (≤100 nmol), which begin to show $P_i$ elution after 1–2 ml of additional washing.

The $P_i$ is eluted with 2 ml of 30 m$M$ HCl. Most of the $P_i$ elutes between 0.5 and 1.5 ml. With larger samples (10–20 $\mu$mol), 3 ml of 30 m$M$ HCl should be collected. The sample is then lyophilized to leave $H_3PO_4$. No isotopic exchange with water or scrambling of the $P_i$ oxygens has been observed to follow lyophilization.

ALTERNATIVE MOLYBDATE EXTRACTION. Some samples may contain compounds (e.g., sugar phosphates) that will elute from the column similarly to $P_i$. Also high salt concentration may require dilution of the sample to an inconvenient volume. In these cases the $P_i$ should first be extracted into an organic solvent as the phosphomolybdate complex, as described earlier. The molybdate can then be removed by Dowex-1 chromatography. This procedure is also useful as a replacement for the Mg precipitations in the previous procedure for purification as $KH_2PO_4$.[2] The precipitations are time consuming and often result in substantial losses of $P_i$.

Sufficient acid (HCl or $HClO_4$) is added so that the free [H$^+$] is approximately 0.5 $N$, and a typical 2-ml sample is then washed with 2 ml of butyl acetate. One-fourth volume of molybdate reagent is added, and the yellow phosphomolybdate complex is extracted into 2 ml of butyl acetate. The lower water layer is further extracted with 1 ml of butyl acetate, and the two butyl acetate fractions are combined and backwashed with 2 ml of 0.5 $N$ HCl. The butyl acetate is extracted with 3 ml of 0.2 $M$ Tris, which destroys the complex and extracts the $P_i$ into the water layer, which is separated, diluted to 15 ml, and loaded onto a 0.5 × 2-cm AG 1-X4 column as before. The column is briefly washed with water and then with 2 × 0.5 ml of 0.1 $M$ KCl followed by a more extensive water wash and the previous elution procedure of 10 and 30 m$M$ HCl.

*Comments.* The phosphomolybdate complex requires an acidity >0.2 $N$ for extraction into the organic phase, and 0.5 $N$ is a useful compromise that assures extraction without use of excess reagents, which may contaminate small $P_i$ samples. This acidity is that commonly used for perchloric acid precipitation of proteins; thus the supernatant can be used directly. Dilute samples can be concentrated by lyophilization before starting the extraction procedure.

The molybdate is highly negatively charged and binds tightly to the AG 1-X4 column along with the $P_i$. When the column is washed with the 10 and 30 m$M$ HCl, the complex reforms and remains so tightly bound that it will not elute even in 1 $N$ HCl. This can result in troublesome, variable losses in yield, particularly with small samples. These can be reduced somewhat by the 0.1 $M$ KCl wash. This neutral salt does not induce the formation of the phosphomolybdate complex and moves the $P_i$ part way down, but not off, the column while the molybdate remains at the extreme top. The complex now cannot reform during the HCl washes, as the molybdate and $P_i$ have been separated.

When preparing large samples (2–10 $\mu$mol) for conversion to $CO_2$, a larger column should be used (0.7 × 5 cm), the 0.1 $M$ KCl wash should be increased to 3 ml, and it requires 5–8 ml of 30 m$M$ HCl to elute the $P_i$. A 3-fold excess of KCl over $P_i$ should be added to the 30 m$M$ HCl fraction; this results in the formation of $KH_2PO_4$ plus excess KCl after lyophilization.

## Measurement of $^{18}O$ in $P_i$ and Other Phosphates

Procedures based on conversions of $P_i$ oxygens to $CO_2$ are given elsewhere.[2,3] Procedures based on conversion to a volatile phosphate compound are given in some detail here. Procedures based on NMR measurements are described more briefly.

### Volatile Phosphate Measurements

*Principle.* The first volatile derivative of $P_i$ to be used in $^{18}O$ analysis was tris(trimethylsilyl) phosphate (TMSP)[14,15] prepared by reaction of $H_3PO_4$ with $N,O$-bis(trimethylsilyl) acetamide in acetonitrile. The derivatization is easily carried out; the acetonitrile can be removed in a stream of $N_2$ and the residue directly introduced into the mass spectrometer. Scaling the original procedure down to the nanomole range, however, proved to be difficult, and a more volatile solvent and reagent were needed. The use of $N$-(trimethylsilyl)diethylamine in $CH_2Cl_2$ overcomes the volatility problem as solvent, and the excess reagent can be removed almost completely before TMSP is lost.

The principal disadvantage of TMSP is that the high natural abundance of $^{29}Si$ and $^{30}Si$ results in considerable spillover of the signal from one [$^{18}O$]$P_i$ species into the peaks at higher molecular weight. This disadvantage is overcome with alkyl esters. Trimethyl phosphate (TMP) has

---

[14] J. Bar-Tana, O. Ben-Zeev, G. Rose, and J. Deutsch, *Biochim. Biophys. Acta* **264**, 214 (1972).

[15] D. H. Eargle, Jr., V. Licho, and G. L. Kenyon, *Anal. Biochem.* **81**, 186 (1977).

much smaller spillover corrections. With electron impact ionization, triethyl phosphate (TEP) and tributyl phosphate give the unsubstituted phosphate ion as the major fragment and thus avoid any spillover corrections. However, their volatility and separation from reagents require a gas chromatograph interface with the mass spectrometer. Suitability of the alkyl esters for accurate and rapid analyses is under current evaluation, and only the trimethylsilyl procedure is described here.

The TMP may be prepared by conventional procedures[9] but because of its less desirable fragmentation pattern, we have been currently working with TEP. This is conveniently prepared by a procedure similar to that described below for TMSP, but with use of 0.5 $M$ triethyloxonium tetrafluoroborate in dry $CH_2Cl_2$, with an equal volume of 0.5 $M$ diisopropylethylamine in dry $CH_2Cl_2$ as a catalyst. However, fragments from reagents (especially $M$ = 105 and 107) and retention of these impurities on columns interfere somewhat. The TEP can also be prepared using diazoethane from $N$-ethyl-$N$-nitroso-$p$-toluenesulfonamide,[16] or from $N$-ethyl-$N'$-nitro-$N$-nitrosoguanidine. Diazoethane is less volatile than diazomethane and concentrated solutions are less readily prepared. Once prepared diazoethane is readily manipulated and excess derivatizing reagent is easily removed by a stream of $N_2$.

Tris(trimethylsilyl) Phosphate (TMSP)

*Materials*

$CH_2Cl_2$ (dried over 4-Å molecular sieves; Linde Co.)

$N$-(Trimethylsilyl)diethylamine (TMSDEA) (Pierce Chemical Co.)

*Procedure.* The $P_i$ sample ($\leq$ 200 nmol) as $H_3PO_4$ from lyophilization of the sample in 30 m$M$ HCl is treated with 50 $\mu$l of dry $CH_2Cl_2$ and 5 $\mu$l of TMSDEA in a Teflon-lined screw-cap vial. The reaction is largely complete after several hours at room temperature with occasional shaking. The samples are stable at room temperature overnight but should be kept cold and dry for prolonged storage.

Calculations of Distributions of $[^{18}O]P_i$ Species

The species of $[^{18}O]P_i$ that may be present are conveniently designated as $P^{18}O_0$, $P^{18}O_1$, $P^{18}O_2$, $P^{18}O_3$, and $P^{18}O_4$, for 0–4 $^{18}O$ per $P_i$, respectively. The mass spectral data give the relative intensities of species with mass M, M + 2, . . . , M + 8, where M is the molecular ion or some fragment containing the four oxygens. These ratios must be corrected for spill-

---

[16] Prepared as described for the methyl derivative, T. J. de Boer and H. J. Backer, *Org. Synth.* **34**, 96 (1954).

over of $^{13}C$, $^{2}H$, etc., to yield the true ratios of $^{18}O$-containing species. This spillover, exclusive of oxygen isotopes, can be calculated[17] from natural abundance data and is summarized in the table.

SPILLOVER CORRECTIONS FOR VOLATILE PHOSPHATE MEASUREMENTS[a]

| Compound | Relative peak heights | | | | |
|---|---|---|---|---|---|
| | M | M + 1 | M + 2 | M + 3 | M + 4 |
| Trimethyl phosphate | 100 | 3.49 | 0.04 | — | — |
| Triethyl phosphate | 100 | 6.94 | 0.20 | — | — |
| Tris(trimethylsilyl) phosphate | 100 | 25.80[b] | 12.93[b] | 2.26[b] | 0.53[b] |
| TMSP minus $CH_3$ | 100 | 24.63 | 12.65 | 2.12 | 0.50 |

[a] Based on natural abundance of $^{13}C$, $^{2}H$, $^{29}Si$, and $^{30}Si$, but not $^{17}O$ or $^{18}O$.
[b] Corrected from earlier values of D. D. Hackney and P. D. Boyer, *Proc. Natl. Acad. Sci. U.S.A.* **75**, 3133 (1978).

With TMP and TEP the spillover into M + 2 is negligible except at very low enrichment, but it is significant with TMSP. The procedure is to start with M and subtract out its spillover into the peaks at higher mass. The corrected value of M + 2 then represents the contribution of the species with one $[^{18}O]P_i$, and the spillover due to this species is subtracted from the peaks at higher mass, etc., until the corrected contributions of all the $^{18}O$-containing $P_i$ species have been determined.

Water that is highly $^{18}O$-enriched is also enriched in $^{17}O$. This becomes a problem when the water (or $P_i$ derived from it) is in the range of 80–99% as the ratio of $^{17}O:^{16}O$ becomes significant. The effect is to shift intensity, which would be at M + 6 to M + 7, for example, and produce an artificially high $P^{18}O_4:P^{18}O_3$ ratio (derived from the M + 8:M + 6 ratio). The way in which the $^{17}O$ is washed out during an exchange reaction is determined by the partition coefficient ($P_c$) of the process. For medium $P_i \rightleftharpoons HOH$ exchange, with a $P_c$ of zero (i.e., oxygens lost one at a time), the effect of $^{17}O$ can be neglected shortly after exchange begins because the $^{17}O:^{16}O$ ratio will drop rapidly owing to the incoming $^{16}O$. When $P_c$ is close to 1 (i.e., all four oxygens are exchanged per encounter), the phosphate that has not undergone exchange will exhibit a high M + 7:M + 6 ratio even after extensive net exchange, whereas the phosphate that has undergone exchange will be mainly $P^{18}O_0$ with a normal M + 1:M ratio. When the experimental data are treated, the contribution of $^{17}O$ can be estimated from the deviations of the intensities of M + 1, M + 3, M + 5,

[17] J. H. Benyon, "Mass Spectroscopy and Its Application to Organic Chemistry." Elsevier, Amsterdam, 1960.

and M + 7 from the spillover values predicted in the table. The $^{17}O$ contribution may then be treated mathematically as though it were $^{16}O$ to calculate the true ratios of $^{18}O$-containing species.

With chemical ionization the molecular ion (at 1 mass unit greater than the true molecular weight; i.e., for TMSP M = 314 . . . M + 8 = 322) is a dominant species and the above considerations apply directly. With ionization by electron impact, significant fragmentation occurs and analysis can be performed on the molecular ion or on more intense fragments. TMSP produces a fragment at M-15 that is due to loss of a $CH_3$ group and is approximately 4-fold more intense than M at 70 eV. There are no other fragments that interfere, and this fragment is preferable for analysis. The spillover corrections for M-15 are also included in the table. TEP produces a fragment at M-27, but analysis is complicated by other fragments at M-29 and M-25. TEP produces the most useful fragment at M-84 by loss of all ethyl groups to leave a completely protonated phosphate ion, which produces a family of intensities for which no spillover corrections are necessary.

### $^{31}P$ NMR Measurements

Cohn and Hu[18] have recently demonstrated that small chemical shifts in the $^{31}P$ NMR of phosphate compounds are induced by substitution of $^{18}O$ for $^{16}O$, and this provides an alternative to the mass spectroscopic techniques for the determination of the distribution of $^{18}O$-containing species. The chemical shift difference is very small, and consequently a high-resolution NMR spectrometer is required. The peaks are well resolved at 145.7 MHz (a magnetic field corresponding to 360 MHz for protons), but at lower field strengths a curve resolver must be used to obtain relative areas. A further advantage of the technique is that it can distinguish bridge from branch localizations of the $^{18}O$ in phosphate esters.[19] It has also been recently demonstrated that high-resolution $^{13}C$ NMR can give similar information about $^{18}O$ substitution at carbon.[19,20]

### Comparison of Methods

The $CO_2$ method has the advantage of high accuracy in the determination of $^{18}O$ content of low-enrichment samples. Its disadvantages are that it requires extensive purification of the $P_i$, high-vacuum techniques, fairly

---

[18] M. Cohn and A. Hu, *Proc. Natl. Acad. Sci. U.S.A.* **75**, 200 (1978).
[19] M. Cohn, in "NMR in Biochemistry" (S. J. Opella, ed.) Dekker, New York, in press.
[20] D. D. Hackney, J. A. Sleep, G. Rosen, R. L. Hutton, and P. D. Boyer, in "NMR in Biochemistry" (S. J. Opella, ed.) Dekker, New York, in press.

large samples (1–10 μmol), and, most important, it yields only the average $^{18}$O content of the $P_i$ and provides no information about the distribution of isotopic $P_i$ species.

Mass spectral analysis of volatile $P_i$ yields the distribution of $P_i$ species and also has the advantage of faster sample purification and preparation. Another major advantage is the small sample size. Samples of <10 nmol can be carried through the complete purification and analysis procedures. This is particularly important in studies that must be performed at or below $K_m$ values and when only limited quantities of enzyme or labeled substrate are available. The accuracy in intensity ratios obtained from a scanning mass spectrometer is not as great as that obtained with $CO_2$ in a mass-ratio spectrometer, but it is quite satisfactory for most purposes and is not limited to low-enrichment samples. Trimethyl and triethyl phosphate are preferable to TMSP when gas chromatographic injection is available, as they do not have the spillover problems associated with $^{29}$Si and $^{30}$Si and are less damaging to the ion source. However, at this stage we obtain better precision with TMSP.

The $^{31}$P NMR method has the advantage of continuous and nondestructive monitoring with no or minimal sample preparation. It has the added advantage that the isotopic composition of phosphate esters, such as glucose-6-P and ATP, can be measured directly without the need for conversion to $P_i$ and information can be obtained about the position of $^{18}$O in the molecule. Thus the $\alpha$, $\beta$, and $\gamma$ positions of ATP can each be followed simultaneously. A further refinement is that bridge positions can be distinguished from branch positions,[19] which is directly useful for scrambling experiments.[10] The major disadvantage is that very large samples are required. Samples of several micromoles require accumulation of thousands of scans in order to obtain a good signal-to-noise ratio needed for accurate ratio measurements. The number of scans required decreases, however, with the square of the sample size, and only a few scans are needed with samples of 50 μmol.

## Some Applications of $^{18}$O Measurements

*General Considerations.* The reversible enzymic hydrolysis of most phosphate compounds (RP; where $P = PO_3$) occurs by one of the two general schemes of Eqs. (1) and (2).

$$RP + E \underset{k_{-1}}{\overset{k_1}{\longleftrightarrow}} E \cdot RP \underset{k_{-2}}{\overset{k_2}{\longleftrightarrow}} \overset{H_2O}{\underset{}{E \cdot \overset{\cdot P_i}{RH}}} \underset{k_{-3}}{\overset{k_3}{\longleftrightarrow}} E + P_i + RH \quad (1)$$

$$RP + E \underset{k_{-1}}{\overset{k_1}{\longleftrightarrow}} E \cdot RP \underset{k_{-2}}{\overset{k_2}{\longleftrightarrow}} \overset{RH}{\underset{}{E - P}} \underset{k_{-3}}{\overset{k_3}{\longleftrightarrow}} \overset{H_2O}{\underset{}{E \cdot P_i}} \underset{k_{-4}}{\overset{k_4}{\longleftrightarrow}} E + P_i \quad (2)$$

Myosin,[21] pyrophosphatase,[22] and probably the ATPases of oxidative and photophosphorylation proceed via Eq. (1) with no evidence for a phosphoenzyme intermediate, whereas acid and alkaline phosphatase,[23] the CaATPase of sarcoplasmic reticulum,[24] and the Na/K ATPase[25] proceed through a phosphoenzyme. The reversible hydrolysis of E · RP or E − P results in the exchange of oxygen between water and bound phosphate. Exchanges that occur during chemical conversion of reactants to products are referred to as intermediate exchanges of the product formed (e.g., and intermediate $P_i \rightleftharpoons$ HOH exchange accompanying ATP hydrolysis), exchanges that occur when a reactant binds, undergoes exchange, and returns to the medium are referred to as medium exchanges (e.g., a medium $P_i \rightleftharpoons$ HOH exchange). An experimentally measured isotopic flux may be produced by a combination of such primary processes.[26] For example, the appearance of $^{18}O$ from HOH into $P_i$ may result from a medium $P_i \rightleftharpoons$ HOH exchange (reversible formation of E · RP from medium $P_i$) plus intermediate $P_i \rightleftharpoons$ HOH exchange accompanying net RP hydrolysis.

Exchange reactions must be clearly differentiated from water oxygen incorporation accompanying net hydrolysis. One water oxygen needs to be incorporated into each $P_i$ formed from RP by hydrolysis, but often "extra" incorporation is observed and each $P_i$ contains more than one water oxygen. Analysis of the distribution of the labeled $P_i$ species both provides a method for assessing the extent of this extra exchange during a medium exchange reaction and provides critical information about the mechanism responsible for the extra exchange, as outlined in the next section.

*Partition Coefficient.* The partition coefficient for a $P_i \rightleftharpoons$ HOH exchange by reactions of Eq. (1) or (2) is defined as the ratio of the rate at which bound $P_i$ loses HOH in the exchange step to the sum of the rate at which bound $P_i$ loses HOH plus the rate of release to the medium. The partition coefficient ($P_c$) approaches 1 as the number of oxygens exchanged before release of bound $P_i$ approaches infinity; $P_c$ approaches zero as the exchange of bound $P_i$ becomes negligible. For example, a medium $P_i \rightleftharpoons$ HOH exchange can be described by the general scheme of Eq. (3), where (EP) is some anhydrous intermediate.

---

[21] J. A. Sleep and P. D. Boyer, *Biochemistry* **17,** 5417 (1978).

[22] C. J. Janson, C. Degani, and P. D. Boyer, *J. Biol. Chem.* **254,** 3743 (1979).

[23] T. W. Reid and I. B. Wilson, *in* "The Enzymes" (P. Boyer, ed.), 3rd ed., Vol 4, p. 373. Academic Press, New York, 1971.

[24] C. Degani, *J. Biol. Chem.*, **248,** 8222 (1973).

[25] C. Degani, A. S. Dahms, and P. D. Boyer, *Ann. N. Y. Acad. Sci.* **242,** 77 (1974).

[26] A more complete discussion of types of oxygen exchange and factors contributing to observed exchange is given in a previous volume.[3]

$$(EP) + H_2O \underset{k_{-2}}{\overset{k_2}{\rightleftarrows}} E \cdot P_i \underset{k_{-3}}{\overset{k_3}{\rightleftarrows}} E + P_i \qquad (3)$$

For this scheme, $P_c$ equals $k_{-2}/(k_{-2} + k_3)$. If $k_{-2} \gg k_3$, then step 2 will reverse many times before $P_i$ is released and up to all four original $P_i$ oxygens could be replaced with water oxygens each time $P_i$ binds to the enzyme. This assumes that rotation of the bound $P_i$ is sufficiently rapid to allow all four $P_i$ oxygens to participate in the exchange. When $k_{-2} \ll k_3$, bound $P_i$ will most often be released without exchange or occasionally with one oxygen exchanged. For intermediate cases the expected extent of exchange can be determined from probability considerations as a function of the partition coefficient. Similar considerations apply to the other possible types of exchanges.

*Transition Probability Functions.* For analysis of $[^{18}O]P_i$ species distributions, the theoretical patterns expected for various cases are essential. A useful set of relationships are the transition probability functions, $P_{mn}$, which give the probability for the scheme of Eq. (3) that, when $P_i$ containing $m$ labeled oxygen binds to the enzyme in unlabeled water, the $P_i$ will be released containing $n$ labeled oxygens. These $P_{mn}$ relationships are as follows[27]:

$P_{00} = 1$
$P_{11} = (1 - P_c)[1 + 3P_c/(4 - 3P_c)]$
$P_{10} = 1 - P_{11}$
$P_{22} = (1 - P_c)[1 + P_c/(2 - P_c)]$
$P_{21} = (1 - P_{22})P_{11}$
$P_{20} = (1 - P_{22})P_{10}$
$P_{33} = (1 - P_c)[1 + P_c/(4 - P_c)]$
$P_{32} = (1 - P_{33})P_{22}$

$P_{31} = (1 - P_{33})P_{21}$
$P_{30} = (1 - P_{33})P_{20}$
$P_{44} = (1 - P_c)$
$P_{43} = (1 - P_{44})P_{33}$
$P_{42} = (1 - P_{44})P_{32}$
$P_{41} = (1 - P_{44})P_{31}$
$P_{40} = (1 - P_{44})P_{30}$

*Medium* $P_i \rightleftarrows HOH$ *Oxygen Exchange.* The value of $P_c$ during exchange of $[^{18}O]P_i$ in unlabeled water by the mechanism of Eq. (3) can be evaluated from the relationship of Eq. (4).[28]

$$P_c = k_{-2}/(k_{-2} + k_3) = (4 - R_4)/3 \qquad (4)$$

where

$$R_4 = \frac{\text{rate of loss of } P^{18}O_4}{\text{rate of loss of average } P_i \text{ enrichment}}$$

The complete solution to the cascade of labeled $P_i$ species during the exchange process can also be derived and is given by[27]

[27] D. D. Hackney, *J. Biol. Chem.*, manuscript submitted for publication.
[28] D. D. Hackney and P. D. Boyer, *Proc. Natl. Acad. Sci. U.S.A.* **75**, 3133 (1978).

$$P^{18}O_4(t) = P_4^0 e^{-kS_4 t}$$
$$P^{18}O_3(t) = (P_3^0 - C_{43})e^{-kS_3 t} + C_{43}e^{-kS_4 t}$$
$$P^{18}O_2(t) = (P_2^0 - C_{42} - C_{32})e^{-kS_2 t} + C_{32}e^{-kS_3 t} + C_{42}e^{-kS_4 t}$$
$$P^{18}O_1(t) = (P_1^0 - C_{41} - C_{31} - C_{21})e^{-kS_1 t} + C_{21}e^{-kS_2 t}$$
$$\qquad + C_{31}e^{-kS_3 t} + C_{41}e^{-kS_4 t}$$
$$P^{18}O_0(t) = P_0^0 + C_{10}(1 - e^{-kS_1 t}) + C_{20}(1 - e^{-kS_2 t})$$
$$\qquad + C_{30}(1 - e^{-kS_3 t}) + C_{40}(1 - e^{-kS_4 t})$$

where $P_j^0$ is the fraction of the phosphate containing $j$ labeled oxygens at time zero and where

$$C_{43} = P_{43}P_4^0/(S_3 - S_4)$$
$$C_{42} = (P_{42}P_4^0 + P_{32}C_{43})/(S_2 - S_4)$$
$$C_{41} = (P_{41}P_4^0 + P_{31}C_{43} + P_{21}C_{42})/(S_1 - S_4)$$
$$C_{40} = (P_{40}P_4^0 + P_{30}C_{43} + P_{20}C_{42} + P_{10}C_{41})/S_4$$
$$C_{32} = P_{32}(P_3^0 - C_{43})/(S_2 - S_3)$$
$$C_{31} = [P_{31}(P_3^0 - C_{43}) + P_{21}C_{32}]/(S_1 - S_3)$$
$$C_{30} = [P_{30}(P_3^0 - C_{43}) + P_{20}C_{32} + P_{10}C_{31}]/S_3$$
$$C_{21} = P_{21}(P_2^0 - C_{42} - C_{32})/(S_1 - S_2)$$
$$C_{20} = [P_{20}(P_2^0 - C_{42} - C_{32}) + P_{10}C_{21}]/S_2$$
$$C_{10} = P_{10}(P_1^0 - C_{41} - C_{31} - C_{21})/S_1$$
$$S_i = 1 - P_{ii}$$

For a rate of loss of the average $^{18}O$ enrichment with a half-life of 1 in arbitrary units, $k$ should be placed equal to $0.693 \times 4/\bar{O}$ where $\bar{O}$ is the average number of $P_i$ oxygens replaced with water oxygen each time a $P_i$ binds to the enzyme and is equal to $4P_c/(4 - 3P_c)$.[29]

*Intermediate $P_i \rightleftharpoons HOH$ Exchange.* This occurs when more than one water oxygen is incorporated into each $P_i$ produced from RP during hydrolysis as depicted by Eq. (1) or (2). The relevant $P_c$ for Eq. (1) is $k_{-2}/(k_{-2} + k_3)$. When labeled RP is hydrolyzed in unlabeled $H_2O$, the average number of water oxygens appearing in $P_i$ will be $\bar{O} = 4/(4 - 3P_c)$, and the distribution of labeled species in the $P_i$ will be given by

$$P^{18}O_3 = P_{33}RP^{18}O_3$$
$$P^{18}O_2 = P_{32}RP^{18}O_3 + P_{22}RP^{18}O_2$$
$$P^{18}O_1 = P_{31}RP^{18}O_3 + P_{21}RP^{18}O_2 + P_{11}RP^{18}O_1$$
$$P^{18}O_0 = P_{30}RP^{18}O_3 + P_{20}RP^{18}O_2 + P_{10}RP^{18}O_1 + P_{00}RP^{18}O_0$$

If reactions governed by $k_{-1}$ or $k_{-3}$ are not negligibly small, the situation is more complex as medium RP $\rightleftharpoons$ HOH and medium $P_i \rightleftharpoons$ HOH exchange can occur and the observed extent of incorporation of water ox-

---

[29] P. D. Boyer, L. de Meis, M. G. C. Carvalho, and D. D. Hackney, *Biochemistry* **16**, 136 (1977).

ygens into $P_i$ will increase with the fraction of hydrolysis. These complications can be removed by analysis of the initial $P_i$ to be formed before significant incorporation of water oxygen into RP or medium exchange of the released $P_i$ can occur. In this case, the observed partition coefficient will be given by $PP_c$ where $P$ is $k_2/(k_{-1} + k_2)$. For the scheme of Eq. (2), the analogous expression for $P$ will be more complex and will be a function of the RH concentration if the release of RH must precede hydrolysis or is rapid relative to hydrolysis of the phosphoenzyme. Measurement of the rate of medium RP exchange with $H^{18}OH$ will allow assessment if release of RP by the $k_{-1}$ step must be taken into account in evaluation of the partition coefficient.

These equations and the ones for intermediate $RP \leftrightharpoons H_2O$ exchange can be readily expanded to deal with the situation in which the water also (or exclusively) contains $^{18}O$. A general computer program for performing such calculations has been developed (D. D. Hackney, unpublished).

*Intermediate $RP \leftrightharpoons HOH$ Oxygen Exchange.* Exchange of water oxygens into RP can occur during synthesis of RP via reversal of Eq. (2). The distribution of labeled species in RP in the absence of exchange will be given by

$$RP'_3 = (0.75)(P^{18}O_3) + P^{18}O_4$$
$$RP'_2 = (0.5)(P^{18}O_2) + (0.25)(P^{18}O_3)$$
$$RP'_1 = (0.75)(P^{18}O_1) + (0.5)(P^{18}O_2)$$
$$RP'_0 = P^{18}O_0 + (0.25)(P^{18}O_1)$$

When $P_c$ is greater than zero, the distribution will be given by

$$RP^{18}O_3 = P_{33}RP'_3$$
$$RP^{18}O_2 = P_{32}RP'_3 + P_{22}RP'_2$$
$$RP^{18}O_1 = P_{31}RP'_3 + P_{21}RP'_2 + P_{11}RP'_1$$
$$RP^{18}O_0 = P_{30}RP'_3 + P_{20}RP'_2 + P_{10}RP'_1 + P_{00}RP'_0$$

The average number of water oxygens present in RP will be $\bar{O} = 3 P_c/(4 - 3P_c)$.

*Medium $RP \leftrightharpoons HOH$ Oxygen Exchange.* Phosphoesters, RP, may also undergo a medium exchange reaction by the general scheme of Eq. (5), where RH may either remain bound or dissociate. When RP is a phos-

$$RP + E \underset{k_{-1}}{\overset{k_1}{\rightleftharpoons}} E \cdot RP \underset{k_{-2}}{\overset{k_2}{\rightleftharpoons}} E \cdot P_i + RH \qquad (5)$$

with $H_2O$ above the middle arrow.

phate ester, only three of the oxygens of RP are exchangeable with water because the fourth is contributed by ROH. For this reaction $P_c = k_2/(k_2 + k_{-1}) = [4(3 - R_3)]/(9 - R_3)$ where

$$R_3 = \frac{\text{rate of loss of } RP^{18}O_3}{\text{rate of loss of average }^{18}\text{O enrichment of 3 terminal oxygens of RP}}$$

The equations describing this exchange cascade are

$$RP^{18}O_3(t) = RP_3^0 e^{-kS_3 t}$$
$$RP^{18}O_2(t) = (RP_2^0 - C_{32})e^{-kS_2 t} + C_{32}e^{-kS_3 t}$$
$$RP^{18}O_1(t) = (RP_1^0 - C_{31} - C_{21})e^{-kS_1 t} + C_{21}e^{-kS_2 t} + C_{31}e^{-kS_3 t}$$
$$RP^{18}O_0(t) = RP_0^0 + C_{10}(1 - e^{-kS_1 t}) + C_{20}(1 - e^{-kS_2 t}) + C_{30}(1 - e^{-kS_3 t})$$

where $RP_j^0$ is the fraction of RP containing $j$ labeled oxygens at time zero and where

$$C_{32} = P_{32} RP_3^0 / (S_2 - S_3)$$
$$C_{31} = (P_{31} RP_3^0 + P_{21} C_{32}) / (S_1 - S_3)$$
$$C_{21} = P_{21}(RP_2^0 - C_{32}) / (S_1 - S_2)$$
$$C_{30} = (P_{30} RP_3^0 + P_{20} C_{32} + P_{10} C_{31}) / S_3$$
$$C_{20} = (P_{20}(RP_2^0 - C_{32}) + P_{10} C_{21}) / S_2$$
$$C_{10} = P_{10}(RP_1^0 - C_{31} - C_{21}) / S_1$$
$$S_i = 1 - P_{ii}$$

Now $k$ should equal $(0.693)(3/\bar{O})$ where $\bar{O}$, the average number of RP oxygens replaced with water oxygen each time RP binds to the enzyme, is equal to $3P_c/(4 - 3P_c)$.

Medium RP $\rightleftharpoons$ H$_2$O exchange is often accompanied by net hydrolysis. Such hydrolysis, if appreciable, will complicate the exchange considerably. The effect of hydrolysis on the average rate of RP $\rightleftharpoons$ H$_2$O exchange has been described.[21,26]

*Calculation of Fluxes.* The fluxes across individual steps of Eq. (3) can be calculated from these relationships. For the medium exchange reaction, the fluxes $\rho_3$ and $\rho_2$ across steps 3 and 2, respectively, are given by

$$\rho_3 = V/\bar{O} = \frac{4 - 3P_c}{4P_c} V$$

$$\rho_2 = \rho_3 \frac{k_2}{k_3} = \frac{4 - 3P_c}{4 - 4P_c} V$$

where $V$ is the observed total flux of isotopic oxygen between water and $P_i$, as can be measured without consideration of the distribution of labeled species.

During irreversible hydrolysis of RP via Eq. (1), the average number of reversals of step 2 per $P_i$ released is given by $R = k_{-2}/k_3$. The total flux, $\rho_2$, across step 2 is a measure of the total amount of water oxygens that have exchanged during the formation of medium $P_i$.

The value for $\rho_2$ is given by

$$\rho_2 = (R + 1)P_i = \frac{3\bar{O}'}{4 - \bar{O}'} P_i$$

where $\bar{O}'$ is the average number of water oxygens present in each $P_i$ molecule released.

This relationship replaces an earlier treatment[2] that did not take all factors into account and leads to underestimation of the total oxygen exchange at higher $P_c$ values. Similar relationships apply to the other exchanges.

*Exchange Patterns with Different Mechanisms.* Comparison of the actual distributions of labeled $P_i$ species observed in an experiment with the preceding theoretical predictions allows an evaluation of whether Eqs. (1) and (2) are an adequate description of the exchange process. Other mechanisms can predict different distributions. This is illustrated in Fig. 1, which gives the theoretical distributions of $^{18}$O-labeled $P_i$ species following hydrolysis of 100% [$^{18}$O]RP in unenriched water. If no extra ex-

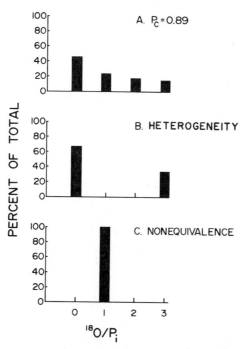

FIG. 1. Theoretical distributions of $^{18}$O-labeled $P_i$ species following hydrolysis of 100% [$^{18}$O]RP in unenriched water. (A) Distribution predicted for model of Eq. (5). (B) Distribution from a limiting model with enzyme heterogeneity. (C) Distribution from a model having nonequivalent hindrance of $P_i$ rotation in E·$P_i$. $P_c$, partition coefficient.

change had occurred, all the $P_i$ should contain 3 $^{18}O$ oxygens, and all the $^{18}O$ can be replaced with $^{16}O$ if the extra exchange is extensive. For all the distributions in the figure, the average extent of exchange is 3 water oxygens incorporated into $P_i$. Panel A gives the distribution predicted for the model of Eq. (5).

The distribution from a limiting model with enzyme heterogeneity is given in panel B. In this model a certain fraction of RP is hydrolyzed with no extra exchange producing $P_i$ with 3 $^{18}O$ oxygens, and the rest is hydrolyzed with extensive exchange producing $P_i$ with no $^{18}O$. The average incorporation is determined by the relative contributions of these two processes. Such a situation can result from actual heterogeneity of the enzyme preparation or may be induced by the binding of effectors that can modify the extent of exchange. This type of distribution can also be produced by a model incorporating enzyme hysteresis in which multiple slowly interconverting forms of the enzyme can exist with different extents of exchange.

The distribution for a model having nonequivalent hindrance of $P_i$ rotation in $E \cdot P_i$ is shown in panel C. In this model one of the four oxygens in $P_i$ is tightly bound and does not participate in the exchange reaction as rapidly as the other three. In the extreme case illustrated here, extensive reversal of the hydrolytic step has resulted in essentially complete exchange of three of the $P_i$ oxygens with water, but the fourth remains resistant to exchange.

*Recent Applications.* Papers from this laboratory making application of $^{18}O$ procedures include demonstration of substrate modulation of intermediate ATP $\rightleftharpoons$ HOH exchange during oxidative phosphorylation and photophosphorylation,[30,31] heterogeneity of ATP cleavage by myosin preparations,[32] characteristics of actin interaction with myosin ATPase,[21,33,34] partition coefficients during medium and intermediate $P_i \rightleftharpoons$ HOH exchange catalyzed by pyrophosphatase,[27,28] and demonstration of alternating site participation in ATP modulation of intermediate $P_i \rightleftharpoons$ HOH exchange by mitochondrial F-1 ATPase.[35]

Applications from other laboratories include use of volatile phosphate techniques to aid studies of oxygen scrambling during ATP $\rightleftharpoons$ HOH ex-

---

[30] D. D. Hackney and P. D. Boyer, *J. Biol. Chem.* **253**, 3164 (1978).
[31] D. D. Hackney, G. Rosen, and P. D. Boyer, *Proc. Natl. Acad. Sci. U.S.A.* **76**, in press, Aug. (1979).
[32] J. A. Sleep, D. D. Hackney, and P. D. Boyer, *J. Biol. Chem.* **253**, 5235 (1978).
[33] J. A. Sleep, D. D. Hackney, and P. D. Boyer, manuscript submitted for publication.
[34] J. A. Sleep and R. L. Hutton, *Biochemistry* **17**, 5423 (1978).
[35] R. L. Hutton, D. D. Hackney, and P. D. Boyer, in "Electrons to Tissues" (P. L. Dutton, J. S. Leigh and A. Scarpa, eds.) p. 494. Academic Press, New York 1978.

change by glutamine synthetase[10] and chloroplasts[36] and application of $^{31}P(^{18}O)$ NMR to study of the P $\rightleftharpoons$ HOH exchange catalyzed by myosin,[37] alkaline phosphatase[16,38] and acid phosphatase.[39]

[36] M. J. Wimmer and I. A. Rose, *J. Biol. Chem.* **252**, 6769 (1977).
[37] M. R. Webb, G. G. McDonald, and D. R. Trentham, *J. Biol. Chem.* **253**, 2908 (1978).
[38] J. L. Bock and M. Cohn, *J. Biol. Chem.* **253**, 4082 (1978).
[39] R. L. Van Etten and J. M. Risley, *Proc. Natl. Acad. Sci. U.S.A.* **76**, 4784 (1978).

## [4] Determination of Heavy-Atom Isotope Effects On Enzyme-Catalyzed Reactions

*By* MARION H. O'LEARY[1]

Heavy-atom isotope effects are extensively used in studies of organic reaction mechanisms.[2-7] Isotope effects are capable of providing detailed information about transition-state structure and other aspects of mechanism not only in organic reactions, but in enzymic reactions as well.[8-10]. Although the first measurement of a heavy-atom isotope effect on an enzyme-catalyzed reaction was made nearly 30 years ago,[11] it is only within the last few years that the technique has gained wider acceptance.

Rate differences resulting from isotopic substitution for atoms other than hydrogen are small. For carbon, oxygen, and nitrogen these rate differences seldom exceed 10%, and they are more commonly in the range of 1–3%. Because the effects are so small, ordinary kinetic techniques are not very useful for their measurement. Instead, the most commonly used method is the competitive method, in which the isotope effect is calcu-

[1] Work in the Author's laboratory was supported by the National Science Foundation.
[2] J. Bigeleisen and M. Wolfsberg, *Adv. Chem. Phys.* **1**, 15 (1958).
[3] L. Melander, "Isotope Effects on Reaction Rates." Ronald Press, New York, 1960.
[4] A. MacColl, *Annu. Rep. Chem. Soc.* **71B**, 77 (1974).
[5] A. Fry, *in* "Isotope Effects in Chemical Reactions" (C. J. Collins and N. S. Bowman, eds.), p. 364. Van Nostrand-Reinhold, Princeton, New Jersey, 1970.
[6] G. E. Dunn, *in* "Isotopes in Organic Chemistry" (E. Buncel and C. C. Lee, eds.), Vol. 3, p. 1. Elsevier, Amsterdam, 1977.
[7] A. V. Willi, *in* "Isotopes in Organic Chemistry" (E. Buncel and C. C. Lee, eds.), Vol. 3, p. 237. Elsevier, Amsterdam, 1977.
[8] M. H. O'Leary, *in* "Transition States of Biochemical Processes" (R. L. Schowen and R. Gandour, eds.), p. 285, Plenum, New York, 1978.
[9] M. H. O'Leary, *in* "Bioorganic Chemistry" (E. E. van Tamelen, ed.), Vol. 1, p. 259. Academic Press, New York, 1977.
[10] M. H. O'Leary, *in* "Isotope Effects on Enzyme-Catalyzed Reactions" (W. W. Cleland, M. H. O'Leary, and D. B. Northrop, eds.), p. 233. Univ. Park Press, Baltimore, Maryland, 1977.
[11] J. A. Schmitt, A. L. Myerson, and F. Daniels, *J. Phys. Chem.* **56**, 917 (1952).

lated from the change in isotopic composition of the starting material or product over the course of the reaction. For carbon, oxygen, and nitrogen it is possible to use the small natural abundance of the heavier isotope in this method. Isotope abundances are then measured with an isotope-ratio mass spectrometer.

In the case of carbon, many isotope effect measurements have been made with $^{14}$C. Isotopic composition is then measured by liquid-scintillation counting. The relationship[12] between $^{13}$C isotope effects and those for $^{14}$C is given in Eq. (1)

$$k^{12}/k^{14} - 1 = 1.9(k^{12}/k^{13} - 1) \tag{1}$$

The radioactivity method is about an order of magnitude less accurate than the stable-isotope method, and current practice favors the latter method.

The purpose of this chapter is to provide experimental details concerning the measurement of heavy-atom isotope effects on enzyme-catalyzed reactions by the stable-isotope method together with guidelines for the interpretation of the results. Measurement and interpretation of carbon isotope effects on enzyme-catalyzed decarboxylation reactions are described in particular detail because of the variety of examples of this reaction that have been studied. We will not consider two newer methods for the determination of heavy-atom isotope effects: the equilibrium perturbation method, which is described elsewhere in this volume,[13] and the direct kinetic method, which has found occasional use in the determination of heavy-atom isotope effects.[14,15]

### Methods

*Theory*

If a substrate containing a mixture of unlabeled (S) and labeled (S*) species is converted into product, the isotopic compositions of the substrate (S*/S) and product (P*/P) change over the course of the reaction because the labeled and unlabeled substrates do not react at exactly the same rate. This is illustrated in Fig. 1 for a decarboxylation reaction in which the labeled substrate reacts more slowly than the unlabeled substrate by 5%; that is, $k^{12}/k^{13} = 1.05$.

Experimentally, the isotope effect can be determined by measuring

---

[12] M. J. Stern and P. C. Vogel, *J. Chem. Phys.* **55**, 2007 (1971).
[13] W. W. Cleland, this volume [5].
[14] C. G. Mitton and R. L. Schowen, *Tetrahedron Lett.* 5803 (1968).
[15] D. G. Gorenstein, *J. Am. Chem. Soc.* **94**, 2523 (1972).

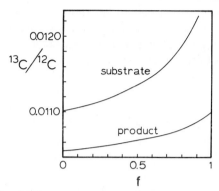

FIG. 1. Isotope ratios ($^{13}C/^{12}C$) for the carboxyl carbon of remaining substrate and for accumulating carbon dioxide product for a decarboxylation showing a carbon isotope effect $k^{12}/k^{13} = 1.05$.

the change in isotopic composition of either the substrate or the product as the reaction proceeds.[2,3] When the substrate is to be analyzed, the isotopic composition of the initial substrate is compared with that of the substrate remaining after partial reaction. The relationship of the isotope effect[16] ($k/k^*$) to the isotopic composition of the substrate prior to reaction ($S_0^*/S_0$) and that remaining after partial reaction ($S_t^*/S_t$) is given by Eq. (2).[17] The fraction reaction ($f$) is given by Eq. (3).

$$\frac{k}{k^*} = \frac{\log(1 - f)}{\log[(1 - f)(S_t^*/S_t)/(S_0^*/S_0)]} \qquad (2)$$

$$f = 1 - \frac{S_t + S_t^*}{S_0 + S_0^*} \qquad (3)$$

The best results are obtained when the fraction reaction is at least 0.5.

When the product is to be analyzed, the isotopic composition of a sample of product obtained after only a small extent of reaction ($P_t^*/P_t$) is compared with that of a sample obtained after complete reaction ($P_\infty^*/P_\infty$). The isotope effect is calculated from Eq. (4).

$$\frac{k}{k^*} = \frac{\log(1 - f)}{\log[1 - f(P_t^*/P_t)/(P_\infty^*/P_\infty)]} \qquad (4)$$

---

[16] Throughout this discussion, an asterisk denotes the position of an isotope label.
[17] The forms of Eqs. (2) and (4) given here are not appropriate for use in studies of hydrogen isotope effects. For a more general description of the mathematics, see Bigeleisen and Wolfsberg.[2] It should be noted that Eqs. (25b) and (26) given by MacColl[4] are incorrect. The correct versions are given here as Eqs. (4) and (5).

Although very high precision is always required in measurements of isotopic composition, it is not important that the extent of reaction in the low-conversion sample be known extremely accurately; a precision of ±1% is quite adequate. However, because of the steep dependence of product isotopic composition on extent of reaction at the end of the reaction (Fig. 1), it is important that the complete-conversion sample be carried all the way to 100% reaction. Correction can be made for incomplete reaction if necessary,[18] but this procedure is best avoided.

To a first approximation the isotope effect is related to the isotopic compositions of the two product samples by Eq. (5)

$$\frac{k^{12}}{k^{13}} = \frac{(P^*_\infty/P_\infty)}{(P^*_t/P_t)} \tag{5}$$

and this provides a useful preliminary estimate of the isotope effect. If the extent of reaction in the low conversion sample is 0%, Eq. (5) is precisely true, and even at 10–15% reaction the error is small.

## Equipment and Materials Required

Isotopic compositions of the required precision can be obtained only with an isotope-ratio mass spectrometer. Such instruments are specifically designed for the high-precision measurement of ratios of isotopic abundances in simple molecules (ratios such as $^{13}CO_2/^{12}CO_2$). Most such instruments have a dual inlet system capable of alternating between a standard of known isotopic composition and the unknown. Because of the necessity for rapid evacuation of the inlet system only volatile substances, such as $CO_2$, $N_2$, $CH_3Cl$, can be measured. The instrument has a pair of collectors so arranged that one collector receives the ion beam from the lighter isotopic species and the other receives that from the heavier isotopic species. The signals from the two collectors are amplified and fed through a bridge circuit for determination of the isotope ratio. The ratios so obtained are relative to some laboratory standard, rather than being absolute abundance ratios. Although with proper calibration it is possible to obtain absolute ratios, this is unnecessary in the determination of isotope effects.

Isolation and purification of materials for isotopic analysis require the use of a good vacuum system incorporating a mechanical pump and a diffusion pump and capable of operating below $10^{-4}$ torr. A Toepler pump is very useful if noncondensable gases (such as $N_2$ and CO) are to be used. A calibrated manometer is useful for measuring yields of volatile products.

---

[18] C. J. Collins and M. H. Lietzke, *J. Am. Chem. Soc.* **81**, 5379 (1959).

Substantially greater quantities of enzyme and substrate are required for determination of heavy-atom isotope effects than are required for steady-state kinetic studies. Each determination of the isotope effect requires about 1 mmol of substrate and sufficient enzyme and cofactor to convert about 20% of this material to product within a few hours. At least a half-dozen measurements of the isotope effect are required in order to demonstrate the reproducibility of the measurement procedure, and a like amount of enzyme and substrate may be required for preliminary experiments.

## Procedure

Carbon isotope effects on enzyme-catalyzed decarboxylation reactions are determined by decarboxylation of two portions of substrate, one to the extent of approximately 10% (we call this the "low conversion" sample) and the other to the extent of 100% (the "complete conversion" sample). The carbon dioxide from each reaction is purified and its isotopic composition is determined on the isotope-ratio mass spectrometer. The isotope effect is then calculated from Eq. 4. A flow diagram that summarizes this procedure is given in Fig. 2. The various steps are described in more detail below.

The scale of the isotope effect experiment is governed largely by the amount of material required for the isotope-ratio measurements. In addition, larger samples reduce errors due to contamination by extraneous carbon dioxide. A reasonable goal is generally 100 $\mu$mol of carbon dioxide per sample. Since only about 10% of the substrate is converted to carbon dioxide in the low-conversion sample, this means that about 1 mmol of substrate is required per experiment.

The substrate solution is made up in buffer containing, if possible, all required cofactors,[19] metal ions, stabilizing agents, etc. However, if necessary, unstable cofactors or other materials can be purified separately and added to the solution just before the enzyme is added. Substrate concentrations in the range 10–20 m$M$ are convenient, although lower concentrations can be used if necessary. The substrate concentration does not influence the observed isotope effect. Volumes larger than about 100 ml per sample are inconvenient to work with. Buffering must be sufficient that the pH does not change substantially even when all the substrate is

---

[19] It is occasionally impractical to use stoichiometric amounts of cofactor or cosubstrate, for reasons of cost, stability, or solubility. In such cases it is necessary to design a system for producing or regenerating the cofactor. See O'Leary and Limburg[20] for an example of this procedure.

[20] M. H. O'Leary and J. A. Limburg, *Biochemistry* **16**, 1129 (1977).

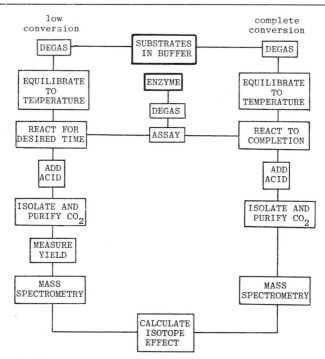

FIG. 2. Flow diagram for determination of carbon isotope effects in enzymic decarboxylation reactions.

decarboxylated (note that below pH 6 most decarboxylations are accompanied by the absorption of a proton from the medium).

This substrate solution is divided into two parts, a large part to be used as the low-conversion sample and a small part to be used as the complete-conversion sample. The division should be made such that the final yields of carbon dioxide in the two samples are approximately the same. Except for amount of enzyme used and reaction time, the two samples are treated identically. It is useful to reserve a small portion of the substrate solution for use in determining the conditions needed to reach the appropriate degree of reaction in the low-conversion sample.

The reaction vessels used for the two samples are shown in Fig. 3. The sidearm allows for a vacuum-tight seal but still permits access to the solution by means of a hypodermic syringe. The top of the flask is initially covered by a serum cap through which is inserted a tube for degassing (to be described shortly). This is later replaced by a glass connector having a stopcock and a joint for attachment to the vacuum line. High-vacuum silicone grease is used on joints and stopcocks.

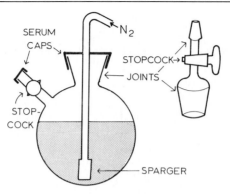

FIG. 3. Reaction vessel for use in determination of carbon isotope effects.

Before the decarboxylation is conducted it is necessary to remove all traces of carbon dioxide present in the solution and in the air space above. In nonenzymic systems this has often been done by evacuating the frozen solution and then subjecting the flask to several thaw–freeze–evacuate cycles. Although this procedure is certainly effective, it is inconvenient in the present case because of the need for continued access to the reaction solution. Instead of evacuating, we bubble carbon dioxide-free nitrogen (prepared by passing tank nitrogen through a column of Ascarite) through the solution for 30 min or more using a fritted disk to disperse the nitrogen. For this purpose the stopcock on the sidearm of the flask (Fig. 3) is opened and the nitrogen is vented out through a syringe needle inserted into the serum cap in the sidearm. Provided that the pH of the solution is below about 6, purging the solution in this manner for 30 min is adequate to remove all traces of carbon dioxide. At higher pH it is necessary to extend the purging time to several hours or else to purge at low pH and then carefully adjust the pH with carbon dioxide-free NaOH (saturated NaOH in water works well). The fritted disk is then removed and quickly replaced with the glass connector shown in Fig. 3. It is useful at this point to flush the flask briefly with nitrogen in order to remove any traces of air that may have been introduced during attachment of the connector. The flask is then placed in a constant-temperature bath.

The enzyme must also be freed of carbon dioxide before use. A convenient apparatus for doing this is shown in Fig. 4. By means of this system we are able to subject the enzyme to gel filtration using carbon dioxide-free buffer. The buffer reservoir above a column of Sephadex G-25 is freed of carbon dioxide by bubbling nitrogen through it, and the column is then washed thoroughly with this buffer. A sample of enzyme is applied to the top of the Sephadex bed by means of the sidearm shown (addition of

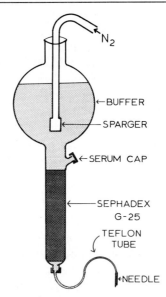

Fig. 4. Column setup for anaerobic desalting of enzymes.

sucrose or glycerol to the enzyme solution ensures that it will layer smoothly on top of the bed) and the column is eluted with buffer. The tube attached to the column outlet is equipped with a small metal tube (a cutoff syringe needle is satisfactory) and the effluent fractions are collected in nitrogen-filled test tubes equipped with serum caps. The carbon dioxide-free enzyme is swirled to ensure homogeneity and then assayed using the small sample of reactant solution that was put aside for this purpose, and the results of the assay are used to calculate the proper reaction time for the low-conversion sample.

After the two reaction samples have reached thermal equilibrium, enzyme is added to each through the sidearm and the contents of the flask are thoroughly mixed. The amount of enzyme added to the low-conversion sample is carefully calibrated to give the desired extent of reaction (usually 5–20%) within a few hours. The exact extent of reaction is later confirmed by measuring the yield of purified carbon dioxide or the amount of unreacted substrate. Excess enzyme is added to the complete conversion sample in order to ensure that the last traces of substrate are decarboxylated. Although correction for incomplete decarboxylation is possible, it is very inconvenient and can be a source of substantial error.

At the appropriate time the decarboxylation in the low-conversion sample is stopped by injection of sufficient concentrated $H_2SO_4$ to inactivate the enzyme and lower the pH below 5. Usual practice is to use enough acid to lower the pH to about 1; this inactivates the enzyme and releases any carbon dioxide tied up in the form of bicarbonate or carbonate ion. However, discretion is required here, as some substrates undergo a slow acid-catalyzed decarboxylation, and such a reaction would provide an undesired source of carbon dioxide. The complete conversion sample is treated in the same manner, although in most cases the reaction time is considerably longer.

After addition of acid, each flask is attached to the vacuum line and the solution is frozen in liquid nitrogen. The stopcock connecting the flask to the vacuum line is then opened, and the nitrogen atmosphere is pumped away. Because a substantial amount of nitrogen is dissolved in the frozen solution it is necessary to subject the solution to a thorough degassing at this point. To accomplish this, the connecting stopcock is closed and the frozen solution is thawed, allowed to outgas, and then refrozen in liquid nitrogen: The stopcock is opened and the liberated nitrogen is pumped away. This thaw–freeze–pump cycle is repeated until all nitrogen has been removed. Two or three cycles are usually required. Failure to remove nitrogen completely slows down subsequent distillation of carbon dioxide.

With the connecting stopcock closed, the solution is again thawed, but this time it is refrozen using Dry Ice and acetone or isopropyl alcohol. The carbon dioxide can then be distilled out of the flask into a temporary receptacle by cooling the receptacle with liquid nitrogen. During the distillation the stopcock connecting the reaction flask to the system should not be left open too long because water has an appreciable vapor pressure at Dry Ice temperature and extensive contamination of the sample by water can occur. After distillation of the carbon dioxide sample, the stopcock connecting the reaction flask to the temporary receptacle should be reclosed and the reaction solution should be thawed and refrozen and a second portion of carbon dioxide distilled out and added to the first. An additional thawing–refreezing cycle is advisable. These additional cycles are necessary in order to obtain a good yield of carbon dioxide and to avoid spurious isotope effects associated with the solubility of carbon dioxide in water.[21]

The isolated carbon dioxide will undoubtedly be contaminated with water at this point. Water has disastrous effects on the mass spectrometer and must be removed before isotope-ratio analysis. This is done by freezing the tube containing the sample in Dry Ice and using liquid nitrogen to distill the carbon dioxide into a second temporary receptacle. This freezing and distillation procedure is repeated about two times.

[21] W. G. Mook, J. C. Bommerson, and W. H. Staverman, *Earth Plan. Sci. Lett.* **22,** 169 (1974).

After purification, the yield of carbon dioxide should be measured. For this purpose a mercury manometer attached to the vacuum system is convenient. Ordinarily it is possible to measure the yield of carbon dioxide and calculate the percent reaction in the low-conversion sample to within at least 1% reaction; such precision is adequate for use in Eq. (4).

However, measurement of yield in the complete reaction sample by this method does not provide an adequate measure of yield within the 95–100% reaction range. An uncertainty of 1% in the extent of reaction will cause a large uncertainty in the calculated isotope effect. The use of an excess of enzyme is ordinarily adequate to ensure that 100% reaction is achieved, but this should be checked in the initial isotope effect experiments by neutralization of the spent acidified reaction mixture and enzymic assay for remaining substrate.

Continued awareness of this issue is required. The first isotope effect experiments conducted in a new system are generally done near the pH optimum of the enzyme. In subsequent experiments different conditions are often used, and the conditions may approach the limits of stability of the enzyme. Slow enzyme denaturation during an isotope effect experiment causes no serious problem provided that proper account is taken of the extent of reaction in both samples. This means using the manometrically measured carbon dioxide yield (rather than the value predicted from the assay) in the low-conversion sample and checking the complete reaction sample carefully to be sure that no substrate remains.

An alternative approach is sometimes appropriate for the complete reaction sample. Since the isotope ratio of the carbon dioxide obtained in the complete reaction sample depends only on the isotopic composition of the starting material, not on the conditions under which the decarboxylation is conducted, it is possible to conduct the complete decarboxylation under more optimal conditions. This procedure is undesirable because it destroys the exact parallel between the preparation of the low-conversion sample and the complete-conversion sample, but if proper precautions are used, successful results can be obtained.

Reversible reactions can also provide complications in measurement of isotope effects. It is usually possible to arrange conditions such that there is little possibility for the reverse reaction to occur in the low-conversion sample, but carrying the complete reaction sample to 100% reaction may require the use of a coupling system. See O'Leary and Limburg[20] for an example of this procedure.

The carbon dioxide is then measured by isotope-ratio mass spectrometry. For determination of carbon isotope effects the isotope ratio is obtained from the intensity ratio of the peaks at $m/e$ 45 and $m/e$ 44. However, before the isotope effect is calculated these ratios must be corrected for the presence of a small contribution due to $^{12}C^{17}O^{16}O$ in the $m/e$ 45 peak.[22] The correction is made by subtracting a value from the 45/44 ratio that corresponds to twice the isotopic abundance of $^{17}O$ in the sample. This correction factor is approximately 0.000740 if the 45/44 ratio is about 0.01100; the isotope effects obtained are quite insensitive to the precise value used for this correction.

*Controls*

Because of the small sizes of heavy-atom isotope effects, particular care must be exercised to guard against errors. The principal sources of error in isotope effect measurements are isotope fractionations occurring during sample preparation, contamination of the product by extraneous carbon dioxide, and contamination of the product with other materials that interfere with the mass spectrometry. In the following paragraphs we describe some of the procedures used to avoid these errors.

After an isotope effect measurement procedure has been designed but before measurements are carried out, the complete procedure should be carried through once with enzyme omitted and once with substrate omitted. In each case, no carbon dioxide should be formed. In practice, the tolerable yield of carbon dioxide for experiments conducted at natural abundance is about 1% of the amount to be formed in the complete experiment.

Contamination of the carbon dioxide by organic compounds that can affect the measured isotope ratios is not a common problem. Contamination can be detected by scanning the mass spectra of the carbon dioxide samples. The presence of water in the samples is indicated by the occurrence of large, slow drifts in the measured isotope ratios as the inlet is alternated between sample and standard. Water can be removed by further distillation of the samples.

In the measurement of isotope effects it is essential that no isotope fractionations intervene between the reaction whose isotope effect is being studied and the isotopic analysis. Many chemical reactions show substantial isotope effects, and if chemical transformation of the sample is required prior to isotopic analysis it must be established that either there is no isotopic fractionation in the conversion or else the conversion must be carried to 100% completion. Many physical processes also show small isotope effects. For example, $^{13}CO_2$ is more soluble in water than is

---

[22] H. Craig, *Geochim. Cosmochim. Acta* **12**, 133 (1957).

$^{12}CO_2$.[21] The boiling point of methanol is increased by $^{18}O$ substitution and decreased by $^{13}C$ substitution.[23] Numerous other such fractionations are known.[24] The most effective deterrent to such problems is to make sure that yields in all physical and chemical processes approach 100% as closely as possible.

A primary criterion for the adequacy of isotope effect experiments is the reproducibility of the measured effect from experiment to experiment. In addition, if all isotope effect experiments are conducted with the same carefully purified batch of substrate, the isotope ratio in the complete conversion sample should be constant, independent of the conditions under which the reaction was conducted.

In separate determinations of the isotope effect it is useful to carry the low-conversion sample to various extents of reaction over the range from as little as 1% reaction to as much as 50% reaction. The isotope ratio for this sample will change markedly as the percent reaction changes, but the calculated isotope effect should be the same for all samples. A carbon dioxide contaminant will contribute to the observed isotope ratio to varying extents, dependent on the amount of reaction. Such contribution is particularly significant at low conversions, and it is for this reason that isotope effects measured at very low conversions may show large errors.

Because of the errors propagated by contamination it is not preferable to measure isotope effects with isotopically enriched materials except in special cases. Errors due to contamination become much more significant in enriched samples than in natural abundance samples. Consider, for example, the error in an isotope ratio caused by the presence of 1% extraneous carbon dioxide. In a natural abundance sample the error in the observed ratio is 1% of the difference between the isotope ratio of the sample and that of the contaminant, and this is a very small error. However, in an enriched sample the error is 1% of the difference between the isotope ratio of the (enriched) sample and that of the (natural abundance) contaminant, and this may be quite a large error. This phenomenon has been responsible for many of the unexplained errors in determination of heavy-atom isotope effects using radioactive isotopes.

Competing chemical reactions can occasionally cause errors in isotope effect measurements. In such cases some of the product may be produced by a nonenzymic reaction, and the isotope effect observed will be a composite of the enzymic and nonenzymic isotope effects, weighted according to their contributions to the overall rate. In studies of the enzymic decar-

---

[23] J. L. Borowitz and F. S. Klein, *J. Phys. Chem.* **75**, 1815 (1971).
[24] J. Bigeleisen, M. W. Lee, and F. Mandel, *Annu. Rev. Phys. Chem.* **24**, 407 (1973).

boxylation of acetoacetic acid, for example,[25] it was necessary to make fresh solutions of acetoacetic acid immediately before use because of the slow spontaneous decomposition of this material. By use of somewhat larger than usual quantities of enzyme we were able to complete the 10% decarboxylation within about 30 min and thus minimize the amount of nonenzymic decarboxylation.

Finally, it is worth emphasizing the need for truly independent experiments. Because of the large number of potential sources of error in isotope effect experiments, one should avoid taking shortcuts that compromise the statistical independence of the separate determinations of the isotope effect. For example, it is bad practice to make up a large volume of substrate or buffer solution and divide it into ten parts in order to make five "determinations" of the isotope effect. Any contaminant in the initial solution will be almost totally masked by the fact that all experiments were conducted with this same material.

*Determination of Heavy-Atom Isotope Effects on Other Reactions*

We have described the protocol for determination of carbon isotope effects on enzymic decarboxylations in detail because this procedure is well understood and illustrates most of the important issues encountered in determination of heavy-atom isotope effects. The applicability of this technique to a number of other reactions is considered in this section.

Heavy-atom isotope effects can be measured for a variety of other reactions if the product can be isolated and converted to a form suitable for measurement by isotope-ratio mass spectrometry. The preceding section should be consulted as to the kind of precautions required in such measurements. Nitrogen isotope effects on amide hydrolysis catalyzed by papain and by chymotrypsin have been determined by the isolation of the ammonia produced and conversion of this material into molecular nitrogen.[26,27] Oxygen isotope effects on hydrolysis of methyl esters have been measured by isolation of the methanol product and conversion of this material into carbon monoxide for isotopic analysis.[28] $^{18}$O-Enriched methyl and ethyl esters have been studied by direct introduction of substrates into a specially designed mass spectrometer.[29,30] Isotope effects on

[25] M. H. O'Leary and R. L. Baughn, *J. Am. Chem. Soc.* **94**, 626 (1972).
[26] M. H. O'Leary and M. D. Kluetz, *J. Am. Chem. Soc.* **94**, 3585 (1972).
[27] M. H. O'Leary, M. Urberg, and A. P. Young, *Biochemistry* **13**, 2077 (1974).
[28] M. H. O'Leary and J. F. Marlier, *J. Am. Chem. Soc.* **101**, 3300 (1979).
[29] C. B. Sawyer and J. F. Kirsch, *J. Am. Chem. Soc.* **95**, 7375 (1973).
[30] C. B. Sawyer and J. F. Kirsch, *J. Am. Chem. Soc.* **97**, 1963 (1975).

the carboxylation of ribulose diphosphate[31] and phosphoenolpyruvate[32] have been obtained through combustion of starting materials and products. However, this procedure is not preferred because it provides only an average isotope effect for all carbons of the starting material.

*The Stable-Isotope Double-Label Method*

The usefulness of the procedures described in the preceding sections is ultimately limited by the mass spectrometer—the isotopic atom must be converted into a simple, volatile form for isotope-ratio analysis, and such conversion is prohibitively difficult for a number of potentially interesting systems. For example, the magnitudes of carbonyl oxygen isotope effects in ester and amide hydrolysis could provide useful information about the intervention of tetrahedral intermediates, but such effects are not readily available by the standard isotope-ratio procedure. We have described a double-label method for the measurement of heavy-atom isotope effects that provides ready access to isotope effects at previously inaccessible sites,[28] and we have shown that by use of this method it is possible to measure carbonyl oxygen isotope effects on the hydrolysis of esters.

Heavy-atom isotope effects are ordinarily obtained by measuring isotope ratios of the site of interest in the substrate or product during the course of the reaction. In the stable-isotope double-label method the same measurements are made, but the comparison is between unlabeled and doubly labeled substrates. One of the labeled positions (the "indicator position") is a site that is easily accessible to isotopic analysis. The other labeled position (the "key position") is the position whose isotope effect is desired. This position may be buried deep within the molecule, as direct isotopic analysis of this position is not necessary. The substrate will be designated by IK, where I is the indicator position and K is the key position. The appropriate equations are then

$$IK \xrightarrow{k} P$$

$$I^*K^* \xrightarrow{k^*} P^*$$

If singly labeled substrates are absent, then measurement of product isotope ratios at the indicator position provides, by use of Eq. (4), an isotope effect $(k/k^*)_0$ that is actually the product of the isotope effect at the indi-

---

[31] J. T. Christeller, W. A. Laing, and J. H. Troughton, *Plant Physiol.* **57**, 580 (1976).
[32] P. H. Reibach and C. R. Benedict, *Plant Physiol.* **59**, 564 (1977).

cator position and that at the key position. The indicator position was chosen because of the ease of isotopic analysis at this position; thus, the isotope effect at this position can easily be measured by the standard single-label method, and the isotope effect at the key position is obtained from the quotient of the two isotope effects. In practice, the situation is slightly more complex than this because of the inevitable presence of small amounts of singly labeled substrate, but if proper isotopic enrichments are used this causes no serious problem.

Use of the double-label method requires synthesis of substrate isotopically enriched at both the key position and the indicator position. To minimize the corrections for singly labeled materials, this synthesis should be carried out with starting materials that are labeled to the extent of at least 90% with the atom of interest, and the use of 97% enriched materials (which are becoming increasingly available) is advantageous. This enriched substrate is then mixed with a roughly equal quantity of natural-abundance substrate, and this material is used for determination of the isotope effect. As we shall see shortly, it is necessary to know the relative abundances of the two singly labeled species relative to the unlabeled and doubly labeled materials in order to correct the observed isotope effect. Provided that 90% or greater enrichment is obtained at each of the two sites in the synthesis of the doubly labeled material, these abundances can be determined with sufficient precision by ordinary mass spectrometry.

The measurement of the apparent isotope effect at the indicator position of this labeled material is conducted as usual, by measuring the isotope ratio of the indicator position at a low fraction reaction and at complete reaction. A first approximation to the isotope effect $(k/k^*)_0$ is calculated from these ratios using Eq. (4). The isotope effect at the indicator position $(k_I/k_I^*)$ must be measured in separate experiments using natural-abundance or singly enriched substrate unless it is clear that there will be no isotope effect at this position. These two isotope effects and the abundances of the singly labeled substrates in the double-label experiment are used to calculate the actual isotope effect at the key position by use of Eq. (6).

$$\frac{k_K}{k_K^*} = \left\{ \left(\frac{k}{k^*}\right)_0 - \left(\frac{k_I}{k_I^*}\right)\left[\frac{1 + (I^*K/I^*K^*)}{1 + (IK/IK^*)}\right] \right\} \Big/ \left\{ \left(\frac{k_I}{k_I^*}\right)\left[\frac{1 + (I^*K/I^*K^*)}{1 + (IK^*/IK)}\right] - \left(\frac{k}{k^*}\right)_0 \left(\frac{I^*K}{I^*K^*}\right) \right\} \quad (6)$$

In this equation, $k_K/k_K^*$ is the isotope effect at the key position, $k_I/k_I^*$ is the isotope effect at the indicator position, which is obtained in a separate experiment. The ratios in parentheses involving the various isotopic species of IK are corrections due to the presence of singly labeled substrates. These ratios are obtained from a carefully determined mass spectrum of the isotopically heterogeneous starting material.

The principal problem with the double-label method lies in the experimental difficulties created by the use of highly enriched isotopic species. Isotopic enrichment appears to be necessary because of the need for correlation between the existence of labels at each of two positions. A slight modification of the procedure overcomes this problem. Along with the increasing availability of highly enriched species containing stable isotopes has come an increasing availability of materials depleted in the heavier isotope. Thus it is possible to synthesize substrate that is substantially free of labeling at the indicator position.[33] This material is then mixed in the ratio of approximately 90:1 with the doubly labelled substrate.[34] The substrate so obtained is very interesting: macroscopically, every atom in the molecule is at very near its natural isotopic abundance; microscopically, there is a near 1:1 correlation between the presence or the absence of an isotopic label at the key position and at the indicator position. The isotope effect experiment can be carried out as above, but most of the hazards associated with studies using isotopically enriched materials have now disappeared.

## Interpretation of Heavy-Atom Isotope Effects

In this section we describe the relationship between experimentally measured heavy-atom isotope effects on enzymic reactions and the rates and isotope effects for individual steps in the reactions. More comprehensive analyses of this issue have been published.[8-10]

If a chemical reaction proceeds in a single step, then the magnitude of a heavy-atom isotope effect on that reaction reflects the structure of the transition state for the reaction. Qualitatively, the magnitude of the isotope effect is related to the change in bonding to the isotopic atom on going from ground state to transition state; the greater the change, the larger the isotope effect. Enzyme-catalyzed reactions proceed by multistep mechanisms, and in addition to transition-state structure, the isotope effect also reflects rates of various steps in the reaction sequence.

---

[33] Although it might seem that this material should also be depleted in the key position, this is not necessary. The desired correlation between positions of labels obtains even if this position is not depleted.

[34] This mixing ratio varies according to what atom is present at the indicator position. Mixing should be done in proportions such that the final isotopic composition at the indicator position is as close as possible to natural abundance.

For example, consider the simple reaction scheme shown in Eq. (7).

$$E + S \rightleftharpoons ES_1 \underset{k_2}{\overset{k_1}{\rightleftharpoons}} ES_2 \xrightarrow{k_3} EP \rightarrow E + P \tag{7}$$

The first step, formation of the Michaelis complex ($ES_1$) is assumed for the present to be at equilibrium. The second step is conversion of this complex to some intermediate ($ES_2$). This intermediate is then converted into the enzyme–product complex (EP), which eventually dissociates to give enzyme and product.

Many heavy-atom isotope effect studies of enzyme-catalyzed reactions can be fitted into a mechanism of the form consistent with Eq. (7) with two useful limitations. In the first place, conversion of $ES_2$ into EP is assumed to be irreversible under the reaction conditions. In the second place, it is often adequate to assume that only $k_3$ shows an appreciable isotope effect. These limitations fit very well with most decarboxylations, for example. In that case $k_3$ is the decarboxylation step and $k_1$ and $k_2$ are not expected to show measurable isotope effects. When these limitations apply, the observed isotope effect is independent of rates or isotope effects for any steps following formation of EP. A more general treatment appropriate for reversible reactions has been given.[8] In general the observed isotope effect in such a system is a function of two variables: the isotope effect on the isotopically sensitive step (called the *intrinsic isotope effect*) and a ratio of rate constants (called the *partitioning factor*), which is related to the extent to which the isotopically sensitive step is rate-determining. The relationship between the observed isotope effect and the intrinsic isotope effect for the mechanism of Eq. (7) is given by Eq. (8).

$$\frac{k}{k^*}(\text{obsd}) = \frac{k_3/k_3^* + k_3/k_2}{1 + k_3/k_2} \tag{8}$$

For the three-step mechanism

$$E + S \rightleftharpoons ES_1 \underset{k_2}{\overset{k_1}{\rightleftharpoons}} ES_2 \underset{k_4}{\overset{k_3}{\rightleftharpoons}} ES_3 \xrightarrow{k_5} E + P \tag{9}$$

in which $k_5$ is the isotopically sensitive step, the isotope effect is given by

$$\frac{k}{k^*}(\text{obsd}) = \frac{k_5/k_5^* + (k_5/k_4)(1 + k_3/k_2)}{1 + (k_5/k_4)(1 + k_3/k_2)} \tag{10}$$

Mechanistic schemes involving additional intermediates have the same general form as Eqs. (8) and (10), but the complex term on the right in denominator and numerator (the partitioning factor) is replaced by a still more complex term involving additional rate constants.

As a matter of practical interpretation, Eqs. (8) and (10) tell us that the observed isotope effect in an enzyme-catalyzed reaction reflects not only the structure of the transition state for the isotopically sensitive step (and thus the magnitude of the intrinsic isotope effect) but the magnitude of the partitioning factor as well. Only if we know the value of the partitioning factor can we obtain information about transition-state structure. Since this factor is not usually available, the more common procedure is to estimate the intrinsic isotope effect by use of appropriate model reactions and use the observed isotope effect to provide information about the partitioning factor.

Heavy-atom isotope effects have been measured for a variety of organic reactions,[1-7] and these effects can often provide at least general guidance for predicting the intrinsic effects to be expected in enzymic reactions. In the case of decarboxylations, for example, a substantial number of examples from organic chemistry[6] lead us to predict intrinsic isotope effects in the range $k^{12}/k^{13} = 1.03-1.07$.

Unfortunately, in a number of cases appropriate data are not available to enable us to predict the intrinsic isotope effect with any confidence. This may occur either because appropriate measurements have not been performed in model reactions or else because evidence from model reactions suggests that intrinsic isotope effects for reactions of a particular type are quite variable. In either case, the enzymologist needs additional sources of guidance in interpreting the observed effects. The alternative to using only the comparison between observed and predicted effects is to measure the isotope effect of interest under a variety of conditions and try to interpret the change in the observed isotope effect with reaction conditions. Useful variables include temperature, enzyme source, pH, substrate structure, cofactor structure, and others.

Examples and Applications

In the table are unpublished data obtained by G. Piazza in the author's laboratory in the course of a study of the pyridoxal phosphate-dependent arginine decarboxylase from *Escherichia coli*. These experiments were intended to probe the temperature dependence of the carbon isotope effect on the decarboxylation, and all determinations were carried out under identical conditions using the same batch of substrate.

The isotope ratio for the complete conversion sample reflects the isotopic composition of the carboxyl carbon of the starting material. This ratio is the same in all samples, within the experimental error of ±0.000003. The isotope ratios given in this table are decade settings for the ratio $m/e$ 45/44 corrected to a constant value of the reference standard, but not corrected for the presence of $^{17}O$. The fact that the ratios are

TABLE

Carbon Isotope Effects on the Decarboxylation of Arginine by Bacterial Arginine Decarboxylase at pH 5.25

| Temperature (°C) | Percent reaction | Isotope ratios[a] × 10⁶ | | $k^{12}/k^{13}$[b] |
|---|---|---|---|---|
| | | Low conversion | Complete conversion | |
| 4.8° | 0.5 | 14143 | 14514 | 1.0279 |
| | 41 | 14227 | 14517 | 1.0284 |
| | 41 | 14229 | 14516 | 1.0281 |
| | 9.1 | 14166 | 14516 | 1.0275 |
| | 8.9 | 14173 | 14515 | 1.0268 |
| | 13 | 14169 | 14515 | 1.0278 |
| | 14 | 14174 | 14514 | 1.0274 |
| | | | | Mean 1.0277 ± 0.0005 |
| 25° | 4.5 | 14324 | 14512 | 1.0142 |
| | 14 | 14317 | 14508 | 1.0153 |
| | 11 | 14330 | 14515 | 1.0145 |
| | 14 | 14327 | 14515 | 1.0150 |
| | 10 | 14333 | 14515 | 1.0142 |
| | 23 | 14334 | 14514 | 1.0152 |
| | 20 | 14339 | 14514 | 1.0145 |
| | 10 | 14329 | 14517 | 1.0147 |
| | 8.2 | 14327 | 14514 | 1.0144 |
| | | | | Mean 1.0147 ± 0.0004 |
| 37° | 19 | 14369 | 14514 | 1.0119 |
| | 16 | 14363 | 14514 | 1.0122 |
| | 13 | 14361 | 14515 | 1.0122 |
| | 11 | 14361 | 14514 | 1.0120 |
| | 11 | 14357 | 14514 | 1.0123 |
| | 10 | 14355 | 14514 | 1.0124 |
| | | | | Mean 1.0122 ± 0.0002 |
| 50° | 46 | 14390 | 14514 | 1.0126 |
| | 13 | 14359 | 14518 | 1.0126 |
| | 5.0 | 14354 | 14521 | 1.0126 |
| | 7.0 | 14351 | 14516 | 1.0126 |
| | 7.4 | 14352 | 14518 | 1.0127 |
| | | | | Mean 1.0126 ± 0.0001 |

[a] Mass spectrometer decade settings for the ratio $m/e$ 45/44 corrected to a constant value of the tank standard, but not corrected for the presence of $^{18}O$.
[b] Calculated from Eq. (4) after correction for the presence of $^{18}O$.

in the range of 0.014, whereas the natural abundance of $^{13}C$ is nearer to 0.011, is due to a slight mismatch in the two amplifiers of the mass spectrometer and does not affect the validity of the isotope effects calculated from these ratios.

At first glance, the variation in the isotope ratio for the low conversion samples appears to be much larger than that for the complete-conversion samples. However, further inspection reveals that this variation is actually caused by a large variation in the percent of reaction. For example, at 4.8° isotope effects were measured at percent of reaction varying from 0.5 to 41. The fact that the calculated isotope effect is not discernibly affected by this variation indicates a remarkable freedom from contamination.

The isotope effects given in the last column of the table are calculated by use of Eq. (4) from the isotope ratios after correction for the presence of $^{17}O$. The standard deviations of the calculated isotope effects range from a low of 0.0001 to a high of 0.0005. The last figure is probably the more realistic of the two in most investigations. Precision of this order is more than adequate for most purposes. The mechanism of decarboxylation of arginine involves formation of a Schiff base between arginine and enzyme-bound pyridoxal phosphate, followed by decarboxylation.[35] Either Schiff base formation or decarboxylation might be the slower step. We have argued previously[9] that intrinsic isotope effects in enzyme-catalyzed decarboxylation reactions near room temperature are expected to be in the range $k^{12}/k^{13} = 1.03-1.07$. If that is true, then at 37° the partitioning factor for decarboxylation of arginine [Eq. (8)] is in the range 1.5–4.8. Thus, neither Schiff base formation nor decarboxylation is entirely rate-determining; instead, both contribute approximately equally to the observed rate.

The isotope effects reported in the table increase substantially as the temperature is decreased. Although theory predicts that isotope effects should increase with decreasing temperature (and this has been borne out in a number of experimental studies), the temperature variation observed here is considerably larger than the expected temperature dependence of the intrinsic isotope effect. This variation must reflect a variation in the partitioning factor with pH. As the temperature is lowered, decarboxylation becomes more and more rate limiting.

Limitations

Two kinds of considerations currently limit the application of heavy-atom isotope effects to problems in enzymology; the first are experimental limitations, and the second are theoretical limitations.

Most of the experimental limitations arise from the need for a volatile material for isotope-ratio analysis. The chemical conversion required in order to convert the isotopic site of interest into a material suitable for mass spectrometry may sometimes be prohibitively difficult. The

[35] E. A. Boeker and E. E. Snell, in "The Enzymes" (P. Boyer, ed.) 3rd ed., Vol. 6, p. 217. Academic Press, New York, 1972.

double-label method described here can alleviate this problem to a considerable degree, but then the analytical problem is replaced by a synthetic one: synthesis of the substrate with the requisite isotope labels may be prohibitively difficult.

Theoretical limitations arise from the need to estimate the magnitude of the intrinsic isotope effect in the reaction being studied. Isotope effects in organic reactions can often provide a useful guide for predicting the magnitude of the intrinsic effect. Certain such predictions (for example, those for carbon isotope effects in decarboxylation and those for the ether oxygen in ester hydrolysis) are clear and relatively unambiguous. In other cases the intrinsic isotope effect may be a complex function of the detailed structure of the transition state, and useful predictions may be impossible. A noteworthy example of this phenomenon is nitrogen isotope effects on amide hydrolysis. Because of the large nitrogen isotope effect on the protonation of amines,[36] the nitrogen isotope effect on amide hydrolysis will be a sensitive indicator of not only the extent of carbon–nitrogen bond breaking but the extent of nitrogen–hydrogen bond making as well. These two factors will work in opposite directions, and it is hard to predict the magnitude of the isotope effect to be expected in such a process.

Some of the theoretical limitations may in the future be overcome by new experimental data. For example, carbonyl carbon isotope effects on enzyme-catalyzed ester and amide hydrolysis (which are now measurable by the double-label method) should provide interesting information about mechanisms of action of proteolytic enzymes. At present there are few data for analogous organic reactions that might be used to predict the intrinsic isotope effects, but the accumulation of an appropriately chosen set of isotope effects in model reactions could change this situation.

Finally we should note that it is possible, at least in theory, to calculate the magnitudes of isotope effects from first principles.[2-4] Although this approach has been widely used in interpreting isotope effects on organic reactions, it has yet to find wide use in enzymic studies. The key difficulty is in assuming a proper structure for the transition state. The assumptions made can have a large impact on the magnitude of the predicted isotope effect. Many organic reactions proceed in a single step, so the observed isotope effect equals the intrinsic isotope effect, and this effect is a direct reflection of the transition-state structure and indicates the adequacy of the assumed transition-state model. No such comparison is possible in most enzymic reactions because of problems caused by partitioning.

---

[36] The ratio of $K_a$ values for $^{14}NH_4^+$ and $^{15}NH_4^+$ is 1.039 at 25° (M. H. O'Leary and A. P. Young, unpublished observation).

Conclusion

The potential for use of heavy-atom isotope effects in studies of enzyme reaction mechanisms is increasing as a result of progress in a variety of experimental studies. New studies of isotope effects in organic reactions provide new models for predicting and understanding isotope effects in enzymic reactions. New studies of isotope effects on enzymic reactions provide an increasingly sound framework on which to base new studies. New methods provide access to information not previously available.

# [5] Measurement of Isotope Effects by the Equilibrium Perturbation Technique

*By* W. WALLACE CLELAND

The classical methods for measuring isotope effects involve direct comparisons of the rates with deuterated and nondeuterated substrates, giving both $V/K$ and $V$ isotope effects, or isotope discrimination experiments for tritium or heavier atom isotope effects, which yield only $V/K$ isotope effects. In 1975, however, Schimerlik, Rife, and Cleland[1] discovered a new and very sensitive method for measuring isotope effects by equilibrium perturbation. The method requires that the reaction be reversible enough that an equilibrium can be established, that there be a color change—or some other monitorable change such as circular dichroism (CD), optical rotation, fluorescence, electron spin resonance (EPR)—during the reaction, and that the labeled compound have a high degree of isotope substitution, but be capable of measuring isotope effects of 1.005 when the colored molecule is involved in the perturbation, and about 1.002 when it is not. It is a cheap and rapid method that has proved to be extremely useful for study of enzymic reactions.

Theory

The method was discovered during attempts to measure the equilibrium isotope effect on the malic enzyme reaction. When enzyme was added to a mixture of TPN, malate-2-D, $CO_2$, pyruvate, and TPNH calculated to be at equilibrium, TPNH disappeared at first and then gradually reappeared until equilibrium was reached (see Fig. 1). This perturbation from equilibrium is caused by more rapid reaction of TPNH, $CO_2$, and

---

[1] M. I. Schimerlik, J. E. Rife, and W. W. Cleland, *Biochemistry* **14,** 5347 (1975).

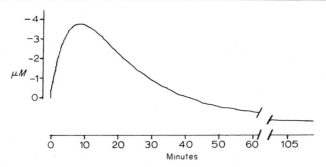

Fig. 1. Equilibrium perturbation with malic enzyme added to mixture of 419 $\mu M$ malate-2-D, 79 $\mu M$ TPNH, 20 mM KHCO$_3$ (3.8 mM CO$_2$), 20 mM MgSO$_4$, 102 $\mu M$ TPN, 3.83 mM pyruvate, pH 7.1. $\alpha = 1.45$; $\alpha\beta = 1.23$, $\beta = 0.85$. The perturbation is normal (TPNH disappears at first).

pyruvate to form malate and TPN than of TPN and malate-2-D to form TPND, pyruvate, and CO$_2$.

$$\text{TPN} + \text{malate-2-D} \xrightarrow{k_1} \text{TPND} + \text{pyruvate} + \text{CO}_2 \qquad (1)$$

$$\text{TPN} + \text{malate} \xleftarrow{k_2} \text{CO}_2 + \text{pyruvate} + \text{TPNH} \qquad (2)$$

That is, $k_2 > k_1$ because of the isotope effect. As isotopic equilibrium is reached, the system comes back to chemical equilibrium as well. The recorder trace is really the difference spectrum between two approximately first-order approaches to equilibrium represented by Eqs. (1) and (2).

*The Equations*

The equations that apply to this method have been published,[1,2] but we shall summarize them here for a typical enzyme with two substrates and two products [mechanism (3)]. A will be the molecule originally containing the deuterium or other heavy isotope (assuming 100% labeling for the moment), and P will contain hydrogen or the other lighter isotope. A will also be the colored molecule in this case, although it does not matter which of the reactants is the colored one. B and Q do not contain the labeled atoms, but participate in the reaction and must be considered. We shall consider an ordered reaction in which A adds before B, and P is re-

---

[2] W. W. Cleland, in "Isotope Effects on Enzyme-Catalyzed Reactions" (W. W. Cleland, M. H. O'Leary, and D. B. Northrop, eds.), p. 153. Univ. Park Press, Baltimore, Maryland, 1977.

leased before Q (we shall consider other possibilities below).

$$E \underset{k_2}{\overset{k_1 A}{\rightleftarrows}} EA \underset{k_4}{\overset{k_3 B}{\rightleftarrows}} EAB \underset{k_6}{\overset{k_5}{\rightleftarrows}} EAB^* \underset{k_8}{\overset{k_7}{\rightleftarrows}} EPQ^* \underset{k_{10}}{\overset{k_9}{\rightleftarrows}} EPQ \underset{k_{12}P}{\overset{k_{11}}{\rightleftarrows}} EQ \underset{k_{14}Q}{\overset{k_{13}}{\rightleftarrows}} E \quad (3)$$

In mechanism (3) the only isotope-dependent steps are those involving $k_7$ and $k_8$ (the chemical change); all others are assumed to be isotope independent. Since we have lumped all non-isotope-dependent steps before and after the catalytic step in $k_5$, $k_6$, $k_9$, and $k_{10}$, this is the most complicated mechanism we normally must look at. An exception occurs when there are two successive chemical steps, such as dehydrogenation and decarboxylation, and we are measuring deuterium and $^{13}C$ isotope effects separately for each step. In this case mechanism (3) must be expanded to include the additional chemical step. For such a case, see Schimerlik et al.[3]

When the net rate constant method of Cleland[4] is used to derive the rate equations for conversion of labeled A to P, and of unlabeled P back to A, we get

$$\frac{-dA_D}{dt} = \left(\frac{k_1 k_3 k_5 k_{7D}(E)}{k_2 k_4 k_6} A_D B - \frac{k_{8D} k_{10} k_{12}(EQ)}{k_9 k_{11}} P_D\right) \Big/ $$
$$\left\{1 + \frac{k_{7D}}{k_6}\left[1 + \frac{k_5}{k_4}\left(1 + \frac{k_3 B}{k_2}\right)\right] + \frac{k_{8D}}{k_9}\left(1 + \frac{k_{10}}{k_{11}}\right)\right\} \quad (4)$$

$$\frac{dA_H}{dt} = \left(\frac{k_{8H} k_{10} k_{12}(EQ)}{k_9 k_{11}} P_H - \frac{k_1 k_3 k_5 k_{7H}(E)}{k_2 k_4 k_6} A_H B\right) \Big/ $$
$$\left\{1 + \frac{k_{7H}}{k_6}\left[1 + \frac{k_5}{k_4}\left(1 + \frac{k_3 B}{k_2}\right)\right] + \frac{k_{8H}}{k_9}\left(1 + \frac{k_{10}}{k_{11}}\right)\right\} \quad (5)$$

The subscripts H and D refer to concentrations of deuterium- or hydrogen-substituted A and P, and rate constants for the corresponding molecules. We will assume first that at equilibrium the levels of B and Q are sufficiently higher than the lowest of A and P concentrations so that B and Q will not change appreciably during the perturbation (we will discuss below the case where this is not true). Then

$$EQ = E(k_{14}Q/k_{13}) \quad (6)$$

Next we must consider the effect of deuterium substitution on equilibrium constants. For dehydrogenases where hydrogen is transferred from a secondary alcohol to DPN to give DPNH, for example, $K_{eq\,H}/K_{eq\,D} = 1.18$, and deuterium becomes enriched in the alcohol. The

---

[3] M. I. Schimerlik, C. E. Grimshaw, and W. W. Cleland, *Biochemistry* **16**, 571 (1977).
[4] W. W. Cleland, *Biochemistry* **14**, 3220 (1975).

table includes a number of equilibrium isotope effects of interest to the enzymologist; some are experimental and others are calculated, but these values can be used with considerable confidence. For careful work, however, one should measure the equilibrium effect for the enzyme under study, and methods for doing this are discussed below. Note that the equilibrium isotope effect for a reaction involving water as one of the perturbants will be different in $H_2O$ and $D_2O$ (footnote $a$ in the table).

In mechanism (3) we can write

$$\frac{K_{eq\,H}}{K_{eq\,D}} = \frac{k_{7H}k_{8D}}{k_{8H}k_{7D}} = \frac{1}{\beta} \quad (7)$$

where $1/\beta$ is the parameter used for the equilibrium isotope effect by Schimerlik, Rife, and Cleland.[1] If the numerator and denominator of Eq. (4) are multiplied by $k_{7H}/k_{7D}$ and Eq. (6) is used to replace EQ, we get

$$\frac{-dA_D}{dt} = \left(\frac{k_1k_3k_5k_{7H}A_DB}{k_2k_4k_6} - \frac{k_{8H}k_{10}k_{12}k_{14}P_DQ}{\beta k_9k_{11}k_{13}}\right)(E)\Big/\left(\frac{k_{7H}}{k_{7D}} + c_f + \frac{c_r}{\beta}\right) \quad (8)$$

where

$$c_f = \frac{k_{7H}}{k_6}\left[1 + \frac{k_5}{k_4}\left(1 + \frac{k_3B}{k_2}\right)\right] \quad (9)$$

$$c_r = \frac{k_{8H}}{k_9}\left(1 + \frac{k_{10}}{k_{11}}\right) \quad (10)$$

Equation (5) can also now be written:

$$\frac{dA_H}{dt} = \left(\frac{k_{8H}k_{10}k_{12}k_{14}P_HQ}{k_9k_{11}k_{13}} - \frac{k_1k_3k_5k_{7H}A_HB}{k_2k_4k_6}\right)(E)\Big/(1 + c_f + c_r) \quad (11)$$

If we now define the observed isotope effect in the forward direction as

$$\alpha = (k_{7H}/k_{7D} + c_f + c_r/\beta)/\Delta \quad (12)$$

where

$$\Delta = 1 + c_f + c_r \quad (13)$$

it is clear that Eq. (11) has $\Delta$ as the denominator, and Eq. (8) has $\alpha\Delta$ as the denominator.

The isotope effect in the reverse direction is $\alpha\beta$, and by multiplying Eq. (12) by $\beta$, we see that it is

$$\alpha\beta = \frac{k_{8H}/k_{8D} + \beta c_f + c_r}{\Delta} \quad (14)$$

It should be noted at this point that Eqs. (12) and (14) are very similar to those for $V/K$ isotope effects, and in many cases $\alpha$ and $\alpha\beta$ will be identical

EQUILIBRIUM ISOTOPE EFFECTS IN AQUEOUS SOLUTION

I. Deuterium[a]

| Structure | Examples[b] | $K_{eq\,H}/K_{eq\,D}$ | Source |
|---|---|---|---|
| —C≡C[H] | $C_2H_2$ | 0.64 | c |
| >C=CH[H] | Phosphoenol pyruvate (3) | 0.78 | d, n |
| [H]COO⁻ | Formate | 0.80 | e |
| —CO—CH$_2$[H] | Pyruvate | 0.84 | f |
| C—C[H]=C | DPN (4), Fumarate | 0.87 | g, h |
| C—CH$_2$[H] | Lactate | 0.88 | i |
| —C[H]=O | Aldehydes | 0.83 | j |
| —CO—CH[H]—C | Ketoglutarate (3) | 0.93 | k |
| C—CH[H]—C | Malate (3), DPNH (4) | 0.98 | l |
| —CO—CH[H]OH | Unhydrated dihydroxyacetone-P | 0.99 | m |
| C—CH[H]OH | Ethanol (1) | 1.04 | m |
| —CO—C[H]OH—C | | 1.10 | m |
| C—C[H]NH$_3^+$—C | Amino acids (2) | 1.13 | m |
| C—C[H]OH—C | Isopropanol (2), cyclohexanol (1), malate (2), isocitrate (2), lactate (2) | 1.16 | g |
| C—C[H]O(P)—C | 2-Phosphoglycerate (2) | 1.19 | n |
| C—C[H]—(OR)$_2$ | Aldehyde hydrates (1) | 1.14 | j |
|  | sugar hemiacetals (1) | 1.24 | g |
| C—S[H] | | 0.43 | o |
| O[H]⁻ | | 0.48 | p |
| H$_2$O[H]⁺ | | 0.69 | o |
| C—NH[H] | | 0.92 | o |
| C—NH$_2$[H]⁺ | | 0.97 | o |
| —COO[H] | | 1.00 | o |
| C—O[H] | Alcohols | 1.00 | o |
| \C/O[H] / \ OR | Aldehyde hydrates, hemicetals, hemiketals | 1.23 | q |

II. Carbon-13[r]

| Compound | $K_{eq\,C-12}/K_{eq\,C-13}$ | Source |
|---|---|---|
| [C]O$_3$H⁻ | 1.009 | s |
| [C]O$_3^{2-}$ | 1.008 | s |
| —[C]OO⁻ | 0.997 | t |
| —[C]OOH | 0.996 | t |
| CH$_3$—[C]H$_2$—CH$_3$ | 0.966 | |
| [C]H$_3$OH | 0.963 | u |
| CH$_3$—CH$_2$—[C]H$_3$ | 0.960 | |

## Equilibrium Isotope Effects (*Continued*)

### II. Carbon-13[r]

| Compound | $K_{eq\ C-12}/K_{eq\ C-13}$ | Source |
|---|---|---|
| $CH_3—[C]H_3$ | 0.956 | |
| $CH_2=[C]H_2$ | 0.954 | |
| $[C]H_4$ | 0.944 | |
| $HC\equiv[C]H$ | 0.939 | |

### III. Nitrogen-15[v]

| Compound | $K_{eq\ N-14}/K_{eq\ N-15}$ | Source |
|---|---|---|
| $[N]H_4^+$ | 1.039 | w |
| $C—CH[N]H_3^+—C$ | 1.058 | x |

### IV. Oxygen-18[y]

| Compound | $K_{eq\ O-16}/K_{eq\ O-18}$ | Source |
|---|---|---|
| $C[O]O$ | 1.041 | z |
| $C—CH[O]H—C$ | 1.033 | aa |
| $C—C=[O]—C$ | 1.031 | bb |
| $C—CH_2[O]H$ | 1.025 | aa |
| $CH_3[O]H$ | 1.017 | aa |
| $H_3[O]^+$ | 1.023 | cc |
| $[O]H^-$ | 0.960 | dd |

[a] The values shown are $K_{eq\ H}/K_{eq\ D}$ values per deuterium for transfer of deuterium from the molecule shown to water. Tritium effects (on which a number of the values are based) are the 1.442 power of deuterium ones. In $D_2O$ where the transfer of deuterium is from the molecule shown to HDO to give $D_2O$, the values in the table are multiplied by 0.945, and in mixed $H_2O$, $D_2O$ solutions, linear interpolation between the $H_2O$ and $D_2O$ values should be used. This factor is K/4, where K is the equilibrium constant for the reaction: $D_2O + H_2O \rightleftharpoons 2HDO$ [L. Friedman and V. J. Shiner, Jr. *J. Chem. Phys.* **44**, 4639 (1966); W. A. Van Hook, *J. Chem. Soc. Chem. Comm.* 479 (1972)]. This correction is important for fractionation factors determined by comparison of $K_{eq}$ values in $D_2O$ and $H_2O$, such as the fumarase and enolase equilibria. For transfer between two molecules listed, divide the value for the donor by the value shown for the acceptor. For example, for transfer of deuterium from a secondary alcohol to DPNH, divide 1.16 by 0.98 to get 1.18 (this is the experimental value found for oxidation of five secondary alcohols by DPN). The table also predicts secondary equilibrium effects. Thus 0.87 for DPN divided by 0.98 for DPNH gives 0.89, the experimental value for the secondary equilibrium isotope effect in conversion of DPN-4-D to DPNH.

Deuterium isotope effects are sensitive mainly to the atoms bonded to the carbon or other atom holding the hydrogen of interest, and replacement of a hydrogen by carbon, nitrogen, or oxygen causes increases in the equilibrium isotope effect by factors of 1.10, 1.15, and 1.18. Thus the value of 1.18 for transfer from a secondary alcohol to DPNH occurs because DPNH has two carbons and a hydrogen on C-4, whereas the alcohols have two carbons and an oxygen in addition to the hydrogen of interest. An exception to this rule is carbonyl compounds, such as acetone or cyclohexanone, where the $CH_2$ group or $CH_3$ group next to the carbonyl carbon shows values that are 5% less per hydrogen than those in the corresponding alcohols or addition compounds, including hydrates (J. M. Jones and M. L. Bender, *J. Am. Chem. Soc.* **82**, 6322 (1960); M. A. Winnick, V. Stoute, and P. Fitzgerald, *J. Am. Chem. Soc.* **96**, 1977 (1974); P. Geneste, G. Lamaty, and J. P. Roque, *Tetrahedron* **27**, 5539 (1971); J. L. Hogg, Ph.D. dissertation, Univ. of Kansas, 1974; unpublished experiments by Dr. P. F. Cook and J. Blanchard in this lab). These secondary inductive effects are not shown by carboxyl groups or by C–C double bonds, but do apply to α-keto acids. Thus the table includes separate entries for $CH_2$ or $CH_3$ groups next to a carbonyl group.

[b] Labeled in the position shown in parentheses.

[c] From the measured $K_{eq}$ of 0.473 in the gas phase for the reaction $C_2H_2 + D_2O \rightleftharpoons C_2D_2 + H_2O$ (J. W. Pyper and F. A. Long, *J. Chem. Phys.* **41**, 1890 (1964)), corrected to liquid water by using the vapor pressure ratio of $H_2O/D_2O$ of 1.152 (W. M. Jones, *J. Chem. Phys.* **48**, 197 (1968)).

[d] I. A. Rose, in "Isotope Effects on Enzyme-Catalyzed Reactions" (W. W. Cleland, M. H. O'Leary, and D. B. Northrop, eds.), p. 225. Univ. Park Press, Baltimore, Maryland, 1977.

[e] W. E. Buddenbaum and V. J. Shiner, Jr., in "Isotope Effects on Enzyme Catalyzed Reactions," loc. cit., p. 11.

[f] H. P. Meloche, C. T. Monti, and W. W. Cleland, *Biochim. Biophys. Acta* **480**, 517 (1977).

[g] From unpublished measurements of equilibria involving DPNH by P. F. Cook in this lab.

*Continued*

Footnotes (*Continued*)

[h] An identical value was seen for trimethoxybenzene in water by A. J. Kresge and Y. Chiang, *J. Chem. Phys.* **49,** 1439 (1968).

The equilibrium isotope effect measured by Dr. P. F. Cook in this lab for conversion of dideuteromalate to dideuterofumarate is 1.45. Since the value for dideuterosuccinate should be 1.45/1.18 = 1.23, the value for fumarate should be that for C-3 of malate divided by 1.11, or 0.88. The value for C—2 of propene calculated by Buddenbaum and Shiner[e] is lower than that for C—2 of propane by a factor of 1.12, which would suggest the value should be 0.98/1.12 = 0.87, in agreement with the value for aromatic CH groups.

[i] Based on the pyruvate value, and the secondary equilibrium isotope effect on the reduction of pyruvate to lactate by DPNH (P. F. Cook and J. Blanchard, unpublished experiments in this lab).

[j] For any given aldehyde, the observed value will be 0.83 times the fraction of free aldehyde, plus 1.14 times the fraction of hydrate. These numbers are calculated from the equilibrium effect on acetaldehyde hydration [$K_D/K_H$ = 1.37; C. A. Lewis, Jr. and R. Wolfenden, *Biochemistry* **16,** 4886 (1977)] and experimental values for dehydrogenase equilibria involving acetaldehyde (P. F. Cook, in this lab) or benzaldehyde and p-methoxybenzaldehyde (J. Klinman, personal communication).

[k] From comparison of glutamate-2,3-$d_3$ with glutamate-2-d in the glutamate dehydrogenase reaction, and malate-3,3-$d_2$ or 3-R-deuteromalate with oxalacetate in the malate dehydrogenase reaction (P. F. Cook and J. Blanchard, unpublished experiments in this lab). Oxalacetate is a mixture of 18% enol, 8% hydrate and 74% keto form, and if 0.87 is used for the enol, 0.98 for the hydrate, and 0.93 for the keto form, the equilibrium mixture should have 0.92, in reasonable agreement with a value of 0.90 reported by Rose.[d]

[l] $K_{eq}$ for the fumarase reaction is 7% less in $D_2O$ than in $H_2O$ (J. F. Thomson, *Arch. Biochem. Biophys.* **90,** 1 (1960)); division of 0.93 by 0.945 gives 0.98 for the value in water.

[m] Calculated from the experimental values for malate and pyruvate by application of the rule that replacement of H with C, N, or O causes increases of 1.10, 1.15, and 1.18 and by the assumption that a $CH_2$ or $CH_3$ group next to a carbonyl group has a value lower by 1.05, and confirmed by the equilibrium isotope effect of 1.06 for oxidation of ethanol by DPN (P. F. Cook, unpublished experiments in this lab). The experimental value for dihydroxyacetone-P determined with triose phosphate isomerase, 1.02 (S. J. Fletcher, J. M. Herlihy, W. J. Albery and J. R. Knowles, *Biochemistry* **15,** 5612 (1976)), reflects the partial hydration of the molecule. The value for a similar ketone will be the fraction of free ketone times 0.99, plus the fraction of hydrate times 1.04.

[n] From the enolase equilibrium; unpublished experiments by V. Anderson in this lab.

[o] R. L. Schowen, *Prog. Phys. Org. Chem.* **9,** 275 (1972). These values are approximate.

[p] K. Henzinger and R. E. Weston, Jr., *J. Phys. Chem.* **68,** 2179 (1964). This apparent value applies to hydrated $OH^-$ in water, which is probably a trihydrate in which the actual fractionation factor for the OH proton is 1.2–1.5, and that for each of the 3 protons hydrogen bonded to $OH^-$ is 0.65–0.70 [V. Gold and S. Grist, *J. Chem. Soc. Perkin II,* 89 (1972)].

[q] J. F. Mata-Segreda, S. Wint, and R. L. Schowen, *J. Am. Chem. Soc.* **96,** 5608 (1974).

[r] Values shown are for transfer of $^{13}C$ from the compound shown to $CO_2$ in aqueous solution at 25°. For $^{14}C$, the values of (isotope effect − 1) are double those shown. Except where shown, the values are from S. R. Hartshorn and V. J. Shiner Jr., *J. Am. Chem. Soc.* **94,** 9002 (1972), corrected for the transfer from $CO_2$(gas) to $CO_2$(aq) [1.001; W. G. Mook, J. C. Bommerson, and W. H. Staverman, *Earth Plan. Sci. Lett.* **22,** 169 (1974)]. It appears that replacing H with C or O on the isotopic carbon increases the value by 1.011 or 1.019, whereas replacing H with C on a carbon bonded to the isotopic carbon increases the value by 1.004. Application of these rules to the compounds in the table should give good estimates for other desired compounds.

[s] The bicarbonate value is experimental (Mook *et al.*[r]). The carbonate value is from Hartshorn and Shiner[r] corrected for transfer of $CO_2$ from gas to aqueous solution.

[t] M. H. O'Leary and C. J. Yapp, *Biochem. Biophys. Res. Commun.* **80,** 155, (1978). Determined from the isocitrate dehydrogenase equilibrium.

[u] The value is actually for $CH_3F$; fluorine behaves like oxygen in these calculations.

[v] The values shown are for transfer of $^{15}N$ from the compound listed to $NH_3$. Since the rules in footnote *r* appear to apply to both $^{13}C$ and $^{18}O$ effects [with (isotope effect − 1) being multiplied by 2 in the latter case], they will probably apply for $^{15}N$ as well.

[w] M. H. O'Leary, unpublished experimental value.

[x] Estimated by use of the rules in footnote *r*; on this basis, methylamine would be 1.050 and primary amines 1.054.

[y] Values are for transfer of $^{18}O$ from the compound shown to water. The rules in footnote *r* appear to apply here, except with (isotope effect − 1) doubled (1.022 for replacement of H with C on the oxygen; 1.008 for replacement of H with C on a carbon attached to the oxygen).

[z] J. R. O'Neil, L. H. Adami and S. Epstein, *J. Res., U.S. Geol. Survey* **3,** 623 (1975).

[aa] The values for $^{18}O$ transfer from methanol to water, and from dimethyl ether to methanol in the gas phase, have been calculated from force fields and geometries to be 1.0216 and 1.022. Thus replacement of H with $CH_3$ gives the same value for $^{18}O$ as it does for $^{14}C$ in the calculations of Hartshorn and Shiner.[r] Replacement of H with C on the carbon attached to the isotopic atom increases the value 1.008 for $^{14}C$ (or 1.004 for $^{13}C$), and application of this rule to $^{18}O$ gives 1.0296 for ethanol and 1.0376 for isopropanol in the gas phase. The value for transfer of $^{18}O$ from $H_2O$(gas) to water of 0.991 presumably reflects largely the effects of hydrogen bonding, and if we assume this to be for an alcohol half what it is for water in aqueous solution, the values for all alcohols are multiplied by 0.9955. The calculated value for isopropanol matches perfectly the experimental one of 1.033 measured for L-malate by John Blanchard from the fumarase equilibrium, and gives considerable support to these calculations and the rules on which they are based.

with $V/K$ isotope effects. In most cases, however, the factors $c_f$ and /or $c_r$ in Eqs. (9) and (10) will include more terms for the equilibrium perturbation isotope effect than for the $V/K$ one. In the present example, for instance, $c_r$ for the $V/K$ isotope effect is the same as that given by Eq. (10), but $c_f$ lacks the last term in Eq. (9). We will have more to say about this later when we discuss the interpretation of results.

We now make the following definitions:

$$K = \frac{P_{H\,eq}}{A_{H\,eq}} = \frac{k_1 k_3 k_5 k_{7H} k_9 k_{11} k_{13} B}{k_2 k_4 k_6 k_{8H} k_{10} k_{12} k_{14} Q} \tag{15}$$

$$k = \frac{k_1 k_3 k_5 k_{7H} B(E)(1 + K)}{k_2 k_4 k_6 \, \Delta \, K} \tag{16}$$

$$P_D = A_0 - A_D \tag{17}$$

$$P_H = P_0 - A_H \tag{18}$$

where $A_0$ and $P_0$ are the initial concentrations of A and P. Substituting these equations into Eqs. (8) and (11) yields

$$\frac{dA_D}{dt} = \frac{Kk}{(1 + K)\alpha} \left( \frac{A_0}{K\beta} - \frac{A_D(1 + K\beta)}{K\beta} \right) \tag{19}$$

$$\frac{dA_H}{dt} = \frac{Kk}{(1 + K)} \left( \frac{P_0}{K} - \frac{A_H(1 + K)}{K} \right) \tag{20}$$

These equations integrate to

$$A_D = \frac{A_0(1 + K\beta \exp - [k(1 + K\beta)t/\alpha\beta(1 + K)])}{1 + K\beta} \tag{21}$$

$$A_H = \frac{P_0(1 - e^{-kt})}{1 + K} \tag{22}$$

Before adding these equations together to give the total concentration of A as a function of time, we will define an apparent isotope effect that determines the size and direction of the perturbation:

$$\text{app } \alpha = \frac{\alpha(\beta + K\beta)}{(1 + K\beta)} = \frac{\alpha\beta(1 + K)}{(1 + K\beta)} \tag{23}$$

---

[bb] Equilibrium isotope effect for transfer of $^{18}O$ from the carbonyl of ketoglutarate to water in the glutamate dehydrogenase reaction (J. E. Rife, Ph.D. Dissertation, Univ. Wisconsin, 1978).
[cc] E. R. Thornton, *J. Am. Chem. Soc.* **84**, 2474 (1962).
[dd] M. Green and H. Taube, *J. Phys. Chem.* **67**, 1565 (1963).

If $t$ is allowed to become infinite in Eqs. (21) and (22), we can calculate the level of A at final chemical and isotopic equilibrium, and since the perturbation returns to the starting point, this equals $A_0$.

$$A_\infty = A_{D\infty} + A_{H\infty} = \frac{A_0}{(1 + K\beta)} + \frac{P_0}{(1 + K)} = A_0 \qquad (24)$$

Thus

$$\frac{P_0}{(1 + K)} = \frac{A_0 K\beta}{(1 + K\beta)} \qquad (25)$$

and if this is substituted into Eq. (22) and Eq. (23) is substituted into Eq. (21), we can add Eqs. (21) and (22) to give

$$A = A_0 + \frac{(A_0 K\beta)}{(1 + K\beta)} (e^{-kt/\text{app}\alpha} - e^{-kt}) \qquad (26)$$

This equation predicts that if app $\alpha > 1$, the perturbation is a normal one (that is, A increases at first), whereas if app $\alpha < 1$, the perturbation is inverse (A decreases at first; the perturbation is away from, rather than toward, the heavier isotope). With primary deuterium isotope effects, both $\alpha$ and $\alpha\beta$ are usually above 1; thus app $\alpha$ is also, and the perturbations are normal ones. For most secondary isotope effects, and for primary ones where the equilibrium isotope effect dominates the kinetic one, however, $\alpha$ may be greater than 1 and $\alpha\beta$ less than 1, or vice versa. In these cases, Eq. (23) predicts that the value of app $\alpha$ (and thus the direction of the perturbation) will depend on $K$. When $K$ is very large, app $\alpha = \alpha$, whereas when $K$ is very small, app $\alpha = \alpha\beta$. Thus, whenever the perturbation is between an element of water (55 $M$ oxygen or 110 $M$ protons) and one of the reactants, one sees only the isotope effect on the reaction toward the water. With fumarase, for example, this means that where the perturbation involves $D_2O$ or $H_2[^{18}O]$ and malate, the primary deuterium or $^{18}O$ effects seen are those for the reaction of malate to give water and fumarate. The isotope effects in the reverse direction must be calculated from $\alpha\beta$.

When $K$ is neither very large nor very small, it is possible that app $\alpha$ will be 1, and no perturbation will be observed. This occurs when:

$$K = \left(\frac{1}{\beta}\right)\frac{(1 - \alpha\beta)}{(\alpha - 1)} \qquad (27)$$

For example, if $\alpha = 1.125$ and $\beta = 0.8$, so that $\alpha\beta = 0.9$, the perturbation will be normal if $K$ is greater than 1 and inverse if $K$ is less than 1, but there will be no perturbation when $K = 1$. In such a situation, one can cause a perturbation to be seen by changing the level of one of the other reactants [B or Q in mechanism (3)] so that $K$ is altered, and in favorable cases one can make the perturbation be either inverse or normal.

Since $K$ is not determined directly during an experiment (it is the ratio of P and A concentrations for the unlabeled reactants at equilibrium), we solve Eq. (25) for $K$ in terms of the experimental values $A_0$ and $P_0$.

$$K = \left\{\left(\frac{P_0}{A_0} - 1\right) + \left[\left(\frac{P_0}{A_0} - 1\right)^2 + \frac{4P_0}{\beta A_0}\right]^{1/2}\right\}\Big/2 \quad (28)$$

and rewrite Eq. (26) as

$$A = A_0 + A_0'(e^{-kt/\text{app}\alpha} - e^{-kt}) \quad (29)$$

where

$$A_0' = 2\Big/\left\{\frac{1}{A_0} + \frac{1}{P_0} + \left[\left(\frac{1}{A_0} + \frac{1}{P_0}\right)^2 + 4\left(\frac{1}{\beta} - 1\right)\left(\frac{1}{A_0}\right)\left(\frac{1}{P_0}\right)\right]^{1/2}\right\} \quad (30)$$

While this equation looks complicated, if $\beta = 1$ it reduces to the reciprocal of sums of reciprocals of $A_0$ and $P_0$, and thus $A_0'$ is approximately equal to the smaller of $A_0$ or $P_0$.

If Eq. (29) is differentiated with respect to time, the maximum point of the perturbation occurs at:

$$t_{\max} = \left(\frac{1}{k}\right)\frac{(\text{app } \alpha) \ln \alpha}{(\text{app } \alpha - 1)} \quad (31)$$

which when app $\alpha$ is close to 1 is approximated by

$$t_{\max} = \left(\frac{1}{k}\right)\frac{(\text{app } \alpha + 1)}{2} \quad (32)$$

which makes $t_{\max}$ only slightly different from $1/k$. Since $1/k$ is given by the point where a reaction mixture near equilibrium containing unlabeled reactants has gone 63% of the way to equilibrium, it is not hard to determine the amount of enzyme needed to put the maximum point of the perturbation at a convenient time (2–5 min).

Substituting Eq. (31) into Eq. (29) gives

$$\frac{A_{\max} - A_0}{A_0'} = (\text{app } \alpha)^{-[1/(\text{app } \alpha - 1)]} - (\text{app } \alpha)^{-[\text{app } \alpha/(\text{app } \alpha - 1)]} \quad (33)$$

which is the equation used to determine the value of app $\alpha$ from the size of the perturbation. A table of values of $(A_{\max} - A_0)/A_0'$ as a function of app $\alpha$ has been published,[2] but it is simpler to use the short computer program given in this chapter, which solves the equation directly by successive approximation.

When the perturbation does not return to exactly the same point where it started, Eq. (33) still applies as long as the starting and final values of A are averaged and used as $A_0$, and the two values differ by less than 20%.

The above equations assume 100% labeling with the heavy isotope. Since perturbation size is directly proportional to the percentage difference in labeling of the two reactants involved in the perturbation,[5] the value of $A_{max} - A_0$ in Eq. (33) must be corrected when this percentage is less than 100. Thus if one reactant contains 90% deuterium and the other one 98%, $A_{max} - A_0$ is divided by 0.08 before Eq. (33) is used to calculate app $\alpha$.

Once app $\alpha$ is available, Eq. (23) can be used to evaluate $\alpha$ by substituting the value of $K$ from Eq. (28). This gives:

$$\alpha = (\text{app } \alpha) z \qquad (34)$$

where

$$z = \left\{ \left(1 - \frac{A_0}{P_0}\right) + \left[\left(1 - \frac{A_0}{P_0}\right)^2 + \frac{4A_0}{\beta P_0}\right]^{1/2} \right\} \bigg/ 2 \qquad (35)$$

The value of $\alpha\beta$ is then obtained by multiplying $\alpha$ by $\beta$.

## B and Q Not Much Greater Than A or P

The equations presented assume that the concentrations of B and Q exceed $A_0'$ sufficiently so that they do not change during the perturbation. If this is not true, as is usually the case for at least one reactant (especially when the perturbation is between noncolored reactants), the concentrations of B and Q will change in a way that tends to reduce the size of the observed perturbation. We then have for a normal perturbation

$$B = B_0 + x = B_0(1 + x/B_0) \qquad (36)$$

$$Q = Q_0 - x = Q_0(1 - x/Q_0) \qquad (37)$$

---

[5] This is not strictly correct unless $\beta = 1$, as can be readily seen by considering a situation where A and P are both 50% labeled to start with. This mixture will not be at isotopic equilibrium if there is an equilibrium isotope effect, and thus a perturbation will be seen. The size of this effect, however, is small [$0.25(\beta - 1)$ times the size of that seen with full labeling for the 50% labeling case, where the effect is maximum], and can usually be ignored. No error is introduced if either A or P contains no deuterium or all deuterium (that is, if P is unlabeled and A is 60% labeled, the perturbation size is 60% of that expected for fully labeled A).

where $x$ is the size of the perturbation. If Eq. (15) is redefined to include $B_0$ and $Q_0$ in place of B and Q, so that the value of $K$ remains constant, $K$ in Eqs. (19) and (20) is multiplied by $(1 + x/B_0)(1 - x/Q_0)$. Since $x^2$ terms will be small, the numerator and denominator of this expression can be multiplied by $(1 - x/B_0)$ to give $1/[1 - x \Sigma(1/C_0)]$, where

$$\Sigma(1/C_0) = 1/B_0 + 1/Q_0 \tag{38}$$

If there are any other reactants (other than A and P) their reciprocal concentrations are also added into $\Sigma(1/C_0)$. Since

$$x = A_D + A_H - A_0 \tag{39}$$

Eqs. (19) and (20) now contain $A_H$ and $A_D$ in both equations so that they must be solved simultaneously, instead of being integrated separately and added together later. This has been done empirically with a computer program for solution of simultaneous differential equations, and it has been found that correct isotope effects will be calculated when $A_0'$ in Eqs. (29) and (33) is replaced with $A_0''$.

$$A_0'' = \frac{1}{(1/A_0') + (1/y)\Sigma(1/C_0)} \tag{40}$$

The value of $y$ depends on the ratio of $1/\Sigma(1/C_0)$ to $A_0'$. For ratios above 0.3, $y$ is 2.2, and it is 2.3 for ratios between 0.16 and 0.3, 2.4 for ratios between 0.08 and 0.16, and 2.5 for ratios below 0.08. The time needed to reach the maximum of the perturbation is also reduced in these cases by roughly the ratio of $A_0''/A_0'$, so care must be taken not to miss the perturbation by running it too rapidly.

*Data Analysis*

The equations presented above are complex enough that their use in practice is difficult. As a result, we include a short computer program, written in simple FORTRAN, which uses the perturbation data to calculate $\alpha$ and $\alpha\beta$. The data for a single perturbation are placed on two cards, and the program will process an unlimited number of sets of data. The program is stopped by a blank card after the last data cards.

*Card 1.* Columns 1–3. I3 format. Number of reactants other than those involved in the perturbation. Maximum number 4, minimum number 1. If water is the only other reactant (or there is no other reactant), put 1 here and 55 M on card 2. For mechanism (3), the number is 2 (B and Q).

Columns 4–20. F17.5 format. The fraction of isotopic labeling. For 60% $^{18}O$, put 0.60, for example. If one reactant is fully deuterated and the other

92% deuterated, put 0.08 (that is, the value to use is the difference in fractional labeling between the reactants involved in the perturbation; see the preceding discussion). If left blank, 1.00 is assumed.

Columns 21–67. Anything put here is printed out as a title.

*Card 2.* Columns 1–10. F10.5 format. The size of the perturbation in micromolar units. If the perturbation is inverse (the reaction moves away from, rather than toward the reactant containing the highest level of the heavier isotope), the value must be preceded by a minus sign.

Column 11–20. F10.5 format. The value of $\beta$ for the reaction in the direction from the reactant with the heavier isotope to the reactant with the lighter isotope. $\beta$ is the reciprocal of the equilibrium isotope effect (i.e., $\beta = K_{eq\,D}/K_{eq\,H}$). The value of $\beta$ *must* be known (determine it separately if it is not known or cannot be accurately calculated).

Columns 21–30. E10.5 format. Concentration of A in molar units. A contains the heavier isotope, or the highest fraction of labeling with it. When water is a reactant, use 110 $M$ when protons are involved in the perturbation, and 55 $M$ when the oxygen is involved.

Columns 31–40. E10.5 format. Concentration of P in molar units. P is the molecule with the lighter isotope, or lowest percentage labeling with the heavier isotope. See previous comment on water as a reactant.

Columns 41–50, 51–60, 61–70, 71–80. E10.5 formats. Concentrations of other reactants in molar units. Place as many values here as are called for in columns 1–3 of card 1 and leave the other fields blank (the order does not matter). When water is a reactant, use 55 $M$.

*Output.* The output will be as follows:

First line: Title preceded by a counting number and $\beta$.

Second line: $1/\Sigma(1/C_0)$, $A_0'$, $A_0''$, $y$, $z$ (see above for meaning of these parameters).

Third line: $\alpha$, $\alpha\beta$. The error in $\alpha$ and $\alpha\beta$ will be the same as that in the perturbation size (that is, if the error in perturbation size is estimated to be ±5%, the error in $\alpha$ is ±5%).

```
      DIMENSION A(4)
      PRINT 100
100   FORMAT(40H CALCULATION OF EQUILIBRIUM PERTURBATION //)
101   FORMAT(I3,F17.5,46H                                              )
1     FORMAT(2F10.5,6E10.5)
      JJ = 0
114   READ 101, NA, FI
      IF (NA) 99, 99, 112
112   READ 1, PERT, BETA, B, R, A
      IF (FI) 80, 80, 81
80    FI = 1.
81    SA = 0
      DO 2 I = 1, NA
2     SA = SA + 1./A(I)
```

```
      AP = 1./SA
      BI = 1./B
      RI = 1./R
      IF(BETA)20,20,21
   20 BETA = 1.
   21 IF (BETA - 1) 3, 4, 3
    4 RP = 1./(BI + RI)
      Z = 1.
      GO TO 5
    3 RP = 2./(RI + BI + SQRT((RI+BI)**2+4.*(1./BETA-1.)*RI*BI))
      BR = B/R
      Z = (1.-BR+SQRT((1.-BR)**2+4.*BR/BETA))/2.
    5 IF (AP/RP - .3) 6, 6, 8
    8 Y = 2.2
      GO TO 15
    6 IF (AP/RP - .16) 9, 9, 11
   11 Y = 2.3
      GO TO 15
    9 IF (AP/RP - .08) 12, 12, 14
   14 Y = 2.4
      GO TO 15
   12 Y = 2.5
   15 RPP = 1./(1./RP + 1./Y/AP)
      JJ = JJ + 1
      PRINT 101, JJ, BETA
      PRINT 102,AP,RP,RPP,Y,Z
  102 FORMAT(6H AP = E14.5,8H    RP = E14.5,9H    RPP = E14.5,
     1 7H    Y = F5.2,7H    Z = F10.5)
      PX = 1.E-06*PERT/RPP/FI
      AA = 1. + 2.72*PX
      DO 22 I = 1,8
      PC = AA**(-1./(AA-1.)) - AA**(-AA/(AA-1.))
   22 AA = AA + (PX-PC)*(AA-1.)**2/LOG(AA)/PC
      ALPHA = AA*Z
      AB = ALPHA*BETA
      PRINT 30, ALPHA, AB
   30 FORMAT(9H ALPHA = F10.5,18H    ALPHA*BETA = F10.5//)
      GO TO 114
   99 PRINT 50
   50 FORMAT(18H PROGRAM COMPLETED /)
      STOP
      END
```

## Experimental Procedures

*Measurement of Equilibrium Constants and $\beta$.* As noted above, it is essential to know the equilibrium isotope effect ($1/\beta$) with a high degree of precision. While it may be estimated from the value for similar reactions, or from the numbers in the table, it is best to determine it experimentally, and to do this requires very accurate determination of $K_{eq}$ values. This is done as follows:

The concentrations of the deuterated and nondeuterated molecules are accurately determined by enzymic end-point analysis (that is, one uses an enzymic assay where the final extent of reaction is limited by the compound being assayed for). The two stock solutions should be similar in concentration, and the assays should be carried out at the same time. Alternatively, the concentrations of the deuterated and nondeuterated molecules can be determined after equilibrium has been reached by assaying a small aliquot enzymically.

Reaction mixtures are made up with all components present except the deuterated or nondeuterated substrate and the product, which will be deuterated or nondeuterated. An aliquot of the deuterated or nondeuterated substrate is then added, and the reaction is allowed to proceed to equilibrium. The initial concentrations should be picked so that the extent of reaction is readily measured, and the changes in the concentrations of reactants present initially are as small as possible (in no case over 50%). Final concentrations are then determined from the initial concentrations and the extent of reaction, and $K_{eq}$ is calculated for the deuterated and nondeuterated samples. If care is taken and replicates run, one should be able to determine $\beta$ to within 1% (that is, ±0.01).

Where water is a reactant, the primary $^{18}O$ equilibrium effect is easily measured by making up reaction mixtures at equilibrium and adding an equal volume of $H_2O$ or $H_2[^{18}O]$. The difference in final equilibrium positions gives the isotope effect (which must be corrected to 100% $^{18}O$). A similar approach can be used with $D_2O$, but since extinction coefficients in $D_2O$ differ from those in $H_2O$, correction for this effect must also be made. Since the equilibrium effect may not be completely linear with $D_2O$ concentration, it also pays to test percentages of $D_2O$ higher than 50%.

*Running the Perturbation.* The amount of enzyme to use can be judged by making up reaction mixtures containing unlabeled reactants that are close to equilibrium and seeing how long it takes for equilibrium to be reached. As noted above, the maximum point of the perturbation will usually occur roughly at the point where an unlabeled mixture has gone 63% of the way to equilibrium. When the concentrations of the reactants other than those involved in the perturbation limit its size, however, the time to reach the maximum may be shortened considerably, so it is better to start with too little enzyme, rather than with too much.

There are two methods of generating reaction mixtures that will return to the starting point and permit accurate measurement of a perturbation. The first is the "two pot" method. Two solutions are prepared with substrates in one and products in the other. Concentrations calculated to be at equilibrium are obtained in the cuvette by adding aliquots from both solutions. Enzyme is added, and the system is allowed to come to final equilibrium. A new equilibrium constant is calculated by correcting concentrations for the extent of reaction observed. The volume of either the substrate or product solution added to the cuvette is then altered on the basis of the corrected equilibrium constant. This method will bring the system very close to equilibrium on the second try.

The second method is the "one pot" method. A single solution of all substrates and products at levels calculated to be at equilibrium is made, and an aliquot is added to the cuvette. Additional concentrated solutions of one of the substrates and one of the products are also made. Enzyme is added to the cuvette and the solution is allowed to come to equilibrium. Depending on which way the system is off from equilibrium, a small aliquot of one of the concentrated solutions is added to the original substrate–product mixture, and it is tested again. This procedure is repeated until the perturbation returns to the starting point. The concentrations of reactants in the final substrate–product mixture may have to be checked by enzymic analysis.

Each of the two methods has its advantages and disadvantages. The two-pot method takes less time, and it is easier to determine the final concentrations being used once the perturbation is obtained. The one-pot method allows more reproducible results; if the concentrations in the final mixture are analyzed, they are known more precisely.

*Comments and Suggestions*

1. Be sure the base line is stable before adding enzyme. Drift, whether from the instrument or from something happening in the cuvette, can give spurious results.
2. The same experiment must be run each time with unlabeled reactants to be sure that any perturbation observed is caused by an isotope effect. This is very important, since some artifacts can produce changes that look like perturbations.
3. After the perturbation reaches its maximum, decrease the chart speed by a factor of 10; it is then much easier to tell when the final equilibrium position has been reached (see Fig. 2).
4. Keep cuvettes tightly stoppered to prevent evaporation. This is especially important if one reactant is volatile ($CO_2$, acetone, etc.).
5. Accurate temperature control is essential, so do not handle the cuvettes. Add enzyme in the smallest possible volume (the exact amount is not critical). Use 25°, or another easily maintained temperature, unless it is essential to use a different value.
6. Enzyme inactivation is not a severe problem as long as turbidity is not generated (loss of activity alters the time scale of events, but does not affect the maximum point of the perturbation).
7. The instability of DPNH at acid pH makes it difficult to do experiments below pH 8 unless they are run rapidly or at low temperature. Even at pH 7, there is a slow rate of breakdown of DPNH, which causes the base line to be skewed.
8. Rapidly equilibrating compounds must be considered a common pool of material and have the sum of their concentrations used in

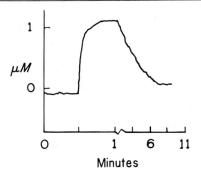

FIG. 2. Equilibrium perturbation with liver alcohol dehydrogenase to determine secondary isotope effect. Cyclohexanone-2,2,6,6-tetra-D, 97 $\mu M$; cyclohexanol, 23.5 $\mu M$; DPN, 200 $\mu M$; DPNH, 18 $\mu M$; pH 5.92. $\beta = 1.24$, $\alpha = 0.9$, $\alpha\beta = 1.12$. The perturbation is inverse (DPNH disappears at first) because of the high ratio of cyclohexanol to cyclohexanone. The chart speed was changed from 1.5 to 0.15 in./min at 1 min. The perturbation did not return exactly to the starting point and has been measured from the average of starting and final values. Enzyme was added at 0.5 min.

the calculations. Examples: $CO_2$ and bicarbonate (add carbonic anhydrase if there is doubt about rapid equilibration). Buffered substrate concentrations (with fructokinase, fructose was equilibrated with sorbitol in the presence of DPN, DPNH, and sorbitol dehydrogenase.[6] The sum of fructose and sorbitol concentrations was treated as a common pool).

## Interpretation of Results

As already mentioned, in many cases the isotope effects calculated by the equilibrium perturbation method are identical with those calculated for one of the $V/K$'s, and thus the interpretations are the same as those for $V/K$ isotope effects. In certain cases there will be differences, however. Equations (12) and (14) apply equally to equilibrium perturbation and $V/K$ isotope effects; the differences lie in the definitions of $c_f$ and $c_r$, and thus we must compare the expressions for $c_f$ and $c_r$ in Eqs. (9) and (10) with the ones we will get for $V/K$. For mechanism 3, $c_r$ in Eq. (10) is the same as one would observe for $V/K_b$ or $V/K_p$ isotope effects, but $c_f$ is different and is in fact the value one observes when measuring apparent $V/K$ isotope effects with A as the variable substrate and B at the same level used in the equilibrium perturbation. For $V/K_b$ or $V/K_p$ isotope effects

$$c_f = k_{7H}/k_6 (1 + k_5/k_4) \qquad (41)$$

[6] F. M. Raushel and W. W. Cleland, *Biochemistry* **16**, 2176 (1977).

The values of $c_r$ in Eq. (10), or $c_f$ in Eq. (41) are not dependent on the levels of any of the reactants, but $c_f$ in Eq. (9) is clearly linear with B, and thus at high B levels the isotope effect will disappear (that is, $\alpha = 1$, and $\alpha\beta = \beta$).

This predicted variation in isotope effect provides a sensitive test for the obligatory order in mechanism (3), since $c_f$ reflects the commitment to catalysis of the ternary EAB complex, and if the mechanism is random and A can dissociate from this complex at a reasonable rate, $c_f$ will not be infinite at infinite B. Thus in mechanism (42)

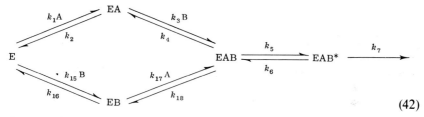

(42)

where again the perturbation is between A and P [we do not show the right side of the mechanism, which we will assume to be the same as in mechanism (3)], the value of $c_f$ will be[7]

$$c_f = \frac{k_{7H}}{k_6}\left(1 + \frac{k_5}{k_4/(1 + k_3B/k_2) + k_{18}}\right) \quad (43)$$

In this case, infinite B does not make $c_f$ infinite, but causes it to become:

$$c_f = \frac{k_{7H}}{k_6}\left(1 + \frac{k_5}{k_{18}}\right) \quad (44)$$

which is the $c_f$ value that applies to the $V/K_a$ isotope effect in mechanism (42). If $k_{18}$ is not extremely small, the isotope effect will not be reduced to 1 at infinite B, and as long as $\alpha$ is significantly larger than 1, the relative size of $\alpha$ and the isotope effect on $V/K_b$ gives some information on the relative sizes of $k_4$ and $k_{18}$.

What one would like is some way to solve for $c_f$ and $c_r$, and then to dissect each of these into its separate partition ratios. This cannot yet be done in the general case, but limits can be put on the various parameters. The method of Northrop[3] of comparing deuterium and tritium isotope effects on $V/K$ can be used to solve for the true isotope effect on the

---

[7] In terms of the net rate constant theory of Cleland,[4] $c_f$ reflects the two possible fates of EAB*; it is the ratio of the rate constant for the isotope-dependent reaction of EAB* to give EPQ* to the net rate constant for backward reaction, ending in release from the enzyme of the varied substrate or the reactant involved in the perturbation. This definition applies equally to $V/K$ or equilibrium perturbation experiments, and when applied in the reverse direction gives $c_r$.

bond-breaking step, or at least put a narrow limit on it. This in turn allows the placing of limits on $c_f$ and $c_r$. The pH variation of the isotope effects can also be used to simplify $c_f$ and $c_r$, since the effective size of the rate constants for reactant dissociation [$k_4$ and $k_{11}$ in mechanism (3)] are very large once one goes past the pK values where the $V/K$ profiles break. Thus at very low or high pH one would expect $c_f$ in Eq. (9) to reduce to $k_{7H}/k_6$, and $c_r$ in Eq. (10) to become $k_{8H}/k_9$.

*Interpretation of Secondary Isotope Effects.* Interpretation of primary isotope effects is not difficult, especially if the true isotope effect has been determined by Northrop's method. Interpretation of secondary isotope effects is less obvious. The true isotope effect on the chemical reaction is now small, often less than the overall equilibrium effect. In fact, the usual situation is that the kinetic isotope effect is normal one way and inverse the other, and both values are less than the full equilibrium effect. For example, if the equilibrium effect is 1.25 ($\beta = 0.8$), one might have true isotope effects in the forward direction of 1.1 and in the reverse direction of 0.88. In such a situation, the apparent isotope effects predicted by Eqs. (12) and (14) will give information on the true secondary kinetic effects only when $c_f$ and $c_r$ are quite small. When large primary isotope effects are observed, and thus one can be sure that $c_f$ and $c_r$ *are* small, this will be the case. What is more common, however, is that the primary effects are small and thus $c_f$ and $c_r$ are large; in these cases, Eqs. (12) and (14) become

$$\alpha = \frac{1/\beta + c_f/c_r}{1 + c_f/c_r} \tag{45}$$

$$\alpha\beta = \frac{\beta + c_r/c_f}{1 + c_r/c_f} \tag{46}$$

and one of these equations can then be used to calculate the ratio of $c_f$ to $c_r$. If the intrinsic primary isotope effect has been determined by Northrop's method, Eq. (12) for the primary isotope effect can then be combined with the value of $c_f/c_r$ to solve for $c_f$ and $c_r$ separately. These methods are clearly in their infancy and have not yet been applied, but it is hoped that in a few years their application will enable us to know the separate commitments in forward and reverse directions for enzymes of interest.

Examples of Use of the Method

The equilibrium perturbation method was first applied to malic enzyme with malate-2-D[1] and gave isotope effects similar to those seen on

$V/K_{\text{malate}}$ (1.5). With $^{13}CO_2$, however, a very large isotope effect of 1.03 was obtained, and by comparison of possible $c_f$ and $c_r$ values for the deuterium and $^{13}C$ effects, and consideration of the true deuterium (measured by Northrop's method) and estimated $^{13}C$ isotope effects on the bond-breaking steps, it was calculated that reverse hydride transfer was 6–8 times the rate of decarboxylation.[3] Decarboxylation is thus the major rate-limiting step in the catalytic process.

With malate dehydrogenase, malate-2-D gave $\alpha$ values of 1.93 at 11 $\mu M$ oxaloacetate and 1.53 at 100 $\mu M$ oxaloacetate, which compare with the $V/K_{\text{malate}}$ of 1.9.[8] The decrease in $\alpha$ at the higher oxaloacetate is caused by $c_r$ being a linear function of oxaloacetate, similar to the form of $c_f$ in Eq. (9).

With isocitrate dehydrogenase and isocitrate-2-D, a normal perturbation is seen. The $\alpha$ value was first reported to be 1.20,[9] but more precise determination by Paul Cook in our laboratory gave 1.15, slightly less than the $1/\beta$ value of 1.17. Cook has also shown that there is no $V/K_{\text{isocitrate}}$ isotope effect at neutral pH, and thus we have the interesting situation that an isotope effect is seen by equilibrium perturbation, but not on $V/K$. The explanation is that $c_f$ for isocitrate is very large because isocitrate dissociates so slowly from the enzyme. In the $V/K$ experiment, $c_r$ is for $CO_2$ and is very small; thus the dominance of $c_f$ in the equation similar to Eq. (9) gives an isotope effect of 1.00. In the equilibrium perturbation experiment, however, $c_r$ is for TPNH, which is released from the enzyme even more slowly than isocitrate. Since $c_r$ for TPNH is larger than $c_f$ for isocitrate, $c_r$ dominates Eq. (9), and $\alpha$ is almost entirely the equilibrium isotope effect, $1/\beta$.

Equilibrium perturbation experiments have been carried out by Cook with cyclohexanol-1-D and liver alcohol dehydrogenase and with isopropanol-2-D and yeast alcohol dehydrogenase. With LADH, $\alpha$ varies from 2.86 at low cyclohexanone to 1.54 at infinite ketone, showing that release of DPNH and cyclohexanone is random, not ordered. With YADH, however, $\alpha$ goes from 3.17 at low acetone to 1.0 at infinite acetone, showing that DPNH dissociates very slowly or not at all from the ternary E–DPNH–acetone complex. For both these enzymes, equilibrium perturbation isotope effects match $V/K$ ones very closely.

The mechanism of fumarase has been investigated by John Blanchard in this laboratory using primary deuterium and $^{18}O$, and secondary deuterium isotope effects. At pH 5, $\alpha$ for the $^{18}O$ isotope effect from the ma-

---

[8] These values have been recalculated from the data in Schimerlik et al.,[1] where the value of $\beta$ was estimated to be 0.76 instead of the correct value of 0.85.
[9] M. H. O'Leary and J. A. Limburg, Biochemistry 16, 1129 (1977).

late side is 1.07, and the same results were obtained using malate containing 92% $^{18}O$, and with water containing 47% $^{18}O$. Since $1/\beta = 1.033$, this is clearly a large kinetic isotope effect in both directions and carbon–oxygen bond breaking appears to be rate limiting. At pH 5 the primary deuterium effect from the malate side is 1.01, and thus C–H bond cleavage is not rate limiting. With dideuterofumarate, or dideuteromalate made from it, the secondary deuterium isotope effects at pH 5 are 1.31 from the malate side and 0.90 from the fumarate side ($1/\beta = 1.45$ from the malate side).

As the pH is raised, the $^{18}O$ effect drops, becoming 1.01 at pH 7 and 1.00 at pH 9 (that is, $\alpha\beta = 0.97$, the full equilibrium isotope effect at pH 9). The primary deuterium isotope effect becomes inverse at pH 7 ($\alpha = 0.92$ from the malate side), suggesting a rapid proton transfer from malate to a nonexchanging catalytic group on the enzyme with a fractionation factor slightly higher than that of water (the equilibrium effect is 0.98 in $H_2O$ and 0.93 in $D_2O$). At pH 9, however, $\alpha$ becomes 1.00 again. The secondary isotope effects become smaller as pH is raised, and $\alpha$ from the malate side reaches 1.00 at pH 9.

We interpret these data to indicate a carbanion mechanism, with C–H bond cleavage much faster than C–O cleavage. At pH 5, where the rates are considerably slower than at pH 7, reactant release has become very fast because of partial protonation of their carboxyl groups, and C–O bond cleavage is the sole rate-limiting step. Carbon–hydrogen bond cleavage has slowed somewhat owing to protonation of the catalytic group (probably carboxyl) which accepts the proton, but C–H cleavage is still probably 67 times, and C–D cleavage 10 times the rate of C–O cleavage.[10] At pH 7, C–H bond breaking is much faster, and $c_r$ has become larger for all three isotope effects (probably as the result of much more rapid C–O bond cleavage relative to conformation changes that permit water and fumarate to dissociate). As a result, we see reduced $^{18}O$ and secondary deuterium effects, and a full inverse primary deuterium effect.

At pH 9, where again the reaction rate has slowed considerably, all three isotope effects show full equilibrium effects from the fumarate side; that is, the full transformation of fumarate to malate is at equilibrium on the enzyme, and the conformation change that permits malate release has become fully rate limiting. This interpretation is supported by the equality

---

[10] Calculated by assuming the true primary deuterium isotope effect to be 7, $1/\beta = 0.92$ for the proton transfer to the enzyme, and $c_f$ to be zero because of the very large size of the $^{18}O$ effect. The value of $c_r$, which is the ratio of reverse proton transfer to C–O bond cleavage, is then 67, and the rate of C–D cleavage is 7 times slower.

at pH 9 of isotope exchanges between malate and either fumarate or the protons or oxygen of water.[11]

The above examples illustrate the wide scope and usefulness of the equilibrium perturbation method. So far it has been used largely in this laboratory, but as soon as its use becomes more widespread, it should become one of the most useful of the available tools for studying isotope effects on enzymic reactions.

[11] K. Berman, E. C. Dinovo, and P. D. Boyer, *Bioorg. Chem.* **1**, 234 (1971).

## [6] Use of Secondary α-Hydrogen Isotope Effects in Reactions of Pyrimidine Nucleoside/Nucleotide Metabolism: Thymidylate Synthetase[1]

*By* THOMAS W. BRUICE, CHARLES GARRETT, YUSUKE WATAYA, and DANIEL V. SANTI

Secondary α-hydrogen isotope effects are useful tools that may aid in ascertaining whether an enzyme-catalyzed reaction or covalent interaction of an inhibitor with an enzyme involves rehybridization of a carbon atom of the ligand at or before the rate-determining step of the reaction.[2] We have used such studies to aid our understanding of the catalytic mechanism of thymidylate (dTMP) synthetase, as well as the interaction of 5-fluoro-2'-deoxyuridylate (FdUMP) with this enzyme; both have been recently reviewed.[3] The currently accepted minimal mechanism of dTMP synthetase is depicted in Fig. 1. A salient feature of this mechanism is that a primary event in catalysis involves addition of a nucleophile of the enzyme to the 6-position of the pyrimidine heterocycle to form transient 5,6-dihydropyrimidine intermediates in which the 6-carbon is rehybridized from $sp^2$ to $sp^3$. FdUMP behaves as a quasi-substrate for this reaction, proceeding through the first two steps of the mechanism depicted in Fig. 1; at this stage, an analog of a steady-state intermediate accumulates,

[1] This work was supported by U.S. Public Health Service Grant CA-14394 from the National Cancer Institute. D. V. S is a recipient of a National Institutes of Health Career Development Award.
[2] W. W. Cleland, M. H. O'Leary, and D. B. Northrop, eds., "Isotope Effects on Enzyme-Catalyzed Reactions." Univ. Park Press, Baltimore, Maryland, 1977.
[3] A. L. Pogolotti, Jr., and D. V. Santi, *in* "Bioorganic Chemistry" (E. E. van Tamelen, ed.), Vol. 1, p. 277. Academic Press, New York, 1977.

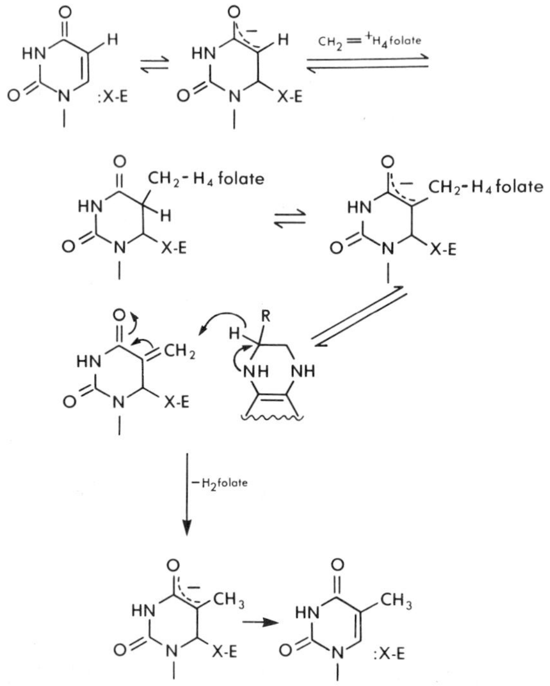

FIG. 1. Minimal mechanism for dTMP synthetase; the 1-substituent of the heterocycle is 5-phospho-2-deoxy-$\beta$-ribofuranose.

the structure of which is depicted in Fig. 2. While this covalent complex is quite stable ($K_d \simeq 10^{-13}$ $M$), it does undergo an enzyme-catalyzed reversal releasing the unchanged reactants, FdUMP and (+), L-5,10-methylenetetrahydrofolate (CH$_2$-H$_4$folate). In addition, dTMP synthetase catalyzes the dehalogenation of 5-bromo-2'-deoxyuridylate (BrdUMP) by a mechanism analogous to the normal enzymic reaction.[4] Here we describe the use of $\alpha$-secondary hydrogen isotope effects in studies of the interaction of FdUMP with dTMP synthetase, as well as in the chemical and enzyme-catalyzed dehalogenation of BrdUrd and BrdUMP, respectively. Using the appropriate [2-$^{14}$C,6-$^3$H]pyrimidine nucleoside/nucleotides, the presence of a tritium isotope effect of 15% or greater provides strong evidence for rehybridization at the 6-carbon and the formation of 5,6-dihydropyrimidine intermediates. Further, certain conclusions may be reached regarding the positioning of the change in rehybridization with respect to the rate-determining step of the reaction.

[4] Y. Wataya and D. V. Santi, *Biochem. Biophys. Res. Commun.* **67**, 818 (1975).

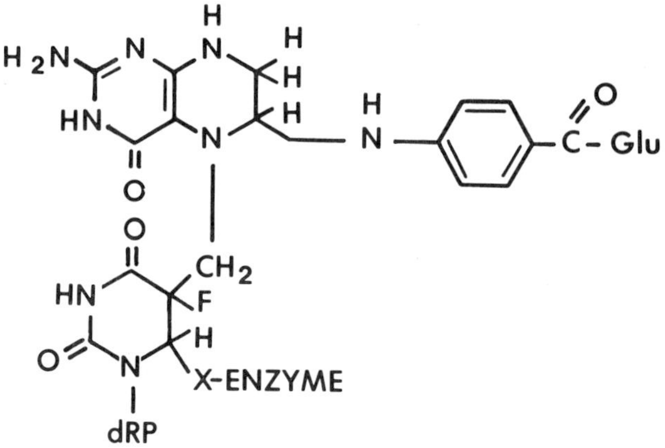

FIG. 2. Structure of the FdUMP-CH$_2$-H$_4$folate-dTMP synthetase complex.

There is good reason to believe that nucleophilic attack at the 6-position of the uracil or cytosine heterocycle may be a common mechanistic feature utilized by many enzymes to enhance the reactivity at various sites of the heterocycle.[5] From what is known thus far, this is almost certainly correct for a variety of enzyme-catalyzed electrophilic substitution reactions occurring at the 5-position of uracil and cytosine heterocycles. These would include the dUMP- and dCMP-hydroxymethylases, the pyrimidine methylases of nucleic acids, pseudo-uridylate synthetase, and a large number of yet uncharacterized enzymes that alkylate the 5-position of minor bases found in tRNA. Indeed, we have recently demonstrated that dUMP hydroxymethylase from SP01-infected *Bacillus subtilis* proceeds by this mechanism.[6] In addition, compared to the unaltered bases, 5,6-dihydrouracil and 5,6-dihydrocytosine derivatives are chemically much more reactive both toward nucleophilic substitution at the 4-position of the heterocycle and toward glycosidic bond cleavage.[5] It is therefore not unreasonable to suggest that at least some of the enzymes that catalyze such reactions might also operate via nucleophilic attack at the 6-position of the heterocycle to achieve saturation of the 5,6-double bond. The use of $\alpha$-secondary hydrogen isotope effects provides a simple tool that may be useful in ascertaining whether these hypotheses are correct. Although dTMP synthetase is used as a paradigm in the studies described here, the general procedures should be useful in detecting dihydropyrimidine intermediates formed in other chemical and enzymic reac-

---

[5] D. V. Santi, Y. Wataya, and A. Matsuda, *in* "Enzyme-Activated Irreversible Inhibitors." (N. Seiler, M. J. Jung, and J. Koch-Weser, eds.), p. 291. Elsevier/North Holland Biomedical Press, Amsterdam, 1978.
[6] M. Kunitani and D. V. Santi, unpublished results.

tions, providing rehybridization of carbon occurs before or at the rate-determining step.

*Materials and Methods*

Thymidylate synthetase from methotrexate-resistant *Lactobacillus casei*[7] was obtained from the New England Enzyme Center and purified as previously described.[8] Radioactive nucleosides were obtained commercially, converted to their corresponding nucleotides using *Escherichia coli* dThd kinase and purified as described[8]; final purification and routine checks of purity were performed by HPLC using Aminex A-27[9] or on Lichrosorb $C_{18}$ using 5 m$M$ ($n$-Bu)$_4$N$^+$HSO$_4^-$ and 5 m$M$ potassium phosphate (pH 7.0). NMM buffer refers to a solution containing 50 m$M$ $N$-methylmorpholine·HCl (pH 7.4), 25 m$M$ MgCl$_2$, 1 m$M$ EDTA, and 75 m$M$ 2-mercaptoethanol. Other materials and methods not described in detail here have previously been reported.[8,10,11]

For analysis of isotope effects, reactants and/or products in reaction mixtures were separated as described for each experiment. After addition of scintillant, $^3$H and $^{14}$C activities were determined by counting samples 3–5 times, with a minimum of $2 \times 10^5$ counts of $^{14}$C collected. Counting efficiencies for each isotope were determined by the external standard ratio method, and disintegrations per minute (dpm) calculations were aided by a tape-fed Hewlett-Packard computer. Standard errors for determinations of $^3$H/$^{14}$C and $^{14}$C dpm were approximately 0.25 and 0.5%, respectively. For kinetic isotope effects, several aliquots of each reaction mixture were analyzed over the course of the reaction. Kinetic isotope effects were calculated using Eqs. (1) and (2),[12] where $R_0$ is the initial $^3$H/$^{14}$C of the substrate, $x$ is the fraction

For product:
$$\frac{k_T}{k_H} = \frac{\log(1 - xR_p/R_0)}{\log(1 - x)} \quad (1)$$

For reactant:
$$\frac{k_T}{k_H} = 1 + \frac{\log(R_s/R_0)}{\log(1 - x)} \quad (2)$$

of substrate reacted at the time of analysis (determined by the $^{14}$C content of reactant and/or product), and $R_s$ and $R_p$ are the $^3$H/$^{14}$C of substrate and product, respectively, at the time of analysis. For dissociation of

---

[7] T. C. Crusberg, R. Leary, and R. L. Kisliuk, *J. Biol. Chem.* **245**, 5292 (1970).
[8] Y. Wataya and D. V. Santi, this series, Vol. 46, p. 307. 1977.
[9] C. Garrett, A. L. Pogolotti, Jr., and D. V. Santi, *Anal. Biochem.* **79**, 602 (1977).
[10] D. V. Santi, C. S. McHenry, and H. Sommer, *Biochemistry* **13**, 471 (1974).
[11] D. V. Santi, C. S. McHenry, and E. R. Perriard, *Biochemistry* **13**, 467 (1974).
[12] L. Melander, "Isotope Effects on Reaction Rates." Ronald Press, New York, 1960.

[2-$^{14}$C,6-$^3$H]FdUMP from the complex, first-order rate constants of the disappearance of bound $^3$H and $^{14}$C also were used to calculate the isotope effect.[10] The latter method, while of limited general utility, has the advantage that the exact fraction of reaction at the time of analysis need not be ascertained; nevertheless, both methods gave identical results. For equilibrium isotope effects of the interaction of [2-$^{14}$C,6-$^3$H]FdUMP with dTMP synthetase, reactants and products were separated after equilibrium was reached, and the isotope effect was calculated using Eq. (3). The values of all isotope effects reported here are presented as mean ± SE of multiple determinations.

$$K_T/K_H = \frac{^3H/^{14}C_{\text{bound FdUMP}}}{^3H/^{14}C_{\text{free FdUMP}}} \quad (3)$$

Secondary α-Hydrogen Isotope Effects in the Interaction of FdUMP and CH$_2$-H$_4$folate with dTMP Synthetase

As described above, FdUMP, CH$_2$-H$_4$folate, and dTMP form a covalent ternary complex having the structure shown in Fig. 2. The *minimal* interactions leading to this complex are depicted in Scheme 1.

FdUMP + CH$_2$-H$_4$ folate + Enz $\rightleftharpoons$ Enz · FdUMP · CH$_2$-H$_4$ folate
         (a)                                            (b)

$\rightleftharpoons$ Enz—FdUMP · CH$_2$-H$_4$ folate $\rightleftharpoons$ Enz—FdUMP—CH$_2$-H$_4$ folate
        (c)                                             (d)

SCHEME 1

Here, the components (*a*) first interact to form a ternary complex (*b*) involving no covalent bonds. A covalent bond is then formed between the enzyme and the 6-position of FdUMP (*c*), and an additional covalent linkage is subsequently formed between the 5-position of FdUMP and the cofactor, CH$_2$-H$_4$folate, to form an isolatable complex (*d*). While the covalent complex (*d*) is relatively stable, all reactions depicted in Scheme 1 are reversible. From the standpoint of utilizing α-secondary isotope effects, it should be noted that the 6-position of the reactant FdUMP is sp$^2$ hybridized and that position of the nucleotide in the covalent complex is sp$^3$ hybridized (Fig. 2). Thus, utilizing [2-$^{14}$C,6-$^3$H]FdUMP, one might expect to observe an equilibrium isotope effect, a secondary kinetic isotope effect ($k_H > k_T$) in dissociation of the complex, and/or an inverse kinetic isotope effect ($k_T > k_H$) in the formation of the complex.

Determination of kinetic isotope effects that occur upon dissociation of the ternary complex were performed essentially as previously described.[10] Briefly, a complex is formed using [2-$^{14}$C,6-$^3$H]FdUMP, excess unlabeled FdUMP is added, and the complex is isolated as the bound radiolabeled FdUMP dissociates and is replaced by unlabeled FdUMP. The $^3$H/$^{14}$C remaining in the complex is determined at intervals, and $k_H/k_T$ is determined as described in Materials and Methods. To measure the equilibrium isotope effect, a moderate excess of [2-$^{14}$C,6-$^3$H] FdUMP is incubated with CH$_2$-H$_4$folate and dTMP synthetase until isotopic equilibrium is achieved between the free and bound ligands. The bound and free [2-$^{14}$C,6-$^3$H]FdUMP are isolated by gel filtration, the $^3$H/$^{14}$C is determined in each, and the equilibrium isotope effect is calculated using Eq. (3).

## Procedures

*Kinetic Isotope Effect upon Dissociation of the [2-$^{14}$C,6-$^3$H] FdUMP–CH$_2$–H$_4$folate–dTMP Synthetase Complex.* A solution (3.4 ml) of the ternary complex was formed[8,10] using 2.25 ml of 2-fold concentrated NMM buffer, 2.0 nmol of dTMP synthetase, 0.23 μmol of (+),L-CH$_2$-H$_4$folate, 30 μmol H$_2$CO and 10.5 nmol [2-$^{14}$C,6-$^3$H]FdUMP (5.25 × 10$^5$ dpm/nmol $^3$H; $^3$H/$^{14}$C dpm = 4.76). Controls omitted CH$_2$-H$_4$folate. To this was added a solution (1.1 ml) containing 1.5 μmol of unlabeled FdUMP. The reaction mixture was kept at 25.0° under nitrogen, protected from light; triplicate aliquots were removed at intervals up to about 35 hr, and the ternary complex was isolated by adsorption on nitrocellulose filters[11]; the aliquots filtered were progressively larger (20–500 μl) with time to obtain sufficient dpm in the isolated complex as the reaction proceeded. The triplicate samples were counted, showing standard errors of $^3$H/$^{14}$C within 0.8% of the mean (see Materials and Methods). From these, the dissociation of the [2-$^{14}$C,6-$^3$H]FdUMP–CH$_2$-H$_4$folate–dTMP synthetase complex was shown to be first order with $k_H = 0.12$ hr$^{-1}$; Fig. 3 shows the data from which $k_H/k_T$ was calculated to be 1.23 ± 0.01 ($n = 12$). Similar experiments using variable amounts of unlabeled FdUMP (10- to 100-fold excess over labeled FdUMP) demonstrated that the rate constant for dissociation and $k_T/k_H$ were independent of the concentration of the free ligand.

*Equilibrium Isotope Effect on the Interaction of [2-$^{14}$C,6-$^3$H]FdUMP with dTMP Synthetase.* The ternary complex was prepared using 0.17 mM(+),L-CH$_2$-H$_4$folate, 1.5 μM dTMP synthetase and 7.5 μM [2-$^{14}$C,6$^3$H] FdUMP and kept under nitrogen at 25.0° protected from light. Nitrocellulose filtration[11] of aliquots demonstrated that at $t = 0$, the $^3$H/$^{14}$C of the complex was identical to that of the FdUMP used, but gradually increased with time until isotopic equilibrium was reached at 45–50 hr. There was no further change up to 90 hr. Excess unlabeled FdUMP was added to aliquots of this solution at 0.5, 60, and 90 hr, and the rate of dissociation was determined over approximately 1.5 half-lives as previously

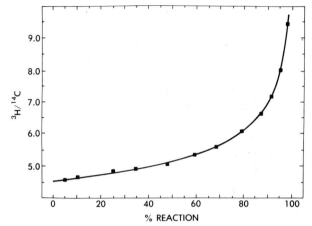

FIG. 3. Secondary α-hydrogen isotope effect upon dissociation of the [2-$^{14}$C,6-$^{3}$H] FdUMP-CH$_2$-H$_4$folate—dTMP synthetase complex. The data points represent the $^3$H/$^{14}$C ratio of [2-$^{14}$C,6-$^3$H]FdUMP bound in the complex isolated over the course of the reaction. The solid line depicts the theoretical isotopic ratios (Eq. 2) for a reaction with $k_H/k_T = 1.229$. Experimental details are described in the text.

described. The facts that the amount of complex was unchanged over this period and that the first-order rate constants for dissociation throughout this period were identical ($k = 2.0 \times 10^{-3}$ min$^{-1}$) indicated that the complex was not modified upon prolonged incubation.

To determine the equilibrium isotope effect, three separate solutions (800 μl) of the complex were prepared as described above and kept under nitrogen at 25°, protected from light, for 60–65 hr. Three 200-μl aliquots of each were passed through a Sephadex G-25 column (1.0 × 19 cm) equilibrated and eluted with 75 m$M$ potassium phosphate–10 m$M$ β-mercaptoethanol (pH 7.4) at 4°; approximately 0.8-ml fractions were collected. The fractions containing bound and free [2-$^{14}$C,6-$^3$H]FdUMP were individually pooled, and aliquots of each were counted. Samples of the free [$^3$H/$^{14}$C]FdUMP were also subjected to HPLC analysis and demonstrated not to have undergone decomposition during the period of incubation and analysis. The equilibrium isotope effect determined by these experiments using Eq. (3) was $k_T/k_H = 1.240 \pm 0.001$ ($n = 9$).

## Comments

The large secondary isotope effect ($k_H/k_T = 1.23$) observed for the dissociation of [2-$^{14}$C,6-$^3$H]FdUMP from the ternary complex and the equilibrium isotope effect ($k_T/k_H = 1.24$) for the interaction are in accord with all data suggesting that the 6-position of FdUMP is sp$^3$ hybridized and covalently bound to the enzyme within the isolatable complex. If covalent bond cleavage in *dissociation* is prerate determining (preequilibrium) the kinetic isotope effect should equal the equilibrium isotope effect, as it does. If this step is rate determining, the kinetic isotope effect would be lower than the equilibrium isotope effect except in the situation where the transition state structure is completely rehybridized to $sp^2$. Thus, from these data we conclude that, in the *association* of FdUMP with the enzyme, covalent bond formation occurs either after the rate-determining step or at the rate-determining step *if* the transition-state structure is $sp^2$ hybridized.

## α-Secondary Isotope Effect in the Dehalogenation of BrdUrd and BrdUMP

dTMP synthetase catalyzes an irreversible thiol-dependent dehalogenation of BrdUMP[4] as shown in Eq. (4). In addition, small amounts of 5-thiol-dUMP are formed.

$$\text{BrdUMP} + 2 \text{RSH} \rightarrow \text{dUMP} + (\text{RS})_2 + \text{HBr} \tag{4}$$

The direct chemical counterpart to this reaction is the dehalogenation of BrdUrd by cysteine and other thiols[13]; as with the enzymic dehalogenation, the chemical reaction is also accompanied by formation of small amounts of 5-thiolated product. Although this reaction was proposed to involve nucleophilic attack of the thiolate at the 6-position of BrdUrd, the presumed 5,6-dihydropyrimidine intermediate is too transient for detection by conventional methods and evidence for its existence rested in large part upon analogy with the bisulfite-mediated dehalogenation which was known to proceed by this mechanism.[14] The general mechanism which is currently believed to occur for both the thiol-mediated chemical and enzymic dehalogenation of BrdUrd and BrdUMP, respectively, is depicted in Fig. 4. We describe here a method for detection of such transient intermediates which could be applicable to both chemical and enzymic reactions of pyrimidines believed to proceed via 5,6-dihydro intermediates. The specific cases illustrated include the thiol-mediated chemical dehalogenation of BrdUrd as well as the dTMP synthetase-catalyzed dehalogenation of BrdUMP.

*α-Secondary Isotope Effect in the Cysteine-Promoted Dehalogenation of BrdUrd.* A solution (330 μl) containing 10 m$M$ [2-$^{14}$C,6-$^3$H]BrdUrd ($^{14}$C, 0.614 μCi; $^3$H, 1.89 μCi) and 0.25 $M$ L-cysteine at pH 7.3 was incubated at

---

[13] Y. Wataya, K. Negishi, and H. Hayatsu, *Biochemistry* **12**, 3992 (1973).
[14] E. G. Sander, *in* "Bioorganic Chemistry" (E. E. van Tamelen, ed.), Vol. 2, p. 273. Academic Press, New York, 1978.

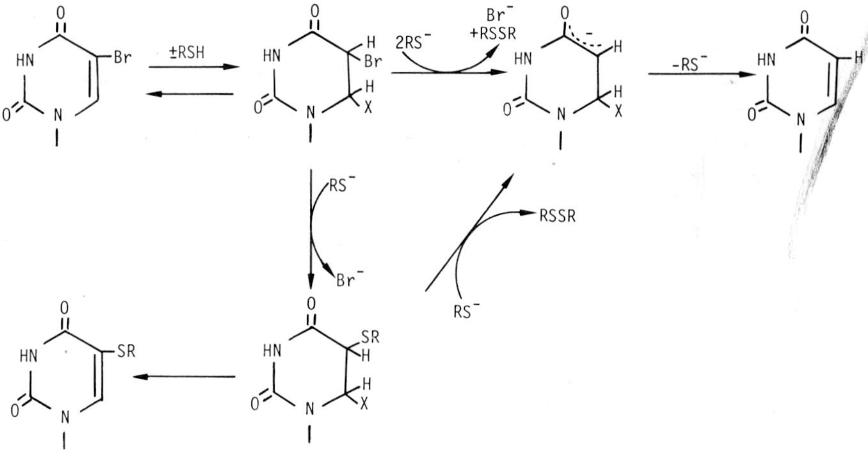

FIG. 4. Proposed mechanisms for thiol-mediated dehalogenation of BrdUrd (X = RS$^-$) and dTMP synthetase-catalyzed dehalogenation of BrdUMP (X = enzyme).

37°. Aliquots (10 μl) were removed at specified intervals, and the extent of debromination was monitored spectrophotometrically[13]; the pseudo-first-order rate constant was $2.2 \times 10^{-2}$ min$^{-1}$. The reactant, BrdUrd, and products, dUrd and $S$-[5-(2'-deoxyuridyl)]cysteine (CysdUrd), were separated by thin-layer chromatography (TLC)[15] as the reaction progressed, and the $^3$H/$^{14}$C ratio of each was determined. Upon completion, dUrd accounted for 92% of the product, the remaining 8% being CysdUrd. The tritium content of both products was enriched at initial stages of the reaction, and approached that of the initial reactant as dehalogenation progressed. Conversely, the $^3$H/$^{14}$C ratio of the reactant decreased as the reaction proceeded. From these data, the calculated $k_T/k_H$ values for formation of dUrd and CysdUrd are $1.187 \pm 0.006$ and $1.156 \pm 0.007$, respectively; $k_T/k_H$ for dehalogenation of BrdUrd is $1.174 \pm 0.005$ ($n = 9$ in each case).

*α-Secondary Isotope Effect in the Thymidylate Synthetase-Catalyzed Debromination of BrdUMP.* A solution (3.5 ml) containing 50 μM [6-$^3$H,2-$^{14}$C]BrdUMP, (5.21 Ci/mol of $^{14}$C, $^3$H/$^{14}$C = 4.360), 1.1 μM dTMP synthetase, 10 mM cysteine, 25 mM MgCl$_2$, 1 mM EDTA, and 50 mM $N$-methylmorpholine·HCl (pH 7.4) was incubated at 37°. Aliquots (0.15 ml) were withdrawn, dUMP and BrdUMP were added as quenching agents and chromatographic markers (approximately 1 mM each) and the samples were immersed in ice water; this procedure was shown to result in rapid and effective quenching of the reaction. Separations of reactants

[15] Y. Wataya and D. V. Santi, *J. Am. Chem. Soc.* **99**, 4534 (1977).

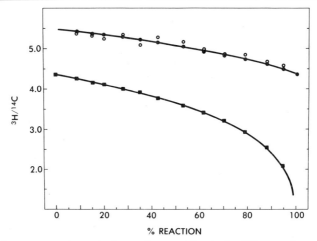

FIG. 5. Inverse secondary α-hydrogen isotope effect in the dTMP synthetase-catalyzed debromination of [2-$^{14}$C,6-$^3$H]BrdUMP. The data points represent the $^3$H/$^{14}$C ratio of dUMP (●), CysdUMP (○), and BrdUMP (■). The lines depict the theoretical isotopic ratios for product (upper line) and reactant (lower line) calculated for a reaction with $k_T/k_H = 1.258$ using Eqs. (1) and (2), respectively. Experimental details are described in the text.

and products were performed on Aminex A-27 as previously described[9] using a buffer containing 0.3 $M$ NH$_4$HCO$_3$ adjusted to pH 9.0 with NH$_4$OH; column temperature was 85°. Dehalogenation in a parallel reaction lacking enzyme was negligible.

As shown in Fig. 5, the $^3$H/$^{14}$C ratios of the reactant (BrdUMP) and the products (dUMP and CysdUMP) change in a manner indicating a more rapid dehalogenation of the 6-tritiated compound. The inverse secondary tritium isotope effects ($k_T/k_H$) are 1.253 ± 0.003 and 1.31 ± 0.03 calculated from the $^3$H/$^{14}$C ratios of dUMP and CysdUMP, respectively ($n = 12$). If, for each determination during the course of the reaction we combine the isotopic content of the products, the mean isotope effect for product formation is 1.256, which agrees well with the value of 1.260 ± 0.003 ($n = 12$) calculated from the decreasing isotopic ratio of the substrate BrdUMP.

## Comments

The magnitude of the inverse α-hydrogen secondary isotope effect ($k_T/k_H = 1.17$) for the cytseine-promoted dehalogenation of [2-$^{14}$C,6-$^3$H]BrdUrd provides strong evidence for sp$^2$ to sp$^3$ rehybridization of the 6-carbon of the heterocycle to form transient 5,6-dihydropyrimidine intermediates in accord with the mechanism depicted in Fig. 4. Kinetic studies[13] indicate that initial attack of thiolate is probably rate determining in this reaction; if so, it may be concluded that the isotope effect observed here is the true kinetic isotope effect of the reaction rather than a preequilibrium isotope effect. Future experimentation and/or calculations may provide knowledge of the transition-state structure of the rate-determining step of this reaction. An even larger inverse $\alpha$-hydrogen secondary isotope effect ($k_T/k_H = 1.26$) is observed in the dTMP synthetase-catalyzed dehalogenation of [2-$^{14}$C,6-$^3$H]BrdUMP. This, together with product analysis and analogy with the aforementioned chemical counterpart verifies the salient features of the mechanism of the enzymic reaction depicted in Fig. 3. However, from available evidence it is not possible to determine whether rehybridization of the 6-carbon of BrdUMP in the enzymic reaction occurs at or before the rate-determining step of the reaction. The above results demonstrate that secondary $\alpha$-hydrogen isotope effects may be of great utility in demonstrating the existence of transient 5,6-dihydropyrimidine intermediates in both chemical and enzymic reactions of pyrimidine nucleosides and nucleotides.

# Section II

# Complex Enzyme Systems

## [7] Cooperativity in Enzyme Function: Equilibrium and Kinetic Aspects

### By KENNETH E. NEET

The term *cooperativity* is widely used in biochemical systems to refer to a variety of interactions with a common feature. In general, cooperativity may be defined as the influence of one molecule (or chemical moiety, or bond formation) upon a succeeding molecule (or chemical moiety, or bond formation) in the process; the overall effect is to increase or decrease the response relative to that of an isolated, independent interaction. The earliest use of cooperativity referred to the intramolecular transitions of macromolecules in which multiple weak bonds were made or broken, in a cooperative and essentially simultaneous fashion, during the transition, e.g., the $\alpha$-helix to random-coil transition of polypeptides and protein segments or the "melting" of the DNA double helix upon heating. In these cases the transition occurs over a much narrower range of temperature or pH than if the process were not cooperative. Cooperativity has been used more recently to describe the association of macromolecular subunits, particularly proteins, into oligomeric structures in which the initial association of two or more subunits gives rise to an increased affinity so that succeeding steps in the association occur more readily.[1] Examples of this behavior that have been studied include octamer formation of hemerythrin,[1] indefinite association of the DNA binding protein of bacteriophage T4,[2] and the requirement for a critical concentration of monomer for the polymerization of tobacco mosaic virus coat protein.[3] Cooperativity of association leads to association modes in which the distribution of oligomers is skewed; for example, in the monomer–octomer system of hemerythrin little, if any, of the protein exists in intermediate oligomers during the dissociation process.

The most common usage of the term cooperativity in biochemical processes is with regard to enzyme kinetics and ligand binding to macromolecules. Positive cooperativity of binding of oxygen to hemoglobin was recognized many years ago, but it was not until the early 1960s that the prevalence of cooperativity in enzyme systems and the potential value of cooperativity in physiological regulation of allosteric systems was fully

---

[1] I. M. Klotz, N. R. Langerman, and D. W. Darnall, *Annu. Rev. Biochem.* **39**, 25 (1970).
[2] R. B. Carroll, K. E. Neet, and D. A. Goldthwait, *J. Mol. Biol.* **91**, 275 (1975).
[3] A. C. H. Durham, *J. Mol. Biol.* **67**, 289 (1972).

appreciated.[4,5] Cooperativity in ligand binding to a protein can either lead to an enhanced binding of succeeding molecules (defined as positive cooperative) and a steeper saturation curve than would be expected from noncooperative binding or lead to a decreased ability of the protein to bind the ligand and a flatter, less-steep saturation curve (defined as negative cooperative). Cooperativity is often referred to as non-Michaelis–Menten (or nonhyperbolic) behavior. Cooperativity used in this sense refers to what is more properly called *homotropic* cooperativity[5] (homotropic interactions) or the influence of a ligand bound to a protein on the further binding or reaction of ligand species of the same kind, e.g., first-oxygen influence on the second, third, and fourth oxygens bound to the tetrameric hemoglobin molecule. Heterotropic cooperativity, in contrast, refers to the influence of a ligand (effector) bound to a protein on the binding to or reaction with the protein of a different type of ligand, e.g., the influence of 2,3-diphosphoglycerate on the binding of oxygen to hemoglobin. *Negative heterotropic* cooperativity refers to an inhibitory action of one ligand on the binding or reaction of another with the protein; *positive heterotropic* cooperativity refers to an activating or enhancing action on the other ligand.

Both homotropic and heterotropic cooperativity imply interactions between ligands in the process of binding to a protein, but it is clear from the many solution and crystallographic studies on these systems that the interaction occurs through the protein structure itself rather than a direct interaction between the ligand molecules themselves. This interaction through the protein structure then requires that the protein, usually in an oligomeric structure, be able to bind several of the ligand molecules and/or be able to undergo conformational changes to transmit the interactions between the ligand binding sites. Many proteins and enzymes have been demonstrated to have these properties. Interactions can be transmitted *spatially* through the protein to different subunits[5,6] or *temporally* to conformations that only slowly relax to the original state.[7,8] These differences will be further distinguished in a later section; this chapter will deal primarily with the consequences and causes of spatial interactions on the cooperativity of oligomeric enzymes; the cooperativity due to temporal interactions will be discussed in Chapter [8] on hysteresis in this volume.

A variation of this type of ligand binding cooperativity is the coopera-

---

[4] J. Monod, J.-P. Changeux, and F. Jacob, *J. Mol. Biol.* **6**, 306 (1963).
[5] J. Monod, J. Wyman, and J.-P. Changeux, *J. Mol. Biol.* **12**, 88 (1965).
[6] D. E. Koshland, Jr., G. Nemethy, and D. Filmer, *Biochemistry* **5**, 365 (1966).
[7] E. Whitehead, *Prog. Biophys. Mol. Biol.* **21**, 321 (1970).
[8] G. R. Ainslie, Jr., J. P. Shill, and K. E. Neet, *J. Biol. Chem.* **247**, 7088 (1972).

tive binding of one macromolecule to another, an example of which is the cooperative binding of certain proteins to DNA. Many of the same considerations apply to this situation, but additional complexities exist because of the interactions that may occur among the protein molecules (i.e., the ligands) or that may occur within the DNA molecule itself. Theoretical analysis of this special case has been made,[9] and interpretation of gene-32 protein of bacteriophage T4 binding to DNA has been presented.[10] Since cooperativity of these systems appears to be of more limited interest than the cooperativity seen in enzymes, we will be concerned in this chapter with small ligand binding to proteins and enzymes.

The question arises as to how common "cooperativity" is among known enzymes or, conversely, what proportion of enzymes deviate from the classical Michaelis–Menten kinetics. This question has been answered in an interesting fashion by a survey of all kinetic studies between 1965 and 1976 reported in the literature.[11] Hill et al.[11] estimate that a minimum of 22% of all *known* enzymes deviate from simple Michaelis–Menten kinetics, about 54% of these (or a total of 420) showing negative cooperativity, positive cooperativity, or a complex, higher degree of mixed cooperativity. The remaining enzymes with non-Michaelis–Menten behavior exhibit substrate inhibition. Since these percentages are minimum estimates and will increase as more enzymes are carefully studied, it is clear that the phenomenon of cooperativity of enzymes is a widespread and important property of enzymes that has mushroomed in visibility and significance in the last 13–15 years.

In this review we do not deal with the myriad of experimental details of cooperativity in individual enzymes, other than to use examples to illustrate points that are being made. Rather, we deal with the practical aspects of analyzing and interpreting data, namely, various ways of measuring and quantitating cooperativity, statistical fitting and evaluation of data, and interpretation of such data in terms of simple and complex models that have been proposed to explain the cooperativity of enzymes.

Measures of Cooperativity

Cooperativity of an enzyme system can be measured from data obtained either from kinetic experiments or from equilibrium binding experiments. Differences in these types of experiments lead to interpretations of mechanisms of cooperativity and will be discussed more fully in a succeeding section. For the present purposes of analyzing the type and ex-

[9] J. D. McGhee and P. H. von Hippel, *J. Mol. Biol.* **86,** 469 (1974).
[10] D. E. Jensen, R. C. Kelly, and P. H. von Hippel, *J. Biol. Chem.* **251,** 7215 (1976).
[11] C. M. Hill, R. D. Waight, and W. G. Bardsley, *Mol. Cell. Biochem.* **15,** 173 (1977).

tent of cooperativity, either relative velocities, $v/V_{max}$, or fractional saturation, $\bar{Y}$ (or the corresponding absolute values of velocity, $v$, or moles of ligand bound, $\nu$) may be used to characterize the system. In the simplest case of rapid-equilibrium binding of all substrates with a pure $K$ system (as defined by Monod, Wyman, and Changeux,[5] in which there is no change in velocities of catalytic sites), the relative velocity is proportional to the fractional saturation. For this section on how cooperativity can be measured and analyzed without invoking a model, we use the velocity or relative velocity, but the results equally apply to the saturation function except where explicitly indicated.

*Methods for Estimating Cooperativity*

*Double-Reciprocal Plots.* Since the Michaelis–Menten assumption of an independent catalytic site leads to a linear double-reciprocal plot, deviation from linearity of such plots is indicative of cooperativity of some kind. (The sole exception to this statement is substrate inhibition, which produces a minimum in the double-reciprocal plot because of decreasing velocities at high substrate concentrations. Such behavior would not be observed in equilibrium binding measurements; we shall not consider those cases of substrate inhibition further in this chapter.) It is interesting to note that Lineweaver and Burk,[12] with brilliant insight in their original 1934 paper, also considered the possibility of complicated mechanisms leading to nonlinear plots and anticipated the development of the concept of allosteric cooperativity by some 30 years.

Positive cooperativity is manifested by double-reciprocal plots that are concave upward; negative cooperativity, by plots that are concave downward (Fig. 1). The extent of curvature is a measure of the degree of cooperativity, and two ways have been suggested to measure the extent of curvature of the double-reciprocal plots.

1. The curvature may be measured as the limit of the second derivative of the $v^{-1}$ vs. $A^{-1}$ plot as $A$ approaches infinity[13,14] ($A^{-1} \to 0$) [Eq. (1)]. A positive sign of the second derivative

$$\Gamma = \lim_{A \to \infty} \frac{\partial^2(v^{-1})}{\partial(A^{-1})^2} \qquad (1)$$

(the slope is increasing) indicates the upward curvature of positive cooperativity whereas, conversely, a negative sign (the slope is de-

---

[12] H. Lineweaver and D. Burk, *J. Am. Chem. Soc.* **56**, 568 (1934).
[13] K. Dalziel and P. C. Engel, *FEBS Lett.* **1**, 349 (1968).
[14] J. Ricard, J.-C. Meunier, and J. Buc, *Eur. J. Biochem.* **49**, 195 (1974).

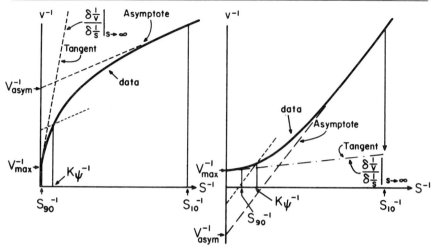

FIG. 1. Illustration of analysis of negative (left) or positive (right) cooperativity from double-reciprocal plots by change of slope ($\Gamma$), ratio of extrapolated velocities ($R_v$), and ratio of substrate concentrations ($R_s$). See Table I. Note that the illustration of the tangent uses the first derivative at $S \to \infty$ whereas the second derivative (i.e., the rate of change of the slope) is needed to calculate $\Gamma$.

creasing) indicates the downward curvature of negative cooperativity (Fig. 1). The treatment using slopes of the derivative curve is analogous to the utilization of the change in slope of $v$ vs. $A$ plots for analysis of complex binding curves.[15]

2. Ainslie et al.[8] used the ratio $R_v$ of the true maximal velocity of the extrapolated from the asymptote to the low-substrate portion of the curve (Fig. 1). A ratio greater than unity indicates the upward curvature of positive cooperativity, and a ratio less than unity indicates the downward curvature of negative cooperativity.

Both these parameters from double-reciprocal plots are useful in determining the extent of cooperativity from the rate equation and rate constants,[8,14] but each has limitations when cooperativity is estimated from experimental data. Obtaining data at high substrate concentration and estimating the rate of change of the slope to calculate $\Gamma$ is difficult and imprecise; two cases frequently occur in which $\Gamma$ has little practical meaning. One is with strong positive cooperativity in which the high substrate concentration region appears flat and there is no change in slope on the double-reciprocal plot. The second is with negative cooperativity such that the double-reciprocal plot appears as two linear segments, and again

---

[15] J. Teipel and D. E. Koshland, Jr., *Biochemistry* **8**, 4656 (1969).

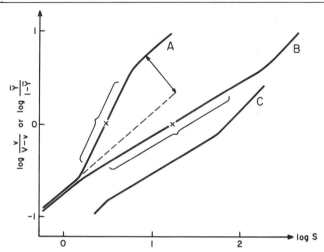

FIG. 2. Possible Hill plots of experimental data. Curve A, positive cooperativity ($n_H \cong 2$). Curves B and C, negative cooperativity ($n_H \cong 0.5$). Curves A and B show symmetrical Hill plots with the maximum or minimum value of the Hill coefficient occurring at the half-saturation position (marked by X). The bracket denotes the region in which experimental data often appear linear. Curve C shows a nonsymmetrical case in which the minimum value of $n_H$ occurs below half saturation. The double arrow indicates where the interaction energy can be measured.

no change in slope will occur at the highest concentrations. In this latter case of two linear segments, the estimation of $R_v$ is more accurate and useful in estimating the cooperativity. However, this method fails in more complicated cases in which the double-reciprocal plot does show continual curvature and estimation of the asymptote is inaccurate.

*Hill Plot and the Hill Coefficient, $n_H$.* The most commonly used expression of cooperativity, and the one most easily understood in terms of ligand binding and interaction, is the Hill[16] coefficient, determined from the slope of a Hill plot of log $v/(V_{max} - v)$ [or log $\bar{Y}/(1 - \bar{Y})$] vs. log $A$ (Fig. 2). The same formal equation has been used in other areas for similar purposes but with different names; Sipp, logit, and Nernst. For the simplest binding situations the Hill plot will have a slope of unity at each extreme end, assuming that the data have been obtained over a wide enough range, because these ends represent interaction of the enzyme with only the first (low concentration) or the last (high concentration) ligand. In theory the maximum Hill coefficient should be obtained at half-saturation by ligand, but in practice the Hill plot is often linear over a range (10–90% saturation) of ligand concentration, and this average slope

[16] A. V. Hill, *J. Physiol.* (*London*) **40**, iv (1910).

is taken as the value for $n_H$. For more complex situations involving mixtures of cooperativity or involving kinetic parameters (see later section) these ideal properties of the Hill plot will not be followed (Fig. 2); however, it is still a useful means of analysis.

One problem in determining Hill coefficients is the estimation of the maximal velocity or $v_{max}$ needed for the plot. $V_{max}$ is best obtained by initially plotting the data in the form of a double-reciprocal plot or a Scatchard plot and extrapolating the data to infinite substrate concentration. For simple positive cooperativity a good estimate of $V_{max}$ can usually be made. Negative cooperativity makes this estimation more difficult, since the experimental concentrations of substrate are usually well below saturation. The slope of the Hill plot is relatively insensitive to the choice of $V_{max}$ if the substrate concentration at which the minimum Hill coefficient occurs gives velocities much less than $V_{max}$.[17] Therefore, estimates of the $n_H$ in the low substrate concentration region can still be made for cases of extreme negative cooperativity. This analysis is equivalent to a direct plot of log $v$ vs. log $A$ and determination of the slope at low ligand concentrations.[18] Alternatively, the problem of determining $V_{max}$ can be resolved by assuming that the Hill plot will be linear and utilizing a computer program with all three parameters, $n_H$ (the slope), $V_{max}$, and $S_{0.5}$ (the substrate concentration for half-saturation) varied to yield the best statistical fit to the data.[19] However, this latter method will not work for cases, now commonly observed, in which the data are not described by the simplest form of the Hill equation, i.e., linear and symmetrical around the half-saturation point.

The Hill coefficient was originally derived[16] for a binding reaction with positive cooperativity of the following form:

$$P + nL = PL_n \qquad (2)$$

where P is the macromolecule, L is the ligand under consideration, $PL_n$ is the liganded complex, $n$ is the stoichiometry of binding in the complex.

Since $n$ is the stoichiometric equivalent of ligand bound and is the exponent of the equilibrium expression, the process as written implies an $(n + 1)$-body collision with no intermediates (i.e., $PL_{n-1}$, $PL_{n-2}$, etc.) formed. Such a situation would occur if the energy of interaction between binding sites was infinitely large so that binding of the first molecule of ligand made the succeeding affinities for ligand exceedingly large. In most physically realistic cases of positive cooperativity this is not the case, and the apparent value of the Hill coefficient, $n_H$, is somewhat smaller than

---

[17] G. R. Ainslie, Jr., unpublished observation.
[18] C. J. Thompson and I. M. Klotz, *Arch. Biochem. Biophys.* **147**, 178 (1971).
[19] G. L. Atkins, *Eur. J. Biochem.* **33**, 175 (1973).

the true number of ligand molecules bound. For example, the binding of four molecules of $O_2$ per tetramer of hemoglobin[20] leads to a Hill coefficient of about 2.8. The value of $n_H$ is now recognized to represent both the number of ligand molecules in the interacting unit and the degree of interaction between them. The closer the value of $n_H$ comes to the value of $n$, the greater the energy of interaction between the sites and the fewer intermediates of partially liganded species exist. For negative cooperativity the situation is different since the value of $n_H$ is always less than unity and is not directly related to the number of sites; indeed, the negative energy of interaction between sites implies that Eq. (2) could not possibly hold and that intermediate species must in fact predominate in this case. Nevertheless, the mathematical relationship

$$\bar{Y} = L^{n_H}/(K + L^{n_H}) \tag{3}$$

[Eq. (3)] still formally holds, and the determination of $n_H$ is also useful in assessing the degree of negative cooperativity.

Wyman[21,22] has shown that another useful quantity can be obtained from Hill plots in certain cases where reliable data are obtained at very low and at very high saturation. Since the slope of the Hill plot is unity at high and low ligand saturation, for simple cooperativity symmetrical around the midpoint the interaction energy between sites can be determined from the perpendicular distance between these asymptotes multiplied by $RT\sqrt{2}$. For a system with identical sites Wyman[22] has interpreted this parameter to represent the average value of the free energy of interaction between the sites during the process of ligand saturation; however, it has been argued that the perpendicular between the asymptotes in fact gives the difference between the free energies of interaction on binding the first and the last ligands and that the average value of the free energy of interaction can be obtained only by assuming a model.[23] A useful improvement in estimation of this energy parameter would be the conversion to a plot of log $\bar{Y}/(1 - Y)(1/L)$ vs. log $L$, which makes the slopes at high and low ligand concentration zero and therefore easier to assess.[24]

The other useful parameter that is probably best estimated from Hill plots is the substrate concentration required for half-maximal saturation, $S_{0.5}$. If there is no cooperativity, the half-saturation point is equivalent to the Michaelis constant, $K_m$, but is more properly termed $S_{0.5}$ in cooperative systems. If simple, symmetric cooperativity exists, this parameter

---

[20] A. Rossi-Fanelli, E. Antonini, and A. Caputo, *J. Biol. Chem.* **236**, 297 (1961).
[21] J. Wyman, *Adv. Prot. Chem.* **19**, 223 (1964).
[22] J. Wyman, *J. Am. Chem. Soc.* **89**, 2202 (1967).
[23] H. A. Saroff and A. P. Minton, *Science* **175**, 1253 (1972).
[24] H. Watari and Y. Isogai, *Biochem. Biophys. Res. Commun.* **69**, 15 (1976).

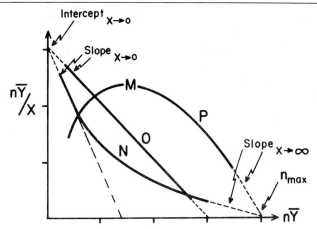

FIG. 3. Examples of Scatchard plots; line $O$, linear (no cooperativity); line $P$, positive cooperativity; line $N$, negative cooperativity or heterogeneity of binding sites. The point $M$ represents the position of the maximum that can be used to calculate the Hill coefficient (Table I). The dashed line from line $N$ extrapolates to a *minimum* estimate of the number of high-affinity sites. The quantity $(n\bar{Y})$ is the experimentally determined moles of ligand bound per mole of protein; $X$ is the concentration of *free* ligand.

corresponds to the average dissociation constant of ligand from protein, but in any case represents the true midpoint of the saturation curve.

$R_s$. The ratio of the substrate concentration required for 90% saturation to the substrate concentration required for 10% saturation has been used by Koshland and co-workers[6,25] as a measure of the cooperativity. Any saturation values could be used, but 10% and 90% were chosen for convenience. For a normal hyperbolic saturation this ratio, $R_s$, is equal to 81, so that any steeper saturation curve for positive cooperativity would have a ratio less than 81, whereas any broad saturation curve indicative of negative cooperativity would have an $R_s$ value larger than 81 (Fig. 1). This ratio is probably most useful in directly indicating the responsiveness of an enzyme to the range of concentrations that it may face in any physiological situation. Furthermore, no mechanism is implied in its utilization. The substrate concentrations needed for estimating the value of $R_s$ can be obtained from virtually any type of plot of the data, including $v$ vs. log $A$, as long as the data span a wide enough range. Obviously the ratio is heavily weighted by the two experimental points at 10 and 90% saturation and is insensitive to the shape of the saturation curve or the precision and accuracy of the experimental data in the intermediate concentration ranges.

[25] M. E. Kirtley and D. E. Koshland, Jr., *J. Biol. Chem.* **242**, 4192 (1967).

*Scatchard Plot.* The Scatchard[26] (or Hofstee) plot of $v/A$ vs. $v$ (or $\bar{Y}/A$ vs. $\bar{Y}$ for binding data) also conveniently describes cooperativity as deviations from linearity; a maximum in the plot indicates positive cooperativity and a curvature in the opposite direction (concave upward) occurs with negative cooperativity (Fig. 3). The extent of positive cooperativity can be estimated from the position of the maximum in the curve.[27] The relationship between the fractional saturation at the maximum, $\bar{Y}_m$, and the Hill coefficient is given by $n_H = (1 - \bar{Y}_m)^{-1}$. Although this method has not yet been extensively used and requires good data around the maximum, it might be useful in some situations. No quantitative estimate of the extent of negative cooperativity can be made.

*Other.* Other suggested measures of cooperativity are the use of three velocities at three substrate concentrations to estimate the shape of the saturation curve[28] and the use of the second moment of the derivative of the saturation function.[29] Neither of these has yet received widespread use.

*Comparison of Methods and Application to a Second-Order Equation*

The five methods of determining the degree of cooperativity are summarized in Table I with the indication of how they vary with negative, no, or positive cooperativity. In addition, the simple second-order saturation equations (4) and (4a) (for binding or rate) are taken as an example, and each cooperativity parameter is calculated in terms of the coefficients of the equation.

$$v = \frac{cA^2 + dA}{1 + aA^2 + bA} \tag{4}$$

$$v^{-1} = \frac{a + b(A)^{-1} + (A)^{-2}}{c + d(A)^{-1}} \tag{4a}$$

Equations (4) and (4a) are applicable to a dimer binding two substrate molecules cooperatively, with the coefficients defined by the particular model chosen; or to a more complex model including changes in kinetic constants; or to a purely kinetic model (see later section). Since $V_{max} = c/a$, Eqs. (4) and (4a) may be rewritten as fractional saturation functions ($v/V_{max}$) or as the fractional binding function ($\bar{Y}$), with the explicit assumption, for the moment, that these expressions are equivalent.

[26] G. Scatchard, *Ann. N. Y. Acad. Sci.* **51**, 660 (1949).
[27] F. W. Dahlquist, *FEBS Lett.* **49**, 267 (1974).
[28] B. I. Kurganov, *Acta Biol. Med. Germ.* **31**, 181 (1973).
[29] T. Sturgill and R. L. Biltonen, *Biopolymers* **15**, 337 (1976).

TABLE I
PARAMETERS USED TO MEASURE COOPERATIVITY AND THEIR INTERRELATIONSHIP FOR A SECOND-ORDER RATE EQUATION

| Parameter measured | Cooperativity | | | Definition of parameter | Parameter in terms of coefficients of Eq. (5) |
|---|---|---|---|---|---|
| | Negative | None | Positive | | |
| $n_H$ | $<1$ | 1 | $>1$ | $\dfrac{d[\log(v/(V_{max}-v))]}{d\log A}$ | $\dfrac{2}{1+\sqrt{\chi}}$ |
| $R_s\,(=81^{1/n_H})$ | $>81$ | 81 | $<81$ | $\dfrac{[A]v = 0.9V_{max}}{[A]v = 0.1V_{max}}$ | $9^{(1+\sqrt{\chi})}$ |
| $R_v$ | $>1$ | 1 | $<1$ | $V_{true}/V_{asym}$ | $1+\dfrac{a}{e^2}(\chi-1)$ |
| $\Gamma$ | $<0$ | 0 | $>0$ | $\lim_{s\to\infty}\dfrac{d^2(V_{max}/v)}{d(A^{-1})^2}$ | $\dfrac{2}{a}(1-\chi)$ |
| $\bar{Y}$ Scatchard max $(=1-1/n_H)$ | Not defined | No max | Max exists | $v$ at $\left(\dfrac{v}{AV_m}\right)_{max}$ | $\dfrac{1-\sqrt{\chi}}{2}$ |
| $\chi$ | $>1$ | 1 | $<1$ | By definition | $\dfrac{e}{a}(b-e)$ |

$$\bar{Y} = \frac{v}{V_{\max}} = \frac{aA^2 + eA}{1 + aA^2 + bA} \tag{5}$$

$$\bar{Y}^{-1} = \frac{V_{\max}}{v} = \frac{a + b(A)^{-1} + (A)^{-2}}{a + e(A)^{-1}} \tag{5a}$$

where $e = (ad)/c$; three constants, $a$, $b$, and $e$, describe the function.

Each of the ways of determining the extent of cooperativity is not independent, but inspection reveals that all five are interrelated by the parameter defined as $\chi$ in Table I. $R_s$ and $\bar{Y}_m$ are directly related to $n_H$, and so interconversions between these can be made if desirable for interpretation or utilization of the parameters. $R_v$ and $\Gamma$ are less directly related to $n_H$, since one or more coefficients of the equation are required in addition to the $\chi$ term. $\chi$ appears to be the most fundamental relationship between coefficients for the dimer, since it is the most simply related to the three coefficients of the second-order equation. For higher-order saturation functions (e.g., a tetramer) the measures of cooperativity in the left-hand column of Table I are still appropriate but are not simply related to the coefficients of the second-order equation.

## Evaluation of Methods of Estimating Cooperativity

*From Experimental Data.* In order to make a reliable estimate of the cooperativity, precise data must be obtained over a wide range of ligand or substrate concentration with no systematic error, particularly at either end of the curve. Often these criteria are not met in data in the literature. With such data available the most obvious demonstration of cooperativity comes from nonlinearity of the double-reciprocal plot or of the Scatchard plot. The Scatchard plot emphasizes curvature at low substrate concentrations whereas the double-reciprocal plot can be used to examine curvature at the higher concentrations. Straightforward plots of $v$ (or $\bar{Y}$) vs. $A$ or vs. log $A$ are sometimes useful in assessing the presence of more complex types of cooperativity (see later section) yielding "bumpy" curves or curves with sharp transitions. Data from cooperative systems need to be plotted in several ways for proper evaluation of the system, and no single method should be relied upon too strongly.

Once the presence of cooperativity has been determined, the utilization of the Hill plot for quantitative assessment is most appropriate. The Hill coefficient as the measure of cooperativity has the advantage that it is the most widely used (and hence the best means of communication) and is also most easily interpreted within the concept of a model, i.e., the relationship (discussed above) between the number of ligand binding sites and the experimental value of $n_H$. Although originally derived to describe a system of the form of Eq. (2), the Hill plot and coefficient is now most frequently used to ascertain the degree of cooperativity rather than to imply a mechanism. In contrast, $\Gamma$ and $R_v$ have no limits and no absolute scale so that the numerical value of either of these parameters, although a perfectly adequate measure of cooperativity, does not present an easily understood concept of a macromolecular model. The Hill coefficient uses a wide range of ligand concentrations for evaluation and therefore is more representative of the data as a whole. Furthermore, for complex mixtures of negative and positive cooperativity (see later) the Hill plot (as opposed to the coefficient) is often the only way of qualitatively assessing what kinds of cooperativity are present. The main disadvantage of the Hill coefficient as the measure of the cooperativity is the relative lack of sensitivity of the parameter to changes in the system. A change of 0.1 to 0.2 in the value of $n_H$ theoretically represents a marked change in cooperativity, but in terms of numerical evaluation it is barely within experimental error in many instances. This reservation is particularly true near $n_H = 1$; i.e., the Hill plot is not very satisfactory as a first means of detecting the presence of cooperativity.

*From Rate or Equilibrium Binding Equations of a Model.* The alternative way to evaluate the cooperativity of a system is to derive the rate equation or the corresponding equilibrium binding expression for a particular mechanism and then to either simulate data based upon assumed constants of the equation or to analytically determine the value of one of the measures of cooperativity, such as $n_H$. The Hill coefficient (at a given $A$ or for the maximal $n_H$) can readily be determined from the expression $n_H = d[\log v/(V_{max} - v)]/d \log A$ for nearly any common saturation expression. Alternatively, the coefficients of the rate equation can be directly related to the Hill coefficient, as was done in Table I for the dimer case. With this information the degree of cooperativity can, in principle, be assessed, but there are several precautions that need to be taken. The cooperativity obtained in this fashion from the rate equation is in fact the maximum cooperativity that is attained at a single given concentration of substrate, whereas the cooperativity measured experimentally is generally an average value over a wide range of concentrations, i.e., the linear portion of the Hill plot. Therefore, the predictions of possible cooperativity from a particular mechanism are not necessarily valid unless more careful analyses or simulations are made.

The first consideration is the ligand concentration at which maximal cooperativity occurs, herein called $S_{n\,max}$. Maximal cooperativity refers either to the highest possible value of $n_H$ in a positively cooperative system or to the lowest possible value of $n_H$ in a negatively cooperative system. For simple, symmetric cooperativity (positive or negative) $S_{n\,max}$

corresponds to $S_{0.5}$ and is often where the Hill coefficient is calculated from experimental data. For any more complex system in which kinetic constants play a role or in which there are mixtures of positive or negative cooperativity or in which there are different degrees of positive cooperativity (or negative) along the saturation curve, $S_{n\,max}$ will not in general coincide with $S_{0.5}$ (cf. Fig. 2). For some mechanisms the maximal cooperativity may occur at very high ligand concentrations, which may not be accessible experimentally. For other mechanisms the $S_{n\,max}$ may occur at very low ligand concentrations, where large experimental errors may obscure the degree of cooperativity. In these cases the maximal cooperativity as judged analytically from the rate or binding equation is misleading and does not necessarily represent what may be expected experimentally. In other, complex instances the region of maximum Hill coefficient may occur over a very narrow range of substrate concentration, i.e., a sharp transition in the plot, and the apparent maximum Hill coefficient from the rate equation will not be observed experimentally. These problems will be illustrated by analyzing the cooperativity of the second-order equation [Eq. (5)] and where it occurs.

Several definitions and parameters, implicit in Eq. (5), are useful to examine and analyze the cooperativity of a system without consideration of the underlying mechanism generating the second-order saturation equation.

$$K_\psi = e/a \qquad V_{asym} = ed/(eb - a) \qquad V_{max}(\text{true}) = 1 = c/a$$
$$\text{Slope asym} = 1/e \qquad\qquad\qquad\qquad v_{n\,max} = e/b$$
$$S_{0.5} = \frac{-(2e - b) \pm [(2e - b)^2 + 4a]^{1/2}}{2a} \qquad S_{n\,max} = [e/a(b - e)]^{1/2} \tag{6}$$

where $K_\psi$ is the $S$ coordinate on the double-reciprocal plot of the intersection point between the experimental data and a line drawn parallel to the asymptote with an intercept halfway between the asymptote and the true intercepts (Fig. 1); $S_{0.5}$ is the substrate concentration at half-maximal saturation; and $v_{n\,max}$ and $S_{n\,max}$ are the velocity and substrate concentration where the value of $n_H$ is maximal. The value of $K_\psi$ will always be smaller than $S_{n\,max}$ for positive cooperativity and larger than $S_{n\,max}$ for negative cooperativity, as is apparent from the shape of the double-reciprocal graphs. The questions of interest are: How does the value of $S_{0.5}$ compare to that of $S_{n\,max}$? and What is the experimentally realizable value for $n_{H\,max}$? It can be shown that $S_{0.5}$ will not be equal to $S_{n\,max}$ unless there is no cooperativity ($n_H = 1$; $\chi = 0$), unless $b = 2e$, or unless $b = e = 0$. In general, $n_H$ will not achieve its maximum value of 2 but will differ by a factor related to how different $\chi$ is from zero (see Table I). As $n_{H\,max}$ approaches 2, $v_{n\,max}$ approaches $V_{max}$ and $S_{n\,max}$ approaches infinity; i.e., the maximum cooperativity of 2 will be achieved only in a greatly asymmetric situation at saturating values of substrate, near the maximal velocity. Two exceptions exist to this general rule.

1. When $b = 2e$, as is the case for at least two of the major cooperativity models extant (see later section), $v_{n\,\text{max}}/V_{\text{max}}$ will be equal to 0.5 and $S_{0.5} = a^{-1/2}$, but the value of the maximal Hill coefficient will be reduced below 2 by a factor equal to $\sqrt{a}/(e + \sqrt{a})$. When $e \ll a$ the Hill coefficient will be near 2; this is equivalent to the second association constant for ligand being much greater than the first.
2. A special case of condition 1 is when $b = e = 0$. The form of Eq. (5) then reduces to that of the Hill equation [Eq. (3)], $v_{n\,\text{max}}/V_{\text{max}}$ remains at 0.5, $S_{0.5} = S_{n\,\text{max}} = a^{-1/2}$, and the value of 2 for $n_{\text{H max}}$ can be attained.

These examples serve to demonstrate the dangers of assuming that a certain mechanism will give large degrees of cooperativity based upon some of the parameters listed in Table I. Other considerations must be made, and the mechanism-dependent relationship between the various coefficients must be examined. Equations of higher degree than two would presumably yield similar analyses, because of the $j - 1$ constraints needed for $j$th order equation to describe a hyperbola; however, they are too difficult to analyze in this simplistic fashion.

Evaluation of Constants of Saturation Functions

*Graphical Methods.*

We have been considering the evaluation of the cooperativity of a ligand binding system through the utilization of the parameters of Table I, but an alternative approach is to evaluate the coefficients of Eq. (4) and assess the cooperativity from their relative values. For the second-order equation [Eq. (4)] this is readily accomplished from the *double-reciprocal plot*. From the definitions of $K_\psi$, $V_{\text{max}}$, $V_{\text{asym}}$, and slope$_{\text{asym}}$ [Eq. (6)] the values of the coefficients $a$, $b$, and $e$ may be determined graphically from the double-reciprocal plot (Fig. 1). These coefficients may then be used to estimate one of the usual cooperativity parameters or, more usefully, may be related to the parameters of a model for cooperativity (see Table II). For higher-order equations, which might still appear to have the same shape on a double-reciprocal or Hill plot, the exact coefficients are not directly determinable from simple plots. One common practice with negative cooperativity is to estimate a "high $K_m$" and a "low $K_m$" from the apparently limiting slopes and intercepts of the high and low substrate con-

TABLE II
RELATIONSHIP BETWEEN GENERAL COEFFICIENTS OF THE SECOND-ORDER EQUATION AND THE PARAMETERS OF SPECIFIC MODELS

| Model | Coefficient of Eq. (5) | | |
|---|---|---|---|
| | $a$ | $b$ | $e$ |
| Adair | $K_1 K_2$ | $K_1$ | $\frac{1}{2}K_1$ |
| Sequential | $K_{tB}^2 K_{XB}^2 K_{BB}$ | $K_{tB} K_{XB} K_{AB}$ | $\frac{1}{2} K_{tB} K_{XB} K_{AB}$ |
| Concerted | | | |
| Exclusive binding | $(1 + L)^{-1} K_R^{-2}$ | $2(1 + L)^{-1} K_R^{-1}$ | $(1 + L)^{-1} K_R^{-1}$ |
| Nonexclusive binding | $(1 + Lc^2)(1 + L)^{-1} K_R^{-2}$ | $2(1 + Lc)(1 + L)^{-1} K_R^{-1}$ | $(1 + Lc)(1 + L)^{-1} K_R^{-1}$ |
| Protomer–oligomer Nonexclusive binding | $2K_E E c^2 (1 + 2K_E E)^{-1} K_1^{-2}$ | $(1 + 4cK_E E)(1 + 2K_E E)^{-1} K_1^{-1}$ | $(1 + 2cK_E E)(1 + 2K_E E)^{-1} K_1^{-1}$ |

[a] The constants are defined in Figs. 6, 7, and 8.

centration regions of the double-reciprocal plots. This procedure is only an approximation and is fraught with potential misinterpretations; it might be useful for comparing similar conditions but should not be applied as a determination of real constants of a system. The "high $K_m$" obtained in this fashion may or may not correspond to the $S_{0.5}$ determined from the Hill plot, depending upon the degree of curvature in the double-reciprocal plot (a longer, linear, high-concentration section will cause the "high $K_m$" to approach the more valid $S_{0.5}$).

The *Scatchard plot* ($n\bar{Y}/X$ vs. $n\bar{Y}$) (Fig. 3) is widely used to graphically evaluate many of the constants in a binding or rate equation, but for cooperative or heterogeneous systems the slopes and intercepts of the plot are complex combinations of the constants of the saturation equation.[30–32] The intercept on the $n\bar{Y}$ axis at high ligand concentration gives the total number of sites (or $V_{max}$ if velocity data are used) but is often difficult to evaluate because of the continual curvature in heterogeneous or negatively cooperative systems (see Klotz[33] for an excellent discussion of the hazards of making this extrapolation). The extrapolation of the high-affinity region (dashed line of Fig. 3) depends upon the cooperativity and gives only a minimum value for the number of high-affinity sites.[32] The intercept on the $n\bar{Y}/X$ axis is $K_1$, the first stoichiometric constant, which is equivalent to the sum of all site binding constants[30,32] (see Klotz[33] for the comparison of stoichiometric constants, used here, and site binding constants). The limiting slope at low free ligand concentrations has been shown to depend only upon the two highest affinity binding constants, whereas the limiting slope at high free ligand concentration is determined by the lowest affinity binding constant[31] (see Fig. 3 for definition of slopes and intercepts). In terms of the second-degree equation [Eq. (5)] these slopes and intercepts are given by Eq. (7).

$$\text{Slope}_{X\to 0} = 2a/b - b \qquad \text{Slope}_{X\to\infty} = 2a/b$$
$$\text{Intercept}_{X\to 0} = 2e \qquad \text{Intercept}_{X\to\infty} = 2 \qquad (7)$$

The interpretation of these slope equations in terms of the parameters of various cooperativity models has been presented[31,32] (see Table II). The application of the slope analysis of Scatchard plots to positive cooperativity is not as useful, since the slopes become very steep and difficult to determine accurately. For these positively cooperative systems, determination of the $S_{0.5}$ and the $n_H$ can be made from Hill plots and translated into some of the constants of the binding equation.[32] The reader is re-

---

[30] I. M. Klotz and D. L. Hunston, *Biochemistry* **10**, 3065 (1971).
[31] Y. I. Henis and A. Levitzki, *Eur. J. Biochem.* **71**, 529 (1976).
[32] F. W. Dahlquist, this series, Vol. 48, p. 270.
[33] I. M. Klotz, *Acc. Chem. Res.* **7**, 162 (1974).

ferred to the review by Dahlquist[32] for a detailed discussion of the utilization of the Hill and Scatchard plots for the analysis of binding equilibria. Thompson and Klotz[18] have presented and discussed the virtues of the log $X$ vs. log $n\bar{Y}$ plot for presentation of binding data and analysis of the binding constants.

*Statistical Fitting of Cooperative Binding Data*

The most appropriate way of analyzing binding or rate data is through a statistical fitting of the data to one (or more) models. The least squares method is now the standard analysis for linear (noncooperative) kinetics (see Volume 63A [6]) and is preferred for cooperativity. In the latter case the statistical methods have not yet been as widely utilized. For the complex, higher-order saturation functions obtained from cooperative systems, statistical methods are preferred because of the desirability of obtaining individual constants, rather than combinations as occurs with most graphical methods, and the objectivity of the process, as compared to graphical estimates. Computer methods are now available to rapidly and easily solve the matrix equations that are a result of the requirements of fitting second-degree and higher equations.

In general, the investigator must initially choose the model(s) that he desires to fit to the data, which may be a general binding equation of the Adair type or a specific model for a particular cooperative model (see later section). A nonlinear least-squares fit involves the minimization of the sum of the squares of experimental deviations from the fitted equation or the closely related statistical parameter $\chi^2$ in order to determine the goodness of fit.[34] In the fitting process it is useful to weight the data, based upon the precision of the experimental data. The weighting factor is usually taken as the inverse of the variance of the observed velocity or fractional binding based upon several measurements of each experimental point. Since negative values of binding constants are not physically realistic, it is convenient to perform the refinement procedure on the logarithm of the parameters or to disallow negative values and reinitiate the search procedure when a parameter goes negative. One problem with minimization procedures for complex functions is the settling of the solution into a local or relative minimum that is different from the true minimum. Some methods need reliable initial estimates of the starting values for the parameters to be refined,[35] whereas others have been designed to allow

[34] P. R. Bevington, "Data Reduction and Error Analysis for the Physical Sciences." McGraw-Hill, New York, 1969.
[35] P. J. Kasvinsky, N. B. Madsen, R. J. Fletterick, and J. Sygusch, *J. Biol. Chem.* **253**, 1290 (1978).

quite incorrect guesses of the starting values.[36] As the number of parameters increases in the fitting equation, i.e., higher-order rate equations, the destinction between different equations (mechanisms) or the necessity for the additional parameters becomes difficult to assess. Other considerations, such as the number of binding sites, must be used to eliminate alternative models. In addition, as more refinable parameters are used, the precision in the experimental data must also be sufficient to obtain a reliable solution, i.e., the minimal mechanism consistent with the data is desired. Conversely, data should be obtained in sufficient density, adequately spaced, and over a wide enough range (at least 25–75% saturation) to define the tested saturation function. Occasionally, the computer program will not converge, indicating unreliable data or a poor choice of mechanism. We will briefly present two fitting procedures here that have been used in the literature with success for analyzing saturation functions greater than second degree.

A sophisticated fitting process has recently been used by Kasvinsky *et al.*[35] and in our laboratory (unpublished), based upon the algorithm of Marquardt for a nonlinear least-squares analysis (see Bevington[34] for FORTRAN programs and a discussion of the fit). The Marquardt algorithm utilizes both an analytical search for the minimum when the solution is close to the minimum in $\chi^2$ and a gradient search when the points are far from the minimum in parameter space. The gradient search examines the slope of the $\chi^2$ function when all parameters are simultaneously varied, and increments the parameters in the direction of steepest descent of $\chi^2$; it is therefore useful when the solution is far from the minimum but is less useful when the points are immediately near the minimum in $\chi^2$. The analytical search, on the other hand, converges rapidly for points near the minimum by linearizing the fitting function through a first-order expansion. The Marquardt process defines a parameter, $\lambda$, which controls the interpolation between the extremes of the analytical or the gradient search by increasing the diagonal of the curvature matrix. The value of $\lambda$ is changed depending upon whether the value of $\chi^2$ is increasing (getting farther from the minimum) or decreasing (approaching the minimum) with successive iterations. The first derivative of the fitting function is required for this algorithm; it can be determined analytically if the function is easily differentiated or it can be numerically evaluated. The standard deviations of the parameters are calculated from the diagonals of the normal error matrix. The Marquardt algorithm, briefly described here, provides a more rapid and efficient method for minimization of nonlinear binding or kinetic functions than other least-squares processes[34,35] and is generally applicable to any functional type of saturation equation.

[36] A. Cornish-Bowden and D. E. Koshland, Jr., *Biochemistry* **9**, 3325 (1970).

Since the Marquardt algorithm requires linearly independent parameters, Kasvinsky et al.[35] have suggested that an examination of the eigenvalues of the normal equation matrix will allow a reduction of the saturation function being fit to one with the minimum number of necessary parameters. Thus, a zero value for an eigenvalue indicates a linear dependency between two parameters, and the mechanism should be reducible to a mechanism with one less refinable parameter. Alternatively, an $F$ test may be made between mechanisms differing by a single parameter.[34] Such analyses could provide a very powerful means of determining the minimal mechanism necessary to fit a particular set of data.

An additional advantage of the Marquardt process is that the refinement program examines all the data in parameter space and thus does not require data to be obtained on straight lines or smooth functions. This property is useful in experiments with inhibitors or activators which yield an array of experimental points at no fixed value or smooth variation of ligand concentration.

Cornish-Bowden and Koshland[36] have presented a fitting procedure for a fourth-order binding equation that assumes the error function to be parabolic in shape. Minimization is then achieved by a gradient search in four-parameter space, solving four simultaneous equations, and iterating all four parameters in the direction to minimize the $\chi^2$. The authors demonstrate that the algorithm approaches the same minimum even though the initial estimates of the parameters were varied over a 3000-fold range; thus this is a very useful program if no good guesses can be made concerning the binding constants of the system. From simulation studies it was also suggested that 15–20 experimental determinations spread over more than 50% of the middle of the saturation curve would be sufficient to yield adequate estimates of the binding constants for a tetrameric protein. A thorough analysis of the limitations and applications of this process to binding data has been presented.[36]

Once the parameters that provide the best fit to a particular model have been obtained, the applicability of the model may be assessed by the value of $\chi^2$, by a visual comparison of the fitted curve to the data to detect unwanted trends, and evaluation of the physical reality of the values of the parameters obtained. The standard deviations of the parameters are also obtained from the least-squares minimization procedures and may be used to assess how precisely the parameters are defined. An equivalent assessment may be made by fixing all but one of the parameters and then determining the variation in that parameter required to produce a particular change, e.g., 10%, in the dependent variable or in $\chi^2$. From the variation required one may obtain not only the absolute dependence upon particular parameters but also the relative sensitivity among the refined parameters.

If a general Adair-type equation was utilized for the fitting, the thermodynamic association constants obtained from the fit may be converted to intrinsic constants, $K_i'$ for the $i$th site with the appropriate statistical factors [Eq. (8)].

$$K_i = \frac{(n + 1 - i)}{i} K_i' \qquad (8)$$

The intrinsic constants may then be interpreted in terms of the models for cooperative binding (see later section). If a general rate equation [cf. Eq. (5)] was utilized for the fitting process, the coefficients may be interpreted in terms of the particular models (see Table II). If a specific rate equation was utilized, the fitted parameters will directly yield the dissociations constants, the catalytic velocities, and the interactions terms characteristic of the specific model chosen.[35]

Kinetic and Equilibrium Aspects of Cooperativity

The practical considerations of measurements of cooperativity in the preceding sections are independent of any model and pertain to all cases of equilibrium measurements of ligand binding and to kinetic data. In this section we will briefly consider those aspects of catalytic kinetics that can alter the apparent cooperativity or generate cooperativity when there is none in binding experiments. There are two basic ways in which kinetic cooperativity may differ from the cooperativity due solely to ligand binding: (a) alteration of catalytic constants as substrate binds with the maintenance of rapid equilibrium binding; and (b) deviations from rapid-equilibrium conditions. In the following section concerning specific equilibrium models, we will point out how kinetic considerations can alter the apparent cooperativity.

Before continuing further, a word must be said about the nomenclature of "cooperativity." Cooperativity, as originally defined, applied to multiple ligand binding to a macromolecule, e.g., hemoglobin, with an enhancement (positive) of *binding* of each subsequent ligand. The sigmoid binding curve and the *mechanism* of producing it were intimately connected. However, *kinetic* cooperativity can be generated in an enzymic system by other mechanisms and is indistinguishable in the kinetic curve. In principle then, when non-Michaelis–Menten behavior is observed in enzyme kinetics, no mechanistic interpretation can initially be made. The author feels that the term "cooperativity" should therefore be applied in a *phenomenological* sense, rather than in a mechanistic sense and should include all forms of deviation from hyperbolic saturation curves (with the exception of simple substrate inhibition, which is clearly different). Mechanistic interpretations may then be applied after more is known about the particular enzyme. This phenomenological usage of the term "cooperativity" now occurs quite generally among enzymologists and is most appropriate.

*Cooperativity Due Solely to Equilibrium Binding of Ligands (Multisite Single-Substrate Enzymes).* The cooperativity observed in initial-rate kinetic studies will be identical to, and the direct result of, the cooperativity of binding when the following two conditions hold.

$$v = k \sum_i^n (\text{ES})_i \tag{9}$$

$$\bar{Y} = v/V_{\max} \tag{10}$$

where the catalytic rate constant, $k$, is the same for all enzyme complexes and $(\text{ES})_i$ represents all forms of enzyme–substrate complexes. These two conditions state that the rate of breakdown of each enzyme–substrate complex is the same regardless of other liganded positions on the enzyme oligomer, and that the saturation of binding sites, $\bar{Y}$, is directly proportional to the fraction of maximal velocity achieved. Each of these two conditions may be invalid in any real case. These restrictions are assumed to hold in the simplest version of cooperative theories when equilibrium binding models (next section) are directly applied to kinetic results.

When the saturation function or rate equation is written for a multisite enzyme, higher power terms of substrate concentration occur, equaling the number of substrate binding sites (or more accurately the number of substrate addition steps in getting to any one enzyme–substrate complex). This fact would suggest that any multisite enzyme should show cooperativity. However, careful examination of the restrictions placed on the enzyme by the independence of sites shows that the derived rate equation can be factored and reduced to a first order expression. Consider a dimer that can bind one molecule of substrate, A, at each of two catalytic sites. With no assumptions about the interactions between the sites, the generalized rate equation may be written:

$$v = \frac{(k_1 K_1 \text{A} + k_2 K_1 K_2 \text{A}^2)}{1 + K_1 \text{A} + K_1 K_2 \text{A}^2} [\text{E}]_t \tag{11}$$

where $k_1$ and $k_2$ are the overall rate constants for breakdown of EA and $\text{EA}_2$, respectively; $K_1$ and $K_2$ are the thermodynamic association constants (or reciprocal Michaelis constants) for formation of EA from E and A, and of $\text{EA}_2$ from EA and A, respectively; and $[\text{E}]_t$ is the total enzyme concentration. Since for independent sites $k_2 = 2k_1$ and $K_1 = 4K_2 = 2K'$ [from statistical factors, Eq. (8)], the numerator and the denominator of Eq. (11) will have a common factor, $(1 + K'\text{A})$, and the rate expression becomes first degree [Eq. (12)].

$$v = 2[E]_t \frac{k_1 K' A}{1 + K' A} \tag{12}$$

The Michaelis–Menten nature of this equation and the fact that $2[E]_t$ is the total number of active sites is obvious. The proof for more than two sites is laborious but has been rigorously given for the multisite case.[37]

*Subunit Interactions Affect Rate Constants of Subsequent Substrate Molecules (Single-Substrate Enzymes).* Since in cooperative models it is assumed that the binding of one ligand to an oligomer can affect the binding of subsequent ligands to the same protein, it is reasonable that the catalytic rate could also be affected when substrate is bound to other sites on the enzyme. This effect can occur even when rapid equilibrium binding of the substrate occurs. Equation (9) is no longer valid and must be replaced by a more general expression.

$$v = \sum_{i}^{n} k_i(ES_i) \qquad \text{where} \qquad k_i \neq k_{i+1} \tag{13}$$

This consideration is a simple extension of each equilibrium model, and we will present the corresponding rate equation, when appropriate. Kinetic curves will deviate from binding curves for the same ligand and more complex possibilities for cooperativity will arise. In other words, the coefficients of Eq. (5) will no longer be simply related, as in certain equilibrium models, but will be independent of each other. Nucleoside diphosphatase[38] catalyzes a reaction in which the ternary complex ($ES_2$), breaks down about six times faster than the binary complex (ES).

*A Single Active Site and a Single Allosteric Site for the Same Substrate (Multisite, Single-Substrate Enzyme).* If the substrate binds at an allosteric site in addition to the catalytic site, the kinetic cooperativity will differ from the binding cooperativity. Enzyme molecules with S bound at both sites (ES, SES, or SE) will be measured by equilibrium methods (in $\bar{Y}$), but kinetic measurements of activity ($v/V_{max}$) will respond only to those complexes with S at the active site (ES or SES but not SE). Differences will also occur depending upon whether or not S is an essential activator at the allosteric site (i.e., whether ES is active). The situation will be further complicated if there are also subunit interactions in an oligomer. Frieden[39] has discussed the kinetic consequences of this type of substrate modifier and the conditions that lead to equivalence between steady-state and rapid-equilibrium assumptions. Kinetic studies cannot

---

[37] J. M. Reiner, "Behavior of Enzyme Systems," 2nd ed., p. 69. Van-Nostrand-Reinhold, Princeton, New Jersey, 1969.
[38] V. Schramm and J. F. Morrison, *Biochemistry* **10**, 2272 (1971).
[39] C. Frieden, *J. Biol. Chem.* **239**, 3522 (1964).

distinguish whether the allosteric site is on the same subunit or on a different one. Isocitrate dehydrogenase[40] and phosphorylase[35] have been analyzed in terms of this kinetic model and appear to have one allosteric–substrate site (isocitrate and glycogen, respectively) for each catalytic site.

*Presence of a Modifier, M, or a Second Substrate, B (Multisite Enzyme).* When a second substrate interacts with the enzyme in a random, rapid-equilibrium system, the saturation kinetic for one substrate (A) will differ from the binding saturation by the dependence on the term $(1 + B/K_B)$ describing the degree of saturation by the second substrate.[41] In other words, the kinetics measure only the amount of EAB whereas the equilibrium experiments would measure the total concentration of EA and EAB. Therefore, the apparent cooperativity measured by the two methods would differ. Similar considerations hold with respect to essential modifiers or activators. Kinetic analysis of isocitrate dehydrogenase[40] and of nucleoside diphosphatase[38] in the presence of allosteric modifiers have demonstrated the complexity of this situation and the differences that will necessarily occur between kinetic and equilibrium binding analyses. If the modifier is an inhibitor that displaces substrate from two active sites simultaneously, cooperativity for the substrate will be observed only in the presence of the inhibitor; this is the ligand exclusion model of Fisher *et al.*,[42] which does not require subunit interactions or conformational changes if the active sites are spatially close.

*Kinetic Models.* Cooperativity may also depend upon the kinetic steps themselves, rather than upon alteration of cooperativity that existed in binding or upon interactions of multiple catalytic or regulatory sites. Monomeric enzymes with no substrate–allosteric site could then produce kinetic cooperativity. The next two paragraphs on random mechanisms and on hysteretic enzymes rely upon particular relationships of rate constants in the steady state to produce cooperativity.

*Kinetic Models: Kinetic Cooperativity through a Random Mechanism (Two-Substrate, Monomeric Enzyme).* A monomeric enzyme (or an oligomer with independent sites) having two substrates (or a substrate and effector) that follow a random, steady-state addition (Bi Bi or Bi Uni) will give rise to a second-order rate equation in A of the form of Eq. (4) (Fig. 4). Each of the coefficients itself is a second-order function of B (Segal[43]). Apparent cooperativity of the kinetics may occur[44] under certain condi-

---

[40] D. E. Atkinson, J. A. Hathaway, and E. C. Smith, *J. Biol. Chem.* **240,** 2682 (1965).
[41] D. L. Purich and H. J. Fromm, *Biochem. J.* **130,** 63 (1972).
[42] H. F. Fisher, R. E. Gates, and D. G. Cross, *Nature (London)* **228,** 247 (1970).
[43] I. H. Segal, "Enzyme Kinetics." Wiley (Interscience), New York, 1975.
[44] B. D. Wells, T. A. Stewart, and J. R. R. Fisher, *J. Theor. Biol.* **60,** 209 (1976).

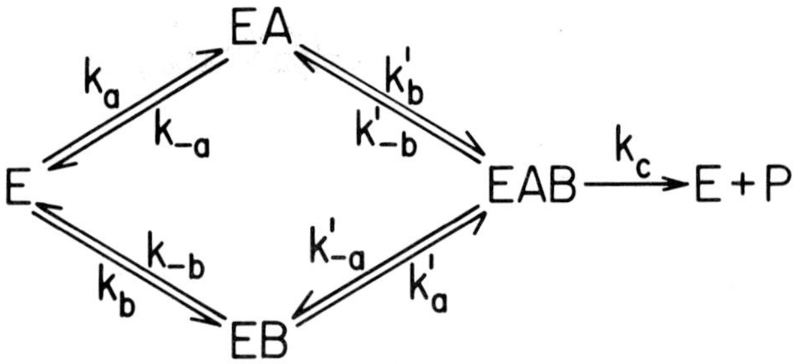

FIG. 4. Generation of apparent kinetic cooperativity by a random, steady-state mechanism. Requirements for cooperativity are that $k_c \sim$ other $k$'s; $K_a \sim K'_a \sim K_b \sim K'_b$; $k_a k'_b \gg$ or $\ll k_b k'_a$.

tions and may be conceptually seen as follows. Positive cooperativity may occur if these conditions hold: (a) the two pathways have approximately the same individual equilibrium constants so that neither pathway is thermodynamically favored; (b) $k_c$ is the same order of magnitude or larger that the other unimolecular rate constants so that rapid equilibrium binding of substrates does not occur; and (c) $k_a k'_b > k_b k'_a$, i.e., the upper pathway is kinetically favored. At a fixed concentration of B and low concentrations of A, the flux will be through the slower E–EB–EAB pathway because of the high proportion of EB. As the concentration of A is raised, the tendency will be for the flux to occur through the faster E–EA–EAB pathway as EA becomes competitive in concentration. The substrate curve for A will therefore be nonhyperbolic, and apparent cooperativity will occur. The cooperativity will depend on the concentration of B and disappear at saturating B concentration. Similarly, negative cooperativity or substrate inhibition may also occur. Simulations of these equations and of similar nonequilibrium random mechanisms have shown that exactly these types of curves may be generated.[44] It must be emphasized that this behavior can be found only in kinetic experiments and requires a rather exact set of rate constants. Its generality, therefore, is unknown.

*Kinetic Models: Hysteretic Cooperativity (One- or Two-Substrate, Monomeric Enzyme).* A monomeric enzyme (or an oligomer with independent sites) which undergoes a slow transition in the presence of substrate will give rise to a second-order rate equation of the same form as Eq. (5), where the coefficients are combinations of the rate constants of the individual steps (Fig. 5). The slow isomerization is termed a hysteretic transition[45] and gives rise to a second substrate addition step and hence a

---

[45] C. Frieden, *J. Biol. Chem.* **245**, 5788 (1970).

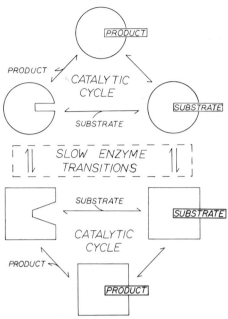

Fig. 5. Generation of apparent kinetic cooperativity through a slow transition (hysteretic) mechanism. Circles and squares represent different active conformations of an enzyme that slowly isomerizes.

second-degree equation.[7,8] With appropriately realistic rate constants, either positive or negative cooperativity may arise, accompanied by either lags or bursts in the kinetic transient.[8] The ramifications of hysteresis and hysteretic cooperativity is discussed more fully in the accompanying chapter in this volume [8]. Conceptually, kinetic positive cooperativity may be seen as follows. At low substrate concentration the enzyme operates mainly in the initial (circle) cycle, since relaxation of the free square to the circle occurs before more substrate adds. At high substrate concentration the addition of substrate to the free square occurs before the transition back to the circle can occur. Depending upon the relative rates and the rates of transition, either negative or positive cooperativity may be observed in the kinetics. It must be emphasized that only one substrate and a monomeric enzyme is required for such a mechanism, but the cooperativity would be observed only in kinetic experiments, not in equilibrium binding measurements (see this volume [8]).

Equilibrium Models for Cooperativity

A molecular model is necessary to better understand the cooperative binding of a ligand to a protein; e.g., the change in affinity of $O_2$ for hemoglobin as $O_2$ is bound. Several general models have been presented to explain the experimental observations of cooperativity. These fall into two main categories: (a) cooperativity in oligomeric proteins that have site–site interactions (or subunit interactions) between the binding sites on each of the identical protomers; and (b) cooperativity in oligomeric proteins that undergo association–dissociation reactions linked to the binding of ligands. These equilibrium models are consistent with the observation that many regulatory enzymes exist as oligomers of relatively few (2–12) numbers of protomers (the identical, repeating unit of the oligomer), each of which possesses a single ligand binding site. In addition, much of the cooperativity observed in enzymes is manifested in the binding of ligands as well as in catalytic activity. We will first look at cooperativity in equilibrium *binding* systems and then consider the additional consequences of *kinetic* factors for each model.

*Site-Site Interactions in Oligomeric Proteins*

The era of allosteric, cooperative transitions of enzymes was initiated by the proposal by Monod, Wyman, and Changeux[5] of a general molecular model to explain the sigmoid binding of ligands to proteins. This was rapidly followed by an alternative model with different assumptions about the molecular changes occurring in the presence of ligand.[6] The succeeding 10 years were occupied with discussion of the differences in these two models and experimental tests to distinguish between them. As we shall see, the argument has abated to a certain extent as more information about specific enzymes has been obtained and extensions of the earliest, simple models have been made. These two general models for site–site interactions with oligomeric proteins will be presented for the conceptual understanding that they provide concerning cooperativity of enzymes. The reader is referred to other reviews of recent years for a more detailed presentation and comparison of the two models.[7,43,46–51]

*The Concerted, Symmetric Model of Monod, Wyman, and Changeux.*[5] The basic assumptions of the concerted model (Fig. 6) are (a) that the oligomeric protein exists in two different states in the absence of ligand, each of which has a different affinity for ligand; (b) that an equilib-

---

[46] D. E. Koshland, Jr., *in* "The Enzymes" (P. Boyer, ed.), 3rd ed., Vol. 1, p. 342. Academic Press, New York, 1970.
[47] A. Levitzki and D. E. Koshland, Jr., *Curr. Top. Cell. Regul.* **10**, 1 (1976).
[48] J. Wyman, *Curr. Top. Cell. Regul.* **6**, 209 (1972).
[49] G. G. Hammes and C.-W. Wu, *Science,* **172**, 1205 (1971).
[50] G. H. Czerlinski, *Currents Mod. Biol.* **2**, 219 (1968).
[51] S. J. Edelstein, *Annu. Rev. Biochem.* **44**, 209 (1975).

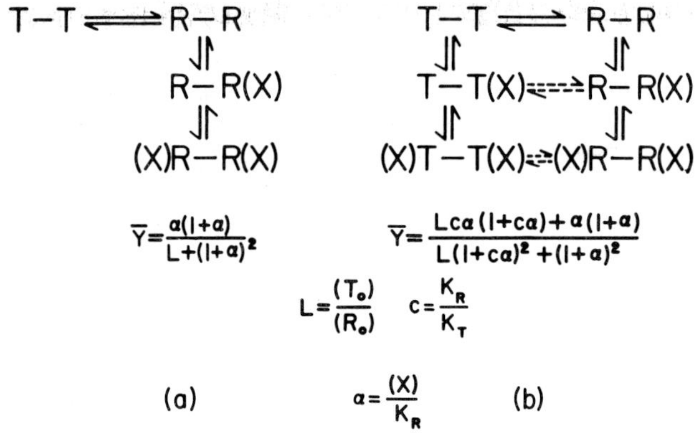

$$T-T \rightleftharpoons R-R \qquad\qquad T-T \rightleftharpoons R-R$$
$$\quad\;\;\Updownarrow \qquad\qquad\qquad\qquad \Updownarrow \qquad\quad \Updownarrow$$
$$R-R(X) \qquad\qquad T-T(X) \rightleftharpoons R-R(X)$$
$$\quad\;\;\Updownarrow \qquad\qquad\qquad\qquad \Updownarrow \qquad\quad \Updownarrow$$
$$(X)R-R(X) \qquad\quad (X)T-T(X) \rightleftharpoons (X)R-R(X)$$

$$\bar{Y} = \frac{\alpha(1+\alpha)}{L+(1+\alpha)^2} \qquad\qquad \bar{Y} = \frac{L c \alpha(1+c\alpha) + \alpha(1+\alpha)}{L(1+c\alpha)^2 + (1+\alpha)^2}$$

$$L = \frac{(T_0)}{(R_0)} \qquad c = \frac{K_R}{K_T}$$

(a) $\qquad\qquad \alpha = \frac{(X)}{K_R} \qquad$ (b)

FIG. 6. The symmetric, concerted model for ligand (X) binding to a dimer. (a) Exclusive binding. (b) Nonexclusive binding. The ligand, X, binds with greater affinity to the R conformation than to the T conformation. Note the preservation of symmetry of the quaternary arrangement of protomers in RR and TT.

rium exists between these conformations that can be shifted by the presence of ligand; and (c) that the symmetry of the oligomeric structure is preserved in the transition between the two states; i.e., each of the protomers is in an identical conformation in each state with no hybrid forms existing. Each protomer in a given state therefore has the same affinity for ligand. Positive cooperativity is readily seen to occur in this system: as the ligand binds and shifts the equilibrium toward the favored binding state, R, more of the high-affinity sites are available to which subsequent ligand molecules may bind. Cooperativity is the result, then, of the concerted transition in a preexisting equilibrium of a symmetrically arranged oligomeric structure brought about by the binding of ligands. In its simplest formulation the binding of ligand occurs exclusively to the high-affinity state, R, but, in general, nonexclusive binding may occur in which there is a quantitative difference in affinities between the T and the R states of the oligomer. The equilibrium binding expression for the concerted model of a protein possessing $n$ protomers, each with one binding site, is

$$\bar{Y} = \frac{Lc\alpha(1 + c\alpha)^{n-1} + \alpha(1 + \alpha)^{n-1}}{L(1 + c\alpha)^n + (1 + \alpha)^n} \tag{14}$$

where $L$ is the equilibrium constant for the concerted transition in the absence of ligand, $(T_0)/(R_0)$; $c$ is the ratio of the intrinsic dissociation constants for the two states, $K_R/K_T$; and $\alpha$ is simply a normalization for the substrate concentration, $\alpha = (S)/K_R$. Cooperativity is quantitatively dependent upon the position of the equilibrium between the R and the T states, i.e., the magnitude of $L$, and upon the relative difference in ligand binding to the two states, i.e., the magnitude of $c$. Equation (14) reduces to the hyperbolic form either when $L$ is very small relative to unity or when $c$ is equal to unity. When exclusive binding to the R form occurs, $c = 0$ and Eq. (14) reduces to the simpler form of Eq. (15).

$$\bar{Y} = \frac{\alpha(1 + \alpha)^{n-1}}{L + (1 + \alpha)^n} \tag{15}$$

The explicit expression for the dimer case is given in Fig. 6.

The concerted model quantitatively describes a ligand binding curve in terms of four independent parameters: $L$, $n$, $K_R$, and $c$ (or $K_T$). Experimental data may be fit to Eq. (14) with a minimization procedure to give the best values of the parameters by the general methods given earlier. In practice some estimate of at least $n$ and $K_R$ is preferred in order to have good initial estimates or for independent verification of the model. For the simple case of nonexclusive binding ($c = 0$) and if $K_R$ is known, a graphical method of determining $n$ and $L$ has been proposed.[52] In more complex instances, when effectors are available to shift the $L$ equilibrium (see below), estimates of the various parameters can be made including the maximum Hill coefficient, $n$, $c$, and $L$.[53,54]

Since the R conformation is considered to bind the substrate more tightly, any effector (modifier) that binds to the R conformation at a separate site from substrate will be an activator, whereas any effector that binds to the T conformation will be an inhibitor. When each form has the same catalytic velocity, a "$K$" system is defined. In general the effectors would also affect the apparent cooperativity of the system since the degree of cooperativity depends upon the position of the initial equilibrium, $L$. However, there are conditions in which $n_H$ is relatively insensitive to changes in other parameters.[48,53] Mathematically, the presence of an effector results in the replacement of $L$ in Eq. (14) by $L'$, where the latter is given by Eq. (16).

$$L' = L \frac{(1 + e\beta)^n}{(1 + \beta)^n} \tag{16}$$

The parameter $e$ is the ratio of the dissociation constants of effector to the two states, $e = K_{eR}/K_{eT}$, and $\beta$ is the concentration of effector relative to its dissociation constant, $\beta = (\text{effector})/K_{eR}$. If $c$ and $e$ are both less than unity, the effector is an activator. For the presence of more than one ef-

---

[52] A. Horn and H. Börning, *FEBS Lett.* **3**, 325 (1969).
[53] M. M. Rubin and J.-P. Changeux, *J. Mol. Biol.* **21**, 265 (1966).
[54] D. Blangy, H. Buc, and J. Monod, *J. Mol. Biol.* **31**, 13 (1968).

fector binding at individual sites, similar terms as in Eq. (16) are added for each.

For a "V" system of the concerted model the catalytic velocities of the R and the T states are different, but the binding of the substrate is the same to both states. The kinetics (or binding of substrate) is therefore hyperbolic. The addition of an effector can shift the equilibrium either toward the more active form (activator) or toward the less active form (inhibitor). Cooperativity in this system occurs solely with the effector, either in terms of the binding or in terms of the activation or inhibition of the enzymic reaction.

More generally, a mixed system can occur in which the affinities and the velocities of the two forms differ. If the higher-affinity R form also has the greater activity, $k_R > k_T$, then extensive cooperativity will occur. If the higher-affinity R form has the lower activity, $k_R < k_T$, then substrate inhibition will occur as the equilibrium is shifted toward the lower activity form at higher substrate concentrations. Effectors that bind preferentially to one form will affect both the maximal velocity and the $S_{0.5}$. The velocity equation for the mixed system is given as

$$v = \frac{nk_T[E]_t L c\alpha(1 + c\alpha)^{n-1} + nk_R[E]_t \alpha(1 + \alpha)^{n-1}}{L(1 + c\alpha)^n + (1 + \alpha)^n} \tag{17}$$

*The Sequential Model of Koshland, Nemethy, and Filmer.*[6] The basic assumptions of the sequential model (Fig. 7) are (*a*) that ligand binding induces a conformational change in the subunit to which it binds; (*b*) that the conformational change affects the interactions with a neighboring subunit such that the next ligand binding-conformational change step occurs more readily or less readily; (*c*) that there is no requirement for symmetry of the oligomeric structure to be maintained, thereby allowing intermediate, hybrid states with protomers of two (or more) conformations in the same oligomer. Cooperativity is then a result of the ease with which a subunit can undergo the conformational transition to the high-affinity form, depending upon the conformational status of its neighboring subunit and the interactions between them. Each association constant, $K_I$, is considered to be a product of the inherent binding constant to the second conformation, $K_{XB}$, and the equilibrium constant for the conformational transition of the protomer in the oligomer; the latter is itself a product of the conformational transition of the isolated subunit, $K_t$, and the equilibrium constant, $K_{AB}$, representing the difference in energy between the AA + B and the AB + A states. In the simplest formulation of the sequential model, the conformational change occurs *only* in the subunit to which the ligand binds [Fig. 7, line (a)], but in a more general extension, the possibility of partial conformational changes in neighboring subunits is also con-

(a) $A\text{-}A \rightleftharpoons A\text{-}B(X) \rightleftharpoons (X)B\text{-}B(X)$

$$K_1' = \tfrac{1}{2}K_1 = K_{1B} K_{XB} K_{AB}$$
$$K_2' = 2K_2 = K_{1B} K_{XB} K_{BB}/K_{AB}$$
$$K_{1B} = \frac{(B)}{(A)}\ ;\ K_{XB} = \frac{(BX)}{(B)(X)}\ ;\ K_{AB} = \frac{(AB)(A)}{(AA)(B)}\ ;\ K_{BB} = \frac{(BB)(A)^2}{(AA)(B)^2}$$

(b) $A\text{-}A \rightleftharpoons C\text{-}B(X) \rightleftharpoons (X)B\text{-}B(X)$

$$K_1' = \tfrac{1}{2}K_1 = K_{1B} K_{1C} K_{XB} K_{BC} \qquad K_{1B} = \frac{(B)}{(A)}\ ;\ K_{1C} = \frac{(C)}{(A)}\ ;$$
$$K_2' = 2K_2 = \frac{K_{1B} K_{XB} K_{BB}}{K_{1C} K_{BC}} \qquad K_{XB} = \frac{(BX)}{(B)(X)}\ ;\ K_{BC} = \frac{(BC)(A)^2}{(AA)(B)(C)}$$
$$K_{BB} = \frac{(BB)(A)^2}{(AA)(B)^2}$$

$$n\bar{Y} = \frac{K_1(X) + 2K_1 K_2(X)^2}{1 + K_1(X) + K_1 K_2(X)^2}$$

FIG. 7. The sequential model for ligand (X) binding to a dimer. (a) The simple sequential model. (b) The generalized sequential model. A, B, and C represent three conformations of the protomer. Note the occurrence of hybrid molecules possessing one A and one B protomer conformation. C can be thought of as an intermediate or partial conformational state between A and B. $K'$ is the intrinsic association constant.

sidered to occur [Fig. 7, line (b)]. The overall binding equation is given by the general Adair equation for binding to multiple sites.

$$n\bar{Y} = \sum_{i=1}^{n} iX^i \prod_{j=1}^{i} K_j \bigg/ \bigg[1 + \sum_{i=1}^{n} X^i \prod_{j=1}^{i} K_j\bigg] \qquad (18)$$

where $X$ is the ligand concentration, $K_i$ is the thermodynamic association constant for the $i$th site, and $n$ is the total number of sites. Each of the association constants, in principle, is defined by the molecular processes of conformational transitions, binding to isolated subunits, and association (or exchange) equilibria of oligomers. The relationship between the Adair constants and the molecular constants is given for the dimer case in Fig. 7. For higher oligomers, such as tetramers, the corresponding relationships are dependent upon the geometrical interaction between subunits ("square," "tetrahedral," etc.). The reader is referred to the original papers[6,25] for these relationships.

Description of the binding of ligand to a dimer using the sequential model requires three independent parameters ($K_1$, $K_2$, and $n = 2$) or four molecular constants ($\overline{K_t K_x}$, $K_{AB}$, $K_{BB}$, and $n = 2$; since $\overline{K_t K_x}$ always occurs as the product). For higher oligomers, an additional Adair constant is required for each binding site, but in the simpler sequential models these can always be related to the same four molecular constants.[6]

Initial fitting of data to the sequential model is best done by the computer fitting methods outlined in an earlier section to obtain the Adair constants. Restrictions may then be applied to the relationships between the intrinsic binding constants[6] to determine whether any of the simple geometric models are applicable. Alternatively, a nomogram method[6] has been suggested as a means of easily determining the molecular constants of a particular sequential model.

The earliest statement of the sequential model[6] was mainly concerned with positive cooperativity and the simplest situation in which the conformation of A did not change until a ligand was bound, but conditions for negative cooperativity and for more complex conformational changes were presaged. In order to explain data on rabbit muscle glyceraldehyde-3-phosphate dehydrogenase, extensions of the original model were made to include negative cooperativity and to include partial conformational changes in the neighboring subunit in addition to those changes in the subunit complexed to ligand [line (b), Fig. 7].[55]

The influence of effectors upon the cooperativity generated by a sequential model is quite general.[25] Any effector may bind to an A or a B conformation and stabilize or destabilize the preferred ligand binding conformation. The main consideration is whether the effector binds competitively with the ligand in question or whether it acts at a true allosteric site. Since the conformational changes in this model are envisaged as being induced by ligand, the effector may induce a new conformation with different properties than those achievable by substrate alone. This possibility gives rise to multiple consequences that are sufficient to explain most observations, but not easily analyzable in terms of simple molecular considerations.

For kinetic analysis the sequential binding model can be extended by consideration of the changes in catalytic rates of adjacent protomers as ligand (substrate) is bound. For rapid equilibrium binding, Eq. (18) can be extended by addition of the catalytic rate constant, $k_I$, of each complex.

$$\frac{v}{[E]_t} = \sum_{i=1}^{n} i k_i X^i \prod_{j=1}^{i} K_j \bigg/ \left[ 1 + \sum_{i=1}^{n} X^i \sum_{j=1}^{i} K_j \right] \tag{19}$$

The relationship among the "Adair" constants of the kinetic equation are now no longer obtainable from the relationship among molecular con-

---

[55] A. Conway and D. E. Koshland, Jr., *Biochemistry* **7**, 4011 (1968).

stants of the geometrical considerations of the sequential model. Description of an enzyme following this case would require a careful comparison of the binding curve and the kinetic saturation curve with proper allowance for differences in protein concentration, other substrates, etc. The author is not aware of a case in which this has been done completely satisfactorily.

*Critical Comparison of the Concerted Model and the Sequential Model.* The original and simplest expositions of each model, concerted[5] and sequential,[6] were truly different models to explain cooperativity, the *major* difference lying in the symmetry requirement of the model of Monod, Wyman, and Changeux and the requirement for hybrid conformational states in the model of Koshland, Nemethy, and Filmer. This difference in the symmetry of the models represents a fundamental difference in viewpoint about forces and constraints holding protomors together in the quaternary structure. The apparent difference as to whether there is a preexisting equilibrium[5] or a ligand-induced change[6] is merely a quantitative consideration with regard to amounts of certain species present and was quickly ameliorated as a distinction between the models (see later section). Both simple models were expanded to include more complexities, such as nonexclusive binding,[47,53,56] additional conformational changes,[47,55,57] kinetic aspects[58] etc. Each model is essentially a limit of a more general model[43,56,48] as illustrated in Fig. 8 for the dimer. In this general model, a conformational change can occur without ligand binding, ligand binding can occur without a conformational change, and hybrid states can occur. The general model may be reduced to each limiting case (Fig. 8), and calculation has been made of the concentrations of intermediates in these cases.[56] Clearly, addition of more conformations than circles and squares (cf. Fig. 7) would make the model even more general. The consequences of the two models may also be compared through their parameters as expressed by the coefficients of the general rate equation for the dimer (Table II; see Segal[43] for the tetramer case).

There are advantages to having simple models and advantages to more complex models. The comparison of the concerted and the sequential models rests on these relative advantages and upon comparison to real data. The concerted model is the simpler model of the two, both conceptually and mathematically. Two states of a protein with a shift of equilibrium between them is more readily understood by the protein chemist than a multitude of states depending upon the degree of saturation and the particular ligand. With the concerted model fewer constants ($n$, $c$, $K_R$, $L$) are

[56] J. E. Haber and D. E. Koshland, Jr., *Proc. Natl. Acad. Sci. U.S.A.* **58** 2087 (1967).
[57] A. Szabo and M. Karplus, *J. Mol. Biol.* **72**, 163 (1972).
[58] K. Dalziel, *FEBS Lett.* **1**, 346 (1968).

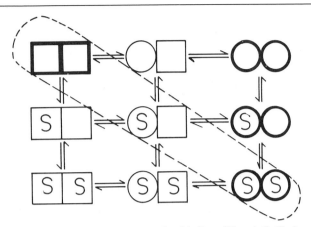

FIG. 8. The general dimer model for cooperative binding of ligand, S. Circles and squares represent different conformations of the protomer. In the most general case, ligand binding can occur without conformational change (vertical columns) to either conformation, or conformational changes can occur in each subunit without ligand binding (horizontal rows). The *simplest concerted* mechanism is shown by the boldface figures. The *concerted mechanism* with nonexclusive binding would correspond to the first and third vertical columns. The *simplest sequential model* is represented by the diagonal within the dashed lines. The *general sequential model* would include an additional partial conformational change of the unliganded square in the center figure, i.e. Ⓢ△.

used to describe the binding curve for a general, multisite enzyme. The constants are thus easier to estimate directly from the data and are easier to compare under different conditions.[54] Although the sequential model has the same number of *molecular* constants ($K_{AB}$, etc.), these are not directly obtainable from the saturation function; thus the number of Adair constants is greater for this model (if $n > 2$), allowing more degrees of freedom in the curve fitting. For the dimer the number of constants is the same in the two models (Table II). On the other hand, since the sequential model is more general, it has the potential to be applicable to more complex situations; if there is anything that the enzymologist has learned, it is that enzymes are more complex than one might expect. With a new enzyme system an investigator would be advised to choose the simpler concerted model for interpretation until data were accumulated to indicate that the model was insufficient.

Negative cooperativity of binding of a ligand can be explained readily by the sequential model, since the interactions that are transmitted to an adjacent subunit can be negative as well as positive with respect to additional ligand binding. The concerted model *cannot* explain negative cooperativity of binding since the requirement of this model is that the equilibrium is shifted toward the state with a higher affinity for ligand. The observation of negative cooperativity is, therefore, prima facie evidence for the sequential model (or at least against the concerted model). However, negative cooperativity and heterogeneity of sites produce the same binding plots and are difficult to differentiate.[32,33] More complex observations of mixed negative and positive cooperativity have been made[59] and can be explained only by the sequential changes of several hybrid states as the affinity decreases and increases.

For positive cooperativity a binding curve is probably insufficient to chose between the concerted and the sequential model. Although it can be shown in principle[36] that different models lead to slightly different saturation curves, practical considerations of real experiments make it difficult or impossible to obtain binding data accurate and precise enough to present a convincing case that only one of the two models is correct without other types of information. Since the two models make different assumptions and predictions about the underlying molecular changes, studies have turned to physical chemical measurements of conformations in solution to distinguish the two models. The simplest sequential model can be distinguished from the simplest concerted model based on the observation that the "state function," $\bar{R}$, should directly coincide with the "saturation function," $\bar{Y}$, if ligand is inducing a conformational change only in the subunit to which it is bound. $\bar{R}$ is a measure of the degree of conformational change and can be measured by a suitable physical probe, such as fluorescence or circular dichroism. In the concerted model the degree of conformational change occurs at lower degrees of saturation (compare the circle to square conversion upon ligand binding of the simplest concerted model in Fig. 6). Many experiments have been directed toward these comparisons in recent years with particular emphasis upon hemoglobin and upon aspartyl transcarbamylase. Conformational data of a very precise nature is required for these comparisons; in addition, there are other problems of interpretation. Comparison of $\bar{R}$ and $\bar{Y}$ and their changes with ligand concentration depend heavily upon the value of $L$ chosen, i.e., the position of preexisting equilibrium of the oligomer before addition of ligand. This value is generally difficult to obtain by physical-chemical methods and thus the comparison of $\bar{R}$ with $\bar{Y}$ becomes hazardous. Second, any one conformational probe of a protein is usually responsive only to a given region of the structure, and therefore more than one measure of the conformational changes occurring must be made. This appears to be true with hemoglobin, wherein some resonance probes appear to favor a direct relationship between the state function and the saturation function[60] whereas others[61] do not. The interpretation of hemoglobin is also complicated by the nonidentity of the $\alpha$ and $\beta$ chains and the

---

[59] A. Cornish-Bowden and D. E. Koshland, Jr., *J. Mol. Biol.* **95**, 201 (1975).
[60] R. T. Ogata and H. M. McConnell, *Biochemistry* **11**, 4792 (1972).
[61] W. H. Huestis and M. A. Raftery, *Biochemistry* **14**, 1886 (1975).

tendency to dissociate to $\alpha\beta$ dimers.[51,62] Last, extensions of the sequential model make the $\bar{R}$ vs. $\bar{Y}$ comparison not very discriminating.[47,55] Figure 7(b) shows the more general sequential model, in which it is clear that the "C" conformation is occurring before ligand is bound, and a physical probe of this protomer conformation would not yield a linear relationship between $\bar{R}$ and $\bar{Y}$. This situation was found for rabbit muscle glyceraldehyde-3-phosphate dehydrogenase, which shows negative cooperativity of binding of NAD (and hence the concerted model cannot apply), but the major conformational change occurs upon binding of the first NAD molecule.[55] At this stage then, there is no general, definitive way of distinguishing between the concerted and the sequential models, until different kinds of data are obtained and are shown to be consistent with one or the other model. At the moment each individual enzyme studied in detail appears to have its own idiosyncrasies, and the general models are only starting points for analyses.

*Extensions of Site–Site Models.* Numerous other interpretations or general models of cooperative binding of ligands through site–site interactions have been published. These may be classified as those studies that have attempted to extend the rigor of the mathematical analysis,[22,63] those that have tried to generalize or extend the concerted or sequential models,[51,64–67] those that have approached the same problem from a different viewpoint,[68,69] and those that have attempted to understand the underlying molecular or energetic aspects of the models.[70–74] These newer developments provide some additional insight into applications to specific enzymes but cannot be adequately treated in the limited confines of this review.

*Catalytic and Substrate Effects on Cooperativity.* Within the last several years other considerations of the site–site interaction models for cooperativity have enlarged our understanding of the potentialities of these systems. One obvious question is whether the introduction of

---

[62] R. G. Shulman, J. J. Hopfield, and S. Ogawa, *Quart. Rev. Biophys.* **8,** 325 (1975).
[63] R. P. Garay, *J. Theor. Biol.* **63,** 421 (1976).
[64] E. Whitehead, *Biochemistry* **9,** 1440 (1970).
[65] J. Herzfeld and H. E. Stanley, *J. Mol. Biol.* **82,** 231 (1974).
[66] J. Herzfeld and P. A. Schlesinger, *J. Mol. Biol.* **97,** 483 (1975).
[67] O. M. Viratelle and F. J. Seydoux, *J. Mol. Biol.* **92,** 193 (1975).
[68] C. J. Thompson, *Biopolymers* **6,** 1101 (1968).
[69] T. R. Chay and C. Ho, *Proc. Natl. Acad. Sci. U.S.A.* **70,** 3914 (1973).
[70] R. W. Noble, *J. Mol. Biol.* **39,** 479 (1969).
[71] A. Cornish-Bowden and D. E. Koshland, Jr., *J. Biol. Chem.* **245,** 6241 (1970).
[72] A. Cornish-Bowden and D. E. Koshland, Jr., *J. Biol. Chem.* **246,** 3092 (1971).
[73] H. A. Saroff and W. T. Yap, *Biopolymers* **11,** 957 (1972).
[74] G. Weber, *Biochemistry* **11,** 864 (1972).

kinetic constants into the concerted model would lead to negative cooperative. The equation [Eq. (17)] for the rapid-equilibrium binding of a substrate with different catalytic rates for the R and the T states of the enzyme has been alleged to produce negative cooperativity.[75] However, it was later demonstrated[28,76,77] that a proper consideration of the experimentally determined $V_{max}$ and the resultant $n_H$ would produce negative cooperativity in only a few limited cases in which the maximum velocity was only a few percent of the potential of the system. The presence of negative cooperativity in kinetics would also eliminate the concerted model from consideration.

Introduction of catalytic constants into the sequential model is more difficult, but the equations have been derived for various tetrameric geometries.[78] In general, the rate equations are not equivalent to the corresponding binding equation, even if rapid equilibrium binding is assumed, because of subunit interaction terms. The $k_i$ of Eq. (19) may be considered as a function of the contribution of subunit interactions to the rate of the conformational change of the subunit within the oligomer and, in general, will not be equal for all $i$ values. Positive and negative cooperativity and substrate inhibition are natural consequences of the steady-state rate equation for the oligomer.

Most discussions of site–site cooperativity have pertained to one-substrate situations or have utilized the assumption that saturating concentrations of a second substrate would make the analysis identical to that of a single substrate case. A recent thorough analysis of the two-substrate case for the concerted model has provided some illuminating information.[79] Equation (20) describes the random binding and rapid equilibration of two substrates before the rate-determining breakdown of the ternary complex in a concerted model.

$$\frac{v}{n[E]_t} = \frac{k_T c e L \alpha \beta (1 + c\alpha)^{n-1}(1 + e\beta)^{n-1} + k_R \alpha \beta (1 + \alpha)^{n-1}(1 + \beta)^{n-1}}{L(1 + c\alpha)^n(1 + e\beta)^n + (1 + \alpha)^n(1 + \beta)^n} \quad (20)$$

where $\beta$ is the ratio of the second substrate to its dissociation constant in the R state, $B/K_{RB}$, and $e$ is the ratio of the dissociation constants of B to the two states, $e = K_{RB}/K_{TB}$, and the other constants have the meaning described in Eq. (14) for the substrate A. This equation [Eq. (20)] differs

---

[75] A. Goldbeter, *J. Mol. Biol.* **90**, 185 (1974).
[76] H. Paulus and J. K. DeRiel, *J. Mol. Biol.* **97**, 667 (1975).
[77] A. Goldbeter, *Biophys. Chem.* **4**, 159 (1976).
[78] J. Ricard, C. Mouttet, and J. Nari, *Eur. J. Biochem.* **41**, 479 (1974).
[79] D. W. Pettigrew and C. Frieden, *J. Biol. Chem.* **252**, 4546 (1977).

from Eq. (16) for effector binding in that here only the ternary complex containing both A and B is considered to be active, whereas in Eq. (16) all complexes containing A contribute to the total binding or activity of the single substrate A. Thus, the kinetic equation is different from one simply putting catalytic rate constants into the binding equation. Examination and simulation studies of Eq. (20) have revealed that several interesting situations can arise. Positive cooperativity can occur in kinetics for one substrate even though that substrate would show hyperbolic equilibrium binding ($k_T = k_R$, $L > 1$, $c < 1$, $e > 1$, $c > e$). If the binding of one substrate affects the $K_m$ of the second (i.e., the dissociation constant is not equal to the kinetically determined half-saturation), positive cooperativity can be observed for that substrate even though it has the same intrinsic binding to both states; this case is tantamount to allowing additional ministates in which the $K_m$ values are allowed to change. Pettigrew and Frieden[79] have concluded that a distinction may be made between all the possible cases if measurements are made on both substrates in kinetics and in binding studies. The use of the conceptually simple concerted model has made these cases easier to analyze but may be an oversimplification of real enzyme systems.

The above treatment is analogous, in some ways, to the earlier treatment of the sequential model with regard to effectors and multiple substrates. Kirtley and Koshland[25] analyzed the effect of a second substrate on the cooperativity of the first for the square geometry of the sequential model. The assumption was made that rapid-equilibrium binding occurred and that the velocity was proportional to the fraction of the appropriately liganded species. Depending upon whether the addition of substrates is ordered or independent, the degree of saturation of the second substrate could alter the apparent cooperativity, half-saturation, or maximally observed velocity of the substrate in question. The authors suggested that, with data available from several kinds of equilibrium, kinetic, and effector experiments, a logical choice may be made among various cooperative models. The application of this analysis in practice is difficult.

*Effect of Products on Cooperativity.* Many considerations of cooperativity are made with respect to *initial* velocities or binding of a single ligand. The influence of products (or other competitive inhibitors, such as substrate analogs) on the cooperativity of enzymic reactions has not received as much attention as it deserves, considering that most catalyses occur *in vivo* in the presence of products as well as substrates. In the presence of product (competitive with substrate) a normal cooperative binding curve for substrate will tend to be "flattened" (lower apparent $n_H$) and shifted to higher concentrations of substrate, as would be expected for a competitive situation.[80] Purich and Fromm[80] have emphasized the regula-

---

[80] D. L. Purich and H. J. Fromm, *Curr. Top. Cell. Regul.* **6**, 131 (1972).

tory role of product and presented simulated curves for a simple Adair-type dimer with exclusive, competitive binding of substrate and product. Conversely, if the sensitivity to inhibition is considered,[81] the velocity of a positively cooperative system, which is more sensitive to changes in substrate, is less sensitive to inhibition by product than a hyperbolic saturation curve. For an enzyme with negative cooperativity, which is less sensitive to changes in substrate concentration, the velocity is more sensitive to inhibition by a product that is capable of inducing the same conformational changes in the enzyme as substrate. Intuitively, product is able to "turn off" more sites than the one it binds to in a negatively cooperative system; in a positively cooperative system, the binding of product at one site enhances the binding of substrate (as well as product) at other sites in the oligomer and thus has less effect than might be expected. The last condition raises the possibility that product (or inhibitor) may produce activation at low concentrations; this has been observed in several enzymes, such as aspartyl transcarbamylase.

Inhibition by substrate analogs can account for the activation observed at low inhibitor concentrations.[82] The general rate equation for the concerted model, assuming that rapid equilibrium binding of $n$ substrates and $n$ competitive inhibitors occurs and that the enzyme may be considered as a single-substrate enzyme, is given by Eq. (21).

$$\frac{v}{n[\mathrm{E}]_t} = \frac{k_\mathrm{T} c\bar{\alpha}(1 + c\bar{\alpha} + r\rho)^{n-1} + k_\mathrm{R}\alpha(1 + \alpha + \rho)^{n-1}}{L(1 + c\alpha + r\rho)^n + (1 + \alpha + \rho)^n} \quad (21)$$

where $\rho = [\mathrm{P}]/K_\mathrm{RP}$; the ratio of product concentration to its dissociation to the R form; $r = K_\mathrm{RP}/K_\mathrm{TP}$, the ratio of the dissociation constants of P to the R and T forms. Although activator and substrate analog bind to the same form of the two-state system, Eq. (21) differs from Eq. (16) because the substrate analog binds to the same site as substrate and leads to inhibition at high concentrations. Application of Eq. (21) to the data for aspartyl transcarbamylase (with minor adjustments to the previously determined allosteric constants) led to a satisfactory fit of the activation at low analog concentration, suggesting that the model and the assumptions are adequate.[83] An analogous expression may be derived for an Adair-type binding equation in which the inhibitor is assumed to bind statistically in the same manner as substrate, with its own set of intrinsic binding constants.

$$\frac{v}{[\mathrm{E}]_t} = \sum_{i=1}^{n} i k_i S^i \prod_{j=1}^{i} K_j \sum_{h=0}^{n-i} (1 + K_h' P^h) \bigg/ \sum_{i=0}^{n} S^i \prod_{j=0}^{i} K_j \sum_{h=0}^{n-i} (1 + K_h' P^h) \quad (22)$$

---

[81] C. J. Lamb and P. H. Rubery, *J. Theor. Biol.* **60**, 441 (1976).
[82] G. D. Smith, D. V. Roberts, and P. W. Kuchel, *Biochim. Biophys. Acta* **377**, 197 (1975).
[83] G. D. Smith, *J. Theor. Biol.* **69**, 275 (1977).

where $K$ = substrate constant; $K'$ = product constant, and $K_0 = 1$; $K'_0 = 0$ by definition.

## Protomer–Oligomer Equilibria

In the site–site interaction models just presented, subunit interactions change upon ligand binding. If this change in interaction is large enough, the actual degree of association of the oligomer might change. Experimentally, a dissociation or an association would be observed in the presence of the ligand. Numerous enzyme and protein systems are known to undergo changes in quaternary structure in the presence of substrate or effector. Nichol et al.[84] and Frieden[85] have pointed out that this process is analogous to the shift in equilibrium envisioned in the concerted oligomer model and should lead to cooperativity of binding of the ligand. The basic assumptions of a simple form of this model are (a) that the protein exists in two states, a protomeric state and a single oligomeric state; (b) that the two states differ in their affinity for the ligand; (c) that the protomer and the oligomer exist in a rapid equilibrium that can be shifted by the presence of ligand; and (d) that the oligomer possesses $n$ subunits with $q$ equivalent binding sites, and that the protomer possesses $p$ equivalent binding sites. The saturation equation for the rapid equilibrium binding of ligand is

$$p\bar{Y} = \frac{p\alpha(1+\alpha)^{p-1} + qac K_E E^{n-1}(1+c\alpha)^{q-1}}{(1+\alpha)^p + nK_E E^{n-1}(1+c\alpha)^q} \qquad (23)$$

where $K_E = E_n/E^n$, $c = K'_1/K'_n$, $\alpha = X/K'_1$, $E_n$ = concentration of oligomer, $E$ = concentration of protomer,

$$K'_1 = \left[\frac{(E)(X)}{(EX)}\right]\left[\frac{p-(i-1)}{i}\right]$$

and $\quad K'_n = \left[\dfrac{(E_n X_{i-1})(X)}{(E_n X_i)}\right]\left[\dfrac{q-(i-1)}{i}\right]$

For the situation in which no sites are gained or lost upon polymerization, $n$ is equal to $q/p$. In many instances $p$ will equal 1; i.e., there will be a single binding site per subunit. The form of Eq. (23) is similar to that derived for the concerted model of site–site interactions within an oligomer, but there is an explicit dependence of the saturation function on the concentration of monomer, $E$. The concentration of monomer may be expressed in terms of the other parameters and the *total* protein concentration, $E_0$ [Eq. (24)]. For the general case ($n > 2$) Eq. (24) is not readily

---

[84] L. W. Nichol, W. J. H. Jackson, and D. J. Winzor, *Biochemistry* **6**, 2449 (1967).
[85] C. Frieden, *J. Biol. Chem.* **242**, 4045 (1967).

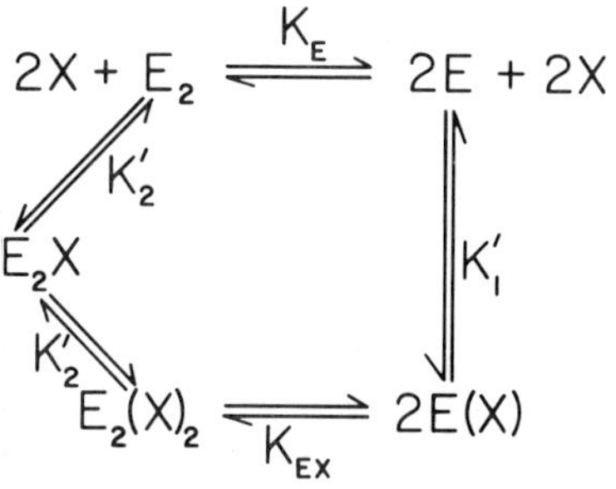

$$2X + E_2 \underset{K_E}{\rightleftharpoons} 2E + 2X$$

$$K_2' \diagdown \qquad \qquad \Big\updownarrow K_1'$$

$$E_2 X$$

$$K_2'' \diagdown \qquad \qquad$$

$$E_2(X)_2 \underset{K_{EX}}{\rightleftharpoons} 2E(X)$$

$$\overline{Y} = \frac{a + 2ac K_E E(1+ca)}{(1+a) + 2K_E E(1+ca)^2}$$

$$E = \frac{1+a}{4K_E(1+ca)^2} \left\{ \left[ 1 + 8E_0 K_E \left( \frac{1+ca}{1+a} \right)^2 \right]^{1/2} - 1 \right\}$$

FIG. 9. Cooperativity due to ligand-linked association–dissociation equilibria of dimeric enzyme, E. Nonexclusive binding is shown; exclusive binding to monomer ($c = 0$) would omit the $E_2(X)_2$ species, exclusive binding to dimer ($c \to \infty$) would omit the E(X) species. Constants are expressed as intrinsic dissociation constants of ligand and molar association constants for dimerization.

solved for $E$ but must be estimated by computer-assisted iterations or simulated to generate appropriate curves.

$$E_0 = E(1 + \alpha)^p + nK_E E^n (1 + c\alpha)^q \tag{24}$$

where $E_0$ is the concentration of the oligomer on a subunit basis (moles of subunit).

The general picture for this type of ligand-influenced association–dissociation (or polysteric[86]) mechanism is exemplified in Fig. 9 for a dimeric enzyme, and the corresponding equations are given. If $E_2$, $E$, and EX are the only species present, then the enzyme is demonstrating a preexisting monomer–dimer equilibrium with exclusive binding of ligand to the monomer, which increases the degree of dissociation. This situation is most analogous to that of the concerted model for site–site interactions in an oligomer without dissociation. Two ways in which this might occur can be imagined for a real protein: (a) that the ligand binding site of the monomer is sterically blocked in the dimer by the neighboring subunit; (b) that a conformational change occurs upon dissociation that allows the ligand binding site to form in the monomer. This exclusive binding case has been called "competitive"[84] because the ligand and an adjacent subunit may compete for the same binding site on another subunit. If $E_2$, $E_2X_2$, and EX are the only species present, then the enzyme is

[86] A. Colosimo, M. Brunori, and J. Wyman, *J. Mol. Biol.* **100**, 47 (1976).

demonstrating a ligand-induced weakening of the subunit interactions, most likely through a conformational change in the dimer (not depicted in Fig. 9). $K_E$ and $K_1$ would both have to be small in this situation.

Conversely, exclusive binding to the dimer could occur and the above arguments would be reversed to coincide with experimentally observed increased association in the presence of ligand. An interesting possibility arises here if the ligand binding site is located on the intersubunit bonding domain, contributed by both subunits, and therefore nonexistent in its entirety on the isolated monomer. Evidence for this has been presented from crystallographic data for a nucleotide binding site (not the active site) in yeast hexokinase.[87]

If the number of ligand binding sites *per subunit* changes between the protomer and the oligomer ($n \neq q/p$), additional complex behavior may occur.[84,88] If the monomer does not bind ligand but the dimer binds only one (0.5 per subunit), the surprising result is that negative cooperativity occurs (at low protein concentration). If the dimer binds two or more ligand molecules, then positive cooperativity results.[88] If the dimer binds one ligand tightly and the monomer binds one ligand more weakly, strong negative cooperativity may occur because of the increase in the total number (2 vs. 1) of ligand molecules bound at high ligand concentration with dissociation to monomer. When the ligand binds to one of two different subunits (R and C) various binding curves can be obtained including ones adequate to explain the positive cooperativity of cyclic-AMP binding to protein kinase.[89]

More generally, the ligand may bind to both monomer and dimer but with different affinities, i.e., the nonexclusive binding or noncompetitive binding case.[84] The extent of cooperativity and the effect of the ligand (association or dissociation) depends on the relative magnitude of the ligand dissociation constants ($K_1$ and $K_2$) and on the relation of the experimental protein concentration, $E_0$, to the dimerization constant, $K_E$. (Note that only four of the five constants of Fig. 9 are independent owing to thermodynamic constraints.) In contrast to the concerted model, negative cooperativity at equilibrium can be observed, with nonexclusive binding. Although the equilibrium is shifted by the ligand, both monomer and dimer, with different affinities, may exist in solution at all times. A thorough analysis and simulation of these different possibilities and the demonstration of conditions for both positive and negative cooperativity have been presented.[90,91]

[87] W. F. Anderson, R. J. Fletterick, and T. A. Steitz, *J. Mol. Biol.* **86**, 261 (1974).
[88] K. C. Ingham, H. A. Saroff, and H. Edelhoch, *Biochemistry* **14**, 4745 (1975).
[89] J. R. Ogez and I. H. Segel, *J. Biol. Chem.* **251**, 4551 (1976).
[90] A. Levitzki and J. Schlessinger, *Biochemistry* **13**, 5214 (1974).
[91] L. W. Nichol and D. J. Winzor, *Biochemistry* **15**, 3015 (1976).

The effect of two or more oligomeric species, e.g., monomer–dimer–tetramer, on cooperative curves has been investigated.[84,91,92] Sedimentation and fluorescence studies of the complex binding curves of anilino-naphthalene-sulfonate binding to human chorionic gonadotropin have demonstrated that the dye binds strongly to the tetramer of the hormone, more weakly to the dimer, and not at all to the monomer. More work needs to be done on this type of system, since it is perhaps more likely to occur with physiological proteins of interest than those in which only a single oligomer exists in equilibrium with monomer.

The most striking behavior of ligand-influenced monomer–oligomer equilibria systems is the dependence of binding and, particularly, cooperativity on the protein concentration used for the experimental studies. Cooperativity of ligand binding will tend to be maximal at intermediate protein concentrations relative to the protein oligomerization constant, but will approach the normal hyperbolic binding curve at extremes of high or low protein concentration. This behavior provides a very valuable tool in the investigation of these systems and a means of distinguishing them from the systems in which cooperativity occurs in a tightly associated oligomer. An additional aid to discerning oligomer equilibrium systems is that they will in general produce Hill plots that are asymmetrical around the half-saturation point.[86]

Modifiers may also influence the protomer–oligomer equilibrium and shift the protein toward the form with higher or lower affinity for the ligand, analogously to the concerted model. Kinetic constants may be introduced to extend Eq. (23) to include the situation of a polymerizing enzyme that has different affinities or different catalytic velocities in the oligomer and the protomer. In the latter case an effector could activate or inhibit the enzyme, concomitant with a change in association state, even without cooperativity in the substrate kinetics.

## Complex Cooperativity

*Mixed Positive and Negative Cooperativity.* The discussion to this point has centered on "pure" positive or negative cooperativity, with a resultant simplification of models. Data have accumulated, however, that more complex situations can arise in which there are mixtures of both positive and negative cooperativity within a single oligomeric enzyme. Such a combination of cooperativity is manifested in binding (or kinetic) curves by "abrupt" transitions[93] or "inflexions"[94] in double-reciprocal plots, by intermediary plateaus (or "bumpy" curves) in direct plots of

---

[92] K. C. Ingham, H. A. Saroff, and H. B. Edelhoch, *Biochemistry* **14**, 4751 (1975).
[93] P. C. Engel and W. Ferdinand, *Biochem. J.* **131**, 97 (1973).
[94] W. G. Bardsley, *J. Theor. Biol.* **65**, 281 (1977).

ligand bound (or velocity) vs. substrate concentration,[15] or by pronounced curvature in Hill plots.[59] These deviations from "normal" cooperativity are clearly a result of binding or rate equations that are of high degree in ligand or substrate. Attention has focused on the mathematical analysis and consequences of these high-degree equations with a lesser emphasis on the molecular basis for them. Teipel and Koshland[15] analyzed the derivatives of Adair-type equations for the changes in slope that must occur to generate the intermediary plateaus. For these bumpy curves to occur, there must be three or more binding sites (third-degree equation or greater) and there must be negative cooperativity followed by positive cooperativity. Bardsley[94] has thoroughly analyzed the third-degree rate equation by the method of ordering functions and determined that there are 26 curve shapes in reciprocal space that are possible. However, these may be reduced to as few as 5 basic types, characteristic of what would normally be termed cooperativity. Complex curves may be explained, conceptually, by a molecular model based upon the sequential changes in intersubunit interactions leading to either positive or negative effects on the next ligand *at each step*. An experimental system may be analyzed by examination of the slopes of Hill plots in each region of ligand binding. Cornish-Bowden and Koshland[59] have demonstrated that this is a sensitive method of determining at least qualitatively the sequence of positive and negative cooperativity resulting from stepwise interactions in ligand binding. A series of simulated Hill plots for a tetramer were presented that should be useful for estimation of the mixture of cooperativity present in experimental data. Abrupt transitions between linear segments appear in double reciprocal plots for glutamate dehydrogenase and have been analyzed by deletion of specific terms in a typical, high-order rate equation.[93] Although the mathematical basis for this analysis is appropriate, the molecular requirements appear to be rather severe in terms of switching on and off specific enzymic species that contribute particular terms to the rate equation. Thus, for glutamate dehydrogenase it requires a tenth-order or a seventh-order equation to describe the series of four linear regions. Too much reliance is placed upon the appearance of straight lines; most binding or catalytic processes are probably continuous rather than discrete functions.[59,94]

A complete analysis of mixed cooperativity in an experimental system is best accomplished by fitting the complete saturation curve and determination of the statistical binding constants.[36] Comparison of these $K'$ values for each step of binding allows a description of the binding in terms of the appropriate mixture of positive and negative cooperativity. Thus, if $K'_1 > K'_2 < K'_3 < K'_4$, the system would have negative cooperativity followed by two steps of positive cooperativity (note that these are association constants) and would generate an intermediary plateau in the $\bar{Y}$ vs. $X$

plot, providing that the magnitudes of the constants were appropriate. Similar analysis of a rate equation would yield the apparent kinetic cooperativity, but with the realization that much greater differences in the constants (and hence in the mixture of cooperativity) may occur since the catalytic rate constant for each site is also involved in the coefficients of the rate equation.

*Half-Site Reactivity.* If the interactions producing negative cooperativity in the reaction of a ligand with a protein are so extreme that the dissociation constant of the last site(s) is (are) very large, then the enzyme will appear to bind fewer ligands than might be expected on the basis of the number of protomers. This is the case of the so-called half-site (or half-of-the-sites) reactivity that has now been observed for numerous enzymes.[47,95,96] A tetramer might bind two molecules of a particular ligand or a dimer might bind one. Although a "partial-site" reactivity might be expected to occur, in all cases the observations appear to be that half of the total number of sites react. Half-site reactivity may be observed in the noncovalent binding of a substrate or ligand; in the covalent reaction with a substrate, pseudosubstrate, or chemical reagent; or in the rapid transient kinetic determination of the maximal amount of covalent enzyme–substrate intermediate.

Glyceraldehyde-3-phosphate dehydrogenase (GPDH) illustrates several half-site phenomena. Only two of the four possible active sites of the tetramer are acylated by the pseudosubstrate, $\beta$-(2-furyl) acryloyl phosphate, but complete loss in activity toward normal substrates occurs concomitantly.[97] Alkylation of the same active site cysteine has been studied with a variety of reagents and may be divided into two classes: those that show half-site reactivity in alkylation (two react more rapidly and cause complete enzymic inactivation), e.g., fluorodinitrobenzene, and those that behave "normally" (four react at the same rate with inactivation proportional to the number of sulfhydryls reacted), e.g., iodoacetate. Yeast and sturgeon muscle GPDH also demonstrate half-site reactivity.[47,95] These results demonstrate the occurrence of cooperativity and half-site reactivity with different ligands or reagents in the same enzyme and the dependence of the nature of the half-site reaction on the particular reagent under observation.

Half-site reactivity could occur in an oligomeric enzyme composed of identical polypeptide chains by one of two mechanisms. Either the enzyme could be symmetrical before addition of the ligand in question and binding *could lead* to the asymmetric final state (Fig. 10, line b), or the

---

[95] F. Seydoux, O. P. Malhotra, and S. A. Bernhard, *Crit. Rev. Biochem.* **2**, 227 (1974).
[96] M. Lazdunski, *Curr. Top. Cell. Regul.* **6**, 267 (1972).
[97] O. Malhotra and S. A. Bernhard, *J. Biol. Chem.* **243**, 1243 (1968).

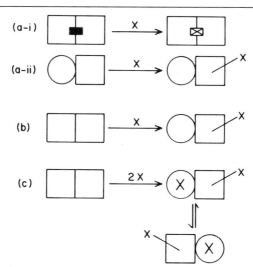

FIG. 10. Half-site reactive and flip-flop dimeric enzymes. The square conformation is reactive toward reagent X, but the circle conformation is not reactive. A covalent bond is indicated by ——X; a noncovalent interaction is indicated by the X inside the circle. (a) Half-site reactivity through a preexisting symmetry: i, occlusion of a site in the dimer; ii, mutual distortion. (b) Half-site reactivity through a ligand-induced conformational change in a symmetric enzyme. (c) A flip-flop mechanism.

oligomeric enzyme could exist in an asymmetric state with different inherent reactivities *before* the addition of the ligand or reagent (Fig. 10, line a). The latter interpretation is favored by Bernhard and co-workers[95] based upon alkylation–acylation experiments of GPDH and the observation of asymmetrical oligomers of human apo-GPDH, yeast hexokinase, coenzyme-bound liver alcohol dehydrogenase, and coenzyme-bound heart muscle malate dehydrogenase. Asymmetrical changes in an *a priori* symmetrical enzyme are favored by Koshland and co-workers[47] as an extension of the sequential model for cooperativity and also based upon alkylation–acylation experiments of muscle and yeast GPDH.[98] More recent observations with a fluorescent analog of NAD[99] and with fluorine magnetic resonance of modified enzyme[100] have indicated that the half-site reactivity is due to ligand-induced changes in the symmetric enzyme.

Half-site reactivity need not lead to cooperativity in the observed kinetics. Lazdunski[96] has demonstrated another manifestation of half-site reactivity in the "flip-flop" mechanism. If an oligomeric enzyme can alter-

---

[98] A. Levitzki, *J. Mol. Biol.* **90**, 451 (1974).
[99] J. Schlessinger and A. Levitzki, *J. Mol. Biol.* **82**, 547 (1974).
[100] J. W. Long and F. W. Dahlquist, *Biochemistry* **16**, 3792 (1977).

nate between two quaternary arrangements of a covalent intermediate (Fig. 10, line c), then a flip-flop may occur between these arrangements that is mediated by the noncovalent binding of ligand on the neighboring subunit. Thus, the enzyme would, in effect, alternate its catalytic cycle between the two protomeric active sites. The enzyme could benefit in terms of catalytic efficiency and the energetics of obtaining the most active conformation. This behavior is best typified in the studies on alkaline phosphatase (AP) from *Escherichia coli,* which has the following properties.

1. AP shows negative cooperativity of inorganic phosphate binding (noncovalent) at pH 8.
2. Negative cooperativity of covalent phosphorylation of the enzyme (2 mol per dimer) by inorganic phosphate has been shown in kinetic and equilibrium measurements with a greater cooperativity at pH 5 than pH 4.2.
3. Covalent phosphorylation by substrates (ATP, AMP) has been shown to be negatively cooperative at pH values below 6 with 2 mol of phosphate incorporated per dimer. However, at pH values above 7 the enzyme demonstrates half-site reactivity with only 1 mol of phosphate per mole of dimer.
4. The $Co^{2+}$ enzyme has somewhat different properties than the $Zn^{2+}$ enzyme with respect to negative cooperativity and half-site reactivity.
5. The overall phosphatase steady-state reaction demonstrates normal Michaelis–Menten kinetics. The rate equation of the flip-flop mechanism for alkaline phosphatase has been shown to generate hyperbolic kinetics and to be consistent with this finding.

Consideration of Induced Fit vs. Preexisting Equilibrium Mechanisms of Ligand Binding and the Utilization of Rapid Relaxation Techniques

When a ligand binds to a protein and a new conformation of the protein is observed, two different pathways may be pictured: a ligand-induced fit or a shift of a preexisting equilibrium by preferential ligand binding. The general scheme for binding of ligand to a monomer is shown in Fig. 11, in which we can assume, for purposes of illustration, that the liganded circle is strongly favored in the presence of ligand. The extreme case of induced fit would occur across the diagonal of the diagram from unliganded square conformation to liganded circle conformation, i.e., no intermediates. In reality, the induced pathway would probably occur by an initial binding of the ligand to the square (lower left), followed by the

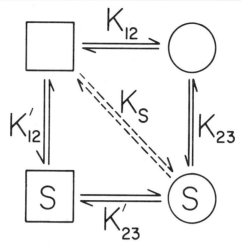

FIG. 11. Comparison of pathways of substrate binding and conformational changes in a monomer. Induced fit in ligand binding would occur through $K'_{12}K'_{23}$ pathway, the limit appearing to occur directly through the diagonal, $K_s$. Ligand-induced shift of preexisting equilibrium would occur through $K_{12}K_{23}$. Constants are written as association constants ($K'_{12}$ and $K_{23}$) or as equilibrium constants for circle/square ($K_{12}$ and $K'_{23}$).

conformational change. The alternative pathway consists of a preexisting equilibrium between square and circle (top line), with the ligand preferentially binding to the circle and pulling the overall equilibrium. The major conceptual difference between these two pathways is simply the amount of the various intermediates either at equilibrium or transiently during the binding process. Thus, if significant amounts of unliganded circles exist in addition to square, then a preexisting equilibrium may be shifted by the ligand binding. If some liganded squares (but no free circles) exist transiently or at equilibrium, the induced-fit pathway would appear to be operating. Thermodynamically, the two pathways are equivalent, since the initial and the final states are the same; hence $K_{12}K_{23} = K'_{12}K'_{23}$.

The equation describing the equilibrium binding will have the same form, no matter which pathway is followed. For example, the simplest sequential model[6] utilizes the concept of induced fit and does not envision the occurrence of unliganded second conformations; however, the mathematics incorporates a term in $K_t$ or $K_{12}$ (always accompanied by a $K_X$, or $K_{23}$, binding term) which is exactly the term for the expression of a preexisting equilibrium. A ligand binding process may, therefore, be conceptually divided into the energy required for the isomerization and the energy required for the intrinsic binding,[78] regardless of the actual pathway used. In considering various mechanisms for binding of ligand to enzymes (e.g., concerted vs. sequential; slow transition vs. mnemonic, see this volume [8], the essential differences will probably not lie in the presence or the absence of a preexisting equilibrium.

The process of binding and isomerization may also be kinetically controlled, the actual pathway being determined by the relationship of the rate constants. The process must be studied by rapid kinetic techniques, such as temperature-jump relaxation or stopped-flow methods, in order to determine the pathway of binding. Even if insignificant amounts of liganded-square exist at equilibrium, the transient presence of this intermediate may be inferred from the dependence of the relaxation time ($\tau$) on the ligand concentration. The dependence of the fast, $\tau_1$, and the slow, $\tau_2$, relaxation on the equilibrium concentrations of ligand, $\bar{S}$, and enzyme, $\bar{E}$, is given for each separate pathway by Eqs. (25)–(29) for a system perturbed from equilibrium in which the binding is rapid relative to the isomerization and the substrate concentration is assumed to be buffered.[49]

Preexisting equilibrium:

$$\tau_1^{-1} = k_{32} + k_{23}[\bar{E} + \bar{L}] \tag{25}$$

$$\tau_2^{-1} = k_{12} + \frac{k_{21}}{1 + K_{23}\bar{L}[1 + K_{23}\bar{E}]^{-1}} \tag{26}$$

Induced-fit:

$$\tau_1^{-1} = k'_{21} + k'_{12}[\bar{E} + \bar{L}] \tag{27}$$

$$\tau_2^{-1} = k'_{32} + \frac{k'_{23}}{1 + (K'_{12})^{-1}[\bar{E} + \bar{L}]^{-1}} \tag{28}$$

where the rate constants are given by

$$K_{12} = k_{12}/k_{21}; \; K_{23} = k_{23}/k_{32}; \; K'_{12} = k'_{12}/k'_{21}; \; K'_{23} = k'_{23}/k'_{32} \tag{29}$$

Both pathways produce a linear increase in the value of the reciprocal relaxation, $\tau_1^{-1}$, with increasing concentration of the ligand. For the slower relaxation process, $\tau_2$, the induced-fit pathway, $K'_{12}K'_{23}$, produces a hyperbolic increase in the reciprocal relaxation time, but the preexisting equilibrium pathway, $K_{12}K_{23}$, produces a decrease in the reciprocal relaxation time with ligand concentration. This difference may be viewed conceptually as follows: in the shift of preexisting equilibrium, the relaxation process at low ligand concentrations is the rapid binding to the appreciable amounts of circle that are already present, but at higher ligand concentrations the relaxation process is limited by the slower isomerization, $K_{12}$. For the induced-fit pathway the relaxation rate will increase as ligand binds to the square form in the first step, but then it will level off as the isomerization, $K'_{23}$, becomes rate limiting. If the two kinetic steps are close enough in time to be coupled, exact analysis becomes more difficult.

Similarly, stopped-flow techniques, which measure the amount of the enzyme–ligand complex or which measure the catalytic activity of the enzyme–substrate complex after rapid mixing, may be used to distinguish the two pathways.[101] An induced-fit pathway will lead to a continually increasing, asymptotic dependence of the reciprocal of the time constant (i.e., the rate of the rapid transient process observed) on the ligand concentration, because the initial step is always a second-order process depending upon the ligand itself. Stopped-flow studies of a preexisting equilibrium pathway, however, will lead to a reciprocal time constant dependence on ligand concentration, which levels off a finite value since the first-order isomerization must always occur before ligand can bind. Furthermore, since the preexisting pathway must always shift the equilibrium toward the higher-affinity species, the pre-steady-state kinetics in a pure $K$ system must always show a lag rather than a rapid burst of activity.[101]

In principle, then, the two pathways for a monomeric enzyme (Fig. 11) may be distinguished if the proper rapid kinetic techniques are applied. For simple enzymes numerous examples have been found of enzyme isomerization in the absence of ligand and of induced-fit pathways of binding and isomerization.[102] Many types of conformational changes occur upon ligand binding, and we are here discussing only those that have some influence on the activity or the equilibrium states of the protein. Other, faster conformational changes may also occur but are not represented in a figure such as Fig. 11, since they would only obscure the picture and not yield additional information.

For more complex enzymes showing cooperativity of ligand binding, the analysis of rapid relaxations is not nearly so simple. The arguments presented above for different pathways of isomerization do not directly distinguish between the concerted and the sequential models for cooperativity, since, as we have seen, the major distinction between the models is related to the preservation of symmetry of the enzyme–ligand complexes. Nevertheless, rapid relaxation techniques have been applied to several allosteric enzymes, notably glyceraldehyde-3-phosphate dehydrogenase, aspartyl transcarbamylase, and homoserine dehydrogenase. In general, utilization of these rapid kinetic techniques allows the determination of rate constants of certain elementary steps in the binding–

---

[101] C. Mouttet, F. Fouchier, J. Nari, and J. Ricard, *Eur. J. Biochem.* **49**, 11 (1974).
[102] G. G. Hammes and P. R. Schimmel, *in* "The Enzymes" (P. Boyer, ed.), 3rd ed., Vol. 2, p. 67. Academic Press, New York, 1970.

isomerization–catalysis process, depending upon what is observable and upon the model chosen to interpret the results. Ideally, one would expect more relaxation times for a sequential model than for a concerted model since there are more intermediate forms present. In practice, these multiple relaxations may be too close together to resolve or may in fact degenerate into nearly the same rate, since they represent ligand–protomer interactions in similar species of enzyme. Indeed, it has been shown[103] that rapid relaxation methods in themselves are insufficient to distinguish definitively between the concerted and the sequential models in a generalized, dimeric enzyme. Equations for the relaxations expected for the various cooperativity models have been presented.[49,50,103–106]

Aspartyl transcarbamylase presents an interesting study of the relaxations associated with ligand binding.[49,102] Briefly, the binding of ATP or CTP and their analogs to the regulatory sites is accompanied by a relaxation that is consistent with binding followed by isomerization. Binding of succinate (a substrate analog) in the presence of the other substrate, carbamyl phosphate, appears to follow the same pathway when binding to the isolated catalytic subunit; when binding to the complete enzyme, an additional slower relaxation rate was observed that decreased with ligand concentration, suggesting a preexisting equilibrium of the enzyme (analogous to pathway $K_{12}K_{23}$ of Fig. 11). Two separate but interdependent relaxations, attributable to isomerizations, were observed when succinate, carbamyl phosphate, and a CTP analog were present. These results clearly show that there are rapid isomerizations associated with the catalytic process as well as at least two slower isomerizations related to the regulatory process. In addition, negative cooperativity in the binding of ATP, CTP, and carbamyl phosphate have been observed. Neither limiting model is exactly sufficient to explain the data, because more relaxations were found than were predicted by the concerted model, but fewer than were predicted by the sequential model. A partially concerted model with *three* conformational changes has been used to interpret the data,[49,102] but a hybrid model with sequential and concerted changes also fits the data.[107] However, it is still useful for many equilibrium type of experiments to interpret the enzyme in terms of a two-state model.[108]

---

[103] G. M. Loudon and D. E. Koshland, Jr., *Biochemistry* **11**, 229 (1972).
[104] J. Janin, *Prog. Biophys. Mol. Biol.* **27**, 77 (1973).
[105] K. Kirschner, E. Gallego, I. Schuster, and D. Goodall, *J. Mol. Biol.* **58**, 29 (1971).
[106] G. G. Hammes and C. W. Wu, *Biochemistry* **10**, 1051, 2150 (1971).
[107] M. Dembo and S. I. Rubinow, *Biophys. J.* **18**, 245 (1977).
[108] G. J. Howlett, M. N. Blackburn, J. G. Compton, and H. K. Schachman, *Biochemistry* **16**, 5091 (1977).

Status and Prospects

*Glyceraldehyde-3-phosphate Dehydrogenase*

The cooperativity of binding of NAD to glyceraldehyde-3-phosphate dehydrogenase (GPDH) from several sources demonstrates a variety of cooperative phenomena and has been interpreted in terms of molecular models. The tetrameric enzyme from rabbit muscle shows strong negative cooperativity,[55] the first two sites having association constants about $10^8$ to $10^{10}$ $M^{-1}$ and the second two about $10^5$ to $10^6$ $M^{-1}$. The GPDH from sturgeon muscle also shows negative cooperativity but the difference between the strong and the weaker sites is only about one to two orders of magnitude.[95,100] The GPDH from rabbit muscle undergoes marked structural changes upon addition of the first NAD to apoenzyme including increased sulfhydryl reactivity, increased susceptibility to proteolysis, and marked changes in ultraviolet difference spectra, fluorescence, and circular dichroism.[55,66,99] The nicotinamide subsite as well as the adenine subsite are necessary for the negative cooperativity.[109] Temperature-jump studies of interconversion of stable forms indicate at least two relaxations for the addition of the second two NAD molecules and an even more complex spectrum for addition of all four coenzymes to the apoenzyme.[49,110] Stopped-flow experiments describe even slower interconversions of enzyme forms, presumably owing to the primary transition to stable enzyme–ligand forms. These data may be interpreted in terms of either a generalized sequential model (cf. Fig 7, line b), in which the major structural change occurs upon binding of the first NAD but with subsequent smaller conformational changes,[49,110] or a hybrid model in which concerted quaternary structural changes occur as well as sequential changes in individual protomers.[66]

The yeast GPDH, in contrast, has been described with either positive cooperativity[105] (particularly at higher pH values and temperatures) or a mixture of positive and negative cooperativity.[111] The latter result has been attributed to enzyme heterogeneity, since the negative interaction is eliminated when the enzyme is highly purified by affinity chromatography; only the positive cooperativity remains.[112] Temperature-jump studies are simpler for the yeast GPDH with only three relaxations, consistent with those expected for the concerted model.[105] Extensive calorimetric studies have shown that the positive cooperativity of NAD binding

---

[109] D. Eby and M. E. Kirtley, *Biochemistry* **15**, 2168 (1976).
[110] G. G. Hammes, P. J. Lillford, and J. Simplicio, *Biochemistry* **10**, 3686 (1971).
[111] R. A. Cook and D. E. Koshland, Jr., *Biochemistry* **9**, 3337 (1970).
[112] L. S. Gennis, *Proc. Natl. Acad. Sci. U.S.A.* **73**, 3928 (1976).

is accompanied by large negative changes in heat capacity and entropy, suggesting a largely hydrophobic effect and a tightening of the protein structure.[113] The binding, rapid kinetic, and calorimetric data for the positively cooperative interactions of NAD with yeast GPDH can be fit either by a concerted model,[105,113] by a generalized sequential model,[113] or by the hybrid model.[66] Consequently, the enzyme from different species may have similar structural changes, but with the benefits of positive or negative cooperative binding related to the needs of the particular organism.

## Combinations of Cooperative Models

Each of the models for cooperativity discussed above represents a single limiting possibility. One may imagine that enzymes, being as predictably complex as they have been found to be, might incorporate one, or more, or all of the different types of cooperativity in their normal functioning. This "worst" case might consist of an enzyme that (a) had a protomer–dimer–oligomer equilibrium; (b) had site–site interactions of the sequential model, with different cooperativities in each of the oligomers; (c) had a different number of ligand binding sites per subunit in one or more of the species; (d) had a different catalytic rate constant in each of the polymeric species; (e) had two (or more) substrates, so that the cooperativity of the ligand in question was dependent upon the second substrate; and (f) had an additional conformational state accessible only through a ligand-induced slow transition (hysteresis) that introduced a new set of kinetically significant states. If this were the case, it is clear that the rate equation would be too complex to write down and analyze in any systematic manner. The investigator would be forced to examine portions of the whole system that were easily accessible (e.g., equilibrium binding). This situation might seem too difficult to conceive, but it appears to the author to be a logical outgrowth of known enzymes and known studies.

One enzyme that appears to have approached this situation is the aspartokinase–homoserine dehydrogenase I of *E. coli,* which has been studied by many laboratories.[114,115] Aspartokinase–homoserine dehydrogenase I is a bifunctional enzyme that is feedback inhibited by threonine. The enzyme possesses (a) a dimer–tetramer equilibrium in which threonine shifts the equilibrium toward the tetramer; (b) site–site interactions of threonine binding in the tetramer; (c) eight threonine bind-

[113] C. W. Niekamp, J. M. Sturtevant, and S. F. Velick, *Biochemistry* **16,** 436 (1977).
[114] P. Truffa-Bachi, M. Veron, and G. N. Cohen, *Crit. Rev. Biochem.* **2,** 379 (1974).
[115] C. F. Bearer and K. E. Neet, *Biochemistry* **17,** 3512, 3517, 3523 (1978).

ing sites per tetramer; (*d*) a different Hill coefficient in the steady-state kinetics of threonine inhibition of the aspartokinase activity than in the binding cooperativity; and (*e*) two separate slow transitions, one for each activity. This type of behavior may be the culmination of eons of evolution in developing the optimal enzyme for a particular function.

## [8] Hysteretic Enzymes

*By* KENNETH E. NEET and G. ROBERT AINSLIE, JR.

### Historical Background and Development of the Hysteretic Concept of Nonrapid Enzyme Transitions

*Hysteresis, Ligand-Induced Slow Transitions, Mnemonic (Memory) Enzymes, Ligand-Induced Conformational Shifts of Equilibria*

The concept that enzymic reactions should demonstrate linear reaction rates and therefore linear progress curves, after a very rapid attainment of the steady state, dates to the early formulations of enzyme kinetics by Michaelis, Menten, Henri, Briggs, and Haldane. Even with the advent in the 1950s of the induced-fit theory and the realization that enzymes would undergo conformational changes (isomerizations) in the presence of substrates, linear progress curves and rapid isomerizations were considered to be proper. Indeed, assay conditions were generally modified until such linearity was obtained. Within the last 8–10 years a somewhat different viewpoint has developed, namely, that not all enzyme transitions need be rapid. The relative slowness of the process could lead to interesting enzymic properties that might prove to be beneficial to the cell (as well as to the enzymologist).

Early suggestions by Witzel[1] and by Rabin[2] were that a slow conformational change or relaxation of an enzyme might produce "cooperative" kinetics or non-Michaelis–Menten behavior for certain special enzyme mechanisms. In 1970 Frieden[3] recognized the prevalence of enzymes that showed a slow response to changes in concentration of substrates or modifiers and coined the term "hysteresis" or "hysteretic enzyme" to describe this type of slow behavior in enzyme assays. The emphasis in this pacesetting work was on the time dependence of the response, the

---

[1] H. Witzel, *Hoppe-Seyler's Z. Physiol. Chem.* **348**, 1249 (1967).
[2] B. R. Rabin, *Biochem. J.* **102**, 22c (1967).
[3] C. Frieden, *J. Biol. Chem.* **245**, 5788 (1970).

slowness of association–dissociation reactions,[4] the possible role of hysteresis in the damping of physiological processes, and the correlation of hysteretic responses with regulatory enzymes. Shortly thereafter, Ainslie et al.[5] extended the considerations of slow processes to a general consideration of the kinetic cooperativity that might be generated by a ligand-induced slow transition of an enzyme. These authors emphasized the analysis of the steady-state kinetics in such a system, the possibilities of producing cooperative kinetics in a monomeric enzyme, the possibility of altering the kinetic cooperativity of an oligomeric enzyme, and the correlation between enzymes showing hysteretic responses and possessing cooperativity in their kinetics. Ricard and co-workers[6] have presented a somewhat different formulation of these concepts with an emphasis on the "memory" of the enzyme and the analysis for a two-substrate (A,B) monomeric enzyme. The latter workers at first[6] neglected the requirement for slowness of the transitions, but in more recent work[7] have recognized the necessity for a slow change in the enzyme. Rübsamen et al.[8] concurred that ligand-induced conformational shifts of equilibria were at the heart of generating kinetic cooperativity in a monomeric enzyme, but missed the realization that cooperativity could not occur in equilibrium binding.

With these different formulations and viewpoints of similar systems, it is necessary that the similarities and true differences between various models be pointed out. We will attempt to present here a general, unified picture of hysteretic enzymes, emphasizing the areas of agreement between different workers and pointing out the details of certain specific models that make them attractive to explain particular enzymes. Two aspects of these systems that will be discussed are the hysteretic (slow) process itself and its relationship to cooperativity in the steady-state kinetics. Although these two aspects are related through the models, they must be remembered as different concepts; hysteresis may occur without cooperativity, but cooperativity (by this mechanism) requires a molecular transition. The terms hysteresis, slow transitions, memory, and mnemonics will hereinafter be used to refer to the same type of enzymes; the term transient will be used for the observed catalytic process, transi-

---

[4] C. Frieden, in "The Regulation of Enzyme Activity and Allosteric Interactions" (E. Kramme and A. Pihl, eds.), p. 59 Academic Press, New York, 1968.
[5] G. R. Ainslie, Jr., J. P. Shill, and K. E. Neet, *J. Biol. Chem.* **247**, 7088 (1972).
[6] J. Ricard, J.-C. Meunier, and J. Buc, *Eur. J. Biochem.* **49**, 195 (1974).
[7] J. Ricard, J. Buc, and J.-C. Meunier, *Eur. J. Biochem.* **80**, 581 (1977).
[8] H. Rübsamen, R. Khandker, and H. Witzel, *Hoppe-Seyler's Z. Physiol. Chem.* **355**, 687 (1974).

tion will refer to the molecular event, and hysteresis will refer in a broad sense to the slow responses.

*General Description of Hysteresis and Cooperativity.* The following points are generally agreed to refer to and define "hysteretic" enzymes (with cooperativity, where appropriate).

1. The key point that distinguishes a hysteretic enzyme is the availability of an additional state of the enzyme that has a different physical form and different kinetic properties. Very rapid or very slow achievement of the state does not result in hysteresis. Only thermodynamically reversible changes of state are germane to this definition of hysteresis, thereby eliminating from consideration systems in which covalent modification occurs, e.g., interconversion of phosphorylase *a* and phosphorylase *b*.
2. Since the transition to the second state is "slow," the term "slow" must be defined; two alternative definitions may be used. First, "slow" may mean that the transition may occur on a time scale, minutes to hours, that is readily observed in assays and is of the same order of magnitude as changes in metabolite concentrations *in vivo*. This time scale can then become important in damping physiological responses to changes in substrate or effector concentrations. Second, "slow" may mean "not fast" when considered on a microscopic basis relative to other rate constants of the system (see later section). Such a rate could still be more rapid than is easily seen in standard spectrophotometers (~1 sec) and would require rapid-mixing techniques.
3. The substrate can either shift the equilibrium toward the additionally available state of the enzyme or can induce a new state which was not previously there. Thus, the question of preexisting equilibrium vs. induced fit is not central to the interpretation of hysteretic enzymes (see this volume [7]), although these alternatives will have an effect on the detailed formalism of particular models. Furthermore, the substrate can shift the enzyme toward the second state either by altering the equilibrium position (changing the free-energy difference between the two states) or by altering the rate of the transition (changing the free-energy barrier between the states).
4. The slow step of the hysteretic transiton can be a conformational change (isomerization), a dissociation–association reaction of the enzyme, or a direct displacement of a bound ligand by a second ligand.[3,5] The authors' bias is that in many situations an underlying slow isomerization, linked to the dissociation or ligand displacement, actually occurs. The importance of conformational changes will be emphasized in this review.
5. Kinetic cooperativity of a monomeric, hysteretic enzyme can occur

with appropriate values of the rate constants. Simulation studies of several related mechanisms using realistic values, of the rate constants, i.e., values consistent with known physical chemical properties of proteins and ligand binding of proteins, have demonstrated the cooperativity. Either positive or negative cooperativity may exist and either burst-type or lag-type slow transitions may occur. Nonhyperbolic saturation curves generated by a slow transition will hereafter be termed hysteretic cooperativity to differentiate them from cooperative curves generated by site–site interactions (this volume [7]; see also a discussion of the terminology of cooperativity in that chapter).
6. Hysteretic cooperativity can occur only in kinetic studies, not in equilibrium techniques, such as direct-binding studies.

## Physiological Rationale for Hysteresis

*Damping.* Hysteretic feedback inhibition by products of various types of pathways may have profound effects on the levels of the metabolites in those pathways and the rate of change of those levels. Frieden[3] has discussed the ways in which hysteretic enzymes could contribute to controlling the flow of metabolites through a sequence or network of enzymes by "damping" effects. If the concentration of any feedback effector reaches an inhibitory level and remains at that level, the concentrations of all intermediates in the pathway and all intermediates and end products of branches of that pathway will decrease to a lower level. However, with hysteretic inhibition the levels of the intermediates and the other end products can continue at their preinhibited levels for a longer time. Thus, the concentration of a feedback inhibitor can go up, due to ingestion of that inhibitor, for example, and a hysteretic response will allow some time for the utilization of the excess feedback inhibitor before decreasing the rates of synthesis of the other intermediates and end products of that metabolic pathway. If the excess inhibitor is used before the hysteretic enzyme can respond, then there would be little or no change in the concentrations of the intermediates and other end products. Hysteretic behavior may have its greatest utility in decreasing the amplitude of oscillations in the concentrations of metabolites in response to oscillations in the concentration of a substrate or effector. For this to be effective, the time constant of the hysteresis would have to be the same order of magnitude as the period of the inherent oscillations of the pathway. The hysteresis of phosphofructokinase is on the order of minutes,[4,9,10] which may allow modulation of the oscillations of similar time period in glycolysis.[11]

[9] P. E. Bock and C. Frieden, *J. Biol. Chem.* **251**, 5630 (1976).
[10] C. Frieden, H. R. Gilbert, and P. E. Bock, *J. Biol. Chem.* **251**, 5644 (1976).
[11] K. Tornheim and J. M. Lowenstein, *J. Biol. Chem.* **250**, 6304 (1975).

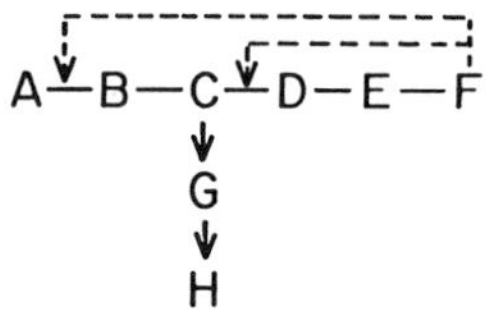

```
          ┌─────────────┐
          ↓    ┌────────↓
        A─B—C─D—E—F
             ↓
             G
             ↓
             H
```

FIG. 1. A to H are intermediates in a metabolic pathway. The solid lines indicate the direction of flow in the pathway. The dashed lines indicate feedback inhibition.

Under certain circumstances the levels of intermediates could actually increase for a time in response to an increase in the level of the feedback inhibitor before decreasing to their final values. If the feedback inhibitor F in a branched pathway (Fig. 1) hysteretically inhibits the A–B conversion and rapidly inhibits the C–D conversion (or D–E if C–D is reversible), the level of C and, therefore, of H will rise temporarily until the hysteretic inhibition becomes effective. Having temporarily raised the levels of H when F is in excess could be beneficial, for example, in the case where H and F are amino acids that must be incorporated into proteins in a specific ratio. The biosynthetic pathway for threonine (F) and lysine (H) in *Escherichia coli* might behave in this fashion since two hysteretic transitions in the bifunctional enzyme aspartokinase–homoserine dehydrogenase I have been reported (see later section).

*Cooperativity.* Another potential physiological benefit of hysteresis is to generate kinetic cooperativity. If an enzyme shows positive or negative cooperativity of kinetics, then it is responsive over a narrower or wider range, respectively, of ligand concentration than it would be for a hyperbolic response. If there is an evolutionary pressure on key regulatory enzymes to have cooperativity, then this could be achieved either with site–site interactions in oligomers or with a slow transition.[12] Hysteresis may contribute, then, by producing cooperativity in a monomeric enzyme or by amplifying or suppressing cooperativity due to site–site interactions. In addition, regulation may occur if an allosteric modifier alters the enzymic hysteresis and hence the kinetic cooperativity. The modifier may itself show kinetic cooperativity of inhibition or activation due to the slow transition.[5]

## Analysis of Kinetic Transients

*Consideration of Artifacts —Is the Transient Real?*

Kinetic transients may be artifactual for several reasons; it is essential that these possibilities be eliminated before one is certain that the

[12] K. E. Neet and G. R. Ainslie, Jr., *Trends Biol. Sci.* **1**, 145 (1976).

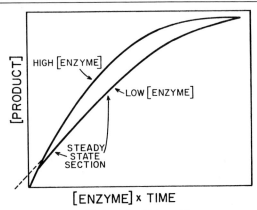

FIG. 2. Progress curves (to equilibrium) for high and low concentrations of an enzyme with a transient burst. The time scales for the two curves have been normalized for comparative purposes, using the enzyme concentrations. The rate of the transition is the same for both concentrations.

kinetic transient is a property of the enzyme. Product accumulation and substrate depletion cause the velocity of a reaction to decrease as it approaches equilibrium. For some reactions this curvature begins almost with the start of the reaction. Substrate concentrations that are at least four or five times the Michaelis constants should be used to minimize this problem. In order to distinguish substrate and product effects from a slow transition in enzyme activity, different enzyme concentrations can be tested. If an enzyme transition is really present, the lower enzyme concentration will have catalyzed a smaller fraction of the reaction (ideally less than 20%) by the time the transient is over (Fig. 2). There will then be a better chance of seeing the transient followed by a more linear section of the progress curve before the reaction rate begins to fall owing to the approach to equilibrium. On the other hand, if the suspected transient is really just an approach to equilibrium, the progress curve, normalized for enzyme concentration, will be the same shape as it was at the higher enzyme concentration. This approach will not work, of course, if the transition is a dissociation itself (phosphorylase[13]) or an isomerization whose rate depends upon the quaternary structure (hexokinase[14]). When possible, adding a coupling enzyme system either to scavenge products or to recycle products to substrates could distinguish approach to equilibrium from a burst (phosphofructokinase[9]). This system would not change the decay rate of an enzyme transition, but it would eliminate a velocity de-

[13] B. E. Metzger, L. Glaser, and E. Helmreich, *Biochemistry* **7**, 2021 (1968).
[14] J. P. Shill and K. E. Neet, *J. Biol. Chem.* **250**, 2259 (1975).

crease that was due simply to product accumulation or substrate depletion. The effect of product can be tested by addition of low concentrations in the initial reaction. If the transient is due to a high sensitivity to low concentrations of product, this should be readily apparent. Product inhibition may also be tested by addition of a second aliquot of enzyme after the reaction has attained a steady state. A slow transient when the second aliquot is added, similar to the first, is indicative of a true transition of the enzyme, not simple substrate or product effects. Although these effects are easy to demonstrate qualitatively, a quantitative analysis of a hysteretic transition in the presence of substrate depletion and product accumulation is difficult (see later section).

An apparent burst of enzyme activity could also be due to enzyme inactivation during the course of an assay. Inactivation can be checked by preincubating the enzyme under assay conditions for varying lengths of time, initiating the reaction with one of the substrates, and determining whether the initial velocity decreases with increasing incubation time. Addition of inert protein (e.g., serum albumin) or coating the glass with silicon or similar substances will sometimes offer protection from a general surface denaturation—inactivation. This process may also be tested for by varying the amount of glass surface with which the enzyme is in contact in different assays. Another test is to start the reaction at less than saturating concentrations of substrate and then add an additional aliquot of substrate in the linear steady state. If the burst were due to partial inactivation of the enzyme, the new steady state (due to higher substrate concentration) should be achieved rapidly, but if the burst were due to a slow transition, the enzyme should go through another burst to the new steady state.

A lag in enzyme activity will be seen if an insufficient concentration of coupling enzyme is used in a linked enzyme assay, i.e., a finite concentration of the coupled product has built up before the maximal rate is observable. The calculation of the amount of coupling enzyme necessary has been presented.[15] A range of concentrations of coupling enzyme should be tested with the enzyme under investigation and its substrates present at the highest concentrations that will be used in the experiments. This condition will be the highest rate of formation of product that the coupling enzyme will have to keep up with. A plot of initial velocity or steady-state velocity vs. coupling enzyme concentration should reach a plateau. A concentration of coupling enzyme that is on the plateau should then be used.

Changes in temperature, buffer, or pH can cause kinetic transients.

---

[15] W. R. McClure, *Biochemistry* **8**, 2782 (1969).

These possibilities can be eliminated by appropriate preincubation of the enzyme and initiation of the reaction with a substrate. If simple dilution causes a slow dissociation of the enzyme and there are differences in activity between the protomer and oligomer, than an assay transient will be observed.

If rapid kinetic methods are being used to measure the reaction, an additional consideration of the number of enzyme turnovers arises. The hysteretic transitions of interest here require hundreds or thousands of catalytic cycle turnovers before the transition to the second state is completed. Thus, the stochiometric burst of product release, as chymotrypsin becomes acylated, is qualitatively different from the multiple turnover during the burst of activity with yeast hexokinase. Comparison of the amount of product formed in the burst to the concentration of enzyme is sufficient to distinguish these possibilities.

## Characteristics of the Transient

All the possible different preincubations of substrates or effectors with enzyme should be tested. Comparison of the transient of a reaction started with enzyme to a reaction started with one of the various substrates may allow determination of the cause of the shift between the initial state and the final steady state of the enzyme. For example, with yeast hexokinase the burst of activity is observed when enzyme is preincubated with MgATP and free $Mg^{2+}$ and the reaction is started with glucose, or when enzyme is preincubated with glucose and free $Mg^{2+}$ and the reaction is started with MgATP. However, the transient burst is not observed when enzyme is preincubated with metal-free ATP plus glucose and the reaction is started with $Mg^{2+}$. Therefore, the conformational transition in yeast hexokinase is caused by glucose and ATP together.[14,16]

One must choose the proper enzyme concentration in order to explore the characteristics of the transient. The amplitude of the transient ($P$ intercept of Fig. 3A) is proportional to enzyme concentration. The assay method must be sensitive enough to be able to see at least half (preferably more) of the amplitude of the transient in order to obtain good values for the initial velocity and the rate of the transition, in particular. The amplitude of a burst is the amount of excess product generated by the enzyme in going from the initial state to the steady state; the amplitude of a lag is the amount of product not generated because of the initially less active form. Effects of pH, temperature, and buffers on the slow transition will also provide information on the molecular causes.

[16] B. A. Peters and K. E. Neet, *J. Biol. Chem.* **252**, 5345 (1977).

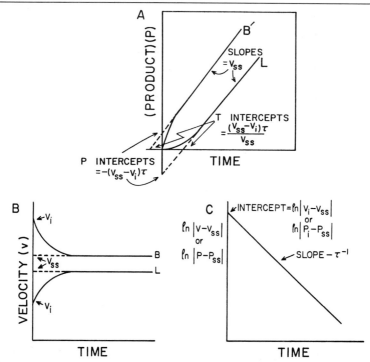

FIG. 3. Product and velocity time courses of a hysteretic enzyme. $P$ and $v$ are the instantaneous product and velocity; $v_i$ and $v_{ss}$ are the initial and steady-state velocities; B' and L indicate a burst or a lag, respectively. (A) Progress curves (solid lines) and extrapolations of the steady-state section (dashed lines). (B) Velocity transients (solid lines) and extrapolations of the steady state (dashed lines). (C) Exponential decay of the transient. Log of the absolute value of the difference between the extrapolated steady-state line and the actual process curve. The line could represent either bursts or lags with either velocity or product data (part A or B).

The time intercept of Fig. 3A (the induction period in the case of the lag, $L$) does not change with enzyme concentration unless there is an association or dissociation of the enzyme or the substrate depletion and/or product accumulation that occur during the decay of the transient are sufficient to change the steady-state rate.[17]

*Evaluation of Parameters*

*Graphical Methods.* In general, three parameters can be obtained from each progress curve in the analysis of a slow transition: the initial

[17] D. J. Bates and C. Frieden, *J. Biol. Chem.* **248**, 7878 (1973).

velocity, the steady-state velocity, and the apparent rate constant for the transition between those two velocities ($\tau^{-1}$).

A transient lag may be analyzed as an exponential increase in velocity to a constant steady state whether or not substrate addition is rapid equilibrium.[3,14,18] The three parameters can be obtained from the following equations for the product concentration (Figs. 3A and 3C).

$$P = v_{ss}t - (v_{ss} - v_i)(1 - e^{-t/\tau})\tau \qquad (1)$$
$$P_{ss} = v_{ss}t - (v_{ss} - v_i)\tau \qquad (2)$$
$$\ln(P - P_{ss}) = \ln[(v_{ss} - v_i)\tau] - t/\tau \qquad (3)$$

where $P$ is the observed product concentration at time $t$, $P_{ss}$ is the product concentration for the steady-state asymptote line, $v_i$ is the initial velocity, $v_{ss}$ is the steady-state velocity, and $\tau^{-1}$ is the observed rate of the transition between the initial state and the steady state. The value of $\tau$ is in general dependent upon all the rate constants for the particular mechanism rather than solely the constants for the slow transition. The three parameters can then be graphically estimated from the slopes of the progress curve (Fig. 3A) and the value of the intercept on either the $t$ axis or the $P$ axis. However, to get a meaningful value of the parameter $\tau$, a semilog plot (Fig. 3C) should be constructed to ensure that the process is truly first order; if it is not, the intercept has no general meaning.

Alternatively, the parameters can be obtained from Figs. 3B and 3C, which are derived from Fig. 3A by taking tangents of the progress curve at various times. Each tangent is the instantaneous velocity, $v$, and the equations that describe the differential form of the progress curve are

$$v = v_{ss} - (v_{ss} - v_i)e^{-t/\tau} \qquad (4)$$

and

$$\ln(v_{ss} - v) = \ln(v_{ss} - v_i) - t/\tau \qquad (5)$$

A burst can be analyzed in the same fashion; the only difference is that the semilogarithmic plots are for $\ln(P_{ss} - P)$ or for $\ln(v - v_{ss})$ versus time since for a burst $P_{ss} > P$ and $v > v_{ss}$.

Multiple exponential decays can also be handled graphically from a plot of $\ln v$ vs. time (Fig. 4). The decrease in velocity of the steady state may be due to product accumulation and the approach of the reaction to equilibrium. In this case, the steady-state velocity can be obtained by extrapolation of the second decay. The validity of a logarithmic extrapolation of a nonlinear steady-state progress curve to zero time has been dis-

---

[18] G. W. Hatfield, W. J. Ray, Jr., and H. E. Umbarger, *J. Biol. Chem.* **245**, 1748 (1970).

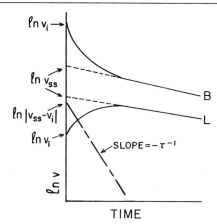

FIG. 4. Analysis of the transient by a first-order plot when the steady state also decays. The solid lines are the logs of the instantaneous velocities, $v$, for a burst, B, or a lag, L. The dashed lines are extrapolated for a decaying steady state. The dotted and dashed line is the log of the absolute value of the difference between the velocities from the solid line and the velocities from the dashed line for B (or for L).

cussed.[19] By subtracting the observed velocities from the corresponding velocities on the extrapolated line and plotting the logarithm of the absolute value of this difference vs. time, the rate of the transient (the first decay) can be obtained. Bursts and lags are handled analogously.

For a double exponential decay or any multiple exponential decay, there must be at least an order of magnitude difference between the rate constants for the decays in order to obtain accurate values of the parameters either by hand or by the above curve-peeling procedure. Otherwise, the fast decay will appear slower and the slow decay will appear faster than they really are.

The Guggenheim method[20] is useful for analyzing exponential decays that are not completely measurable to the final steady state. If not enough of the steady state were observable in Fig. 3A, then the dotted line could not be accurately drawn and good values of $(P - P_{ss})$ could not be obtained. The Guggenheim method is applied by taking the differences in concentrations of substrates at a fixed time interval. An exponential decay can be described by the equation

$$A = A_0 e^{-t/\tau} \tag{6}$$

[19] C. Zervos, *J. Theor. Biol.* **50**, 253 (1975).
[20] D. P. Shoemaker and C. W. Garland, "Experiments in Physical Chemistry," p. 222. McGraw-Hill, New York, 1962.

where $A$ is the substrate concentration, $A_0$ is the substrate concentration at zero time, $k$ is the rate constant for the decay, and $t$ is the time. If a series of points $A_i$ and $A_{i+\Delta}$ are taken with a constant time interval, $\Delta t$, it can be shown that

$$A_i - A_{i+\Delta} = A_0 e^{-t/\tau}(1 - e^{-\Delta t/\tau}) \tag{7}$$

Since the expression in parentheses is constant, a plot of $\ln(A_i - A_{i+\Delta})$ vs. time will be linear with a slope of $-\tau^{-1}$. An analogous treatment holds true for multiple exponential decays.[21] The equation is true for any time interval, but ordinarily $\Delta t$ is taken to be 30–50% of the period of time that the decay is observed. If a linear steady state is obtained, the Guggenheim method must be applied to the instantaneous velocities.

*Statistical Methods.* A more sophisticated method of data analysis involves smoothing of digitized data, computerization, and iterative fitting procedures. The most convenient methods for following a reaction give a continuous output with time, e.g., following the reaction with a spectrophotometer, fluorometer, or pH stat. The values of the product concentration at various times can be obtained either by digitizing the analog curves by hand or by monitoring the output with an analog-to-digital converter and collecting data at appropriate time intervals. Digitizing by hand is laborious; satisfactory values for the parameters have been obtained by using about 50 points.[14] Using an analog-to-digital converter, especially one interfaced with a computer or storage device, makes it practical to use more points for the calculations.[22]

Data can be smoothed and velocities calculated for the values of product concentrations and associated times by a sliding five-point fit. Five data points are fit to a second-order polynomial and the slope of the tangent to these five points at the center point is calculated from the parameters found for the polynomial.[14,23] The second-order polynomial will be a good approximation for the curvature of the data as long as curvature, over the small range of the five points, is not too extreme. Next, the first data point in the set of five is dropped, an additional data point is added, and the process is repeated; points 1–5 give the velocity at point 3, points 2–6 give the velocity at point 4, etc.

If the computer in which the product concentration and associated

---

[21] C. F. Bearer, Doctoral dissertation, Case-Western Reserve Univ., Cleveland, Ohio (1977).

[22] In our laboratory we routinely use a Gould 6000 Data Logger and manually transfer the magnetic tape to an interface into a DEC PDP 11/40 Computer. However, any small microprocessor would suffice.

[23] P. R. Bevington, "Data Reduction and Error Analysis for the Physical Sciences." McGraw-Hill, New York, 1969.

times have been stored is interfaced with a plotter or video terminal, then a graphical display of the log velocity vs. time plot may be made. From this curve the computer program can calculate a weighted least-squares fit to a straight line for the steady-state region, from a starting point chosen for linearity. The logarithm of the absolute value of the difference between the velocity and the extrapolated steady-state velocity line at the various earlier times is calculated and fit to an appropriately weighted straight line. From these lines, estimates of the parameters are computed.

The more straightforward and statistically more proper procedure is to use the parameters to fit the product vs. time data rather than to fit the velocities that were derived from those data. The general forms of the product equations are

$$P = C_1 - C_2 e^{-C_3 t} + C_6 t \tag{8}$$

for a single exponential decay to a steady state and

$$P = C_1 - C_2 e^{-C_3 t} - C_4 e^{-C_5 t} \tag{9}$$

if the steady state decays as well. The relation of the parameters in these equations to the desired parameters can be obtained by differentiation.

$$v_i = C_2 C_3 + C_6 \quad \text{for Eq. (8)} \tag{10}$$
$$\text{or} \quad = C_2 C_3 + C_4 C_5 \quad \text{for Eq. (9)}$$
$$v_{ss} = C_6 \quad \text{for Eq. (8)} \tag{11}$$
$$\text{or} \quad = C_4 C_5 \quad \text{for Eq. (9)}$$
$$\tau^{-1} = C_3 \quad \text{for either Eqs. (8) or (9)} \tag{12}$$

These equations are suitable when substrate depletion, product increase, or product inhibition are not major factors, i.e., when the second phase does not decay at all [Eq. (8)] or when the time of the experiment is less than one half-time ($C_5 t < 1$) of the second decay [Eq. (9)]. These approximations are reasonably good in many situations. In the event that product accumulation and substrate depletion have a substantial effect on the steady-state rate, it may be necessary to fit the progress curves to a more complex rate equation for a specific mechanism.[17] For example, at low substrate concentrations the transient is usually not prominent and it may not be clear whether the initial or the steady-state velocity is observed. Fitting the full time course of the progress curve may be particularly useful in this situation.[17,24]

The appropriate equation [Eq. (8), Eq. (9), or see Bates and Frieden[17]] may be fitted to the product–time data by an iterative, nonlinear least-squares procedure that minimizes the statistical parameter $\chi^2$. The parameters from the log $v$ vs. time data may be used as initial estimates. De-

[24] D. J. Bates and C. Frieden, *J. Biol. Chem.* **248**, 7885 (1973).

tailed descriptions, including Fortran programs, of the procedures that can be used to search for parameters to minimize $\chi^2$ for functions that are not linear in their parameters are available.[23] The procedure (Marquardt algorithm) that converges most rapidly is a cross between using a gradient search and an approximate analytical function that can be treated as being linear in its parameters. The Fortran programs for obtaining velocities from the product vs. time data according to Eq. (8) or Eq. (9), fitting the velocities to get preliminary parameters, and improving the values of these parameters with the Marquardt fitting procedure are available.

## Models for Hysteresis and Kinetic Cooperativity

Three ranges of rates for the transitions among various states of an enzyme may occur: slower than the slowest steps in the catalytic sequence, the same order of magnitude as the rate-limiting steps in the catalytic sequence, and faster than the rate-limiting steps in the catalytic sequence.

In the event that the transitions are slow, i.e., slower than the rates for substrate or product release from the enzyme or the rates of the catalytic interconversion steps, the enzyme forms within a state will have reached a steady state with respect to each other before the interconversion of states has reached a steady state. The *initial* velocity of such a system is the sum of the velocities of the separate states; in other words, the system can be treated as though multiple enzymes catalyzing the same reaction were present. For the special case in which the velocities for $n$ different states are all hyperbolic functions of substrate concentration the initial velocity is:

$$v_{\text{initial}} = \sum_{i=1}^{n} \frac{V_i A}{K_i + A} \qquad (13)$$

where A is the varied substrate. The initial velocity is a hyperbolic function of A if all the $K_i$ values are equal and has the same appearance as negative cooperativity if they are not. When the final steady state among states of the enzyme is reached, the *steady-state* velocity can be either positively or negatively cooperative or not cooperative at all. Slow transitions could be important as metabolic controls, because of the hysteresis, or because of the cooperativity they cause, or both.

For transitions of intermediate speed, when the range of rates for the transitions is the same order of magnitude as the rate-limiting step in the catalytic sequence of the reaction, there is no simply defined initial velocity for the reaction. During the time a steady-state distribution is being reached among the forms of an enzyme within all the states, a steady-state distribution is being reached among the states as well. Once again the steady-state velocity can be positively or negatively cooperative or not cooperative at all. Transitions of this speed would not produce hysteresis except on the time scale of the catalytic reaction. However, these transitions could be important for metabolic control due to the cooperativity they can produce.

For transitions that are an order of magnitude faster than the fastest steps of the catalytic sequence, equilibrium always exists between states of the enzyme, during the time that a steady state is being reached within each separate state. In this case the transition can cause neither hysteresis nor cooperativity (see later section).

*Conditions for Catalytic Transients (Hysteresis)*

There are two questions to consider for a model that allows one or more transitions between enzyme states: first, Is there hysteresis? and second, Is there cooperativity? Hysteresis can be present in any model in which a change in the concentration of a ligand, i.e., a substrate or an effector, can shift enzyme between states that have different catalytic activity. The type of transition we are considering is not one that is obligatory to each turnover of the enzyme, but rather one involved in a side branch from the catalytic path or an alternate reaction path. Catalytic transients involving the first turnover of a substrate with an enzyme are excluded.

One type of model with a transition that can cause hysteresis but never cause cooperativity has a deadend branch off the catalytic cycle. Two examples (Figs. 5C and 5D) can be obtained from the general mechanism of Fig. 5 by allowing appropriate rate constants to be zero. Mechanism 5C is the simplest example. If a significant amount of enzyme is initially in the unprimed form, then increasing the substrate concentration will give a lag-type transient as the enzyme is pulled into the primed form. In the other example (Fig. 5D), the rate of the $E \rightleftharpoons E'$ transition is so slow that there is no flux between E and E' during the course of the catalytic reaction and EA is catalytically inactive. If the equilibrium lies in favor of E in the absence of A and in favor of E'A with high A present, addition of A will shift enzyme from the sink (E + EA) into the catalytically active primed form. If the $EA \rightleftharpoons E'A$ transition is slow there can be hysteresis, a lag-type transient. Bursts and substrate inhibition may occur when EA is favored at high concentration. However, after the hysteresis is over, the steady-state distribution of the E and EA enzyme forms is at equilibrium. Derivation of the rate equation with these conditions shows that the velocity can only be a hyperbolic function of the concentration of A. This equilibrium condition holds true for any mechanism with a deadend branch.

In mechanism 5, hysteresis will be observed if one state of the enzyme

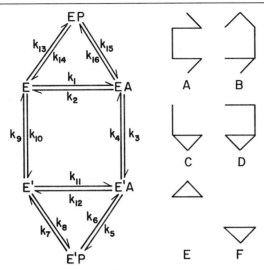

FIG. 5. The ligand-induced slow-transition mechanism for a monomeric, single-substrate enzyme existing in two conformations, E and E'. Parts A–F represent some of the pathways for interconversions among the enzyme forms (see text) Left-hand side reprinted, by permission, from G. R. Ainslie, Jr., J. P. Shill, and K. E. Neet, *J. Biol. Chem.* **247**, 7088 (1972).

has a different affinity for substrate than the other and the transitions between those states are slower than the maximum velocity of the catalytic reaction. For example, if the primed enzyme form binds substrate more strongly than the unprimed enzyme form, then enzyme will be shifted from the unprimed state to the primed state. If the primed state is more catalytically active, then there will be a slow transition to a higher velocity, a lag, whenever the substrate concentration is raised. Alternatively, if the primed state is less catalytically active, then there will be a slow transition to a lower velocity, a burst. The hysteresis is a slow transition whether the reaction velocity is higher or lower in the steady state.

For any slow transition to be observed, a kinetically significant amount of enzyme must be shifted between states. For example, if the primed enzyme form (Fig. 5) has a higher affinity for substrate and a higher catalytic activity but 99% of the enzyme is in the primed form in the absence of substrate, the transition of enzyme between the states will not be observed. Even if the entire 1% of the enzyme initially in the unprimed state were shifted, only an insignificant change in the rate of the reaction would occur.

The rate of the transient is a function of the substrate concentration. For any monomeric, two-state mechanism (e.g., Fig. 5 or Fig. 8) the function is the ratio of two polynomials, second order in substrate concentration.[3,5,7,18] For more complex mechanisms the polynomials will be $n$th order; $n$ is the number of places in the mechanism where substrate combines with enzyme. These functional forms hold true whether or not the mechanism generates cooperativity. The rate of the transition in general approaches constant values at high and low substrate concentration with potentially complex behavior at intermediate concentrations.

In summary, any model that purports to produce hysteresis must satisfy the following criteria:

(1) The transition that produces the hysteresis must be slower than the maximum velocity of the reaction.
(2) The transition must lie in an alternate path or side branch.
(3) A kinetically significant amount of enzyme must be shifted by addition of ligand to enzyme.
(4) The different states must have different catalytic activities at the concentration of substrate tested.

*Conditions for Hysteretic Cooperativity in a Monomeric Enzyme*

The second question to consider in any model involving a transition between enzyme states is whether or not cooperativity can occur. In any mechanism with alternate paths for substrate addition and product release, the highest power of each substrate or product concentration that can potentially appear in the reaction velocity equation is equal to the number of different places in the mechanism in which that substrate or product combines with the enzyme. The power of a concentration term can be lower than the maximum for several reasons (see later).

The single-substrate mechanism (Fig. 5), which also represents a two-substrate system when the second substrate is saturating, is capable of generating positive or negative cooperativity.[5] The rate equation is second order in substrate A because of the two points of addition of substrate

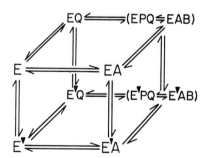

FIG. 6. The general ligand-induced slow-transition mechanism for a monomeric, two-substrate enzyme existing in two conformations, E and E'. A and B are substrates, P and Q are products. An ordered pathway is shown.

in the mechanism (EA and E'A). Since most regulatory enzymes have more than one substrate, we will discuss in more detail the bisubstrate, slow-transition mechanism.

The general ordered mechanism for two substrates, two products, and two states of the enzyme, E and E', is shown in Fig. 6. Equation (14) describes the behavior of the steady-state velocity for an enzyme with this mechanism.[25]

$$v_{ss} = \frac{\begin{array}{l}AB(1 + A + B + AB + P + Q + PQ + AP + BP + BQ + ABP \\ + BPQ) - PQ(1 + A + B + AB + P + Q + PQ + AP + BP \\ + BQ + ABP + BPQ)\end{array}}{\begin{array}{l}\text{constant} + A + B + AB + A^2 + B^2 + A^2B + AB^2 + A^2B^2 + P \\ + Q + PQ + BQ + AP + AQ + BP + P^2 + Q^2 + ABP + BPQ \\ + APQ + ABQ + A^2P + B^2P + B^2Q + P^2Q + AP^2 + PQ^2 \\ + BQ^2 + BP^2 + A^2BP + ABPQ + AB^2P + B^2PQ + AB^2Q \\ + ABP^2 + BP^2Q + BPQ^2 + AP^2Q + P^2Q^2 + B^2Q^2 + A^2P^2 \\ + A^2B^2P + AB^2PQ + ABP^2Q + BP^2Q^2 + B^2P^2Q + A^2BP^2 \\ + B^2PQ^2 + AB^2P^2 + B^2P^2Q \\ + A^2B^2P^2 + B^2P^2Q^2\end{array}} \quad (14)$$

Two points of addition for each substrate and product are present so that the equation in its most general form contains the second power of the concentration of each of these species. The only differences between the general form of the equation for the initial velocity (not shown) and that of the steady-state velocity are the absence of an $AB^2P$ term and a minus $BP^2Q$ term in the numerator and the absence of $B^2P$, $BP^2$, $AB^2P^2$, and $B^2P^2Q$ terms in the denominator in the case of the initial velocity equation. The coefficients of the individual terms in the steady-state equation are all different from those of the initial-velocity equation.

In the absence of products and with a rearrangement the equation for the steady-state velocity can be put in the following form [Eq. (15)].

$$\frac{1}{v_{ss}} = \frac{\text{coeff}_1(1/A)^2 + \text{coeff}_2(1/A) + \text{coeff}_3}{\text{coeff}_4(1/A) + \text{coeff}_5} \quad (15)$$

The coefficients of each term of Eq. (15) are quadratic functions of B. If sets of data were collected with varying A concentration at a number of changing fixed concentrations of B, the intercepts and asympotote slopes $(AS_A)$ (see this volume [7]) could be cooperative functions of B.

$$\frac{1}{V_{app}} = \frac{\text{coeff}_3}{\text{coeff}_5} = \frac{\text{const}_1(1/B)^2 + \text{const}_2(1/B) + \text{const}_3}{\text{const}_4(1/B) + \text{const}_5} \quad (16)$$

---

[25] Each term in the equation has a different coefficient. The coefficents are omitted to clarify the subsequent discussion.

$$AS_A = \frac{\text{coeff}_1}{\text{coeff}_4} = \frac{\text{const}'_1(1/B)^2 + \text{const}'_2(1/B) + \text{const}'_3}{\text{const}'_4(1/B) + \text{const}'_5} \tag{17}$$

Since Eq. (15) is symmetrical with respect to A and B, it could be rearranged to give a parallel dependence of the coefficients of $1/v$ versus $1/B$. The question of whether or not cooperativity will be observed for either A or B depends upon the relationships among the coefficients of the 2/1 functions of Eqs. (15), (16), and (17).

Support for a mechanism involving a slow transition can be gleaned from product inhibition studies. The full equation for the steady-state velocity [Eq. (14)] can be rearranged to demonstrate that, with P as the inhibitor and A as the varied substrate, the P inhibition of the slope of the asymptote line ($AS_A$) on the plot of $1/v$ vs. $1/A$ has the form of a 2/1 function of P concentration.

$$AS_A = \frac{\text{coeff}''_1(P)^2 + \text{coeff}''_2(P) + \text{coeff}''_3}{\text{coeff}''_4(P) + \text{coeff}''_5} \tag{18}$$

Note that because there is no $B^2P^2$ term in the denominator of the general equation there is no $B^2$ term in the coefficient of the $P^2$ term in the numerator of Eq. (18). A plot of $AS_A$ versus P may show cooperativity and the asymptote slope ($AS_P$) would have the following form:

$$AS_P = \frac{\text{coeff}_1}{\text{coeff}_4} = \frac{\text{const}'''_1(1/B)^2 + \text{const}'''_2(1/B)}{\text{const}'''_3(1/B) + \text{const}'''_4} \tag{19}$$

If $AS_P$ were plotted versus B, it would approach zero at high concentrations of B. A simple ordered Bi Bi mechanism would produce similar behavior with respect to P.

Mechanism 6 can be simplified by allowing only one state of the enzyme to catalyze the reaction (Fig. 7), thereby becoming an example of the mnemonic enzyme concept of Ricard et al.[6] (cf. Fig. 8). Raising the concentration of P or Q will keep a higher proportion of the enzyme in one state, the unprimed state in these cases. The product inhibition patterns of P and Q are all linear functions of P and Q, and P becomes a linear uncompetitive inhibitor of A rather than a noncompetitive one at infinite concentrations of B. Both A and B can still be cooperative even if the other substrate is saturating. As pointed out,[6] the presence of Q will increase the negative cooperativity or decrease the positive cooperativity of A.

*Conditions for the Absence of Cooperativity.* Whether cooperativity will be observed or not is a complex function of all the rate constants. Some general relations among the rate constants will eliminate cooperativity. Rapid-equilibrium addition of substrates will eliminate cooperativity due to a slow transition. The condition necessary for rapid equilibrium is that the rate constant for dissociation of the substrate from the

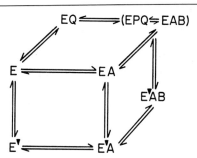

FIG. 7. A slow-transition mechanism with only one catalytically active form, EAB. (See Fig. 6.)

enzyme–substrate complex must be an order of magnitude or more faster than the maximum velocity of the reaction. Since in general the maximum velocity of the reaction contains the rate constants for the steps involved in the catalytic interconversion and the rate constants for the dissociation of products, the unimolecular rate constant for substrate dissociation must be an order of magnitude faster than whichever of those unimolecular rate constants is rate limiting. Cooperativity is lost for mechanism 5 by allowing the rate constants $k_2$ and $k_{12}$ in the equation that describes this mechanism [Eq. (1) of Ainslie et al.[5]] to become very large compared to $k_{14}$, $k_{15}$, $k_{16}$, $k_5$, $k_6$, and $k_7$. Alternatively, rederiving the velocity equation using the conditions

$$\text{EA} = k_1(\text{A})(\text{E})/k_2 \quad \text{and} \quad \text{E}'\text{A} = k_{11}(\text{A})(\text{E}')/k_{12} \quad (20)$$

results in a hyperbolic dependence on A.

Cooperativity will also be eliminated in a mechanism containing a transition if the transition itself is fast enough to be at equilibrium. Cooperativity is lost for mechanism 5 by allowing $k_3$, $k_4$, $k_9$, and $k_{10}$ to become large compared to the constants making up the maximum velocity expression.[5] Alternatively, the equation can be rederived using the equilibrium constraints

$$\text{E}' = k_{10}(\text{E})/k_9 \quad \text{and} \quad \text{E}'\text{A} = k_3(\text{EA})/k_4 \quad (21)$$

The resulting equation for the velocity is a hyperbolic function of A.

The loss of cooperativity when the rate of the transition is rapid can be illustrated also with the mnemonic mechanism[6] (Fig. 8). A hyperbolic equation results if it is assumed that the rhombus and circle forms of Fig. 8 interconvert rapidly enough to be in equilibrium during the course of the reaction; i.e., $k_4$ (rhombus) = $k_{-4}$ (circle).

Cooperativity is eliminated if the flux through the nonequilibrium catalytic pathways of the mechanism is insignificant. The distribution of en-

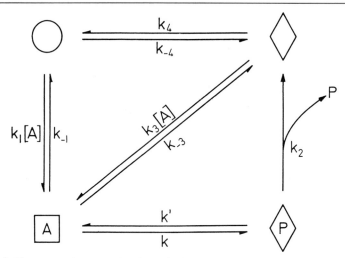

FIG. 8. The mnemonic mechanism for a single-substrate, monomeric enzyme. The circle and rhombus represent two conformations of the free enzyme. If $k_1 > k_3$ negative cooperativity may be observed; if $k_1 < k_3$, positive cooperativity may be observed. Reprinted, by permission, from J. Ricard, J.-C. Meunier, and J. Buc, *Eur. J. Biochem.* **49**, 195 (1974).

zyme among pathways may be determined from the King–Altman patterns (the equilibrium distribution when no reaction is occurring can also be obtained from the patterns). For any mechanism there are two types of King–Altman patterns[26] that make up the distribution equations for the fraction of enzyme present in each enzyme form as a function of the rate constants, substrate concentrations, and product concentrations. One type of pattern does not include the irreversible step of product release, whereas the other does.

The first type of pattern in a complete distribution equation is related to the equilibrium among the enzyme forms (in the absence of reaction). For example, in mechanism 5 one pattern of this type would be Fig. 5A, describing the pathway EP = EA = E = E′ = E′A = E′P. All the steps are reversible and this pattern will contribute to the distribution equation of each enzyme form, either when considering equilibrium or when considering the nonequilibrium steady state present during catalytic reaction. The second type of pattern contains the irreversible step. For example, Fig. 5B describes the pathway E′(EP′) = E′A = EA = EP → E. Due to the irreversible step EP → E + P, this pattern can only contribute to the

[26] E. P. Whitehead, *Biochem. J.* **159**, 449 (1976).

distribution equation for the E enzyme form and would not contribute at all when considering equilibrium among the enzyme forms.

If the second type of term makes a negligible contribution to the amount of enzyme present in the various enzyme forms, then there will be no cooperativity. The reason for this is that the velocity of the reaction will be the rate constant for the irreversible step times the concentration of the enzyme form that precedes that step. If that enzyme form is in equilibrium with the other enzyme forms, there is no cooperativity of binding, and there can be no cooperativity for the velocity of the reaction. Whitehead[26] suggested ways to isolate the second type of term in order to determine more readily whether there will be any cooperativity for a given mechanism and set of rate constants.

*How Does the Cooperativity Originate?* Cooperativity occurs when there are different catalytic paths and the proportion of the total velocity of the reaction due to each separate path changes as the substrate concentration is changed. For example (Fig. 9), cooperativity will be observed in mechanism 9A if at low substrate concentration the path 9B and the path 9C both contribute significantly to the flux of reaction, since at high substrate concentration only the path 9C contributes. The change in relative importance (with changing substrate concentration) among the paths catalyzing the overall reaction produces the cooperativity. These paths are the same ones that are involved in the nonequilibrium patterns in the exposition of Whitehead.[26]

To illustrate in a different way, if the rate constants in mechanism 5 were such that substrate addition were at rapid equilibrium, then the two paths 5E and 5F would both contribute to the total rate of reaction in constant proportions over the full range of substrate concentration. Since no shift in proportion from one path to the other would occur, there would be no cooperativity.

In summary, for a mechanism involving a transition to give cooperativity three things are important:

1. The transition can be slower than (or the same speed as) the maximum velocity of the reaction, but it cannot be an order of magnitude faster than the maximum velocity of the reaction and still be responsible for cooperativity.
2. The transition cannot produce cooperativity if the addition of the substrate(s) is (are) rapid equilibrium.
3. The rate constants for the transition (and for the other steps of the reaction) must be such that raising the substrate concentration shifts the proportion of the overall reaction that goes through two (or more) of the alternative catalytic paths.

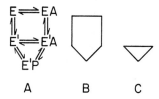

A  B  C

FIG. 9. The ligand-induced slow-transition mechanism for a monomeric, single-substrate enzyme with only one catalytically active form, E'A. Diagrams B and C represent two pathways by which the catalytic reaction can occur.

## Hysteresis in Oligomers

*Models*

More complex situations arise if hysteresis occurs in oligomeric enzymes that also show site–site cooperativity or association–dissociation reactions.[5] In the former case the cooperativity due to the slow transition will either increase the cooperativity inherent in the site–site interactions or tend to counteract it and decrease the overall cooperativity. The complete rate equation will be $n$th order in substrate, when $n$ is the number of oligomeric sites times the number of states due to slow transitions. For a dimer with two states of each protomer the rate equation will be fourth order. The equation for the half-time of the transient decay will also be $n$th order in substrate. Complex cooperativity involving a mixture of positive and negative cooperativity will also be possible in the kinetics (see this volume [7]). The contribution of cooperativity from the slow transition will be eliminated for the same reasons as those discussed for the monomeric enzymes, i.e., rapid equilibrium of substrate binding, rapid equilibration of the isomerization steps, and insignificant flux through the nonequilibrium pathways of the mechanism.

Hysteresis in the concerted, site–site interaction model has been analyzed assuming rapid equilibrium of binding; as expected, the cooperativity in the steady state is solely due to the parameters of the concerted mechanism.[27] No analysis has been made of this system in the true steady state without the assumption about the rapid binding of substrate. The influence of subunit interactions on the substrate response of the transient rate has been presented for specific concerted and sequential models.[28,29]

[27] B. I. Kurganov, A. I. Dorozhko, Z. S. Kagan, and V. A. Yakovlev, *J. Theor. Biol.* **60**, 247 (1976).
[28] C. Mouttet, F. Fouchier, J. Nari, and J. Ricard, *Eur. J. Biochem.* **49**, 11 (1974).
[29] K. Kirshner, *J. Mol. Biol.* **58**, 51 (1971).

When association–dissociation occurs, this process itself may be the slow transition. The most interesting situation arises when the rate of the association–dissociation reaction is on the same order of magnitude as the enzymic velocity and nonlinear progress curves are observed. Two effects of the association–dissociation are possible: a dissociation due simply to dilution, which can be eliminated by prior incubation of the enzyme under assay conditions; and a dissociation or association that is influenced by the binding of the substrate (or other ligand) to one form of the enzyme. In both cases, the half-time of the assay transient will be dependent upon the enzyme concentration, but only in the latter case will the half-time be dependent upon the substrate concentration. Equations have been derived for these dependencies assuming rapid-equilibrium binding.[30,31] Graphical analysis of the dimeric case allows a determination of the first-order rate constant for dissociation, the second-order rate constant for dimerization, and the equilibrium constant.[31] These parameters have been determined for phosphorylase and aspartic $\beta$-semialdehyde dehydrogenase. Interestingly, a cooperative effect of substrate upon the initial rate of enzyme dissociation may occur under certain conditions.

### *Discrimination between Hysteretic Cooperativity and Site–Site Cooperativity*

*Binding Curves.* Since hysteretic cooperativity is a purely kinetic phenomenon, the most direct way of determining the relative contribution is to compare kinetic and equilibrium data. If the cooperativity from kinetic assays is different than the cooperativity for the same ligand–enzyme interaction measured by an equilibrium process (such as equilibrium dialysis or fluorescence quenching), one way infer that a kinetic component contributes to the enzymic cooperativity.

The most convincing case for hysteretically caused differences in cooperativity is for a monomeric enzyme to show cooperativity in kinetics (but not in equilibrium binding), as observed with wheat germ hexokinase[32] and liver glucokinase.[33] However, a requirement for this conclusion is to demonstrate the continued monomeric status of the enzyme during catalysis, using a technique such as reacting enzyme sedimentation.

Differences between kinetic and equilibrium cooperativity are a neces-

---

[30] B. I. Kurganov, A. I. Dorozhko, Z. S. Kagan, and V. A. Yakovlev, *J. Theor. Biol.* **60**, 287 (1976).
[31] B. I. Kurganov, *J. Theor. Biol.* **68**, 521 (1977).
[32] J.-C. Meunier, J. Buc, A. Navarro, and J. Ricard, *Eur. J. Biochem.* **49**, 209 (1974).
[33] A. C. Storer and A. Cornish-Bowden, *Biochem. J.* **165**, 61 (1977).

sary, but not a sufficient, condition to implicate hysteretic cooperativity. The kinetic component may not only be due to a hysteretic process, but other factors may also be important.

1. If $v/V_{max}$ is not proportional to the fractional saturation of bound substrate in an oligomeric enzyme, then the kinetic cooperativity will differ from the binding cooperativity (see this volume [7]).
2. Conditions in kinetic experiments and in binding experiments often differ, e.g., protein concentration, presence of a second substrate, time of analysis, and temperature.

*Variations in the Transient.* Investigation of a kinetic transient (by rapid-mixing techniques, if necessary) is useful for determining that a certain kinetic cooperativity is due to hysteresis. The rate constants for the transitions ($k_3$, $k_4$, $k_9$, and $k_{10}$ of Fig. 5) appear in both the expression for the overall rate of the transient and in the coefficients of the steady-state rate equation.[5]

Therefore, correlation of the changes in the kinetic transient with changes in observed cooperativity as experimental conditions change is circumstantial evidence that the two processes are mechanistically related. For example, yeast hexokinase shows negative cooperativity with respect to MgATP and a slow burst at pH 6.5; both the burst and the negative cooperativity are abolished if the reaction is run at pH 8, if citrate is added to the pH 6.5 assay, or if MgGTP is used as substrate.[14,16] Such a parallel change in parameters could either be indicative of hysteretic cooperativity or simply reflect separate responses to a common, underlying alteration in protein structural changes. Conversely, a change in one of the properties, e.g., the slow kinetic transient, *without* a concomitant change in the other, i.e., cooperativity, would suggest that the two processes are independent of each other. Studies of the kinetic transient per se are therefore suggestive of the underlying mechanism for cooperativity but are not definitive proof of the particular mechanism. This approach is equivalent to the "desensitization" studies of allosteric enzymes used to demonstrate the physical separation of the allosteric and the catalytic sites.

*Rate Constants.* The most definitive evidence for the existence of hysteretic cooperativity is the determination of the various individual rate constants, some by independent methods, and the verification of the rate equation (using either initial velocities or the full progress curve[17]) for the particular mechanism in question. The most thorough treatment has been for wheat germ hexokinase, in which the results from steady-state kinetics,[32] stopped-flow initial rates,[7] and temperature-jump fluorescence studies,[34] have been interpreted to be consistent with a two-substrate, mnemonic (slow transition) mechanism. Steady-state kinetics, rapid kinetics, and structural studies will all be necessary to show that a kinetic mechanism that links hysteresis and cooperativity is the most probable mechanism.

Classification of Hysteretic Enzymes

In early papers[3,5] discussing hysteretic enzymes, all known enzymes that showed a slow transition or that showed hysteresis linked to kinetic cooperativity were listed. Further subdivision was made into those that showed a lag or a burst with substrate or with effector or those that showed negative cooperativity or positive cooperativity.[27] Although this is a logical way to classify hysteretic enzymes, it does not provide much insight into their role or behavior. We propose a classification based on the physiological function attributed to the slow transition of the enzyme (Table I). Such a procedure requires more detailed information about the mechanism of the enzyme.

*Class I. Incidental to Function.* Some enzymes may show hysteresis or hysteretic cooperativity with certain substrates under certain conditions but with no relevance to the actual functioning of the enzyme. This may simply reflect a "slower" transition induced by the unusual substrate and/or a coincidental combination of rate constants that leads to hysteretic cooperativity. Two digestive enzymes appear to fall into this cate-

TABLE I
TENTATIVE CLASSIFICATION OF SOME HYSTERETIC ENZYMES

| Enzyme | Class | Comments (reasoning) |
|---|---|---|
| Chymotrypsin[a] | I | No apparent physiological significance for the transient or the cooperativity |
| Ribonuclease[b] | I | No apparent physiological significance for the transient or the cooperativity |
| Phosphofructokinase[c,d] (bovine liver) | II | Hysteretic interconversion between inactive dimer and active tetramer ($t_{1/2} \simeq 0.5$ min) on the time scale of glycolytic oscillations; cooperativity in tetramer (see text) |

(*Continued*)

[34] J. Buc, J. Richard, and J.-C. Meunier, *Eur. J. Biochem.* **80**, 593 (1977).

TABLE I (*Continued*)

| Enzyme | Class | Comments (reasoning) |
|---|---|---|
| Phosphorylase[e,f] (rabbit muscle) | II | Hysteretic interconversion between inactive tetramer and active dimer ($t_{1/2} \simeq 2$ min); no cooperativity |
| Chorismate synthase[g,h] (*Neurospora crassa*) | II | Hysteresis in dimer ($t_{1/2} \simeq 10$ min), but no cooperativity |
| Glutamate dehydrogenase[i] (*N. crassa*) | II or IV[j] | Hysteretic isomerization in hexamer ($t_{1/2} \simeq 3$ min) probably under physiological conditions |
| Glutamate dehydrogenase[k,l,m] (beef) | II | Hysteretic displacement of inhibitory GTP ($t_{1/2} \simeq 2$ sec) and changes in the state of aggregation ($t_{1/2} < 1$ sec) on the time scale of the catalytic reaction |
| Pyruvate kinase[n] (red blood cell) | II or IV[j] | Hysteresis ($t_{1/2} \simeq 0.5$ min) and cooperativity in tetramer; activator FDP eliminates cooperativity but not transient, so class II more probable |
| Pyruvate carboxylase[o,p] (sheep kidney) | II or IV[j] | Hysteresis ($t_{1/2} \simeq 0.3$ min) and cooperativity in tetramer |
| Acetyl CoA carboxylase[q,r,s] (rat liver) | II or IV[j] | Hysteresis ($t_{1/2} \simeq 1$ min) in oligomer and cooperativity |
| Adenyl cyclase[t,u] (rat liver) | II or IV[j] | Hysteresis ($t_{1/2} \simeq 0.5$ min) with effectors ITP IDP and cooperativity; slow response to hormone proposed |
| Hexokinase[v] (yeast) | II, III, or IV | Slow enough ($t_{1/2} \simeq 1$ min) to be physiologically important; cooperativity in monomer and dimer from reacting enzyme sedimentation |
| Phosphoenolpyruvate carboxykinase[w] | III | Hysteresis ($t_{1/2} \simeq 0.5$ min) and cooperativity in monomer; reacting enzyme sedimentation indicates only monomer |
| Hexokinase[x,y] (wheat germ) | III | Hysteresis ($t_{1/2} \simeq 2$ min) and cooperativity in monomer (see text) |
| Glucokinase[z] (rat liver) | III | Cooperativity in monomer; kinetics interpreted in terms of slow transition |
| Aspartokinase–homoserine (*Escherichia coli*) dehydrogenase I[α] | II and IV | Both hysteretic cooperativity and site–site cooperativity in tetramer with kinase activity; a second, slower transition indicates class II, also (see text) |
| Rhodanase[β] (human liver) | IV | Cooperativity in dimer; no report of hysteresis in assay but fluorescence shows a hysteretic isomerization on same time scale as reaction |

(*Continued*)

TABLE I (*Continued*)

| Enzyme | Class | Comments (reasoning) |
|---|---|---|
| Threonine deaminase[γ] (*Bacillus subtilis*) | IV (II?) | Hysteresis ($t_{1/2} \simeq 1$ min) and cooperativity in dimer; simulation shows hysteresis can account for observed cooperativity |
| Glyceraldehyde-3-phosphate dehydrogenase[δ] (rabbit muscle) | IV | Cooperativity and hysteresis ($t_{1/2} \simeq 0.1$ sec) in tetramer; hysteresis is too fast to be physiologically important. |

[a] G. R. Ainslie, Jr. and K. E. Neet, *Mol. Cell. Biochem.* **24,** 183 (1979).
[b] H. Rübsamen, R. Khandker, and H. Witzel, *Hoppe-Seyler's Z. Physiol. Chem.* **355,** 687 (1974).
[c] C. Frieden, in "The Regulation of Enzyme Activity and Allosteric Interactions" (E. Kramme and A. Pihl, eds.), p. 59. Academic Press, New York, 1968.
[d] P. M. Lad, D. E. Hill, and G. G. Hammes, *Biochemistry* **12,** 4303 (1973).
[e] B. E. Metzger, L. Glaser, and E. Helmreich, *Biochemistry* **7,** 2021 (1968).
[f] B. E. Metzger, E. Helmreich, and L. Glaser, *Proc. Natl. Acad. Sci. U.S.A.* **57,** 994 (1967).
[g] F. H. Gaertner and K. W. Cole, *J. Biol. Chem.* **248,** 4602 (1973).
(1971).
[h] G. R. Welch, K. W. Cole, and F. H. Gaertner, *Acta Biochim. Biophys.* **165,** 505 (1974).
[i] B. Ashby, J. C. Wootton, and J. R. S. Fincham, *Biochem. J.* **143,** 317 (1974).
[j] The possibility of a link between hysteresis and cooperativity has not yet been investigated.
[k] H. Eisenberg, R. Josephs, and E. Reisler, *Adv. Prot. Chem.* **30,** 101 (1976).
[l] J. M. Jallon, A. di Franco, and M. Iwatsubo, *Eur. J. Biochem.* **13,** 428 (1970).
[m] C. Frieden and R. F. Colman, *J. Biol. Chem.* **242,** 1705 (1967).
[n] J. A. Badway and E. W. Westhead, *J. Biol. Chem.* **251,** 5600 (1976).
[o] R. Bais and B. Keech, *J. Biol. Chem.* **247,** 3255 (1972).
[p] H. G. Wood and R. E. Barden, *Annu. Rev. Biochem.* **46,** 385 (1977).
[q] M. D. Greenspan and J. M. Lowenstein, *J. Biol. Chem.* **243,** 6273 (1968).
[r] C. A. Carlson and K.-H. Kim, *Acta Biochim. Biophys.* **164,** 490 (1974).
[s] K. Bloch and D. Vance, *Annu. Rev. Biochem.* **46,** 263 (1977).
[t] J. Wolff and G. H. Cook, *J. Biol. Chem.* **248,** 350 (1973).
[u] M. Rodbell, L. Birnbaumer, S. L. Pohl, and H. M. J. Kraus, *J. Biol. Chem.* **246,** 1877 (1971).
[v] J. P. Shill and K. E. Neet, *J. Biol. Chem.* **250,** 2259 (1975).
[w] Yu-Bin Chaio, Doctoral dissertation, Case-Western Reserve Univ., Cleveland, Ohio (1975).
[x] J.-C. Meunier, J. Buc, A. Navarro, and J. Ricard, *Eur. J. Biochem.* **49,** 209 (1974).
[y] J. Buc, J. Ricard, and J.-C. Meunier, *Eur. J. Biochem.* **80,** 593 (1977).
[z] A. C. Storer and A. Cornish-Bowden, *Biochem. J.* **165,** 61 (1977).
[α] C. F. Bearer and K. E. Neet, *Biochemistry* **17,** 3512, 3517, 3523 (1978).
[β] R. Jarabak and J. Westly, *Biochemistry* **13,** 3233, 3237, 3240 (1974).
[γ] G. W. Hatfield, W. J. Ray, Jr., and H. E. Umbarger, *J. Biol. Chem.* **245,** 1748 (1970).
[δ] K. Kirshner, *J. Mol. Biol.* **58,** 51 (1971).

gory. Ribonuclease has been demonstrated to show a small degree of positive cooperativity with the substrate, 2′,3′-cyclic CMP.[8] Since ribonuclease is monomeric and undergoes a conformational change in the absence of substrates (rate constant about $10^3$ sec$^{-1}$),[35] the model proposed to explain the cooperativity[8] is both reasonable and analogous to the more general slow transition model. Another monomeric enzyme, δ-chymotrypsin, also undergoes a slow transition in the alkaline region that has been well studied.[36] A small degree of negative cooperativity[37] has recently been demonstrated for this enzyme with the artificial substrate N-acetyltryptophan nitrophenyl ester, and its kinetics have been interpreted in terms of the slow transition mechanism. Both these enzymes are extracellular, not under any apparent regulatory control, and do not appear to require any hysteresis for their function.

*Class II. Physiological Damping.* A potentially important role that hysteresis may play in the living organism is that of damping out a cellular response to some rapid change in substrate or effector response (see preceding discussion). Such a "damping" might prevent the cell from going into wildly fluctuating changes when some particular metabolite either increases or decreases suddenly in concentration. Proving this role for any enzyme is difficult, since study of the isolated enzyme cannot determine its temporal interactions with other enzymes of the system. One must rule out contributions to cooperativity (Classes III and IV) and provide a rationale for the intracellular hysteresis in order to rule out Class I. Studies of reconstituted enzyme systems or of *in situ* systems may provide the best evidence. Nevertheless, there are two enzymes that have been interpreted in terms of a damping response role: phosphofructokinase and the second transition of aspartokinase.

The tetramer of phosphofructokinase reversibly dissociates to an inactive dimer upon dilution into slightly acidic conditions (pH < 7) with a half-time between 10 and 90 min.[4,38] The rate and extent of dimerization are promoted by lower pH and lower temperature and depend upon enzyme concentration.[9] The inactivation is either biphasic or a single, first-order process, depending upon conditions, and has been interpreted as a sequence of protonation, isomerization to the inactive tetramer, and a slow dissociation ($E_4 \rightarrow HE_4 \rightarrow HE_4' \rightarrow HE_2'$). ATP and citrate shift the equilibria toward the protonated forms whereas fructose 6-phosphate and AMP favor the active form.[39] If the enzyme is preincubated with citrate

---

[35] T. F. French and G. G. Hammes, *J. Am. Chem. Soc.* **87**, 4669 (1965).
[36] J.-R. Garel and B. Labouesse, *Eur. J. Biochem.* **39**, 293 (1973).
[37] G. R. Ainslie, Jr. and K. E. Neet, *Mol. Cell Biochem.* **24**, 183 (1979).
[38] P. M. Lad, D. E. Hill, and G. G. Hammes, *Biochemistry* **12**, 4303 (1973).
[39] P. E. Bock and C. Frieden, *J. Biol. Chem.* **251**, 5637 (1976).

plus ATP and catalysis is initiated with fructose 6-phosphate, a slow (minutes) activation occurs.[40] Interestingly, liver phosphofructokinase shows a slow dilution-inactivation that does not appear to alter molecular weight.[41] The allosteric and cooperative properties of muscle enzyme all occur within the tetramer, suggesting that the slow dissociation may play a role in the damping of metabolite fluctuations, particularly when the muscle pH has become more acidic during anaerobiosis.

*Class III. Cooperativity in a Monomeric Enzyme.* An important function for hysteresis of an enzyme would be to confer kinetic cooperativity upon a monomeric enzyme that could not attain similar cooperativity through site–site interactions. Identification of cooperativity of the kinetics of an enzyme and a concomitant slow transition is relatively easy, but the difficult part is demonstrating, unequivocally, that the enzyme is monomeric under conditions either of assay or of physiological functioning. Two enzymes, wheat germ hexokinase[32] and rat liver glucokinase,[33] appear to satisfy the criteria.

The mnemonic mechanism has been applied to explain the hysteresis and negative cooperativity of glucose kinetics with wheat germ hexokinase.[6,7,32,34] The "mnemonical enzyme" concept describes a mechanism for cooperativity in which the enzyme (monomer) form released by dissociation of the last product of a reaction can isomerize to another state, and both states of the free enzyme can combine with substrate (Fig. 8). The glucose 6-phosphate–enzyme complex dissociates leaving the enzyme in the rhombus form; this form can then either slowly isomerize to the circle form or combine with glucose; both the circle-glucose and rhombus-glucose forms can isomerize to a common square form; the latter form can then combine with ATP, catalyze the reaction, release ADP, and release glucose 6-phosphate. During the catalytic reaction or upon the release of products, the enzyme is returned to the rhombus state.

Support for this mechanism rests on a number of observations: Glucose binding by equilibrium binding measurements is hyperbolic; the burst observed in the absence of products can be converted into a lag by preincubation with glucose 6-phosphate; hysteretic processes of about the same rate ($10^{-3}$ to $10^{-2}$ sec$^{-1}$) can be observed in both the catalytic reaction and the binding of glucose; the rate of the transient increases to a plateau with increasing glucose concentration and decreases to a plateau with increasing glucose 6-phosphate concentration; the first substrate, glucose, is negatively cooperative at pH 8.5 and must therefore combine with the enzyme at two or more points in the mechanism; the second,

---

[40] G. Colombo, P. W. Tate, A. W. Girotti, and R. G. Kemp, *J. Biol. Chem.* **250**, 9404 (1975).
[41] A. Ramaiah and G. A. Tejwani, *Eur. J. Biochem.* **39**, 183 (1973).

ATP, is not cooperative at pH 8.5; and preincubation with glucose 6-phosphate makes glucose more negatively cooperative.

*Class IV. Alteration of Cooperativity of an Oligomeric Enzyme.* Oligomeric enzymes may generate cooperativity through site–site interactions between the ligand binding sites on different subunits so that it is not immediately obvious why they should also have a hysteretic effect that alters this cooperativity. That such an alteration of cooperativity can occur has been demonstrated theoretically[5] by consideration of the "hybrid" mechanisms in which both slow transitions and site–site interactions occur. One enzyme, aspartokinase–homoserine dehydrogenase I (AK-HSDH), of *E. coli* has been carefully studied and interpreted in terms of a mechanism that provides a more general rationale for the hybrid type of mechanism.[42]

AK-HSDH has been shown to undergo two threonine-induced hysteretic transitions[42] (Fig. 10). The first transition (R → T) is associated with the binding of threonine to the four high-affinity threonine binding sites and has a time constant on the order of 0.1 sec. This first transition has been demonstrated to be an isomerization and has been studied by stopped-flow assay, stopped-flow fluorescent and temperature-jump techniques; it produces a threonine–enzyme complex which is 80–90% inhibited for both AK and HSDH activities. The second transition (T → U) is an isomerization that occurs on a much slower time scale (on the order of minutes), is correlated with the binding of threonine to the second low affinity set of four sites, does not affect the HSD activity any further, but reduces the AK activity to zero. The cooperativity of binding to the high-affinity set of sites at equilibrium is identical with the cooperativity of inhibition of the HSDH activity ($n_H$ = 2.3) under a wide range of conditions, suggesting that the hysteretic transition has not altered the kinetic cooperativity for this activity. In contrast, the cooperativity of the inhibition of the AK activity is greater ($n_H$ = 4) than the cooperativity of equilibrium binding to threonine to either the first or the second set of sites ($n_H$ = 2.3), suggesting that there is a kinetic component superimposed upon the cooperativity due to site–site interactions. The kinetic component has been interpreted as a reflection of the hysteretic cooperativity from the first transition (R → T).

The second, slower transition (T → U), which affects only the AK activity, does not appear to influence cooperativity but may well be important for the damping of oscillations of fluctuations of metabolites within the *Escherichia coli* (i.e., a Class II hysteresis). Thus, with a single set of threonine binding sites (the high-affinity ones) the enzyme is able to provide a different kinetic cooperativity of inhibition of the two activities

---

[42] C. F. Bearer and K. E. Neet, *Biochemistry* **17**, 3512, 3517, 3523 (1978).

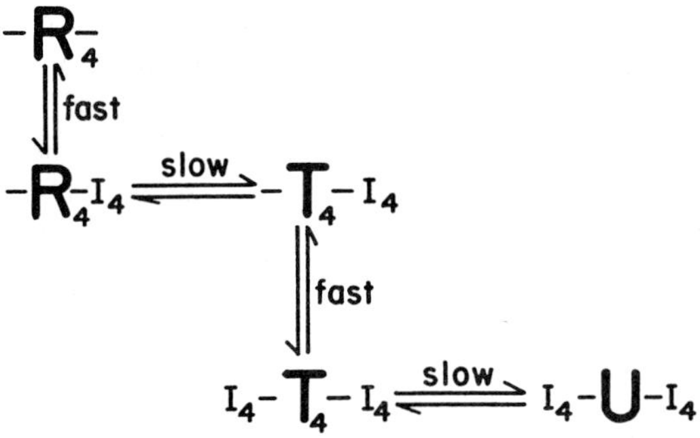

$$-\underset{\substack{\Updownarrow \text{fast}}}{R_4^-}$$

$$-R_4^- I_4 \underset{\text{slow}}{\rightleftharpoons} -T_4^- - I_4$$

$$\Updownarrow \text{fast}$$

$$I_4 - T_4^- - I_4 \underset{\text{slow}}{\rightleftharpoons} I_4 - U - I_4$$

FIG. 10. A model for threonine binding to aspartokinase–homoserine dehydrogenase I. R, T, and U represent three conformational states. R is active and binds substrates. T is 80–90% inhibited for both activities. U is 100% inhibited for aspartokinase activity. The enzyme is a tetramer and binds eight molecules of the inhibitor, I (threonine).

comprising the bifunctional enzyme. Superimposition of the hysteretic cooperativity upon the subunit interaction cooperativity for one activity (AK), but not for the other (HSDH), allows an effective alteration in their relative regulatory properties. Such a complex combination of hysteresis in an oligomeric enzyme might be a widespread solution of enzymes to regulatory problems.

Molecular Aspects of Hysteresis

*Possible Structural Basis for Slow Conformational Change.* A slow conformational change of a protein differs from a more rapid one only in that the former must have a much more difficult barrier (high energy) to pass through in getting to the new state. No precise detail is known for any slow conformational transition, but several possibilities exist (Fig. 11).

1. The first possible type of movement would be a rotation or sliding of two α-helices with respect to each other. Two α-helices in the 75 amino acid N-terminus of muscle phosphorylase make contact with the analogous region in the intersubunit contact region, but move inward to make a more compact structure upon substrate binding.[43] Whether or not this is related to the slow transition observed in solution is unknown.[13]
2. A second possibility is the cis-trans isomerization of proline bonds that is known to occur, but slowly.[44] Both the slow step in refolding of ribonuclease[44] and the "locking" of the conformation of concan-

[43] R. J. Fletterick, personal communication.
[44] J. F. Brandts, H. R. Halvorson and M. Brennan, *Biochemistry* **14**, 4953 (1975).

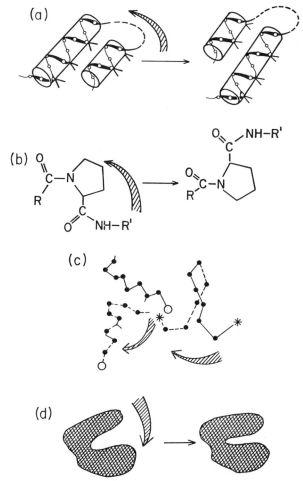

FIG. 11. Possible molecular basis for slow conformational changes. (a) Rotation or sliding of α-helices with respect to each other. (b) Cis-trans isomerization about a proline-imide bond. (c) Segmental movement of a large portion of a chain (taken approximately from chymotrypsin). (d) Hinged movement of lobes (domains) (taken approximately from hexokinase).

avalin A in the presence of divalent metal cations[45] have been interpreted in terms of an isomerization between cis and trans conformations of a proline bond (but see also Hagerman and Baldwin[46]).

3. A third possibility could be called a "segmental rotation" of a relatively large portion of the protein and is exemplified by chymo-

[45] R. D. Brown, III, C. F. Brewer, and S. H. Koenig, *Biochemistry* **16**, 3883 (1977).
[46] P. J. Hagerman and R. L. Baldwin, *Biochemistry* **15**, 1462 (1976).

trypsin.[36,47] The conformational change which occurs slowly at high pH values is thought to be the same as the one that occurs upon activation of chymotrypsinogen to chymotrypsin. The change is well described in terms of the three-dimensional structure and can be viewed as a rotation about a segment of polypeptide chain involving Met-192 with an associated movement of about 5–6 Å of Ile-16 and Asp-194 to form a salt linkage.

4. A fourth possibility for a slow conformational change could be called a "hinged movement" of domains with respect to each other. This appears to occur in yeast hexokinase[48] upon binding of glucose but is probably not the slow transition itself, since the latter requires both substrates to occur. A related movement of the hinge, however, may occur slowly.

These four possibilities share one or more of the expected properties of the barrier to a conformational change: the making or breaking of many bonds simultaneously; charge repulsion or steric hindrance in the course of the movement; or unfavorable rotation about a single bond. The combination of X-ray crystallographic information and solution studies of these processes will be necessary to determine the details of hysteretic changes.

*Activation Energetics of Slow Conformational Changes.* Study of the energetics of the isomerizations of enzymes can provide insight into the molecular changes that are occurring and, possibly, into the role of the isomerization in enzyme function (Table II). The slower enzyme transitions are characterized by a larger free energy of activation, but the contribution of enthalpy of activation and entropy of activation might differentiate various aspects of the enzyme isomerizations. Table II summarizes rate constants, $\Delta H^*$ and $\Delta S^*$, for the limited amount of data available. The enthalpies of activation for ribonuclease and concanavalin A are consistent with the corresponding parameter measured in proline model compounds (Table II), but the entropy of activation is quite different. This may reflect the contribution of other portions of the protein to the isomerization or the contribution of changes in solvation upon isomerization. The more rapid isomerization of ribonuclease observed by temperature-jump experiments occurs with an activation enthalpy that is quite a bit lower than that of refolding of the protein.

The transition of homoserine dehydrogenase I in the presence of threonine has been studied by a number of techniques (Table II). There is

---

[47] S. T. Freer, J. Kraut, J. D. Robertus, H. T. Wright, and N. H. Xuong, *Biochemistry* **9**, 1997 (1970).

[48] T. A. Steitz, C. Anderson, W. Bennett, R. McDonald, and R. Stenkamp, *Biochem. Soc. Trans.* **5**, 620 (1977).

TABLE II
ACTIVATION ENERGETICS OF CONFORMATIONAL TRANSITIONS

| Enzyme[a] | Method | $k$ (sec$^{-1}$) | $\Delta H^*$ (kcal/m) | $\Delta S^*$ (cal/m-deg) |
|---|---|---|---|---|
| Ala-Pro[44] (cis-trans) | H+ titration | 0.0035 (25°) | 19.2 | −5.1 |
| RNase folding[44,b] | Spectral | 0.02 (25°) | 17.4 | −8.4 |
| Con A[45] | NMR | 0.001 (5°) | 21.1 | +3.2 |
| AK-HSD[c,d] | Spectral | 2 (20°) | 21.4 | +15.7 |
| AK-HSD[42] | Assay transient | 6.7 (20°) | 16.7 | +2.0 |
| WGHK[34] | Spectral | 0.001 (35°) | 9.2 | −42 |
| YHK[14] | Assay transient | 0.033 (25°) | 4.8 | −49.6 |
| CHT[e]: Forward | Spectral | 0.55 (25°) | 18.3 | +2 |
| Reverse | Spectral | 1.9 (25°) | 25.4 | +29 |
| RNase[35,f]: Forward | Spectral | 446 (25°) | 6.0 | −25.5 |
| Reverse | Spectral | 1823 (25°) | 3.0 | −33.2 |

[a] RNase, ribonuclease; Con A, concanavalin A; AK-HSD, aspartokinase–homoserine dehydrogenase; WGHK, wheat germ hexokinase; YHK, yeast hexokinase; CHT, chymotrypsin; Ala-Pro, alanyl-proline dipeptide. Superscript numbers refer to text footnotes.
[b] T. Y. Tsong, R. L. Baldwin, and E. L. Elson, *Proc. Natl. Acad. Sci. U.S.A.* **69**, 1809 (1972).
[c] J. Janin and G. N. Cohen, *Eur. J. Biochem.* **11**, 520 (1969).
[d] J. Janin and M. Iwatsubo, *Eur. J. Biochem.* **11**, 530 (1969).
[e] A. R. Fersht, *J. Mol. Biol.* **64**, 497 (1972).
[f] This would be a "rapid" conformational transition.

a remarkable agreement between the $\Delta H^*$ for the spectroscopically observed isomerization and that obtained from stopped-flow measurements of the transient during activity. This correspondence of values probably indicates that the rate-determining step is the same. Differences in $\Delta S^*$ values could be interpreted in terms of substrates (which are present only in the activity measurements) affecting either solvation or orderedness in the enzyme–substrate complex.

The energy barrier for some processes is primarily entropic in nature (large negative value of $\Delta S^*$) whereas others are assisted by the increase in entropy in the activated complex; others occur with relatively little effect of entropy in the transition state. Large positive enthalpy changes are associated with all the isomerizations, indicating that a number of noncovalent bonds must be broken in going to the transition state, with a concomitant release of thermal energy. Whether these differences in activation energetics may be associated with particular kinds of structural changes or whether they are due to specific changes in each protein will not be clear until more data of this type are correlated to structural changes.

## [9] The Kinetics of Immobilized Enzyme Systems

*By* KEITH J. LAIDLER and PETER S. BUNTING

Most kinetic studies on enzyme-catalyzed reactions are made with both the enzyme and the substrate in free solution. During the past few years there has also been interest in systems in which either the enzyme or the substrate, or both, are immobilized on a solid support. We have elsewhere[1] reviewed some of the earlier kinetic work on such systems, and some more recent and more detailed reviews have been presented.[2,3] In the present chapter we emphasize some of the more important kinetic principles that apply to immobilized enzymes and discuss some of the applications of these principles to systems that have been investigated experimentally.

There are several reasons why it is important to investigate the kinetics of immobilized enzymes. In the first place, enzymes in nature are frequently immobilized in cells or tissues, and kinetic studies in free solution, although providing an important basis for understanding, do not give the entire answer as to how enzymes function in the living organism. Second, chemical engineers are becoming increasingly interested in the use of enzymes as catalysts for industrial processes. If an enzyme is to be used in this way there is considerable waste of material if a batch process is employed, in which case the enzyme must be removed at the end of the run in order for it to be used for another batch. It is much more economical to attach the enzyme to a solid support, for example to the interior of a tube or at the surface of beads which are packed into a tube, and to cause the reactant solution to flow through the tube. By the use of this technique the same enzyme can be used many times over.

Third, immobilized enzymes lead to important procedures in clinical medicine. Suppose, for example, that a patient suffers from a condition in which an undesirable substance has accumulated in his blood stream. An enzyme that can remove this substance, for example by hydrolyzing it, could be attached to the interior of a tube. The patient's blood could then be pumped through this tube (an extracorporeal shunt) and back into his body. In this way the undesirable substance has been removed without

---

[1] K. J. Laidler and P. S. Bunting, "The Chemical Kinetics of Enzyme Action." Oxford Univ. Press (Clarendon), London and New York, 1973.
[2] L. Goldstein, this series, Vol. 44, p. 397.
[3] J. M. Engasser and C. Horvath, *in* "Applied Biochemistry and Bioengineering" (L. B. Wingard, Jr., E. Katchalski-Katzir, and L. Goldstein, eds.), Vol. 1, p. 127. Academic Press, New York, 1976.

any enzyme having entered the patient's body. The alternative procedure of injecting the enzyme directly into the blood stream is less economical in enzyme and can also be harmful, since many enzymes are toxic and lead to antigen–antibody reactions that can have a serious effect on the patient. Another important clinical application is the automated analysis of important metabolites, biological fluid being pumped into one end of a tube and an analysis made of the effluent.

There are various techniques by which an enzyme may be attached to a solid support. For details the reader is referred to various articles on this topic in Vol. 44 of this series. Immobilization may occur by physical adsorption, by physical inclusion within a matrix, or by covalent chemical bonding. The present review is concerned with the kinetic behavior observed when an enzyme is included within a solid matrix that is in contact with a solution containing the substrate, and also with the situation in which the enzyme is attached to the interior surface of a tube through which substrate solution is flowing. Emphasis is placed on the main features of the theoretical treatment of these systems and on the applications of the theories to the experimental results. Our particular concern will be the kinetic effects of substrate concentration, flow rate, temperature, and pH.

Basic Kinetic Principles of Immobilized Enzyme Systems

There are several reasons why an enzyme attached to a solid may behave differently from one in free solution. In the first place, the immobilization may cause the enzyme molecules to assume a different *conformation*, and therefore be kinetically different. Second, the immobilized enzyme exists in an *environment* different from that when it is in free solution, and this can have a significant effect on the kinetics. Third, there is a *partitioning* of substrate between the solution and support, with the result that the substrate concentration in the neighborhood of the enzyme may be significantly different from that in the bulk solution. Finally, *diffusional* effects will play a more important role with immobilized enzymes, where there is often a very considerable degree of diffusional control.

*Enzymes Embedded in Solid Matrices*

The principles relating to an enzyme supported within a matrix, immersed in substrate solution, are illustrated in Fig. 1. The case of the enzyme and substrate both in free solution is shown in Fig. 1a. If the total enzyme concentration is $[E]_0$, and the substrate concentration is $[S]$, the rate for many systems is given by the Michaelis–Menten equation

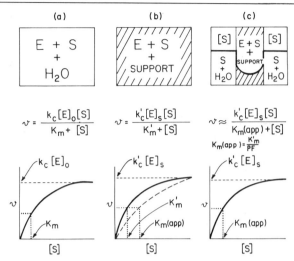

FIG. 1. Three enzyme–substrate systems. In (a) enzyme and substrate are present in aqueous solution; in (b) they are present in a solid support. In (c) the enzyme is embedded in a solid, which is immersed in a solution of substrate. The variation of rate with substrate concentration is shown.

$$v = \frac{k_c[E]_0[S]}{K_m + [S]} \quad (1)$$

where $k_c$ is the catalytic constant and $K_m$ the Michaelis constant. Both these constants are functions of the individual rate constants for the particular mechanism.

Suppose that instead of being in free solution the enzyme and substrate are trapped in a solid, such as a gel. The situation, illustrated in Fig. 1b, is now different in various respects. Because the enzyme and substrate may now exist in different *conformations,* the rate constants may be altered. In addition, the reaction is now occurring in a different *environment.* As a result of these two effects, the Michaelis parameters will be different from those in free solution, and we will now denote these parameters as $k'_c$ and $K'_m$. If these were the entire effects, the result would be as shown by the bold line in Fig. 1b. However, there is also the possibility that in the solid support there will be some degree of diffusion control; this effect sometimes occurs with enzyme systems in aqueous systems, but only for exceedingly fast reactions, and can usually be ignored (see also Vol. 63[10]), The effect of diffusion in the solid will be to lower rates, as shown by the dashed line in Fig. 1b. The limiting rate, $k'_c[E]_0$, will be the same as in the absence of diffusion control, because this limiting rate corresponds to complete saturation of enzyme by the substrate, and diffusion is then not involved. At lower substrate concentrations, however, the diffusion will have a significant effect and will cause the apparent Michaelis constant, $K_m$(app), to be greater than $K_m'$.

A situation with which we are often concerned in practice is represented in Fig. 1c. Here the enzyme is distributed throughout the support, which may be in the form of a disk, and the substrate is present in the external solution. The substrate therefore has to diffuse into the solid in order to come into contact with the enzyme molecules. In the situation shown in Fig. 1c, the substrate can flow into the support from two sides of a cylindrical disk. The rates of diffusion are therefore important, as in the previous case (Fig. 1b). However, there is now an additional factor, the partitioning of the substrate between the solution and the support. For example, the substrate might contain nonpolar (hydrophobic) groups, and the solid might also contain hydrophobic groups; the substrate would then be more soluble in the support than in the aqueous solution. In other words, the partition coefficient $P$ would be greater than unity, and this would have the effect of increasing the rates. In the concentration profile shown in Fig. 1c the partition coefficient is less than unity, so that the concentration of substrate immediately inside the support is less than that immediately outside. This has the effect of diminishing the rate.

After the steady state has become established, the variation of substrate concentration throughout the support is as shown in Fig. 1c. The substrate diffuses from each of the two opposite surfaces, and the concentration passes through a minimum, the substrate being converted into product under the action of the enzyme. An exact solution of the steady-state equations for this system cannot be obtained, but a useful approximate treatment has been given by Sundaram, Tweedale, and Laidler[4] and Thomas et al.[5] have dealt with the case of diffusion from only one surface. The treatment of Sundaram et al. is concerned with diffusional effects within the membrane and assumes that diffusion in the aqueous phase is sufficiently rapid so that there is no diffusional control. This assumption is satisfactory for stirred systems, but if there is little stirring it is necessary to consider diffusional effects in the substrate solution; this has been discussed by Goldstein.[2]

The treatment of Sundaram et al.[4] leads to the conclusion that the rate equation again is approximately of the Michaelis–Menten form

$$v = \frac{k_c'[E]_s[S]}{K_m(\text{app}) + [S]} \tag{2}$$

where $[E]_s$ is the concentration of enzyme within the solid support. The

---

[4] P. V. Sundaram, A. Tweedale, and K. J. Laidler, Can. J. Chem. **48**, 1498 (1970).
[5] D. Thomas, G. Brown, and E. Selegny, Biochimie **54**, 229 (1972).

apparent Michaelis constant $K_m(\text{app})$ is related to the $K'_m$ value for the immobilized enzyme (which does not involve partitioning and diffusional effects) by the equation

$$K_m(\text{app}) = K'_m/PF \tag{3}$$

Here $P$ is the partition coefficient (ratio of surface concentration of substrate in the support to that in the solution) and $F$ is a function given by

$$F = \frac{\tanh \gamma l}{\gamma l} \tag{4}$$

where

$$\gamma = \frac{1}{2}\left(\frac{k'_c[E]_s}{DK'_m}\right)^{1/2} \tag{5}$$

The function $F$ is the Thiele function, which commonly appears in catalytic problems where diffusion is important. In Eq. (5), $D$ is the diffusion coefficient for the substrate in the support.

The form of the function $F$ is very significant in the theory and is shown in Fig. 2. We see that when $\gamma l$ is small the function $F$ is close to unity; in that case $K_m(\text{app})$ is equal to $K'_m/P$, and the $K'_m$ value is therefore modified by the partitioning but not by diffusional effects, which are now unimportant. It follows from Eq. (5) that the product $\gamma l$ tends to be small under the following conditions:

1. When $l$ is small, which means that the substrate has ready access to the enzyme; in the limit of $l = 0$ the enzyme is in free solution.
2. When $D$ is large, when again the substrate can easily reach the enzyme molecules.
3. When the enzyme concentration $[E]_s$ within the solid is small, which means that the catalysis is sufficiently slow that diffusion can keep up with it.
4. When $k'_c/K'_m$ is small; again, this means slow catalysis, the chemical process being rate limiting rather than the rate of diffusion.

At higher $\gamma l$ values the function $F$ becomes significantly less than unity, and it is an interesting property of the Thiele function that when $\gamma l$ is larger than about 2 the value of $F$ is given quite accurately by

$$F = 1/\gamma l \tag{6}$$

This region of $\gamma l$ values where $F$ is significantly less than unity corresponds to some measure of diffusion control of the reactions. Table I summarizes the conditions that lead to diffusion control.

This variation of $F$ with $\gamma l$ has an important consequence as far as the

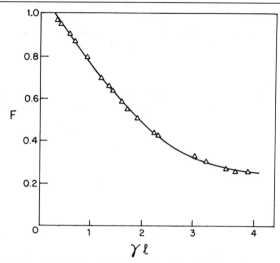

FIG. 2. The function $F (= \tanh \gamma l/\gamma l)$ plotted against $\gamma l$. The points are derived from an experimental study of acetylcholinesterase in polyacrylamide disks. From T. T. Ngo and K. J. Laidler, *Biochim. Biophys. Acta* **377**, 303 (1975).

dependence of the rate on the enzyme concentration is concerned. It follows from Eqs. (2) and (3) that if $F$ is approximately unity the rate should be proportional to $[E]_s$, and for sufficiently thin slices (low $\gamma l$ values) $F$ is approximately unity. In general, however, Eqs. (2) and (3) show that at low substrate concentrations the rate is given by

$$v \propto [E]_s F \tag{7}$$

and when $\gamma l$ is large enough ($>2$), the function $F$ is inversely proportional to $\gamma l$; thus, under these conditions, with the use of Eq. (5)

$$v \propto [E]_s/\gamma l \propto [E]_s^{1/2} \tag{8}$$

In other words, as $\gamma l$ increases from a small to a large value, which will occur as the thickness $l$ is increased, there will be a transition from a dependence of rate on the first power of the enzyme concentration to a dependence on the square root. An example of this behavior will be seen later.

The above relationships, which apply to enzyme immobilized in a cylindrical support, also apply with some modification to other particle shapes. For example, Hinberg et al.[6] have developed a treatment for an enzyme immobilized in spherical particles. For such systems there enters a similar parameter $F$, related to the kinetic and diffusional parameters

[6] I. Hinberg, R. Korus, and K. F. O'Driscoll, *Biotechnol. Bioeng.* **16**, 943 (1974).

TABLE I
SUMMARY OF CONDITIONS FOR DIFFUSIONAL CONTROL
OF IMMOBILIZED ENZYME SYSTEMS

| System | Conditions |
|---|---|
| Solid matrix | High enzyme concentration, $[E]_s$ |
| | Low substrate concentration, $[S]$ |
| | Large membrane thickness, $l$ |
| | Low substrate diffusion constant, $D$ |
| | High enzyme catalytic constant, $k_c'$ |
| | Small Michaelis constant, $K_m'$ |
| Tubes | Low substrate flow rate, $v_f$ |
| | Low substrate concentration, $[S]$ |
| | Large tube radius, $r$ |
| | Large tube length, $L$ |
| | Low substrate diffusion constant, $D$ |

and to the radius of the sphere, and the resulting conclusions are very similar to those for the disks.

*Enzymes Attached to the Inner Surfaces of Tubes*

A detailed theoretical treatment of the kinetics of a system where the enzyme is attached to the inner surface of a tube, through which the substrate solution is flowing, has been developed by Kobayashi and Laidler.[7] The treatment is quite complicated, and only a very brief account of the main conclusions can be given here. The system is represented in Fig. 3. The enzyme is present in a unimolecular layer at the surface, so that diffusional effects within the support are not involved. However, diffusion in the substrate solution, which is flowing through the tube, has to be taken into account and in some circumstances has a very important effect on the kinetics. It is convenient, as a simplification, to think in terms of a diffusion layer established in the solution at the inner surface of the tube, as shown in Fig. 3. A typical substrate concentration variation is as shown in the figure; the concentration is constant in the central part of the tube, but falls to a lower value in the diffusion layer, since at the surface the substrate is being removed when it comes into contact with the enzyme.

The treatment of Kobayashi and Laidler[7] leads to the conclusion that two factors have an important effect on the flow kinetics of enzymes immobilized on tubes: (a) the mass transfer of the substrate to the surface;

[7] T. Kobayashi and K. J. Laidler, *Biotechnol. Bioeng.* **16,** 99 (1974).

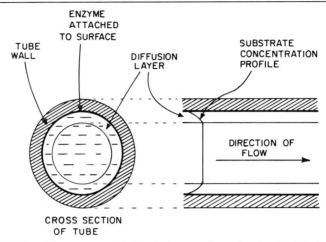

FIG. 3. A tube with enzyme attached to the inner surface, showing the diffusion layer and the concentration profile.

(b) the rate of diffusion within the diffusion layer at the surface. The thickness of the diffusion layer depends on the magnitude of the mass-transfer coefficient $k_L$, which is a measure of the rate at which substrate diffuses through the layer. This coefficient is given by

$$k_L = \frac{3}{2} \frac{(12)^{1/3}}{\Gamma(1/3)} \left(\frac{D^2 v_f}{rL}\right)^{1/3} \qquad (9)$$

where $D$ is the diffusion coefficient of the solute, $v_f$ the flow rate, $r$ the radius of the tube, and $L$ its length. The gamma function $\Gamma(1/3)$ has the value 2.67, and Eq. (9) becomes

$$k_L = 1.29(D^2 v_f/rL)^{1/3} \qquad (10)$$

We see from Eq. (10) that increasing the flow rate increases the rate of diffusion of substrate to the surface. This effect arises because there is a reduction in the effective thickness of the diffusion layer with increasing flow rates. At very high flow rates there is therefore little diffusional control; the substrate reaches the surface rapidly, and the rate of formation of product depends on the chemical interaction between enzyme and substrate at the surface. At low flow rates, on the other hand, a substantial diffusion layer is established, and the substrate is slow to reach the surface. There is therefore considerable diffusion control and the kinetic behavior deviates from that obtained with the enzyme in free solution. These and other conclusions are summarized in Table I.

The theory leads to the conclusion that to a good approximation the

Michaelis–Menten equation, Eq. (2), is obeyed. The value of $K_m(\text{app})$ is given by the expression

$$K_m(\text{app}) = K'_m + 0.39\, k'_c\, [\text{E}]_s \left(\frac{rL}{D^2}\right)^{1/3} v_f^{-1/3} \tag{11}$$

where $K'_m$ and $k'_c$ are the kinetic parameters for the immobilized enzyme in the absence of diffusional effects, and $v_f$ is the rate of flow of the substrate solution through the tube. We see from Eq. (11) that $K_m(\text{app})$ should, according to the theory, vary with the reciprocal cube root of the flow rate. At very high flow rates the second term in Eq. (11) is negligible and $K_m(\text{app})$ is equal to $K'_m$; under these conditions the thickness of the diffusion layer is very small and there is little effect of diffusion. At lower flow rates, on the other hand, the diffusion layer becomes thicker; the apparent Michaelis constant is then larger, and the rates are smaller. Examples of this effect will be seen later.

Other important results predicted by the Kobayashi–Laidler treatment relate to the product concentrations at the tube exit and the overall rates of reaction within the tube. If the conditions are such that there is little diffusional control (e.g., high [S], low tube diameter; see Table I), the rate of formation of product within the tube is equal to the internal surface area of the tube, $2\pi rL$, multiplied by the rate $v$ of the reaction per unit surface area; this rate is given by Eq. (2). The rate of product formation within the tube, $v_0$, is thus

$$v_0 = 2\pi rLv \tag{12}$$

and is independent of flow rate. The linear rate of flow of substrate solution through the tube is $v_f$, and multiplication by the cross-sectional area gives the volume rate of flow, $\pi r^2 v_f$. The product concentration at the tube exit is the rate $v_0$ divided by the volume of solution that emerges in unit time, and is thus given by

$$[\text{P}]_0 = \frac{v_0}{\pi r^2 v_f} = \frac{2L}{rv_f} v = \frac{2Lk'_c[\text{E}]_s[\text{S}]}{rv_f(K'_m + [\text{S}])} \tag{13}$$

If, on the other hand, the conditions are such that there is complete diffusional control (e.g., low [S]; see Table I), the theory leads to the result that the rate of product formation within the tube is given by

$$v_D = 8.06(v_f D^2 r^2 L^2)^{1/3}[\text{S}] \tag{14}$$

The product concentration $P_D$ at the tube exit is now

$$[\text{P}]_D = 2.56(DL/r^2 v_f)^{2/3}[\text{S}] \tag{15}$$

The rate $v$ is now not involved, but the diffusion coefficient is involved, and there is a different dependence on flow rate. We shall see that Eqs. (13) and (14) provide a convenient way of determining the extent of diffusional control.

## Analysis of Data, with Examples

The equations given in the preceding section lead to a number of useful methods for testing the experimental data with respect to the theoretical treatment. It is of particular interest to deduce, from the data, the extent of diffusion control. This knowledge is important for the design of preparative enzyme reactors and for the application of immobilized enzymes in analytical procedures. The following is a summary of the various procedures that can be used in testing the theories and determining the extent of diffusion control.

### Enzymes Immobilized in Solid Matrices

*Lineweaver–Burk Plots.* It is very often found that Lineweaver–Burk plots of $1/v$ against $1/[S]$ show curvature, being convex to the $1/[S]$ axis; an example is shown in Fig. 4. This type of curvature results from the fact that the extent of diffusion control is greater at lower substrate concentrations. As we have seen, diffusion control has the effect of lowering the rates and increasing the $K_m(app)$ values; the points at lower substrate concentrations thus correspond to abnormally high $1/v$ values as compared with those at higher substrate concentrations. The result is curvature convex to the $1/[S]$ axis. The results shown in Fig. 4 relate to $\beta$-galactosidase trapped in polyacrylamide gel.[8] A similar analysis of data has been made by Regan et al.[9]

*Determination of F Values.* A more quantitative procedure is to determine $F$ values from the observed $K_m(app)$ values, and see how they vary with $\gamma l$. With very thin disks $F$ is close to unity and $K_m(app)$ is therefore equal to $K'_m/P$ [Eq. (3)]. Measurements of $K_m(app)$ at greater thicknesses therefore allow $F$ values to be calculated. The curve in Fig. 2 represents the theoretical variation of $F$ with $\gamma l$, as predicted by Eqs. (3) and (4). The points were obtained by Ngo and Laidler[10] for acetylcholinesterase in polyacrylamide. The excellent agreement between theory and experiment shows that the theory is along the right lines and gives a quantitative index, in terms of $F$ values, of the extent of diffusion control at the various membrane thicknesses.

[8] P. S. Bunting and K. J. Laidler, *Biochemistry* **11**, 4477 (1972).
[9] D. L. Regan, M. D. Lilly, and P. Dunnill, *Biotechnol. Bioeng.* **16**, 1081 (1974).
[10] T. T. Ngo and K. J. Laidler, *Biochim. Biophys. Acta* **377**, 303 (1975).

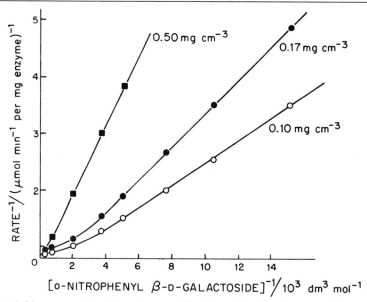

FIG. 4. Lineweaver–Burk plots for immobilized $\beta$-galactosidase in polyacrylamide disks. The curvature indicates diffusion control, becoming more important as the substrate concentration is lowered. From P. S. Bunting and K. J. Laidler, *Biochemistry* **11**, 4477 (1972). Copyright by the American Chemical Society.

*Influence of Enzyme Concentration.* Another indication of the extent of diffusion control is provided by studying the effect of enzyme concentration on the kinetic behavior at low substrate concentrations ($[S] \ll K_m$). For diffusion-free kinetics the rate is directly proportional to the enzyme concentration. For full diffusion control, on the other hand, the rate is proportional to the square root of the enzyme concentration, as shown by Eq. (8). Figure 5 shows results obtained for acetylcholinesterase in polyacrylamide slices.[10] With the thin slices the rate is proportional to the first power of the enzyme concentration, indicating absence of any diffusion control. With the thicker slices the rate is proportional to the square root of the enzyme concentration, a result that is consistent with full diffusion control.

*Temperature Dependence.* This dependence also provides evidence with regard to diffusion control, a topic that is considered later.

## Enzymes Attached to Tubes

*Lineweaver–Burk Plots.* As with enzymes supported in disks, plots of $1/v$ against $1/[S]$ may be convex to the $1/[S]$ axis, because of the increasing extent of diffusion control at the lower substrate concentrations,

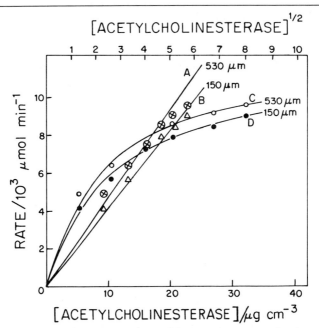

FIG. 5. Plots of rate against the square root of the enzyme concentration (curves A and B) and against the enzyme concentration (curves C and D) for acetylacholinesterase in polyacrylamide slices, 150 and 530 μm in thickness. There is little diffusion control with the thin slices, but considerable diffusion control with the thick ones. From T. T. Ngo and K. J. Laidler, *Biochim. Biophys. Acta* **377,** 317 (1976).

where there is an increase in $K_m$(app). Figure 6 shows an example for L-asparaginase attached to the inner wall of nylon tubing.[11] This plot shows that an increase in the flow rate $v_f$ decreases the diffusion effect by reducing the thickness of the Nernst diffusion layer; there is then a decrease in the slope of the Lineweaver–Burk plot, $K_m$(app) becoming smaller and approaching $K'_m$.

*Influence of Flow Rate on $K_m$(app).* Equation (11) predicts that $K_m$(app) varies linearly with $v_f^{-1/3}$, becoming $K'_m$ when $v_f^{-1/3}$ approaches zero. A plot of $K_m$(app) against $v_f^{-1/3}$, for acetylcholinestarase attached to nylon tubing,[12] is shown in Fig. 7. The extrapolated value of $1.7 \times 10^{-4}$ M, corresponding to $K'_m$, is not far from the $K_m$ for free solution ($1.2 \times 10^{-4}$ M), where diffusion effects do not apply.

*Influence of Flow Rate on Product Concentration.* According to Eq. (13), a plot of the logarithm of the product concentration against the loga-

---

[11] P. S. Bunting and K. J. Laidler, *Biotechnol. Bioeng.* **16,** 119 (1974).
[12] T. T. Ngo and K. J. Laidler, *Biochim. Biophys. Acta* **377,** 317 (1975).

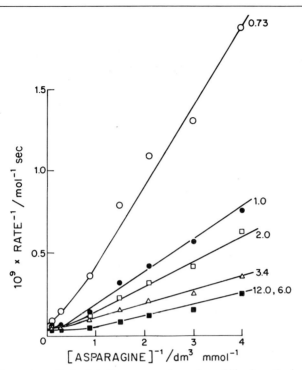

FIG. 6. Lineweaver–Burk plots for L-asparaginase immobilized on the inner wall of a nylon tube. The diffusion control, indicated by the curvature, becomes less as the flow rate is increased from 0.73 to 12.0 cm sec$^{-1}$ From P. S. Bunting and K. J. Laidler, *Biotechnol. Bioeng.* **16,** 119 (1974).

rithm of the flow rate will have a slope of −1 if there is no diffusion control. If there is full diffusion control, which will occur at low substrate concentrations and low flow rates, a double-logarithmic plot will have a slope of −0.67 [Eq. (15)]. Figure 8 shows a typical plot, for β-galactosidase attached to nylon tubing.[13] This plot shows that there is very extensive diffusion control at the low substrate concentrations and low flow rates. The slopes never became −1, showing that there are diffusion effects under all the experimental conditions employed.

Alternatively, we can make double-logarithmic plots of the rate of product formation within the tube against the flow rate. If there is no diffusion control, there is no dependence on flow rate [Eq. (12)] and the slope of the double-logarithmic plot is zero. If there is full diffusion control, Eq. (14) applies and the slope of the double-logarithmic plot is 1/3.

[13] D. Narinesingh, T. T. Ngo, and K. J. Laidler, *Can. J. Biochem.* **53,** 1061 (1975).

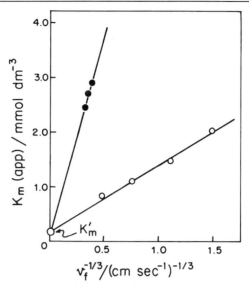

FIG. 7. Plots of $K_m$(app) against $v_f^{-1/3}$, where $v_f$ is the flow rate, for acetylcholinesterase attached to the inner surface of nylon tubing. ○, length = 10 cm; ●, length = 50 cm. From T. T. Ngo and K. J. Laidler, *Biochim. Biophys. Acta* **377**, 317 (1975).

Intermediate slopes give some indication of the extent of diffusion control.

*Use of Dimensionless Parameters.* The extent of diffusion control can be determined by the use of two dimensionless parameters defined by Kobayashi and Laidler[7] [Eqs. (16) and (17)].

$$\phi = \frac{[P]}{[S]} \left(\frac{v_f r^2}{DL}\right)^{2/3} \tag{16}$$

$$\rho = \frac{K_m(\text{app})}{[S]} \tag{17}$$

where [P] is the concentration of product at the tube exit. A plot of $\phi$ against $\rho$ can be split into regions, as shown in Fig. 9, corresponding to different degrees of diffusion control. Region 1 corresponds to little diffusional control (<5%), region 3 to considerable (>60%) diffusional control, and region 2 is intermediate. Figure 9 shows experimental points for acetylcholinesterase attached to nylon tubing.[12] Note that two of the points, corresponding to high flow rates and high substrate concentrations, are in the region of little diffusion control.

The advantage of this type of plot is that it can be used to estimate the extent of diffusion control on the basis of a single experiment, in which product concentration [P], initial substrate concentration [S], and flow

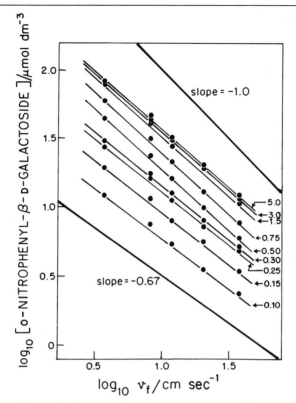

FIG. 8. Double-logarithmic plots of product concentration against flow rate for β-galactosidase attached to the inner surface of nylon tubing. The substrate concentrations are shown. There is more diffusion control at the low substrate concentrations and flow rates. From D. Narinesingh et al., Can. J. Biochem. 53.

rate $v_f$ are measured. On the basis of this one experiment the conditions can be modified in order to alter the extent of diffusion control.

*Temperature Dependence.* This also provides evidence for diffusional effects in tubes, a topic that is considered later.

## Two-Substrate Systems

A few investigations have also been made on the flow kinetics of enzymes involving two substrates. For example, Daka and Laidler[14] have attached rabbit muscle lactate dehydrogenase to the inner surface of nylon tubing and have made a study of the flow kinetics for the reaction

[14] N. J. Daka and K. J. Laidler, *Can. J. Biochem.* **56**, 774 (1978).

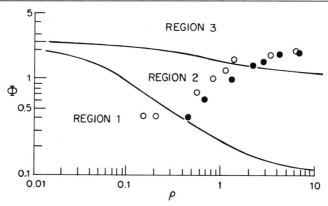

Fig. 9. Plots using the dimensionless parameters $\phi$ and $\rho$ [see Eqs. (16) and (17)]. The lines separate the results into three regions: region 1, little diffusional control (<5%); region 2, moderate diffusional control (5–60%); region 3, considerable diffusional control (>60%). The experimental points are for acetylcholinesterase attached to nylon tubing. $\bigcirc$, $v_f = 4.24$ cm sec$^{-1}$; $\bullet$, $v_f = 0.45$ cm sec$^{-1}$. Two of the points, corresponding to high flow rates and high substrate concentrations, are in the region of little diffusional control. From T. T. Ngo and K. J. Laidler, *Biochim. Biophys. Acta* **377**, 317 (1975).

between pyruvate and reduced nicotinamide adenine dinucleotide (NADH). In one series of measurements the pyruvate was held in excess and the NADH concentration varied; in another the NADH was in excess and the pyruvate concentration was varied. Application of all the tests showed that there was a considerable degree of diffusion control under all the conditions employed. The apparent Michaelis constants again varied linearly with $v_f^{-1/3}$, and the values extrapolated to infinite flow rate approached the values for the enzyme in free solution. It thus appears that the Kobayashi–Laidler treatment applies satisfactorily to two-substrate systems under these limiting conditions. There is still need for a general theoretical treatment of two-substrate systems, for both membranes and tubes.

## Temperature Effects

We have elsewhere[1] given a general treatment of temperature effects in enzyme kinetics, and the reader is also referred to Vol. 63[10] for further background information. Of particular interest are the temperature coefficients in the limiting cases of low and high substrate concentrations, for both free and immobilized enzymes.

For the case of enzymes immobilized in matrices immersed in substrate solution, the rate $v$ at *low substrate concentrations* is given by

$$v_l = \frac{k'_c[\text{E}]_s[\text{S}]}{K_m(\text{app})} \tag{18}$$

[see Eq. (2)], and use of Eq. (3) gives

$$v_l = \frac{k'_c[\text{E}]_s[\text{S}]PF}{K'_m} \tag{19}$$

If $\gamma l$ is small, $F$ is approximately unity and there is no diffusion control. If $k'_c$ and $K'_m$ vary with temperature according to Eqs. (20) and (21),

$$k'_c = A'_c \exp(-E'_c/RT) \tag{20}$$

and

$$K'_m = A'_m \exp(\Delta E'_m/RT) \tag{21}$$

the temperature dependence of the rate under these conditions will be given by

$$v_l \propto \exp[-(E'_c + \Delta E'_m)/RT] \tag{22}$$

and the observed activation energy will correspond to $E'_c + \Delta E'_m$. If, on the other hand, $\gamma l$ is greater than about 2, $F$ is given quite accurately by $1/\gamma l$, and

$$v_l = \frac{k'_c P[\text{E}]_s[\text{S}]}{K'_m \gamma l} = \left(\frac{4k'_c[\text{E}]_s D}{K'_m l^2}\right)^{1/2} P[\text{S}] \tag{23}$$

If the temperature-dependence of $D$ is given by

$$D = A_D \exp(-E_D/RT) \tag{24}$$

the temperature-dependence of the rate, under these conditions, is given by

$$v_l \propto \exp[-(E'_c + \Delta E'_m + E_D)/2RT] \tag{25}$$

At *high substrate concentrations,* on the other hand, the rate is simply given by

$$v_h = k'_c[\text{E}]_s = [\text{E}]_s A'_c \exp(-E'_c/RT) \tag{26}$$

and the observed activation energy is equal to $E'_c$.

Figure 10 shows some results obtained by Ngo and Laidler[15] for acetylcholinesterase trapped in polyacrylamide. The activation energy of 9.9 kcal mol$^{-1}$ found at the higher substrate concentration and in the lower temperature range is close to that for the free enzyme (9.4 kcal mol$^{-1}$). This value is presumably $E'_c$ for the breakdown of the Michaelis complex.

[15] T. T. Ngo and K. J. Laidler, *Biochim. Biophys. Acta* **525**, 93 (1978).

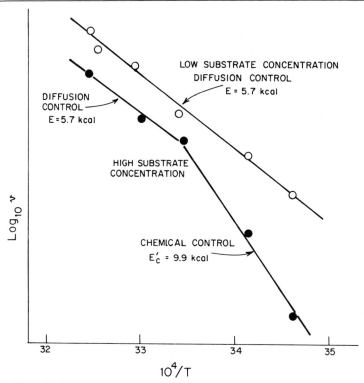

Fig. 10. Arrhenius plots for acetylcholinesterase immobilized in acrylamide. Different scales are used for the rates at high and low substrate concentrations. From T. T. Ngo and K. J. Laidler, *Biochim. Biophys. Acta* **525**, 93 (1978).

The lower value (5.7 kcal mol$^{-1}$) obtained at higher temperatures suggests that low-[S] behavior is now being observed. At the higher temperatures the enzyme more rapidly converts the substrate to product, and the enzyme within the disk will have a low-[S] environment. The same low value of 5.7 kcal mol$^{-1}$ is found with the substrate at low concentrations and is attributed to diffusion control. If Eq. (25) applies, the observed activation energy is $\frac{1}{2}(E'_c + \Delta E'_m + E_D)$. With $E'_c = 9.9$ kcal mol$^{-1}$ it follows that $\Delta E'_m + E_D = 1.4$ kcal mol$^{-1}$. The energy $E_D$ is expected to be about 2 kcal mol$^{-1}$, and $\Delta E'_m$ will be small and may be negative or positive. Somewhat similar results were obtained by Buchholz and Rugh[16] for immobilized trypsin.

The treatment of temperature effects for enzymes attached to tubes is

[16] K. Buchholz and W. Rugh, *Biotechnol. Bioeng.* **18**, 95 (1976).

quite complicated, but some limiting cases are of interest and can be applied to the experimental results. At high flow rates and high substrate concentrations $K_m(\text{app})$ approaches $K'_m$ and the rate is given by Eq. (13); the activation energy will be $E'_c$, corresponding to the breakdown of the enzyme–substrate complex. At the other extreme (low flow rates and low substrate concentration), the rate is given by Eq. (14), which involves diffusion, but not the enzyme-catalyzed reaction. The observed activation energy will then be two-thirds of the value for the diffusion process.

Results obtained by Ngo and Laidler[15] for acetylcholinesterase attached to nylon tubing are consistent with this interpretation. At low substrate concentration and low flow rates the activation energy was 1.7 kcal mol$^{-1}$. Higher values were obtained at higher substrate concentrations and flow rates.

An alternative procedure for enzymes trapped in matrices was employed by Ngo and Laidler,[10] and involves comparing low-substrate concentration rates for thin and thick disks. From Eqs. (2) and (3) the rate $v_l$ at low substrate concentrations is given by

$$v_l = \frac{k'_c[E]_s[S]PF}{K'_m} \tag{27}$$

For thin disks $F$ is approximately unity, and thus

$$v_l(\text{thin}) = \frac{k'_c[E]_s[S]P}{K'_m} \tag{28}$$

For thick disks $F$ is approximately $1/\gamma l$, and thus

$$v_l(\text{thick}) = \frac{k'_c[E]_s[S]P}{K'_m \gamma l} \tag{29}$$

$$= \left(\frac{4k'_c D[E]_s}{K'_m l^2}\right)^{1/2} P[S] \tag{30}$$

using Eq. (5). It follows from Eqs. (28) and (30) that

$$\frac{[v_l(\text{thick})]^2}{v_l(\text{thin})} = \frac{4DP[S]}{l^2} \tag{31}$$

In this expression only $D$ is temperature-dependent, and a plot of the logarithm of the left-hand-side of Eq. (31) against $1/T$ will therefore give the activation energy for the diffusion process. In this way Ngo and Laidler[10] obtained a value of 3.7 kcal for the activation energy of the diffusion process for the acetylcholinesterase trapped in polyacrylamide. A procedure for obtaining the activation energy for diffusion in the case of flow through tubes was described by Ngo and Laidler.[12]

## Influence of pH

The general theory of pH effects in enzyme kinetics has been presented previously[1], and is also dealt with in Vol. 63 [9]. With immobilized enzymes, pH effects are of interest from various points of view. The partitioning of hydrogen ions between the solution and the supported enzyme has an important influence on the dependence of rate on pH. Also, if the enzyme reaction produces or consumes acid, some special effects may be observed.

Figure 11 shows a plot of the logarithm of the rate of an enzyme-catalyzed reaction against the pH. At low substrate concentrations the variations in rate with pH depend upon the p$K$ values for the free enzyme. At high substrate concentrations the behavior depends on the ionizations of the enzyme–substrate complex or some subsequently formed complex.

If there are charged groups on the support, the p$K$ values will be altered. Consider, for example, the ionization of a group $-B-H$, present as part of the active center of the enzyme:

$$-B-H \xrightleftharpoons{K_a} -B^- + H^+$$

If there is a negatively charged group in the neighborhood of the B–H group the ionization can be represented as follows:

$$\begin{array}{c} B-H \\ | \\ COO^- \end{array} \xrightleftharpoons{K_a'} \begin{array}{c} B^- \\ | \\ COO^- \end{array} + H^+$$

The dissociation constant $K_a'$ will be less than $K_a$, since the $-COO^-$ group attracts the leaving protons; p$K_a'$ is thus *larger* than p$K_a$. The $\log_{10}$ (rate) vs. activity curves will therefore be displaced to the right, as seen in the dashed curve shown in Fig. 11. A positively charged support will give displacements to the left. Goldstein[17] has collected a number of examples of this type of behavior.

Systems have also been studied in which the immobilized enzyme releases acid[15,18,19] or base,[20] and there may then be remarkable changes in the pH profiles. Suppose, for example, that acid is produced in a reaction brought about by an enzyme immobilized in a membrane. The acid diffuses out and the buffer in the aqueous solution diffuses in, and as a result

---

[17] L. Goldstein, *Biochemistry* **11,** 4072 (1972).
[18] R. Goldman, O. Kedem, H. I. Silman, S. R. Caplan, and E. Katchalski, *Biochemistry* **7,** 486 (1968).
[19] H. I. Silman and A. Karlin, *Proc. Natl. Acad. Sci. U.S.A.* **58,** 1664 (1965).
[20] D. Thomas, G. Brown, and E. Selegny, *Biochimie* **54,** 229 (1972).

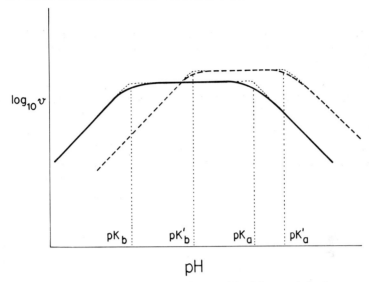

FIG. 11. Schematic plots of $\log_{10}$ (rate) against pH. The full curve is for free enzyme and shows the $pK_a$ and $pK_b$ values corresponding to the inflection points. The dashed curve, displaced to the right, is for enzyme immobilized in a negatively charged support. This behavior is also found for reactions producing acid that has to diffuse out of the support. With enzymes immobilized in positively charged supports, or producing base, the displacement is to the left.

the pH in the solution will be greater than that in the vicinity of the enzyme. The pH profiles are therefore shifted to the right, as shown in Fig. 11. If the enzyme reaction produces base, as is the case with urease,[20] the apparent pH optimum is shifted to a lower value. Ngo and Laidler[15] have given a quantitative interpretation of this type of behavior, in terms of Fick's law of diffusion, and have shown that very considerable pH shifts can result from this effect. Other workers have observed similar effects.[2,3]

Enzymes *in Vivo*

Kinetic studies of enzyme–substrate systems in free solution provide very important background information, but enzymes *in vivo* are frequently attached to cells and tissues. Kinetic studies of immobilized enzymes provide valuable information about the physiological behavior of enzymes.

An interesting example, discussed by Bunting and Laidler[8] and by Ngo and Laidler,[10] is myosin or adenosine triphosphatase (ATPase), which is an important constituent of muscle. The density of muscle is

TABLE II
DIFFUSION CONTROL IN MUSCLE SYSTEMS
ATPase: $\gamma = 3.5 \times 10^4$ cm$^{-1}$

| Component | Diameter ($\mu$m) | $\gamma l$ | F | Diffusion control |
|---|---|---|---|---|
| Filaments | 0.1 | 0.35 | 0.96 | Little |
| Fibrils | 2 | 7 | 0.14 | Some |
| Fibers | 5 | 17.5 | 0.006 | Almost complete |

about 1.0, and the fraction of it that is myosin is about 0.1. Since the molecular weight of myosin is about $2 \times 10^7$, the concentration of myosin in muscle is about $5 \times 10^{-6}$ mol dm$^{-3}$. Use of the kinetic parameters *in vitro* for ATPase[21] and with $D = 10^{-6}$ cm$^2$ sec$^{-1}$ leads to a $\gamma$ value of ~$3.5 \times 10^4$ cm$^{-1}$. The diameters of muscle filaments, fibril, and fiber are approximately 0.1, 2, and 5 $\mu$m, respectively.[22] Table II shows the corresponding $\gamma l$ values calculated from Eq. (5), and the F values calculated from Eq. (4). An F value of close to unity means that $K_m$(app) is hardly modified by diffusional effects, whereas a low value means a very large diffusional influence on $K_m$(app). At low substrate concentrations, where $K_m$(app) is involved in the rate equation, the diffusion control will therefore be as shown in Table II; we see an interesting range of behavior, from hardly any diffusion control in the filaments to almost complete diffusion control in the fibers. At higher substrate concentrations the diffusion control will tend to disappear.

We saw in Vol. 63[10] that diffusional effects are rarely significant with enzymes in free solution. For enzyme reactions under physiological conditions, on the other hand, diffusion control is a distinct possibility.

[21] L. Ouellet, K. J. Laidler, and N. F. Morales, *Arch. Biochem. Biophys.* **39,** 37 (1952).
[22] B. Katzy, "Muscle and Synapse," p. 162. McGraw-Hill, New York, 1966.

# [10] Subsite Mapping of Enzymes: Application to Polysaccharide Depolymerases

## By JIMMY D. ALLEN

It has been shown for several depolymerizing enzymes that substrate reactivity is influenced by polymer units distant from the bond actually undergoing reaction. This observation led to the proposal that monomer residues remote from the point of catalysis interact with the enzyme. Consequently, the substrate binding region of the enzyme can be envisioned as an array of tandem subsites with each subsite geometrically complementary to and interacting with a single monomer residue of the substrate. Such a subsite model has been invoked to account for the enzymic properties of proteinases,[1] nucleases,[2] and carbohydrases.[3]

This chapter deals with subsite mapping, the term applied to the experimental determination of the number of subsites comprising the binding region, the energetics of interaction of each subsite with a monomer residue, and the hydrolytic rate coefficients. The value of the generated subsite map lies in its ability to correctly predict the action pattern of the enzyme. This development will be restricted to subsite mapping of polysaccharide depolymerases; however, the general principles developed here are also applicable, within certain limitations, to other polymer-acting enzymes.

Theory

The symbols and notations employed are listed in Table I.

The interactions of maltotetraose with the binding region of a hypothetical five-subsite amylase are depicted in Fig. 1. If the substrate binds so as to expose a susceptible bond to the catalytic amino acids on the enzyme (binding modes indexed IV,4, V,4, and VI,4 in Fig. 1), the complex is productive, and bond cleavage can occur. The remaining positional isomers are nonproductive in the sense that they cannot lead to bond scission. The rates of bond scission of the substrate depend on the energetics of binding of the productive positional isomers and on the respective hydrolytic rate coefficients.

*Equations for the Subsite Model*

The subsite model is best characterized by conceptually dissecting the enzymic processes into microscopic and macroscopic phenomena. A microscopic coefficient characterizes one particular positional isomer of an $n$-mer substrate. Hence, each positional isomer has an attendant microscopic association constant, and each productive positional isomer has an attendant microscopic hydrolytic rate coefficient. We cannot normally directly measure these microscopic constants, but they can be related to measurable macroscopic parameters, which are a function of all

---

[1] R. C. Thompson and E. R. Blout, *Proc. Natl. Acad. Sci. U.S.A.* **67**, 1734 (1970).
[2] F. G. Walz, Jr. and B. Terenna, *Biochemistry* **15**, 2837 (1976).
[3] D. M. Chipman and N. Sharon, *Science* **165**, 454 (1969).

## TABLE I
### Symbols and Notations Used in Subsite Mapping

| | |
|---|---|
| A | Substrate |
| b.c.f. | Bond cleavage frequency |
| $c$ | Index of the position of the catalytic site, specifying the subsite to the right of the position of bond cleavage |
| E | Enzyme |
| G | Measured radioactivity for a sugar after chromatographic separation |
| $\Delta G$ | Unitary free energy of binding |
| $\Delta G_a$ | Acceleration factor |
| $h$ | General binding-mode index, specifying the real or virtual subsite occupied by the reducing-end glycosyl unit |
| $i$ | Subsite index |
| $j$ | Maximum chain length substrate for which experimental data are available |
| $K$ | Microscopic dissociation constant |
| $K'$ | Microscopic association constant |
| $K_i$ | Inhibition constant |
| $K_m$ | Michaelis constant |
| $K_{int}$ | Microscopic dissociation constant for a binding mode in which the entire binding region is occupied |
| $k'$ | The first-order rate constant for enzymic hydrolysis |
| $k_{+1}$ | Microscopic association rate constant |
| $k_{-1}$ | Microscopic dissociation rate constant |
| $k_{+2}$ | Hydrolytic rate coefficient |
| $l$ | Number of real subsites comprising the binding region of an enzyme |
| $m$ | Chain-length index for product |
| min | Minimum value |
| $n$ | Chain-length index for substrate |
| P | Product |
| $Q$ | Normalized sum of the weighted-squared residuals |
| $R$ | Gas constant |
| $r$ | Specific binding-mode index, specifying the real or virtual subsite occupied by the reducing-end glycosyl unit |
| $T$ | Absolute temperature |
| $t$ | Sugar chain length |
| $V$ | Maximum velocity |
| $v_0$ | Measured velocity |
| $W$ | Weighting factor |
| $X$ | An experimental parameter, $\tilde{K}_{m,n}$, $\tilde{V}_n$, or b.c.f. |
| [ ] | Concentration |
| $\sim$ | Measured or apparent value |
| $\cdot$ | Time derivative |
| ○ | α-D-Glucopyranoside unit |
| ⌀ | Reducing α-D-glucopyranoside unit |
| ● | Reducing radioactively labeled α-D-glucopyranoside unit |
| ○—○ | $\alpha(1 \rightarrow 4)$ bond joining two α-D-glucopyranoside units |

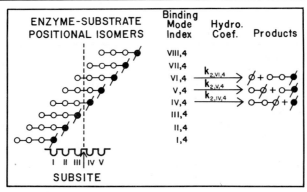

FIG. 1. Positional isomers of maltotetraose on a five-subsite enzyme. U, subsite on the enzyme; ↑, position of the catalytic site; the other notations are given in Table I. The broken line shows the position of bond cleavage. The enzyme binding region topography depicted is $l = 5$ (number of subsites), $c = 4$ (index number of the subsite to the right of the catalytic site). The binding mode index is a Roman numeral indicating the substrate holding the reducing unit followed by an Arabic number indicating the chain length of the substrate. When the substrate extends beyond the right of the binding region, virtual subsites are used to designate the binding mode index. The hydrolytic rate coefficients (Hydro.Coef.) are of the form $k_{2,r,n}$ for chain length $n$ in binding mode $r$. From J. D. Allen and J. A. Thoma, *Biochem. J.* **159**, 105 (1976).

positional isomers of the $n$-mer substrate. Hence, the task of subsite mapping is to determine the relationship between subsite binding energies and the microscopic constants, then to relate these microscopic constants to experimentally accessible macroscopic parameters.

*Relation of Subsite Binding Energies to Microscopic Parameters.* The free energy of binding at each subsite is assumed to be an intrinsic constant, unaffected by binding or absence of binding at any other subsite. Therefore, the sum of the binding energies of the occupied subsites for each positional isomer of an $n$-mer is related to the microscopic association constant by

$$\sum_{i=r-n+1}^{r} \Delta G_i + 2400 = -RT \ln K'_{r,n} \qquad (1)$$

where $r$ indexes the specific positional isomer and is the subsite occupied by the reducing end of the substrate; $\Delta G_i$, the unitary free energy of binding a substrate monomer unit in subsite $i$, is summed over all occupied subsites. Subsites lying beyond the ends of the binding region, namely, virtual subsites, have zero free energy of binding. The 2400 cal/mol is the cratic free energy contribution to binding, i.e., the entropy of mixing at

$25°$,[4] and $K'_{r,n}$ is the microscopic association constant for a substrate of chain length $n$ in binding mode $r$.

*Relation of Microscopic Parameters to Macroscopic Parameters.* The scheme for binding and hydrolysis of a positional isomer in binding mode $r$ is

$$E + A_n \underset{k_{-1,r,n}}{\overset{k_{+1,r,n}}{\rightleftarrows}} EA_{r,n} \xrightarrow{k_{+2,r,n}} E + P_{r,n-m} + P_{r,m} \quad (2)$$

where E is enzyme, $A_n$ is a substrate of chain length $n$, $P_{r,n-m}$ and $P_{r,m}$ are products of chain length $n-m$ and $m$ arising from the nonreducing end and reducing end, respectively, of the substrate molecule. The microscopic rate coefficients $k_{+1,r,n}$ and $k_{-1,r,n}$ describe binding of the substrate, and $k_{+2,r,n}$ is the microscopic hydrolytic rate coefficient.

When deriving the Michaelis expression from this scheme, making the usual steady-state approximations, the conservation expression for an enzyme with $l$ subsites is

$$[E]_0 = [E] + \sum_{h=1}^{l+n-1} [EA_{h,n}] \quad (3)$$

where $h$ indexes all possible potential isomers and is the subsite occupied by the reducing end of the substrate. By defining a microscopic Michaelis constant as

$$K_{r,n} = \frac{k_{-1,r,n} + k_{+2,r,n}}{k_{+1,r,n}} = \frac{[A_n]\left([E]_0 - \sum_{h=1}^{l+n-1}[EA_{h,n}]\right)}{[EA_{r,n}]} \quad (4)$$

the rate of formation of product arising from a specific positional isomer, indexed $r$, is

$$\frac{d[P_{r,m}]}{dt[E]_0} = \frac{[A_n]k_{+2,r,n}/K_{r,n}}{1 + [A_n]\sum_{h=1}^{l+n-1} 1/K_{h,n}} \quad (5)$$

Equation (5) has the form of typical competitive inhibition where the positional isomers not leading to production of $P_{r,m}$, that is, binding modes where $h \neq r$, act as competitive inhibitors. The measurable macroscopic velocity, $\bar{v}_{0,n}$, is the sum of the microscopic velocities for all production positional isomers or

---

[4] R. W. Gurney, "Ionic Processes in Solution," p. 80, McGraw-Hill, New York, 1953.

$$\frac{\tilde{v}_{0,n}}{[E]_0} = \sum_{h=c}^{c+n-2} \frac{d[P_{h,m}]}{dt[E]_0} = \frac{[A_n] \sum_{h=c}^{c+n-2} k_{+2,h,n}/K_{h,n}}{1 + [A_n] \sum_{h=1}^{l+n-1} 1/K_{h,n}} = \frac{[A_n]\tilde{V}_n/[E]_0}{A_n + \tilde{K}_{m,n}} \quad (6)$$

where $c$ identifies the position of the catalytic site and is the index number of the subsite shown in Fig. 1 to the right of the position of bond scission.

The inverse of the macroscopic Michaelis constant, $\tilde{K}_{m,n}$, is the sum of the inverses of the microscopic Michaelis parameters, and in the limiting case where $k_{-1,r,n} \gg k_{+2,r,n}$ the microscopic Michaelis constant, $K_{h,n}$, approaches a dissociation constant so that

$$\frac{1}{\tilde{K}_{m,n}} = \sum_{h=1}^{l+n-1} \frac{1}{K_{h,n}} = \sum_{h=1}^{l+n-1} K'_{h,n} \quad (7)$$

The macroscopic maximum velocity is

$$\frac{\tilde{V}_n}{[E]_0} = \sum_{h=c}^{c+n-2} \frac{k_{+2,h,n}}{K_{h,n}} \bigg/ \sum_{h=1}^{l+n-1} 1/K_{h,n} = \sum_{h=c}^{c+n-2} k_{+2,h,n} K'_{h,n} \bigg/ \sum_{h=1}^{l+n-1} K'_{h,n} \quad (8)$$

The first-order rate constant for $n$-mer hydrolysis, $\tilde{k}'_n$, measured at low substrate concentrations where $A_n \ll \tilde{K}_{m,n}$, can be seen from Eq. (6) to be

$$\tilde{k}'_n = \frac{\tilde{V}_n}{\tilde{K}_{m,n}[E]_0} = \sum_{h=c}^{c+n-2} \frac{k_{+2,h,n}}{K_{h,n}} \quad (9)$$

which contains terms for productive binding only. Hence, the measurable parameters $\tilde{K}_{m,n}$, $\tilde{V}_n$, and $\tilde{k}'_n$ are related to the microscopic association constants by Eqs. (7)–(9), and the microscopic association constants are related to the subsite binding energies by Eq. (1).

Substrates that can cover the entire binding region of the enzyme ($n \geq l$) can be used to measure the sum of the subsite binding energies. Binding modes in which the subsites are all filled will have the same microscopic dissociation constant, $K_{int}$. Equation (7) is written in terms of $K_{int}$ for $n \geq l$ by partitioning it as

$$\frac{1}{\tilde{K}_{m,n}} = \sum_{h=1}^{l-1} \frac{1}{K_{h,m}} + (n - l + 1)\frac{1}{K_{int}} + \sum_{h=n+1}^{n+l-1} \frac{1}{K_{h,n}} \quad (10)$$

where summation from 1 to $l - 1$ accounts for binding modes where subsites depicted on the right side of the enzyme in Fig. 1 are vacant, summation from $n + 1$ to $n + l - 1$ accounts for binding modes where subsites depicted on the left side of the enzyme are vacant, and the remaining term accounts for the $n - l + 1$ binding modes where all of the subsites are occupied. Equation (10) has the form of a straight line where the slope of a plot of $n - l + 1$ ($n \geq l$) against $1/\tilde{K}_{m,n}$ is a measure of $1/\tilde{K}_{int}$. However, a weighted least squares fit to the reciprocal of Eq. (10), which is a hyperbola, is preferred because it does not necessitate an inversion of $\tilde{K}_{m,n}$ with attendant distortion of the error functions.

There is yet another experimentally accessible parameter that will yield information about the binding region. As depicted in Fig. 1, each productive binding mode gives rise to a characteristic product. The rate of appearance of each product is a function of the population of the positional isomer giving rise to that product and the associated hydrolytic rate coefficient. From Eq. (5), we can see that the ratio of the rate of formation of products from an $n$-mer binding in two adjacent binding modes is

$$\frac{[\dot{P}_{r,m}]}{[\dot{P}_{r+1,m+1}]} = \frac{k_{+2,r,n}/K_{r,n}}{k_{+2,r+1,n}/K_{r+1,n}} \tag{11}$$

where $[\dot{P}_{r,m}] = d[P_{r,m}]/dt$. Since $k_{+2}$ is zero for nonproductive complexes, bond cleavage frequencies serve only as a probe of productive complexes. The rates of bond cleavage can be normalized by dividing the rate of formation of a product by the rate of formation of all products from an $n$-mer. These normalized rates are referred to as bond cleavage frequencies (b.c.f.'s). Equation (11) is expressed in terms of subsite binding energies by using Eq. (1) to give

$$RT \ln \frac{[\dot{P}_{r,m}]}{[\dot{P}_{r+1,m+1}]} = \Delta G_{r+1} - \Delta G_{r-n+1} + RT \ln \frac{k_{+2,r,n}}{k_{+2,r+1,n}} \tag{12}$$

Hence, bond-cleavage-frequency ratios provide a measure of the difference between two subsite binding energies and the ratio of the hydrolytic rate coefficients.

As this development reveals, an analysis of the action of polysaccharide depolymerases on homologous oligosaccharides provides the information necessary to map the binding region of these enzymes. The experimentally accessible parameters that are useful in subsite mapping are Michaelis parameters, $\tilde{K}_{m,n}$ and $\tilde{V}_n$; the first-order rate constants, $\tilde{k}'_n$; and the bond cleavage frequencies measured as a function of substrate chain length. The remainder of this chapter will be devoted to the methods of collection of these data and to the methods of application of these data to generate a subsite map.

Practical Aspects

*Enzyme*

Polysaccharide depolymerases that degrade homologous oligosaccharides can be mapped using the procedures in this chapter. The depolymerase must be scrupulously purified, since a trace-contaminating enzyme that also acts on the substrates can profoundly influence the measured parameters.

*Substrates*

To generate the data necessary for subsite mapping, a series of homologous substrate oligosaccharides of varying chain lengths must be prepared and purified. To obtain bond cleavage frequencies, the oligosaccharides must be labeled in a terminal unit so that the substrate bond cleaved can be identified by means of the labeled product.

*Preparation of Unlabeled Substrates.* Unlabeled oligosaccharides can be obtained from the controlled acid or enzymic hydrolysis of the corresponding polysaccharides.[5] For example, exhaustive treatment of starch with *Pseudomonas* isoamylase[6] or pullanase[7] will result in hydrolysis of the 1 → 6 linkages and give a reasonable yield of maltooligosaccharides without further treatment. Debranched polysaccharides remaining in the digest can be degraded to oligosaccharides using an endoamylase.[7] Maltooligosaccharides can also be obtained from the acid hydrolysis of cyclic dextrins.[8]

*Preparation of End-Labeled Substrates.* Maltooligosaccharides with the reducing-end unit radiolabeled are readily prepared by the action of cyclodextrin glucanotransferase[9] [EC 2.4.1.19] on cyclohexaamylose[10] and radiolabeled glucose. This enzyme condenses cyclohexaamylose to the labeled glucose, then, through further transglycosylation reactions on the resulting linear oligosaccharides, generates a series of maltodextrin oligosaccharides with the label preserved in the reducing end.

Maltooligosaccharides labeled in the nonreducing terminus can be prepared using a phosphorylase. With appropriate conditions,[11] labeled glucose from glucose 1-phosphate is added to the nonreducing end of the primer oligosaccharides. For subsite mapping of dextranases, isomaltooligosaccharides labeled in the reducing or nonreducing end can be prepared using exodextranase and D-glucanotransferase.[12]

*Purification of Substrates.* Oligosaccharides can be purified using col-

---

[5] J. H. Pazur, *in* "The Carbohydrates" (W. Pigman, D. Horton, and A. Herp, eds.), 2nd ed., Vol. 2A, p. 69. Academic Press, New York, 1970.
[6] J. D. Allen and J. A. Thoma, *Carbohydr. Res.* **61**, 377 (1978).
[7] J. F. Robyt and D. French, *J. Biol. Chem.* **245**, 3917 (1970).
[8] Y. Nitta, M. Mizushima, K. Hiromi, and S. Ono, *J. Biochem. (Tokyo)*, **69**, 567 (1971).
[9] D. French, this series, Vol. 5, p. 148.
[10] D. French, this series, Vol. 3, p. 17.
[11] J. H. Pazur, *in* "Starch: Chemistry and Technology" (R. L. Whistler and E. F. Paschall, eds.), Vol. I, p. 133. Academic Press, New York, 1965.
[12] G. J. Walker, *Carbohydr. Res.* **30**, 1 (1973).

umn or preparative paper chromatography. For large-scale preparations oligosaccharides can be fractionated using cellulose, carbon, or polyacrylamide column chromatography.[5,13] For small-scale preparations, paper chromatography[14] using multiple ascents is the most convenient method. The number of ascents necessary to fractionate the sugars will depend on the chain lengths. The progress of fractionation of unlabeled oligosaccharides is followed using an alkaline silver nitrate reagent[15] to visualize the sugars on a strip cut off from the side of the chromatogram. Cyclic dextrins can be detected by the yellow or brownish-yellow complex formed when the paper is placed in a chamber saturated with iodine vapors or by the alkaline silver nitrate reagent after they have been hydrolyzed by spraying the paper with *Aspergillus oryzae* α-amylase. Radiolabeled oligosaccharides are detected by autoradiography. The fractionated oligosaccharides are eluted from the paper with water and assayed for carbohydrate content.[16]

*Collection of Data*

*Bond Cleavage Frequencies (b.c.f.'s)*. As discussed under "Limitations", the b.c.f.'s should be determined over a range of substrate concentrations from approximately 0.1 $K_m$ to 10 $K_m$ to test for complicating multimolecular substrate reactions. Since the products of initial bond scission can be further degraded by the enzyme, i.e., secondary attack, care must be taken to assure that initial products are detected. Aliquots are removed during the course of hydrolysis of the end-labeled oligosaccharide with most of the samples taken before the substrate has reached 50% hydrolysis. The enzyme in each aliquot is immediately denatured by a method that does not affect the sugars nor substantially increase the salt concentration. The aliquots are concentrated, if necessary, and spotted along with standards on chromatography paper. The paper is developed using multiple ascents to obtain resolution of the oligosaccharides. The radiolabeled oligosaccharides are detected by autoradiography, and the resulting autoradiogram is used as a guide in cutting out the paper containing the radiolabeled sugars. When a product is not present in sufficient quantities to darken the film, the standards are used as a guide. The products are quantitated by scintillation counting in a toluene cocktail.

---

[13] M. John, G. Trénel, and H. Dellweg, *J. Chromatogr.* **42,** 476 (1969).
[14] J. A. Thoma and D. French, *Anal. Chem.* **29,** 1645 (1957).
[15] J. C. Dittmer and M. A. Wells, this series, Vol. 14, p. 528.
[16] M. Dubois, K. A. Gilles, J. K. Hamilton, P. A. Rebers, and F. Smith, *Anal. Chem.* **28,** 350 (1956).

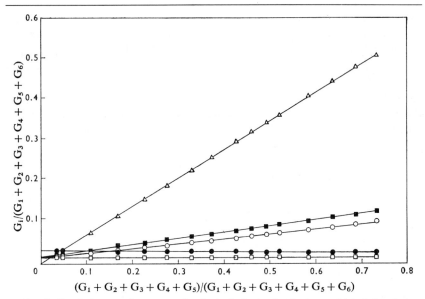

FIG. 2. Bond cleavage frequency plot for hydrolysis of reducing end-labeled maltohexaose by *Aspergillus oryzae* α-amylase. □, glucose; ○, maltose; △, maltotriose; ■, maltotetraose; ●, maltoheptaose. From J. D. Allen and J. A. Thoma, *Biochem. J.* **159**, 121 (1976).

It has been shown[17] that the radioactivity counting data for $n$-mer hydrolysis is best analyzed by plotting (Fig. 2)

$$\sum_{t=1}^{n-1} G_t \bigg/ \sum_{t=1}^{n} G_t \quad \text{against} \quad G_i \bigg/ \sum_{t=1}^{n} G_t$$

for each product $i$-mer, where $G_t$ and $G_i$ are the measured radioactivities for sugars of chain length $t$ and $i$, respectively. The sample background, arising from contamination of the substrates by radiolabeled impurities, is not subtracted from the raw counting data since it influences only the intercept of this b.c.f. plot. The slope of the plot, determined from a weighted least-squares fit, is the bond cleavage frequency for the bond that upon cleavage yields a product of chain length $i$. Curvature in the plot is diagnostic of a change in bond cleavage frequencies, which is due to either a shift in mechanism as the substrate is depleted or to a secondary attack on the products. Consequently, the initial slope is used in determining bond cleavage frequenices.

*Michaelis Parameters.* Michaelis parameters for the series of oligosaccharides are determined by measuring the rate of enzymic hydrolysis of the substrate at concentrations with a range of at least 0.2–5 times $K_m$.

[17] J. D. Allen, *Carbohydr. Res.* **39**, 312 (1975).

The extent of hydrolysis must be kept low enough to ensure that initial velocities are measured.

Bond cleavage in an oligosaccharide can be detected as an increase in the reducing value due to the freed hemiacetal group. The neocuproine assay[18] provides a sensitive measure of reducing value. When the substrate oligosaccharides are themselves reducing sugars, the relative increase in reducing value generated by a low extent of hydrolysis is small, which introduces a large error into the velocity measurements. This difficulty can be overcome by using nonreducing substrates such as the α-methyl maltooligosaccharides.[19] Radiolabeled oligosaccharides[20] provide an alternative method of measuring initial velocities that also eliminates this problem. The extent of hydrolysis of these substrates is readily quantitated, after fractionation of the products using paper chromatography, by radioactivity counting. In fact, it is possible to determine initial velocities and bond cleavage frequencies simultaneously using this technique.

## Examples

### Application of the Subsite Model

There are a number of different approaches that can be used in applying the subsite model to polysaccharide depolymerizing enzymes. Some specific examples of subsite mapping are presented below, and the difficulties in the procedures are pointed out. The summary at the end of the chapter gives the most reliable overall procedure for subsite mapping.

*Mapping of Glucoamylase and β-Amylase.* Hiromi and co-workers[21,22] applied the subsite model to exoenzymes by using the first-order rate constants for hydrolysis of a series of oligosaccharides. Since there is only one productive binding mode for an $n$-mer on an exoenzyme as shown in Fig. 3, Eq. (9) simplifies to

$$\bar{k}'_n = k_{2,r,n} K'_{r,n} \tag{13}$$

insofar as the inverse of the Michaelis constant approximates an association constant. The ratio of first-order rate constants for $n$-mer and $n + 1$-mer expressed as subsite binding energies [Eq. 1)] is

$$\frac{\bar{k}'_n}{\bar{k}'_{n+1}} = \frac{k_{2,r,n}}{k_{2,r+1,n+1}} \exp(\Delta G_{n+1}/RT) \tag{14}$$

---

[18] S. Dygert, L. Li, D. Florida, and J. A. Thoma, *Anal. Biochem.* **13**, 367 (1965).
[19] J. A. Thoma, G. V. K. Rao, C. Brothers, J. Spradlin, and L. H. Li, *J. Biol. Chem.* **246**, 5621 (1971).
[20] J. D. Allen and J. A. Thoma, *Biochemistry* **17**, 2345 (1978).
[21] K. Hiromi, *Biochem. Biophys. Res. Commun.* **40**, 1 (1970).
[22] K. Hiromi, Y. Nitta, C. Numata, and S. Ono, *Biochim. Biophys. Acta* **302**, 362 (1973).

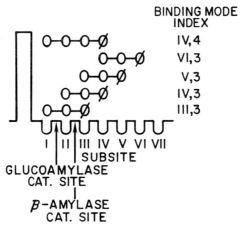

FIG. 3. Binding region of an exoamylase. The notations are given in Fig. 1 and Table I. The barrier to the left of subsite I normally prevents binding of internal glucosyl residues in subsite I.

Thus, if the hydrolytic rate coefficients are equal, first-order rate constants provide a direct estimate of the binding energies of subsites indexed $c + 1$ and larger. The apparent binding energies for subsites III–VII of *Rhizopus delemar* glucoamylase, calculated from the first-order rate constants for maltose through maltoheptaose in Eq. (14), are given in Table II.

The binding energy for maltose, calculated from the Michaelis constant [Eqs. (1) and (7)], indicates tighter binding than can be accounted for by binding within subsites III–VII (Table II). Consequently, Hiromi *et al.*[22] assume that only binding modes $r$ and $r + 1$ where subsites I and II are occupied contribute significantly to the binding energy of the oligosaccharides, so that Eq. (8) can be expressed in terms of subsite binding energies to give

$$\exp(-\Delta G_{n+1}/RT) = \left\{\frac{k_{+2,r,n}}{\bar{V}_n/[E]_0} - 1\right\} \exp(-\Delta G_I/RT) \quad (15)$$

Hence, if the above assumptions are valid, and $k_{+2,r,n}$ is a constant, a plot of $\exp(-G_{n+1}/RT)$ against $[E]_0/\bar{V}_n$ will be linear and the intercepts will yield $k_{+2}$ and $\Delta G_I$. Hiromi and co-workers[22] use the apparent binding energies for subsites III–VI and experimental maximum velocities for $n = 2$–7 and propose that the resultant plot is linear with $k_{+2}$ and $\Delta G_I$ at 77 sec$^{-1}$ and 0 cal/mol, respectively. However, considering that the apparent subsite binding energies are calculated assuming a constant hydrolytic rate coefficient and that only binding modes $r$ and $r + 1$ are considered, normal experimental scatter could obscure nonlinearity of such a plot.

TABLE II
APPARENT SUBSITE BINDING ENERGIES FOR SOME MAPPED AND PARTIALLY MAPPED ENZYMES

| Enzyme | Apparent subsite binding energies (cal/mol) | | | | | Catalytic site → | | | | | |
|---|---|---|---|---|---|---|---|---|---|---|---|
| *Bacillus amyloliquefaciens* (BLA-N) α-amylase[a] | −1070 | −2440 | −160 | −1010 | −2280 | 3300 | −3440 | −1720 | −960 | 1260 | |
| *Rhizopus delemar* glucoamylase[b] | | | | | | 0 | −4820 | −1590 | −430 | −220 | −110 |
| Wheat bran β-amylase[c] | | | | | | >1400 | 1850 | −5890 | −1090 | 570 | −790 |
| *Aspergillus oryzae* α-amylase[d] | | | | −400 | <−2700 | −7210 | | −3120 | −1130 | | −100 |
| *B. amyloliquefaciens* (BLA-F) α-amylase[e] | −1100 | −2400 | 0 | −600 | −2400 | — | — | −1200 | | | |
| *B. amyloliquefaciens* (BLA-D) α-amylase[e] | −1300 | −2800 | −100 | −1200 | −2700 | 2670 | −3650 | −1000 | | | |
| *Streptococcus mutans* KI-R dextranase[f] | 170 | −410 | −1110 | −1920 | −1650 | — | — | −640 | −280 | | |
| Lysozyme[g] | | | −1800 | −2900 | ≤−5700 | 3000–6000 | ~−4000 | −1000 | | | |

[a] The apparent binding energies listed contain an acceleration factor of 370 cal/mol. From J. D. Allen and J. A. Thoma. *Biochem. J.* **159**, 121 (1976).
[b] The binding energy of the subsite to the right of the catalytic site is the average of two determinations. From K. Hiromi, Y. Nitta, C. Numata, and S. Ono, *Biochim. Biophys. Acta* **302**, 362 (1973).
[c] The sum of the two subsites adjacent to the catalytic site is calculated to be −1330 cal/mol. From M. Kato, K. Hiromi, and Y. Morita. *J. Biochem. (Tokyo)* **75**, 563 (1974).
[d] From J. D. Allen and J. A. Thoma. *Biochem. J.* **159**, 121 (1976).
[e] The sum of the two subsites adjacent to the catalytic site are estimated to be 3100 and 3800 cal/mol for BLA-F and BLA-D, respectively. From S. Iwasa, H. Aoshima, K. Hiromi, and H. Hatano, *J. Biochem. (Tokyo)* **75**, 969 (1974). The proportioning of the energies of the two subsites adjacent to the catalytic site of BLA-D was calculated by J. A. Thoma and J. D. Allen, *Carbohydr. Res.* **39**, 303 (1975).
[f] From A. Pulkownik, J. A. Thoma, and G. J. Walker, *Carbohydr. Res.* **61**, 493 (1978).
[g] From D. M. Chipman and N. Sharon, *Science* **165**, 454 (1969).

The binding energy of subsite II is calculated using two different methods.[22] Again considering only binding modes $r$ and $r + 1$, the Michaelis constant for each $n$-mer is a measure of the sum of subsite binding energies, which can be used to calculate the binding energy of subsite II since the binding energies of all other subsites have been estimated. Alternatively, Eq. (13) in conjunction with Eq. (1) provides a measure of the sum of subsite binding energies by using the value of the hydrolytic rate coefficient estimated above. The binding energy for subsite II calculated by the two different methods is in good agreement. The final subsite map for the gluocoamylase is given in Table II. It is questionable whether subsites V–VII and possibly IV are real or whether the low binding energies attributed to these subsites arise as a result of the assumptions of the model and experimental scatter.

The Michaelis parameters calculated from this subsite map and the experimental values agree within experimental error except for the values for a maltodextrin of average chain length 15.5, where the experimental $\tilde{K}_m$ and $\tilde{V}$ are lower than predicted by the subsite map. This discrepancy may be attributable to nonproductive binding, not accounted for by the model, where subsite I is occupied by a glucosyl residue other than the nonreducing end glucosyl residue so that the nonreducing "tail" of the substrate extends beyond subsite I.[22,23]

Although $\beta$-amylase potentially has two productive binding modes for each $n$-mer substrate (e.g., III,3 and IV,3 in Fig. 3), the only detectable bond scission produces maltose from the nonreducing end. Hence, if the hydrolytic rate coefficients are equal, the only significantly populated positional isomer is the one in which the substrate nonreducing end occupies subsite I. Therefore, $\beta$-amylase is analogous to gluocoamylase except that the hydrolyzable unit is maltose rather than glucose. Using a procedure similar to that employed for glucoamylase, Kato et al.[24] calculated the apparent subsite binding energies for wheat bran $\beta$-amylase given in Table II.

*Mapping of Bacillus amyloliquefaciens (strain BLA-N)$\alpha$-amylase.* Bond cleavage-frequency ratios can be used to measure the size of the binding region, to locate the catalytic amino acids within the binding region, and to calculate apparent binding energies for some of the subsites.[25] Equation (12), which relates b.c.f. data to subsite binding energies, contains a ratio of hydrolytic rate coefficients that cannot be readily evaluated. However, this equation can be simplified by writing it in terms of apparent subsite binding energies, $\Delta \tilde{G}_i$, as

[23] J. A. Thoma and D. E. Koshland, Jr., *J. Am. Chem. Soc.* **82**, 3329 (1960).
[24] M. Kato, K. Hiromi, and Y. Morita, *J. Biochem. (Tokyo)* **75**, 563 (1974).
[25] J. A. Thoma, C. Brothers, and J. Spradlin, *Biochemistry* **9**, 1768 (1970).

$$RT \ln \frac{[\dot{P}_{r,m}]}{[\dot{P}_{r+1,m+1}]} = \Delta \tilde{G}_{r+1} - \Delta \tilde{G}_{r-n+1} \tag{16}$$

so that the rate ratio becomes integrated into $\Delta \tilde{G}_i$. If the hydrolytic rate coefficients are the same for all positional isomers, the apparent and actual subsite binding energies are identical ($\Delta \tilde{G}_i = \Delta G_i$); so that, bond cleavage frequencies provide a direct measure of subsite binding energies.

Considering the enzyme–substrate complexes depicted in Fig. 1, the ratios [●]/[○—●] and [○—●]/[○—Ȯ—●] provide a direct measure of the differences in apparent binding energies of subsites I and V and subsites II and VI (a virtual subsite), respectively. By using substrates of increasing chain length, the differences in apparent binding energies for subsites farther removed from the cleavage site can be calculated. So that an oligosaccharide of chain length $n$ provides binding energy information for $n - 1$ subsites on either side of the catalytic site. The b.c.f. data for a series of oligosaccharides provide multiple measures of the apparent binding energies for subsites nearer the catalytic site that can be used as a check for internal consistency of the model.

As a starting point in the analysis, a subsite is chosen as a reference subsite and assigned a binding energy of zero ($\Delta G_{ref} = 0$). Apparent binding energies are then calculated relative to this reference subsite. For example, Thoma and co-workers[25] determined the *B. amyloliquefaciens* $\alpha$-amylase b.c.f.'s for $\alpha$-methyl maltooligosaccharides ($n = 3$–12) radiolabeled in the $\alpha$-methylglucosyl unit. With subsite IX as the reference subsite, the relative apparent subsite energy histogram shown in Fig. 4 (open bars) is generated using Eq. (16). The multiple measures of the subsite energies, which are the same within 140 cal/mol, are averaged to give the histogram shown.

Note that the relative binding energies on the periphery of the binding region (subsites $-IV$ through $-I$ and XI and XII) appear at an approximately constant level of 1100 cal/mol. Virtual subsites, that is subsites lying beyond the ends of the binding region, have no interaction with the substrate and consequently must have a binding energy of zero. This apparent constant interaction on the periphery of the binding region is due to assigning subsite IX, the reference subsite, a zero binding energy. When these peripheral, virtual subsites are adjusted to a zero binding energy, the apparent binding energies shown as filled bars in Fig. 4 are established. The bond cleavage frequencies predicted by this subsite map closely conform to the experimental values.

Since subsites VI and VII (Fig. 4) are occupied in all productive positional isomers, b.c.f. data do not yield any information about these subsites. However, Michaelis parameters can be used to predict the binding energies of these two subsites. The sum of the binding energies for subsites VI and VII can be seen from Eq. (1) to be

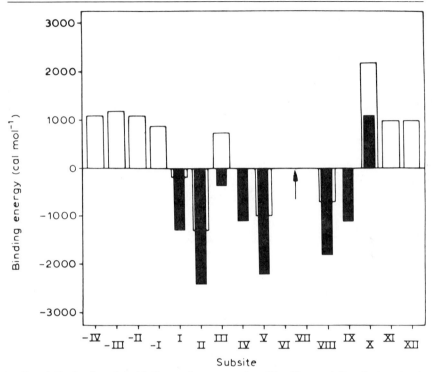

FIG. 4. Evaluation of the binding-region topography of *Bacillus amyloliquefaciens* (strain BLA-N) α-amylase from bond-cleavage frequencies. ↑, Position of the catalytic site; ☐, apparent binding energies with $\Delta G_{ref} = \Delta G_{IX} = 0$; ■, apparent subsite binding energies with $\Delta G_{IX} = -1100$ cal/mol. From J. A. Thoma and J. D. Allen, *Carbohydr. Res.* **48**, 105 (1976).

$$\Delta G_{VI} + \Delta G_{VII} = -RT \ln K'_{int} - 2400 - \sum_{\substack{i=1 \\ (i \neq VI, VII)}}^{X} \Delta G_i \quad (17)$$

Although $K'_{int}$ can be calculated using $K_{m,n}$ data ($n \geq l$) in Eq. (10), the only estimates of the binding energies of subsites I through V and VIII through X are the apparent binding energies, $\Delta \tilde{G}_i$. If, as a starting point in the analysis, $k_{+2}$ is taken to be constant so that $\Delta \tilde{G}_i = \Delta G_i$, then the sum of the binding energies of subsites VI and VII can be calculated using Eq. (17). There is no direct way to calculate the contribution of subsites VI and VII to their sum. However, the binding energy sum can be systematically partitioned between the two subsites until the model is best able to account for the Michaelis parameters.

Thoma and co-workers[19] used this analysis with the Michaelis parameters for α-methyl maltooligosaccharides to determine the binding energies for subsites VI and VII of *B. amyloliquefaciens* α-amylase. However, the resulting subsite map has a systematic error in fitting the experimental data that is larger by an order of magnitude or more than expected from experimental scatter. From the trends in the $\tilde{K}_m$ data, Thoma et al.[19] proposed that the poor fit is due to the assumption that the hydrolytic rate coefficient is constant and that a better approximation of $k_{+2,r,n}$ can be obtained by assuming that filling each subsite lowers the activation-energy barrier, enhancing the rate of bond cleavage. Accordingly, the larger oligosaccharides in binding modes where most of the subsites are occupied would be cleaved at a faster rate than the smaller oligosaccharides, bound so that only a few subsites are occupied. This approximation is expressed as

$$k_{+2,r,n} \propto \exp\left(\sum_{i=r}^{r-n+1} \Delta G_{a,i}/RT\right) \quad (18)$$

where $\Delta G_{a,i}$ is the contribution of the $i$th real subsite occupied by a substrate monomer unit to acceleration of bond scission. Equation (12) can then be partitioned as

$$RT \ln \frac{[\dot{P}_{r,m}]}{[\dot{P}_{r+1,m+1}]} = (\Delta G_{r+1} - \Delta G_{a,r+1}) - (\Delta G_{r-n+1} - \Delta G_{a,r-n+1})$$
$$= \Delta \tilde{G}_{r+1} - \Delta \tilde{G}_{r-n+1} \quad (19)$$

showing that apparent binding energies measured by bond cleavage frequencies must be increased by the acceleration factor to give the true thermodynamic subsite binding energy.

There is no direct way of measuring the acceleration factor for each subsite. However, if the acceleration factor is approximated as a constant for each subsite, a value can be estimated by varying the binding energies for subsites VI and VII and the acceleration factor to achieve the best possible fit to the experimental data. Thoma and co-workers[19] found that setting $\Delta G_a \cong 450$ cal/mol with subsites VI and VII set at 3000 and $-2640$ cal/mol, respectively, brought the experimental and calculated values into good agreement.

The application of the subsite model outlined above has some flaws. In using the b.c.f. analysis to calculate apparent subsite binding energies, a rigorous statistical analysis is not employed, so that disproportionate error is introduced into certain subsites. A subjective judgment as to when the relative subsite binding energies on the periphery of the enzyme become constant is necessary in evaluating the number of subsites. Further, weighting factors are not used to take into account differences in experimental precision.

A computer simulation of the depolymerase model developed by Allen and Thoma[26] provides a statistically sound method of transforming exper-

---

[26] J. D. Allen and J. A. Thoma, *Biochem. J.* **159**, 105 (1976).

imental data into a subsite map. In mathematical terms subsite mapping reduces to adjusting the number of subsites, the position of the catalytic site, and the subsite binding energies until the "best" subsite map to account for the experimental data is generated. The goodness-of-fit between experimental and calculated parameters is defined as

$$QX = \sum_j W_j(X_{j,\text{exptl}} - X_{j,\text{calcd}})^2 \qquad (20)$$

where $j$ is summed over the experimental data collected for parameter $X$, which may be Michaelis parameters or b.c.f.'s. The weighting factor, $W_j$, is the reciprocal of the experimental variance. Hence, the subsite map that minimizes $Q$ to give $Q_{\min}$ is the optimum subsite map. The depolymerase computer model[27] offers a means of rapidly and efficiently optimizing the subsite map by using a conjugate gradient minimization routine to obtain $Q_{\min}$.

The use of the depolymerase computer model to assess the binding region topography by using bond cleavage frequencies is demonstrated for *B. amyloliquefaciens* $\alpha$-amylase[28] in Table III. Subsites are added to or deleted from the ends of the binding region, and the subsite energies for each new binding topography are adjusted by the computer model to establish $Q_{\min,\,\text{b.c.f.}}$. When the model has an adequate number of subsites, $Q_{\min,\,\text{b.c.f.}}$ is not further significantly lowered by allowing additional subsites. As shown in Table III, the 10-subsite map (minimization 3) is required to account for the data. Deleting subsite I (minimization 2) or subsite X (minimization 1) results in a 26- or 64-fold poorer fit, respectively. Addition of another subsite does improve the fit (minimization 5), but this improvement in fit was found[28] to be insignificant. The optimized apparent subsite energies generated in minimization 3 are given in Table IV.

Once the binding region topography has been established and apparent binding energies have been estimated from bond cleavage frequencies, the next step is to use the computer model to obtain the best possible fit to all the data ($Q_{\min,\,\text{total}}$) without allowing the hydrolytic rate coefficient to vary ($\Delta G_a = 0$). It is most efficient to first optimize the two subsites adjacent to the catalytic site using ($Q_{\min,\,K_m} + Q_{\min,\,V} + Q_{\min,\,K_{\text{int}}}$) with the other subsites constrained as determined by $Q_{\min,\,\text{b.c.f.}}$. At that stage all the subsites are near their final values and the computer model can rapidly establish $Q_{\min,\,\text{total}}$.

Care must be exercised when optimizing using $\bar{K}_m$ and $\bar{V}$ data since

---

[27] A complete listing of the depolymerase computer model is given in J. D. Allen, Ph.D. thesis, University of Arkansas (1975). The author will lend a computer tape of the model, which can be copied.

[28] J. D. Allen and J. A. Thoma, *Biochem. J.* **159**, 121 (1976).

TABLE III
EVALUATION OF THE NUMBER OF SUBSITES ON *Bacillus amyloliquefaciens* α-AMYLASE[a,b]

| Minimization no. | -I | I | II | III | IV | V | Catalytic site VI↓VII | VIII | IX | X | XI | $Q_{min,b.c.f}$ |
|---|---|---|---|---|---|---|---|---|---|---|---|---|
| 1 |  | X | X | X | X | X |  | X | X |  |  | 12.9 |
| 2 |  |  | X | X | X | X |  | X | X | X |  | 5.2 |
| 3 |  | X | X | X | X | X |  | X | X | X |  | 0.2 |
| 4 |  |  | X | X | X | X |  | X | X | X | X | 0.2 |
| 5 | X | X | X | X | X | X |  | X | X | X |  | 0.1 |
| 6 | X | X | X | X | X | X |  | X | X | X | X | 0.1 |

[a] From J. D. Allen and J. A. Thoma, *Biochem. J.* **159**, 121 (1976).
[b] The experimental bond cleavage frequencies were used to optimize the binding energy of selected subsites (designated by X), and the remaining subsite binding energies were constrained at zero.

local minimums are sometimes established with $Q_{min}$ much larger than the absolute $Q_{min}$. Hence, in contrast to b.c.f. data, the error surface established by Michaelis parameters has local minimums, and a minimization routine simply establishes the lowest point of whatever valley it starts in. Local minimums can be avoided by using different initialized values for the subsite binding energies so that the local minimum valley is not encountered. If a wide range of starting values leads to the same minimum, it is safe to assume that the absolute minimum has been established.

When subsites VI and VII of *B. amyloliquefaciens* α-amylase are optimized to obtain ($Q_{min, K_m} + Q_{min, V} + Q_{min, K_{int}}$) with the other subsites constrained as determined by the b.c.f. analysis, the results shown in Table IV are obtained with $Q_{min, total} = 11.3$. When all the subsite binding energies are allowed to vary to obtain the overall best fit overall without an acceleration factor, $Q_{min, total}$ is reduced to 4.7 (Table IV). Finally, when the acceleration factor is also allowed to vary, the fit is improved to give $Q_{min, total} = 2.4$ with the subsite binding energies and acceleration factor listed in Table IV. Applying the $F$ test shows that the improvement in fit upon allowing the acceleration factor is significant at the 99.5% confidence level.

The Michaelis parameters calculated from the model with $G_a$ set at 0 and 370 cal/mol are compared to the experimental Michaelis parameters in Table V. In the absence of an acceleration factor the ratios of experimental to calculated parameters do not approach a random distribution about unity but rather show a trend, which is diagnostic of a poor model.

TABLE IV
OPTIMIZED SUBSITE MAPS FOR *Bacillus amyloliquefaciens* α-AMYLASE[a]

| Subsite no. | $\Delta G_a$ constrained at zero | | | $\Delta G_a$ allowed to vary[d] |
|---|---|---|---|---|
| | $Q_{bcf}$ minimized | $(Q_{K_m} + Q_V + Q_{K_{int}})$ minimized | $Q_{total}$ minimized | $Q_{total}$ minimized |
| | Apparent subsite binding energies (cal/mol) | | | |
| I | −1160 | | −890 | −1070 |
| II | −2350 | | −2120 | −2440 |
| III | −240 | | −130 | −160 |
| IV | −1040 | | −890 | −1010 |
| V | −2310 | | −1950 | −2280 |
| VI | [b] | 6090 | 4970 | 3300 |
| VII | [b] | −2290 | −2880 | −3440 |
| VIII | −1740 | | −1500 | −1720 |
| IX | −1020 | | −880 | −960 |
| X | 1250 | | 1440 | 1260 |
| | $\Delta G_a$ (cal/mol) | | | |
| | 0 | 0 | 0 | 370 |
| $Q_{min}$ | Normalized sum of residual errors squared | | | |
| b.c.f. | 0.2 | 0.2 | 0.6 | 0.3 |
| $\tilde{K}_m$ | [c] | 50.4 | 24.7 | 9.9 |
| $\tilde{V}$ | [c] | 31.3 | 6.6 | 6.0 |
| $\tilde{K}_{int}$ | [c] | 0.003 | 0.3 | 0.02 |
| Total | | 11.3 | 4.7 | 2.4 |

[a] From J. D. Allen and J. A. Thoma, *Biochem. J.* **159**, 121 (1976). -------, position of the catalytic site.
[b] Subsites adjacent to the catalytic site cannot be estimated from bond cleavage frequencies.
[c] Michaelis parameters and $K_{int}$ cannot be computed without a complete subsite map.
[d] The binding energies listed are apparent energies and contain an acceleration factor of 370 cal/mol.

It is apparent that allowing $\Delta G_a$ to vary improves the fit significantly and more randomly distributes the errors.

A $Q_{min}$ value larger than is accountable for by experimental error[26] is diagnostic that the model is not a true representation of the enzyme. $Q_{min, b.c.f.}$ for *B. amyloliquefaciens* α-amylase is within the value predicted for experimental error. However, the lowest values of $Q_{min, K_m}$ and $Q_{min, V}$ (Table IV) are, respectively, 4.7- and 4.3-fold higher than predicted from experimental scatter alone. The model bias is possibly due to the approxi-

TABLE V
COMPARISON OF EXPERIMENTAL AND COMPUTED MICHAELIS PARAMETERS FOR *Bacillus amyloliquefaciens* α-AMYLASE[a]

| Substrate chain length | $K_m$ or $K_i^b$ (M) | | | $V$ (relative)[c] | | |
|---|---|---|---|---|---|---|
| | Experimental | Experimental/calculated | | Experimental | Experimental/calculated | |
| | | $\Delta G_a = 0$ cal/mol | $\Delta G_a = 370$ cal/mol | | $\Delta G_a = 0$ cal/mol | $\Delta G_a = 370$ cal/mol |
| 1 | $6.0 \times 10^{-1}$ | 2.0 | 2.7 | — | — | — |
| 2 | $2.2 \times 10^{-2}$ | 0.7 | 0.7 | $2.4 \times 10^{-4}$ | 3.1 | 3.7 |
| 3 | $1.9 \times 10^{-2}$ | 2.3 | 1.6 | $2.0 \times 10^{-3}$ | 5.2 | 0.9 |
| 4 | $1.7 \times 10^{-2}$ | 8.0 | 0.7 | $6.3 \times 10^{-3}$ | 11.8 | 0.9 |
| 5 | $1.9 \times 10^{-2}$ | 53.7 | 2.2 | $3.4 \times 10^{-2}$ | 11.6 | 1.0 |
| 6 | $9.9 \times 10^{-3}$ | 28.0 | 1.3 | $2.5 \times 10^{-1}$ | 8.3 | 1.4 |
| 7 | $5.2 \times 10^{-3}$ | 14.8 | 1.3 | $4.1 \times 10^{-1}$ | 1.6 | 0.8 |
| 8 | $1.5 \times 10^{-3}$ | 4.5 | 1.2 | 1.0 | 1.2 | 1.1 |
| 9 | $8.8 \times 10^{-4}$ | 3.0 | 1.3 | 1.0 | 1.1 | 1.1 |
| 10 | $5.6 \times 10^{-4}$ | 1.9 | 0.8 | 1.0 | 1.1 | 1.0 |
| 11 | $4.9 \times 10^{-4}$ | 1.7 | 0.8 | 1.0 | 1.0 | 1.0 |
| 12 | $8.6 \times 10^{-4}$ | 3.1 | 1.4 | | | |

[a] From J. D. Allen and J. A. Thoma, *Biochem. J.* **159**, 121 (1976).
[b] The value for glucose is an inhibition constant; all others are $K_m$ values.
[c] Normalized to a substrate chain length of 12 maximum velocity.

mation that $\Delta G_{a,i}$ values are equal; since, it is unlikely that each subsite on the enzyme contributes equally to the hydrolytic coefficient.

*Mapping of Asperigillus oryzae α-amylase.* The chain length dependence of Michaelis parameters can be used as a probe of the size of the binding region.[8] As can be seen from Eq. (8), $\tilde{V}_n/[E]_0$ will become constant, insofar as the hydrolytic coefficient becomes constant, whenever productive complexes dominate nonproductive complexes. Nitta *et al.*[8] assume that productive complexes dominate whenever the substrate exceeds the size of the binding region. From the break point in a plot of log $\tilde{V}_n/[E]_0$ as a function of substrate chain length they estimated that *A. oryzae* α-amylase has seven subsites. However, it has been pointed out[29] that productive complexes can dominate nonproductive complexes whenever the substrate chain length is shorter than the site of the binding region, so that the break point in such a plot can only set a lower limit on the number of subsites.

It has been shown[30,31] that *A. oryzae* α-amylase exhibits significant transglycosylase activity and that, at least for maltotriose, Michaelis–Menten kinetics are not applicable.[20] Consequently, the Michaelis parameters reported for this enzyme are probably influenced by second-order substrate reactions to some unassessed extent and are not suitable for subsite mapping. Hence, bond cleavage frequencies, determined at low substrate concentrations where multimolecular substrate reactions are insignificant, are the only available data for subsite mapping of this enzyme. Using the depolymerase computer model and b.c.f. data for $n = 3-12$ in an analysis as with *B. amyloliquefaciens* α-amylase (Table III) reveals that the binding region of *A. oryzae* α-amylase has eight subsites with the catalytic amino acids located between subsites III and IV. The apparent subsite binding energies are given in Table II; the ability of this map to account for the experimental bond cleavage frequencies is shown in Table VI.

Because of the inaccessibility of Michaelis parameters, the binding energies of subsites III and IV and the acceleration factor for this enzyme cannot be estimated. However, the partial subsite map predicts a dissociation constant that is much smaller than the measured inhibition constants for glucose and maltose. This is diagnostic that the calculated apparent binding energies contain a rate-coefficient term and are consequently too low by the factor $\Delta G_{a,i}$.

*Subsite Mapping of B. amyloliquefaciens (Strains BLA-F, -D) α-*

---

[29] J. A. Thoma and J. D. Allen, *Carbohydr. Res.* **48**, 105 (1976).
[30] J. D. Allen and J. A. Thoma, *Biochemistry* **17**, 2338 (1978).
[31] T. Suganuma, M. Ohnishi, R. Matsuno, and K. Hiromi, *J. Biochem. (Tokyo)* **80**, 645 (1976).

TABLE VI
EXPERIMENTAL AND MODEL-CALCULATED BOND CLEAVAGE FREQUENCIES
FOR *Aspergillus oryzae* α-AMYLASE[a]

| Substrate | Cleavage frequencies | | | | | | | | | | Source |
|---|---|---|---|---|---|---|---|---|---|---|---|
| o—o—● | | | | | | | | 0.00 | 1.00 | | Experimental |
| | | | | | | | | 0.00 | 1.00 | | Calculated |
| o—o—o—● | | | | | | | 0.00 | 0.99 | 0.01 | | Experimental |
| | | | | | | | 0.00 | 0.99 | 0.01 | | Calculated |
| o—o—o—o—● | | | | | | 0.01 | 0.78 | 0.20 | 0.00 | | Experimental |
| | | | | | | 0.00 | 0.78 | 0.22 | 0.00 | | Calculated |
| o—o—o—o—o—● | | | | | 0.00 | 0.16 | 0.72 | 0.12 | 0.00 | | Experimental |
| | | | | | 0.00 | 0.16 | 0.73 | 0.11 | 0.00 | | Calculated |
| o—o—o—o—o—o—● | | | | 0.00 | 0.34 | 0.17 | 0.41 | 0.08 | 0.00 | | Experimental |
| | | | | 0.00 | 0.34 | 0.18 | 0.42 | 0.06 | 0.00 | | Calculated |
| o—o—o—o—o—o—o—● | | | 0.00 | 0.18 | 0.39 | 0.11 | 0.27 | 0.05 | 0.00 | | Experimental |
| | | | 0.00 | 0.21 | 0.39 | 0.11 | 0.25 | 0.04 | 0.00 | | Calculated |
| o—o—o—o—o—o—o—o—● | | 0.00 | 0.14 | 0.27 | 0.30 | 0.08 | 0.19 | 0.03 | 0.00 | | Experimental |
| | | 0.00 | 0.15 | 0.28 | 0.28 | 0.08 | 0.18 | 0.03 | 0.00 | | Calculated |
| o—o—o—o—o—o—o—o—o—● | 0.00 | 0.11 | 0.22 | 0.21 | 0.22 | 0.06 | 0.14 | 0.02 | 0.00 | | Experimental |
| | 0.00 | 0.12 | 0.22 | 0.22 | 0.22 | 0.06 | 0.14 | 0.02 | 0.00 | | Calculated |
| o—o—o—o—o—o—o—o—o—o—● | 0.00 | 0.11 | 0.20 | 0.20 | 0.17 | 0.16 | 0.04 | 0.10 | 0.02 | 0.00 | Experimental |
| | 0.00 | 0.10 | 0.18 | 0.18 | 0.18 | 0.18 | 0.05 | 0.12 | 0.02 | 0.00 | Calculated |
| o—o—o—o—o—o—o—o—o—o—o—● | 0.00 | 0.10 | 0.18 | 0.19 | 0.18 | 0.14 | 0.15 | 0.04 | 0.09 | 0.01 | 0.00 | Experimental |
| | 0.00 | 0.08 | 0.15 | 0.15 | 0.15 | 0.15 | 0.15 | 0.04 | 0.10 | 0.02 | 0.00 | Calculated |

[a] From J. D. Allen and J. A. Thoma, *Biochem. J.* **159**, 121 (1976).

*Amylase*. Iwasa *et al.*[32] estimate from plots of log $\bar{V}_n$, $(1/\bar{K}_{m,n})$, and $\bar{k}'_n$ as a function of substrate chain length that the binding region of two strains of *B. amyloliquefaciens* α-amylase are comprised of eight substrates. From the predominant modes of cleavage of reducing-end labeled maltoheptaose by strain BLA-F α-amylase and of unlabeled oligosaccharides up to a chain length of seven by strain BLA-D α-amylase, these authors propose that the cleavage site is located between subsites VI and VII. However, it should be noted that the use of unlabeled substrates complicates the problems of subsite mapping since it is impossible to tell whether the products arise from the reducing or nonreducing end of the substrate.

[32] S. Iwasa, H. Aoshima, K. Hiromi, and H. Hatano, *J. Biochem.* (*Tokyo*) **75**, 969 (1974).

Assuming a constant hydrolytic rate coefficient, estimated from the plateau region of a log $V_n$ vs. $n$ plot, Iwasa et al.[32] have calculated the sum of the apparent binding energies for subsites VI and VII using the first-order rate constant for maltose hydrolysis with Eqs. (1) and (9). In order to evaluate the binding energies of the remaining subsites, a series of equations obtained by dividing $\bar{k}'_n$ ($n = 3-8$) by $\bar{k}'_2$ are generated. For example when $n = 3$ the resulting equation is

$$\frac{\bar{k}'_3}{\bar{k}'_2} = \frac{k_{+2,\text{VII},3}}{k_{+2,\text{VII},2}} \exp(-\Delta G_\text{V}/RT) + \frac{k_{+2,\text{VIII},3}}{k_{+2,\text{VIII},2}} \exp(-\Delta G_\text{VIII}/RT) \qquad (21)$$

If the hydrolytic rate coefficients are equal, they cancel from Eq. (21); so that six nonlinear equations are generated that have the six subsite binding energies as unknowns. Iwasa et al.[32] propose to solve this set of nonlinear, simultaneous equations by an iterative technique to generate the apparent subsite energies given in Table II. However, it has been shown[29] that the interative technique for solving these equations does not always converge to a unique solution.

Thoma and Allen[29] have further examined the subsite map for strain BLA-N of *B. amyloliquefaciens* α-amylase. Starting with the apparent subsite energies generated from the first-order rate constants and optimizing subsites VI and VII, the model-predicted and experimental Michaelis parameters differ by two orders of magnitude, on the average. However, by assuming a constant acceleration factor of 630 cal/mol, the fit is dramatically improved. Hence, the data collected for this enzyme seem to support the conclusion that rates of bond cleavage are not equal.

*Mapping of Dextranases.* Pulkownik et al.[33] have used bond cleavage frequencies for nonreducing-end labeled isomaltose oligosaccharides ($n = 4-9$) with the depolymerase computer model to predict the binding topography of some dextranases and to partially map the binding region of *Streptococcus mutans* K1-R dextranase as given in Table II.

*Prediction of the Time Course of Oligosaccharide Degradation.* Equation (5) in an integrated form can be used to predict the distribution of saccharides at any time during the degradation of the substrate oligosaccharide. Torgerson and co-workers,[34] starting with a subsite map for a strain of *B. amyloliquefaciens* α-amylase, used numerical integration of Eq. (5) to predict the saccharide distribution of maltononanose degradation as shown in Fig. 5. The agreement between experimental and calculated values for a number of other oligosaccharides is similar, attesting to the predictive value of the model.

---

[33] A. Pulkownik, J. A. Thoma, and G. J. Walker, *Carbohydr. Res.* **61**, 492 (1978).
[34] E. M. Torgerson, L. C. Brewer, and J. A. Thoma, *Arch. Biochem. Biophys.* **196**. 13 (1979).

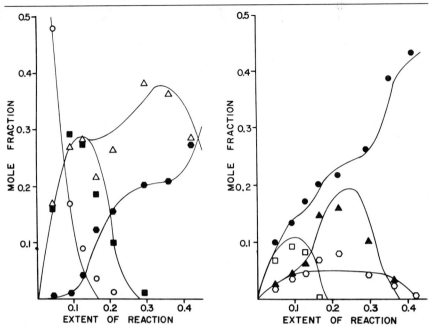

FIG. 5. Time course of hydrolysis of uniformly radiolabeled maltononanose. The data points are the experimentally determined distribution of saccharides in the digestion of uniformly labeled maltononanose by a strain of *Bacillus amyloliquefaciens* α-amylase. ○, $G_9$; □, $G_7$; ■, $G_6$; ▲, $G_5$; ◐, $G_4$; △, $G_3$; ●, $G_2$; ◯, $G_1$. The lines are the distribution of saccharides predicted from the subsite model as described in the text. From E. M. Torgerson. L. C. Brewer, and J. A. Thoma, *Arch. Biochem. Biophys.* **196**, 13 (1979).

## Features of the Mapped Enzymes

An examination of the enzymes that have been mapped reveals certain characteristics that have mechanistic significance. The subsite depicted to the left of the catalytic site (Table II) shows unfavorable energetics of interaction with the substrate monomer unit. This unfavorable binding energy is proposed to assist in straining or distorting the stable chair form of the pyranose ring into a half-chair form, approaching the transition state.[3] The actual distortion energy may be larger than the measured value if some favorable interactions also occur upon substrate binding at this subsite.

Strain BLA-N of *B. amyloliquefaciens* and *A. oryzae* α-amylases also have subsites distant from the catalytic site interacting unfavorably with the substrate monomer unit (Table II). These subsites are termed barrier subsites, since they act as a barrier to substrate binding. As reflected in the b.c.f. data for *A. oryzae* α-amylase (Table VI), binding so as to avoid the barrier subsite, resulting in labeled maltotriose, is preferred over binding in the barrier subsite to give labeled maltotetraose. However, since favorable binding at the subsite shown on the right of the barrier subsite more than compensates for barrier subsite binding, the largest b.c.f. for substrates equal to or larger than the size of the binding region occurs when all the subsites are occupied. However, *B. amyloliquefaciens* α-amylase has a barrier subsite on the end of the binding region so that the highest b.c.f. is a result not of all subsites being occupied, but of avoidance of the barrier subsite. An exoenzyme may be thought of as having such a large barrier subsite that binding across the barrier subsite is energetically excluded.

## Limitations

### Complicating Reactions

The subsite model, as developed here, requires that only a single substrate and enzyme molecule interact at any given time and that the enzyme catalyze only a single bond scission of the substrate per encounter. A cursory examination of the action pattern of an enzyme may not reveal reactions that can complicate the interpretation of the data used in subsite mapping. We will discuss here two types of complicating reactions and propose some methods of testing for these reactions.

*Multimolecular Substrate Reactions.* A simple hydrolytic event involves only one substrate molecule and is, thus, unimolecular in substrate. If more than one substrate molecule is involved in the catalytic event, the enzyme can be said to catalyze multimolecular substrate reactions. As the availability of highly purified enzymes has increased and more quantitative studies have been carried out, carbohydrases such as amylases that are classified as hydrolases have been shown to exhibit significant nonhydrolytic, multimolecular substrate catalytic activity.[30]

Multimolecular substrate reactions can take a number of different forms. Condensation, the reverse of hydrolysis, is the formation of a new glycosidic bond. Transglycosylation is the rearrangement of glycosidic bonds; after bond cleavage, the glycosidic residue from the nonreducing end of the substrate is transferred to the nonreducing end of another saccharide rather than to water. Shifted binding is the binding of a second substrate molecule in a nonproductive mode within the subsites so as to sterically shift the otherwise more energetically favorable binding of a productively bound substrate. Porcine pancreatic,[7] *B. subtilis*,[35] and *A.*

---

[35] H. Fujimori, M. Ohnishi, M. Sakoda, R. Matsuno, and K. Hiromi, *J. Biochem. (Tokyo)* **82**, 417 (1977).

*oryzae*[30,31] α-amylases have been shown to use more than one of these mechanisms to a significant extent in the degradation of oligosaccharides.

One way of testing for multimolecular substrate reactions is to determine the bond cleavage frequencies of end-labeled oligosaccharides over a wide concentration range.[30] As the substrate concentration is increased the probability of multimolecular substrate reactions is enhanced, generally leading to different apparent bond cleavage frequencies. If the enzyme carries out multimolecular substrate reactions, bond cleavage frequencies must be determined at low enough substrate concentrations so that the occurance of multimolecular reactions is insignificant; only then can the b.c.f. data be used to determine the apparent binding energies of the subsites. However, care must be used in interpretation of apparent Michaelis parameters generated by a multimolecular-substrate enzyme.

*Repetitive Attack.* Some depolymerases have been found to hydrolyze several glycosidic bonds (i.e., repetitively attack the substrate) during the lifetime of a single enzyme–substrate encounter, which complicates the interpretation of bond cleavage frequencies as well as Michaelis parameters. Various degrees of repetitive attack has been observed for enzymes such as sweet potato β-amylase[36] and porcine pancreative, human salivary, and *A. oryzae* α-amylase.[6,37]

The degree of repetitive attack can be assessed by using uniformly labeled oligosaccharides.[6] In the absence of repetitive attack (and multimolecular reactions) enzymic degradation of uniformly labeled $n$-mer will produce equal quantities of the products of chain length $n-m$ and $m$. However, repetitive attack will generate more of the shorter chain length products at the expense of the longer chain length products.

Alternatively, two series of oligosaccharides, one labeled in the nonreducing end and one labeled in the reducing end, can be employed to determine the degree of repetitive attack.[38] In the absence of repetitive attack (and multimolecular reactions), the products of the hydrolyses of the two series of oligosaccharides will be consistent with the same bond cleavage pattern. However, when repetitive attack occurs, one of the series will have reduced amounts of the longer chain length products in the digest. Assuming an absolute polarity of repetitive attack, the products of the other series of oligosaccharides still reflect the first hydrolytic event and can be used in subsite mapping. The interpretations of Michaelis parameters from a repetitive-attack enzyme have been discussed[39] and are beyond the scope of this chapter.

[36] J. M. Bailey and D. French, *J. Biol. Chem.* **226**, 1 (1957).
[37] J. A. Thoma, *Biopolymers* **15**, 729 (1976).
[38] J. F. Robyt and D. French, *Arch. Biochem. Biophys.* **138**, 662 (1970).
[39] J. A. Thoma, *Biochemistry* **5**, 1365 (1966).

*Assumptions in the Model*

In development and application of the depolymerase model a number of assumptions are made that, judging from the ability of the model to predict the experimental data, must indeed be largely justified. Nevertheless, some of the assumptions discussed below can only be approximations of the physical system.

*Additivity of Intrinsic Subsite Binding Energies.* The unitary free energy of binding a substrate monomer unit in a subsite is assumed to be an intrinsic constant unaffected by binding or absence of binding at other subsites. Further, these subsite binding energies are assumed to be additive in calculation of the microscopic association constants. However, a recent study[40] of the binding of a number of oligosaccharides to lysozyme reveals that the additivity of subsite contributions to the binding energies does not adequately explain the data for this enzyme.

In addition, since the end monomer residues of an oligosaccharide are different chemical entities than the internal residues, they are likely to exhibit different energetics of subsite binding. In the maltooligosaccharide series, the internal monomer units are constrained in the $\alpha$ configuration; whereas, the reducing-end unit exists as an equilibrium of the $\alpha$ and $\beta$ anomeric forms. It is likely that these two anomeric forms bind differently in the subsites. In fact, it has been shown[3] that the two anomeric forms of $N$-acetylglucosamine bind in the same subsite of lysozyme but have different orientations. Also, the two anomeric forms of D-glucose have different energetics of interaction with glucoamylase.[41] By using the $\alpha$-methyl maltooligosaccharides, Thoma and co-workers eliminated the problem of the two anomeric forms at the reducing end of the oligosaccharides, but the methyl group itself is likely to influence subsite binding. In light of the above discussion, the additivity of subsite energies can only be an approximation.

*One–One Complex Formation.* The subsite model assumes that a single-substrate molecule interacts with a single-enzyme molecule at any given time to give a one–one complex. The binding of a second substrate molecule to give a two–one complex might influence the productive binding modes and, consequently, be easily detectable as a shifted multimolecular reaction as delineated above. Alternatively, a second substrate molecule might bind so that the productive binding modes are not affected. In such a case, two–one binding may not influence the b.c.f. but will be reflected in the Michaelis parameters. Such a two–one complex has been

---

[40] M. Schindler, Y. Assaf, N. Sharon, and D. M. Chipman, *Biochemistry* **16**, 423 (1977).
[41] K. Hiromi, M. Kawai, N. Suetsugu, Y. Nitta, T. Hosotani, A. Nagao, T. Nakajima, and S. Ono, *J. Biochem. (Tokyo)* **74**, 935 (1973).

observed for lysozyme.[42] The extent to which two–one complexes complicate the subsite model is difficult to assess precisely. It has been proposed[43] that, in calculation of $V_n$, two–one binding can be partially accounted for by considering only nonproductive binding modes that sterically impede binding in a productive mode.

*Estimations of the Relative Hydrolytic Rate Coefficients.* At the conception of the subsite model, the hydrolytic rate coefficient was assumed to be a constant, irrespective of the substrate binding mode or chain length. This assumption seems to be adequate to explain the action of some enzymes, but it does not adequately account for the action of *B. amyloliquefaciens* α-amylase. A better approximation for this enzyme is to assume that the hydrolytic rate is enhanced as the number of filled subsites increases. Although this approximation significantly improves the fit, the model still does not account for the experimental data within error limits. The test of the validity of this assumption will have to await further experimental developments.

*Approximation of $\tilde{K}_m$ as a Dissociation Constant.* The dissociation constant is approximated by $\tilde{K}_m$ only if the rate of dissociation of the enzyme–saccharide complex is much faster than the rate of forward reaction to give products (i.e., $k_2 \ll k_{-1}$). The absence of repetitive attack implies that this is a good approximation, since the substrate dissociates from the enzyme faster than it can rearrange and be reattacked. However, for an enzyme that catalyzes a high degree of repetitive attack, several catalytic events can occur before dissociation, implying that the rate of reaction is not insignificant as compared to the dissociation rate. Sweet potato β-amylase has been shown to exhibit a high degree of repetitive attack; the derivation of the subsite map for wheat bran β-amylase given in Table II also may well be complicated by the interpretation of $\tilde{K}_m$ for a repetitive attack enzyme.

Summary

*Prefered Protocol for Subsite Mapping*

We can now outline the best overall procedure for mapping of the binding region of a polysaccharide depolymerase.

1. *Test for complicating reactions.* Test the enzyme, which has been purified to homogeneity, for complicating reactions such as multimolecular substrate reactions or repetitive attack. If the enzyme catalyzes such reactions, establish experimental conditions where

[42] E. Holler, J. A. Rupley, and G. P. Hess, *FEBS Lett.* **40,** 25 (1974).
[43] Correction to K. Hiromi, M. Ohnishi, and S. Shibata, *J. Biochem. (Tokyo)* **74,** 397 (1973).

they are insignificant or modify the depolymerase model to account for the complications.

2. *Use bond cleavage frequencies to establish the binding region topography and to measure the apparent subsite binding energies.* Determine bond cleavage frequencies for a series of end-labeled oligosaccharides up to a chain length that exceeds the size of the binding region, typically to chain length 12. Examine the b.c.f. data to estimate the binding region topography, then test the estimate in the depolymerase computer model by establishing the least number of subsites that adequately accounts for the data. With this binding region topography, use the depolymerase computer model to calculate the apparent binding energies of the subsites.

3. *Use $\bar{K}_m$ and $\bar{V}$ measured as a function of substrate chain length to complete the subsite map.* Measure $\bar{K}_{m,n}$ and $\bar{V}_n$ for oligosaccharides up to a chain length that exceeds the size of the binding region by several units so that $K_{int}$ can be evaluated. Use $K_{m,n}$, $V_n$ and $K_{int}$ in the depolymerase computer model to optimize the two subsites adjacent to the catalytic site; then use all the experimental data to obtimize all the subsites. If the measured parameters are predicted within experimental scatter the subsite map is complete. Otherwise, test the improvement in fit on allowing the hydrolytic rate coefficient to vary.

# [11] The Kinetics and Processivity of Nucleic Acid Polymerases[1]

*By* WILLIAM R. MCCLURE and YVONNE CHOW

The first question that should be answered in a kinetic study of a nucleic acid polymerase is: How processive is it? In other words, after formation of a phosphodiester bond, what is the probability that the polymerizing enzyme will translocate along the template to incorporate yet another nucleotide into the nascent chain rather than dissociate to bind to some other suitable polymerization site? Mechanistic studies on this class of enzymes are motivated by *in vivo* observations of high-fidelity replication and repair of DNA and highly specific initiation, elongation, and termination of RNA transcription and by the belief that positive and negative regulators of both these aspects of gene expression must act by altering in

---

[1] Supported by National Institutes of Health Grant GM 21052 and by the Milton Fund of Harvard University.

some fashion the underlying reaction sequence catalyzed by the polymerases involved. Kinetic studies in parallel with product analysis and structural information are an essential component of a complete biochemical description of these enzyme reaction mechanisms. However, owing to some novel complexities of enzymic polymerization, a complete and unambiguous interpretation of the steady-state kinetics for any of these enzymes has not been achieved. The processivity is one of the interesting complexities referred to above.

We show in this chapter how the processivity of a polymerase affects the interpretation of other kinetic parameters. To illustrate these arguments we have included a derivation of the steady-state equation for a generalized nucleic acid polymerase. In addition, we describe a new method for quantitatively determining the processivity of a DNA polymerase and illustrate the method with data obtained employing *Escherichia coli* DNA polymerase I. Finally, we discuss processivity in connection with other kinetic factors that affect the fidelity of nucleic acid polymerization.

Theory

*General Mechanism for Polymerase Reactions in the Steady State*

In Fig. 1 we show a steady-state scheme applicable in slightly altered form to most nucleic acid polymerases. The rate equation for this iso-ordered mechanism is

$$\frac{v}{E_t} = \frac{V_1(Pn)(N) - V_2(K_n/K_{ppi})(Pn)(PP_i)}{K_{i,Pn}K_n + K_n(Pn) + K_{pn}(N) + (Pn)(N)}$$
$$+ \frac{K_n}{K_{pp_i}} K'_{pn}(PP_i)' + \frac{K_n}{K_{pp_i}} (Pn)(PP_i) + \frac{K_n}{K_{pp_i}} K_{i,n}(Pn)(N)(PP_i) \quad (1)$$

where the species Pn and Pn + 1 correspond to the polynucleotide template and nascent product strand of length n or n + 1, and N and $PP_i$ refer to the triphosphate substrate and inorganic pyrophosphate product, respectively. The individual kinetic constants have been subscripted according to the substrate or product to which they correspond. All other nomenclature is according to Cleland.[2]

The essential difference between this general mechanism and the standard Ordered Bi Bi mechanism is the possibility for translocation from the elongated E·Pn + 1 state to the E·Pn complex, shown as $k_9$ in Fig. 1. In addition, although we can distinguish between the translocated and untranslocated complexes, there is no difference between Pn + 1 and

---

[2] W. W. Cleland, *Biochim. Biophys. Acta* **67**, 188 (1963).

```
            E
    k₇ ↗  ↖ k₁Pn
     ↙ k₈Pn  k₂ ↘
              k₉
    EPn+1   ⇌   EPn
              k₁₀
    ↗ ↖ k₆PPᵢ     k₄ ↗ ↘
   k₅ ↙              ↙ k₃N

    (EPn+1 PPᵢ ⇌ EPnN)
```

FIG. 1. Steady-state kinetic scheme for a generalized nucleic acid polymerase. The addition of polynucleotide template (Pn) and triphosphate (N) are shown as ordered. The release of pyrophosphate ($PP_i$) results in an EPn + 1 complex that can either dissociate or translocate. The effect of this latter branching pathway on the overall kinetics is discussed in the text.

Pn in solution. For that reason, we have defined kinetic constants corresponding to the different enzyme polymer complexes, but have not differentiated between the dissociated polymer species (Pn is used throughout for polynucleotide concentration). Indeed, polymers of any length remain at unit activity through all elongation steps.

In addition to the kinetic constants used in Eq. (1), two additional measurable parameters are required properly to characterize a steady-state polymerase mechanism. These are the forward and reverse processivities, $P_f$ and $P_r$, respectively,

$$P_f = k_9/(k_7 + k_9) \qquad (2)$$
$$P_r = k_{10}/(k_2 + k_{10}) \qquad (3)$$

Defined in this fashion, the processivity of a nucleic acid polymerase is seen to vary from 0 (i.e., distributive synthesis) to 1 (i.e., strictly processive). Intermediate values of $P_f$ and $P_r$ express the probability that, having catalyzed the incorporation of a nucleotide into a growing primer strand, the enzyme will translocate rather than dissociate.

In Table I we list modifications of the basic mechanism and indicate the changes that result thereby in the denominator of the rate equation. As noted above, when $k_9 = k_{10} = 0$, the standard Ordered Bi Bi equation results. The properties of this mechanism as applied to *E. coli* DNA polymerase I and herpesvirus polymerase have been discussed.[3,4] When the steps corresponding to either $k_7$ and $k_8$ or $k_1$ and $k_2$ are omitted (modifications 3 and 4), the Michaelis constants for polynucleotide approach

---

[3] W. R. McClure and T. M. Jovin, *J. Biol. Chem.* **250**, 4073 (1975).
[4] S. S. Leinbach, J. M. Reno, L. F. Lee, A. F. Isbell, and J. A. Boezi, *Biochemistry* **15**, 426 (1976).

TABLE I
THE MODIFICATIONS IN EQ. (1) RESULTING FROM CHANGES
IN THE GENERAL MECHANISM SHOWN IN FIG. 1

| Mechanism | Modification | Resultant change | PP$_i$ inhibition pattern[a] Vary Pn | Vary N | $V_1/E_t K_{pn}$ | |
|---|---|---|---|---|---|---|
| 1 | None | None | NC → UC | NC | $k_1 =$ | $\dfrac{k_1 k_9}{k_7} + \dfrac{k_8 k_9}{k_7}$ |
| 2 | $k_9 = k_{10} = 0$ | None | NC → UC | NC | $k_1$ | |
| 3 | $k_7 = k_8 = 0$ | $K_{pn} = 0$ | NC → UC | NC | $k_1$ | $\dfrac{k_{10}}{k_9}$[b] |
| 4 | $k_1 = k_2 = 0$ | $K'_{pn} = 0$ | UC | NC | $k_8$ | $\dfrac{k_9}{k_7}$ |
| 5 | $k_9 = k_{10} \to \infty$ and $k_7 = k_8 = 0$ or $k_1 = k_2 = 0$ | $K_{pn} = 0$ $K'_{pn} = 0$ | UC → E | C | $\infty$ | |
| 6[c] | $k_2$ and $k_7 \to 0$ | $K_{i,pn} \to 0$ $K_{pn} \to 0$ $K'_{pn} \to 0$ | — | NC | $\infty$ | |

[a] The product inhibition patterns are abbreviated: NC, noncompetitive; UC, uncompetitive; C, competitive; E, eliminated. Pattern alterations on saturation with the nonvaried substrate are indicated with an arrow.

[b] In this case the ratio is $V_1/K'_{pn}$.

[c] In the strictly processive mechanisms, as $k_2$ and $k_7$ approach zero (e.g., $<10^{-3}$ sec$^{-1}$) the kinetic constants corresponding to template also approach zero (e.g., $<10^{-10}$ M).

zero. For example, if $k_7 = k_8 = 0$, saturation with nucleotide decreases the steady-state level of E·Pn to zero so that dissociation cannot occur from that complex. Another formal possibility would be a further modification of mechanism 3 or 4 to allow translocation to occur in concert with chemical reaction or PP$_i$ release, thus making $k_9 = k_{10} \to \infty$ (mechanism 5). The prediction here is that both Michaelis constants for polynucleotide would tend to zero with saturating substrates.

The above discussion has dealt with polymerases where $0 \leq P_f < 1$. For a few enzymes, $P_f$ is for all practical purposes equal to one. In these cases both $k_2$ and $k_7$ can be set equal to zero. The effect on the steady-state kinetics is 2-fold. First, $K_{pn} = K'_{pn} = K_{i,pn} = 0$. The more realistic way to view such an instance is that the polynucleotide site is required in equal concentration to enzyme. In other words, very tight binding or essentially titration behavior is observed in the interaction between enzyme and polynucleotide. Enzymes of this class also typically display an initiation phase followed by elongation. DNA-Dependent RNA polymerases have been characterized as members of this sixth class. The second major feature of the steady-state kinetics of strictly processive enzymes is that the triphosphate saturation and $PP_i$ product-release steps are all totally independent of the enzyme–polynucleotide interaction. In other words, if the initiation phase of synthesis can be separated from the steady-state conversion of triphosphates into polymer, the latter process can be analyzed with normal kinetic equations in which $E \cdot Pn$ is employed as the catalytic species equivalent to E in standard steady-state formalism.

To some extent this separation of enzyme–polynucleotide interaction from triphosphate binding can be achieved for enzymes belonging to the first five classes of Table I by saturating with the polymer template. At least the *patterns* of triphosphate saturation conform to simple analysis; however, evaluations of discrete rate constants from steady-state kinetic constants can be rather complex. As an example of this problem, we show in the last column of Table I the quotient of $V_1/E_t K_{pn}$ for each mechanism. In the simple Ordered Bi Bi reaction this calculation yields a lower estimate of $k_1$. In all other cases the result is $k_1$ multiplied by a term containing rate constants associated with the dissociation and/or translocation of the enzyme–polynucleotide complex. Clearly, an estimate of the values of $k_9$ and $k_{10}$ in combination with the polynucleotide dissociation steps is required for a full description of the enzyme template kinetics. In addition, the kinetic constants associated with triphosphate and pyrophosphate also contain contributions from the elementary steps in the enzyme–template interaction. Saturation with template can yield interpretable data on the triphosphate kinetics, but information on the coupling between the two major phases of polymerization is of necessity lost in such an approach. Another minor problem in the interpretation of $K_n$'s arises when a template with more than one nucleotide base is employed.

Consider a strictly processive enzyme or a polymerase synthesizing at saturating concentrations of a dissociable template. The initial-velocity equation, Eq. (1), reduces to a one substrate–one product expression when only one triphosphate is required as substrate (e.g., homopolymer templates). For alternating copolymer templates; a modified Ping Pong equation describes the saturation, and two Michaelis constants are defined for the two triphosphate substrates. In reciprocal form, the initial-velocity equation is

$$\frac{1}{v} = \frac{K_{n1}}{V_1(N_1)} + \frac{K_{n2}}{V_1(N_2)} + \frac{1}{V_1} \tag{4}$$

In general, the equation for any template with fractional base composition $F_{ni}$ is

$$\frac{1}{v} = \sum_i \frac{F_{ni} K_{ni}}{V_1 (N_i)} + \frac{1}{V_1} \tag{5}$$

The sum extends over all the different template bases. For natural DNAs $i = 1-4$ and the four $N_i$ correspond to A, G, C, and T; $K_{ni}$ correspond to the four equivalent homopolymer Michaelis constants for the substrate triphosphates. Equation (5) was derived by Hyman and Davidson[5] and shown by Rhodes and Chamberlin[6] to be applicable to a large number of templates for RNA polymerase. In both the examples cited, the initiation phase (i.e., enzyme–template association) was experimentally bypassed. The Ping Pong nature of Eq. (5) implies that triphosphate binding and incorporation is not affected by DNA sequence and that the kinetically significant pathway in polymerization entails binding of only one triphosphate at a time. This conclusion has also been verified for DNA polymerase I[2] although contradictory claims that several prealigned triphosphates may occur in DNA polymerization can be found.[7-9] The naive expectation for kinetically significant prealignment of triphosphates would be an interaction between them corresponding to a sequential mechanism and thus an intersecting pattern on a reciprocal plot in which one triphosphate was varied at fixed concentrations of the others. The parallel pattern actually observed in all cases corresponds to Ping Ping, i.e., noninteractive, binding of triphosphates. Although Ping Pong steady-state kinetics does not totally disprove the notion of triphosphate prealignment, severe restrictions are thereby placed on such models. For example, the dissociation constants for the putative prealigned nucleotides must be well below 1 $\mu M$.

## How To Determine the Processivity of a DNA Polymerase

The new method we describe here is quantitative, accurate, technically convenient, and versatile enough to allow a substantial variation of reaction conditions (e.g., solution composition, substrate concentrations, temperature, and template-primer compositions). The technique relies upon the resolution of discrete product lengths on a high percentage acrylamide gel in 7 $M$ urea. The theory employed to analyze the resulting product length distribution data is a simple application of the Kuhn distribution law.

Consider a poly(dA) template with oligothymidylate primers at a con-

---

[5] R. W. Hyman and N. Davidson, *J. Mol. Biol.* **50**, 421 (1970).
[6] G. Rhodes and M. J. Chamberlin, *J. Biol. Chem.* **249**, 6675 (1974).
[7] N. Battula, D. K. Dube, and L. A. Loeb, *J. Biol. Chem.* **250**, 8404 (1975).
[8] E. C. Travaglini, A. S. Mildvan, and L. A. Loeb, *J. Biol. Chem.* **250**, 8647 (1975).
[9] J. P. Slater, I. Tamir, L. A. Loeb, and A. S. Mildvan, *J. Biol. Chem.* **247**, 6784 (1972).

centration such that the ratio of primer 3'-OH groups to enzyme is ≥ 1000. We now allow polymerization to occur for an average of ~ 100 turnovers. Two extremes in product length can be imagined. First, if the enzyme were strictly processive and never dissociated from an elongating primer we expect to find primers elongated about 100 nucleotides. The molar concentration of elongated primer would equal the molar enzyme concentration. The second extreme mode of synthesis is distributive, where the enzyme dissociates after each nucleotide incorporation event. In this case we predict a Poisson distribution of primer lengths. At early times in the reaction only the original primer elongated by one nucleotide would be found because reinitiation on previously elongated primers would not have occurred. In this case the elongated primers would be found at 100 times the enzyme concentration.

The extremes described above have been reported, but we also need to analyze data corresponding to intermediate processive synthesis. We consider that a primer chain of length $m$ ($R_m$) that is elongated an additional $n$ nucleotides ($R_{m+n}$) would have originated as follows: the enzyme would associate with the template-primer and after each incorporation step translocate with probability $P$ or dissociate with probability $1 - P$. Therefore, each elongated primer is the result of $n - 1$ translocation steps and 1 dissociation step. Assuming that the probability of translocation remains constant throughout the elongation reaction, the distribution of product lengths is then,

$$\%R_{m+n} = KP^{n-1}(1 - P) \qquad (6)$$

where $\%R_{m+n}$ is the mole percentage of each product length and where $K$ is a normalization constant. The probability in Eq. (6) can be evaluated by taking logarithms

$$\log[\%R_{m+n}] = (n - 1) \log P + \log K(1 - P) \qquad (7)$$

Thus, if we plot the log of the mole percentage of each product chain length vs. $n - 1$, the slope will be log $P$ and the intercept will yield a value of $K$ and an internal check on $P$. The constant $K$ has been evaluated analytically and can be found elsewhere.[10] The probability of translocation discussed above corresponds to the processivity defined in Eq. (2).

Equation (7) predicts a linear curve from an analysis that includes all the product synthesized. Therefore, deviations from linearity can be considered diagonostically as having arisen from experimental problems, such as product recovery or radioactivity quantitation, or from more complex origins, such as different enzyme synthetic modes. The illustrative

---

[10] Y. Chow and W. R. McClure, submitted for publication.

## The Coupling between Triphosphate Binding and Processivity

We have argued in the section on general mechanism that, by experimentally separating the enzyme–polymer interaction from the triphosphate binding steps, meaningful kinetic constants could be assigned in the overall reaction mechanism. We have also shown in the preceding section how the processivity of an enzyme can be determined by an analysis of the product length distribution. We now consider explicitly the coupling between triphosphate binding and the processivity of a DNA polymerase. The mechanism depicted in Fig. 1 indicates that the enzyme can dissociate from either the Pn + 1 or Pn complexes. Thus, although we have defined $P_f$ only in terms of $k_7$ and $k_9$, it is clear that we could expect a decrease in the observed processivity that would be inversely proportional to the probability of the enzyme's continuing at the EPn stage. The observed processivity will then be the product of $P_f$ and the probability of continuing on the same primer at the triphosphate binding step; we have termed this probability, $P_t$. Therefore,

$$P_{obs} = P_f P_t \tag{8}$$

Where, $P_{obs}$ is the observed processivity, $P_f$ is the forward processivity as defined in Eq. (2), $P_t$ is the probability of the binary EPn complex incorporating another triphosphate rather than dissociating.

The contribution of $P_t$ to the forward processivity can be seen as follows. If, after a nucleotide has been incorporated, a fraction of the EPn + 1 complex translocates to form EPn, we want to know what additional fraction of these newly formed EPn complexes will react further to yield EPn + 1 again. There are three ways for the EPn complex to be converted to other enzyme forms.

$$-(d[\text{EPn}]/dt) = (k_3[\text{N}] + k_2 + k_{10})[\text{EPn}] \tag{9}$$

To a first approximation only the $k_3[\text{N}]$ route guarantees another nucleotide incorporation event. Both $k_2$ and $k_{10}$ can or do lead to dissociation and an end of synthesis on a given primer. As the above three steps are competing (pseudo) first-order reactions, the ratio of one to the sum of all three will be a probability of reacting along the given pathway. We, therefore, define in the following expression the probability of incorporating another nucleotide after translocation to the EPn complex.

$$P_t = \frac{k_3[\text{N}]}{k_2 + k_{10} + k_3[\text{N}]} \tag{10}$$

As defined initially, $P_t$ is the probability of the EPn complex binding and incorporating another triphosphate compared to the sum of the three possible steps available to that enzyme species. We next define

$$C_n = (k_2 + k_{10})/k_3 \tag{11}$$

Dividing Eq. (10) by $k_3$, and using Eq. (11) yields on substitution into Eq. (8)

$$P_{obs} = P_f[N]/(C_n + [N]) \tag{12}$$

The constant, $C_n$, corresponds to the concentration of triphosphate required to make $P_t$, the probability of continuing, equal to 0.5. The prediction of Eq. (12) is that the observed processivity should approach $P_f$ asymptotically as the nucleotide concentration increases. The constants corresponding to this saturation behavior can be evaluated by employing Eq. (13), the double-reciprocal form of Eq. (12).

$$1/P_{obs} = C_n/P_f(1/[N]) + 1/P_f \tag{13}$$

The forward processivity, $P_f$, is obtained from the intercept of a plot of $1/P_{obs}$ vs. $1/[N]$ and the slope divided by the intercept yields $C_n$. The form of Eq. (12) is analogous to the Michaelis–Menten equation. The value of $C_n$, however, is only distantly related to $K_n$, the triphosphate Michaelis constant. Indeed, the expression for $C_n$ in this mechanism is simpler than that for $K_n$. As shown in Fig. 4 and discussed later, the two constants are not always numerically the same.

The relation defining $P_t$, Eq. (10), is actually somewhat more complex than shown. We have included only $k_3[N]$ into the fraction of EPn complexes that remain processive. Actually, only a portion of the EPn complexes that translocate back to EPn + 1 (i.e., the $k_{10}$ step) will dissociate; the remaining fraction will once again translocate to yield EPn. But we know this latter fraction to be $P_f$; thus a more precise term in the denominator of Eq. (10) would be $k_{10}(1 - P_f)$ rather than simply $k_{10}$. We could also consider the reversibility of triphosphate binding to diminish $k_3[N]$. This alteration and the simpler modification of $k_{10}$ affect only the expression for $C_n$. The operational form of Eq. (12) remains identical. We should emphasize, however, that a determination of $C_n$ cannot be used to determine $k_2$, $k_{10}$, or $k_3$ without additional information. For example, if polymer release or translocation were rate limiting in the reverse reaction, the processivity, $P_r$, could be used with $C_n$ to determine $k_3$. Equation (10) was derived here using the oversimplified rate constants for heuristic reasons and because inclusion of the known complexities would not enhance the final equation used to determine $P_f$.

As a numerical illustration of the formalism in the Theory section,

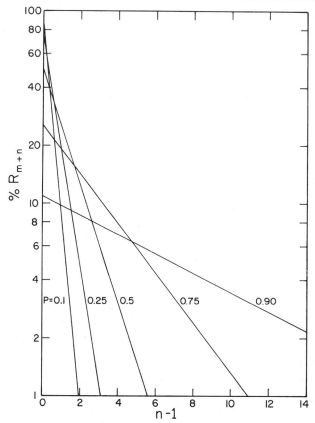

FIG. 2. Calculated product distribution pattern for a DNA polymerase. The mole percentage of each primer elongated $n$ nucleotides is plotted on a logarithmic ordinate vs. $n - 1$ according to Eq. (7). Each line corresponds to the processivity ($P$) indicated.

consider a DNA polymerase whose initial velocity kinetics can be described by Eq. (1) and where $k_7 = k_9$. In this case $P_f = 0.5$. The enzyme has an equal probability of translocating or dissociating after the incorporation of each nucleotide; at saturating triphosphate concentrations, (e.g., $[N] \geq 100 C_n$) we calculate from Eq. (12) that $P_{obs} \cong 0.5$. The product distribution generated by this enzyme would consist of one-half of the product primers elongated one nucleotide, one-quarter elongated two nucleotides, one-eighth elongated three nucleotides, etc. This hypothetical distribution when plotted according to Eq. (7) yields the pattern shown by the middle line in Fig. 2. The same enzyme synthesizing at a concentration of nucleotide equal to $C_n$ would still dissociate or translocate with equal probability at the EPn + 1 stage, but in addition significant dissociation would occur from the EPn complex. In this case,

$P_t = 0.5$ and the overall $P_{obs} = 0.5 \times 0.5 = 0.25$. The calculated product distribution for this example is shown in Fig. 2. We also show representative product distribution curves for other values of $P$.

## Materials and Methods

Deoxythymidine triphosphate, oligo(dT)$_{10}$, and poly(dA)$_{380}$ (P. L. Biochemicals); [$\alpha$-$^{32}$P]TTP (New England Nuclear); acrylamide (Sigma); $N,N'$-methylenebisacrylamide (Eastman Kodak); and TEMED (Bio-Rad) were all purchased from the sources indicated. Buffers and other salts were reagent grade or better.

Unless otherwise indicated, the standard conditions for the DNA polymerase I reaction were a final reaction volume of 0.10 ml containing 50 m$M$ KP$_i$, pH 7.4; 7.5 m$M$ MgCl$_2$; 30 $\mu M$ poly(dA) (DNA phosphorus); 30 $\mu M$ (pT)$_{10}$ (3 $\mu M$ 3'OH); 50 $\mu M$ [$\alpha$-$^{32}$P]TTP (400 cpm/pmol). The reactions were run at 25° and quenched with a 3-fold molar excess of EDTA. The samples were then depurinated in 50% HCOOH for 2 hr at 65° and lyophilized to remove formic acid. The samples were dissolved in 25 $\mu$l of 5% glycerol, 0.03% bromophenol blue, and 0.1 concentration of the electrophoresis running buffer. The electrophoresis was run according to Maniatis et al.[11] in a 12% acrylamide–0.6% bisacrylamide gel containing 7 $M$ urea, and TBE buffer (90 m$M$ Tris, 90 m$M$ boric acid, 2.5 m$M$ EDTA, pH 8.3). A lower percentage stacking gel (5% acrylamide, 0.04% bisacrylamide, 7 $M$ urea, 50 m$M$ Tris-Cl pH 6.7) enhanced resolution when larger sample volumes were run ($\geq 50$ $\mu$l). The 40 $\times$ 16 $\times$ 0.3 cm gel was run at 250 V for 24 hr at room temperature using TBE buffer in the reservoirs. The individual product lengths were located after autoradiography (Kodak XR-5 film) and quantitated either by scintillation counting (Cerenkov) of individual slices from the corresponding region of the gel or by integrating a densitometer tracing of the autoradiogram. For both methods, base lines were drawn assuming a linear gradient of background radioactivity throughout the product region on the gel. The molar amount of each product ($R_{m+n}$) was calculated from the specific activity of TTP and the number of added nucleotides ($n$). These values were summed, and the mole percentage (% $R_{m+n}$) of each product was then computed. The logarithms of these values were plotted vs. $n - 1$ according to Eq. (7). The summation required by Eq. (6) extends to infinity; we were obliged to omit from our analysis those longer product peaks that were about equal to or less than the background radioactivity. In most cases, these longer species were present at less than 1 mol%. The error resulting from this truncation was therefore considered to be insignificant.

[11] T. Maniatis, A. Jeffrey, and H. van de Sande, *Biochemistry* **14**, 3787 (1975).

## Results and Discussion

The only requirement for applying our method of determining the processivity of a polymerase is recovery and resolution of individual product species. As shown in Fig. 3, the resolution on gels between products varying in length by one nucleotide is very good. Both densitometer tracing and gel slicing yielded identical results in the analysis shown in Fig. 3C. For convenience, we have generally used the densitometer method when many samples are analyzed. There are many methods for integrating such data; we have found cutting out peaks and weighing the paper to be the most convenient. Occasionally, the gel electrophoresis resulted in unsymmetric bands, so that densitometer tracing was not reliable. In this case, gel slicing can often salvage the rather misshapen data. We adjust the specific activity of the TTP so that a 6-hr exposure will be sufficient for adequately exposure of the autoradiogram. This minimizes spreading of the product bands. As shown in Fig. 3C, this method of analysis is a rather "forgiving technique." The data are precisely linear to nearly 1 mol% of product. In Fig. 4, some of the data extend well below 1 mol%. Some individuals might exercise themselves about how the peaks in Figs. 3A and 3B should be divided among the various products. Because this is an integral type of analysis, errors in apportioning an area for a given peak tend to be compensated by the effect on the adjacent peaks. Errors in quantitation lead primarily to poorer linearity rather than to an incorrect slope.

The coupling between triphosphate saturation and processivity predicted by the model in Fig. 1 was demonstrated by the data in Fig. 4. At the lowest triphosphate concentrations employed, the product distribution approached that predicted by Poisson's equation (i.e., distributive synthesis). As the triphosphate concentration was increased, $P_{obs}$ approached a limiting value in accordance with Eq. (12). The constants that characterize the coupling between triphosphate saturation and processivity were evaluated from the reciprocal plot shown in Fig. 4B. The limiting processivity was found to be 0.64, and $C_n = 13$ $\mu M$. The initial velocities of product formation for the same data yielded $K_n = 12$ $\mu M$. In this case the two constants that characterize triphosphate saturation were comparable; however, a different value for $P_f$ could yield unequal $C_n$ and $K_n$ values.

In this short chapter we have included only those results that illustrate our method of determining processivity. It should be clear from the data in Figs. 4A and 4B that, properly defined, the processivity of a DNA polymerase is not a constant characteristic. Indeed, we have found that many reaction variables have a large effect on the processivity. In the work described here, we have employed only the DNA polymerase I fragment

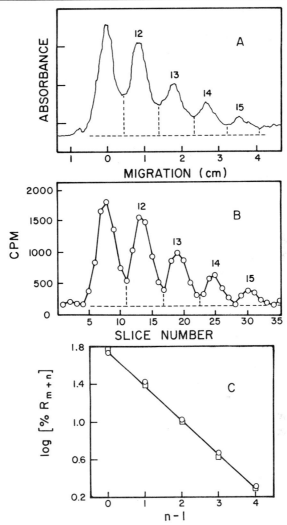

FIG. 3. Product distribution and processivity analysis for *Escherichia coli* DNA polymerase I fragment enzyme. Reactions were run as described in Methods with poly[dA·(dT)$_{10}$] as the template-primer; enzyme was 2.9 n$M$. The $\alpha$-$^{32}$P-labeled elongated products were electrophoretically separated on an acrylamide gel and autoradiographed. The densitometer tracing of the autoradiogram is shown in panel A. The abscissa represents distance from the bromophenol blue dye front. In panel B, we have plotted counts per minute in each gel slice taken from the region of the gel corresponding to the densitometer tracing in panel A. The baselines and divisions between each of the $R_{m+n}$ products are indicated with dashed lines in both A and B. The logarithmic product distribution function, log[% $R_{m+n}$], is plotted vs. $n - 1$ in C. The two symbols correspond to data taken from the integrated intensities in A (○) or from the summed counts per minute in B (□). A linear least-squares fit to all the points yielded a value of $P = 0.43$.

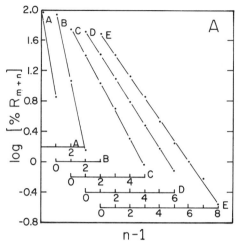

FIG. 4A. Product distribution patterns for *Escherichia coli* DNA polymerase I fragment enzyme at various TTP concentrations. Reactions were run under standard conditions at 2.9 n$M$ enzyme except that the TTP concentration was varied as follows: Curves A, 1.0 $\mu M$; B, 4.8 $\mu M$; C, 20 $\mu M$; D, 51 $\mu M$; E, 95 $\mu M$. The $\log[\% R_{m+n}]$ is plotted vs. $n - 1$, but each curve is displaced to the right an equal distance. The letter on each abscissa corresponds to the set of data designated above. The $P_{obs}$ values were 0.08, 0.13, 0.43, 0.49, and 0.54 for curves A–E, respectively.

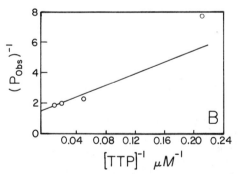

FIG. 4B. Processivity saturation plot for *E. coli* DNA polymerase I fragment enzyme. The observed processivities from Fig. 4A are plotted vs. TTP concentration in reciprocal form. The intercept and slope were evaluated using a weighted least-squares analysis devised by G. N. Wilkinson [*Biochem. J.* **80**, 324 (1961)] and yielded values of $P_f = 0.63$ and $C_n = 13$ $\mu M$ according to Eq. (13) in the text.

enzyme. The processivity of intact DNA polymerase I is more complex because two modes of synthesis are apparent.[10]

A variety of other methods have been employed to determine the processivity of nucleic acid polymerases. Most of these methods have been applied to *E. coli* DNA polymerase I. In Table II we have compiled a sampling of representative data corresponding to several enzymes and several template-primer combinations. In most cases the processivity was expressed as an average number of nucleotides incorporated per enzyme–template association event. We have recalculated the processivities according to our Eq. (2). For example, if $n$ was the average number of nucleotides incorporated, the processivity in our nomenclature, $P_{obs} = (n - 1)/n$. Although it was not always clear that number average lengths had been determined, as required for our processivity calculation, in most cases upper or lower estimates of $P_{obs}$ could be calculated.

The template challenge method can be useful when the processivity is high enough so that average enzyme–template association times are long compared to the mixing and sampling time. Because this method relies on observing a perturbation in the polymerization on one template by the addition of the challenging template, the relative affinity of the enzyme for both templates must be determined in control or independent experiments.

A new method devised by Bambara *et al.*[12] measures a cycling-time perturbation during synthesis. The concept is similar to the template challenge approach except that steady-state rate measurements are used to deduce the average time spent by an enzyme on the template of interest. Thus, the time resolution demanded by the direct template challenge experiment is not a requirement of the cycling time measurement. Because this method relies on a knowledge of the steady-state partitioning of the enzyme between template and inhibitor complexes, the experimenter is vulnerable to the criticism that the enzyme may be present in additional complexes including those nonspecifically bound to double-stranded DNA, nonproductively bound at 3'-OH ends, or indeed even free in solution. This method is also not applicable to homopolymer templates.

Careful work employing the template-challenge and cycling-time perturbation methods has minimized the disadvantages referred to above and yielded relatively consistent measurements of the processivity of several enzymes. However, as indicated above, both techniques require important kinetic information about the perturbing templates. The necessary

---

[12] R. A. Bambara, D. Uyemura, and T. Choi, *J. Biol. Chem.* **253**, 413 (1978).

TABLE II
PROCESSIVITY OF NUCLEIC ACID POLYMERASES

| Enzyme source | Template and primer | Processivity | Method[a] | Comments | Reference |
|---|---|---|---|---|---|
| 1. *Escherichia coli* Pol. I | Poly(dA·oligo T) | <0.98 | TC | 35° | b |
| 2. *Escherichia coli* Pol. I | Poly[d(AT)][c] | <0.10 | TC | 4° | d |
| 3. *Escherichia coli* Pol. I | Poly[d(AT)] | <0.97 | TC | 37° | |
| 3. *Escherichia coli* Pol. I | λ "Sticky ends" | >0.90 | PA | 6° and 22° | e |
| 4. *Escherichia coli* Pol. I | fd single strand | <0.10 | PA | 30° | f |
| 5. *Escherichia coli* Pol. I | Poly[d(AT)] | 0.99 | CP | 37° | g |
| 6. *Escherichia coli* Pol. I | Poly[d(AT)] | 0.94 | CP | 4° | |
| 6. *Escherichia coli* Pol. I | Poly(dA·oligo T) | 0.5 | PA | 25° | h |
| 7. *Escherichia coli* Pol. II | fd single strand | >0.99 | PA | DNA binding protein was present, 30° | f |
| 8. AMV[c] reverse transcriptase | Poly(rA·oligo T) | <0.96 | TC and PA | 37° | i |
| 9. Calf thymus 3.4 S | Poly(dA·oligo T) | <0.10 | PA | 20° | c |
| 10. Calf thymus 6.8 S | Poly(dA·oligo T) | <0.10 | PA | 20° | c |
| 11. Human KB cell | Calf thymus DNA | <0.10 | CP | 37° | g |
| 12. *Escherichia coli* RNA polymerase | Poly(dAT) | >0.999 | TC and PA | 37° | j |
| 12. *Escherichia coli* RNA polymerase | Poly(dG)·poly(dC) | 0.99 | TC and PA | | |

[a] The methods are abbreviated as follows: TC, template challenge; PA, product (size) analysis; CP, cycling-time perturbation.
[b] L. M. S. Chang, *J. Mol. Biol.* **93**, 219 (1975).
[c] Other abbreviations used here: AMV, avian myeloblastosis virus; Poly[d(AT)], the alternating copolymer of dA and dT; Pol., polymerase.
[d] W. R. McClure and T. M. Jovin, *J. Biol. Chem.* **250**, 4073 (1975).
[e] D. Uyemura, R. Bambara, and I. R. Lehman, *J. Biol. Chem.* **250**, 8577 (1975).
[f] L. A. Sherman and M. L. Gefter, *J. Mol. Biol.* **103**, 61 (1976).
[g] R. A. Bambara, D. Uyemura, and T. Choi, *J. Biol. Chem.* **253**, 413 (1978).
[h] Y. Chow and W. R. McClure, submitted for publication.
[i] D. K. Dube and L. A. Loeb, *Biochemistry* **15**, 3605 (1976).
[j] N. F. Neff and M. J. Chamberlin, *J. Biol. Chem.* **253**, 2455 (1978).

data can be independently obtained in each case, but the real interest in measuring processivity is to determine its relationship to the overall mechanism. Thus, a complete characterization must include measurements done as a function of solution composition, substrate concentration, temperature, template-primer composition, etc. The perturbation methods therefore require a characterization of the perturbing molecule interaction with enzyme over the same range of experimental variables.

Because of the disadvantages of the indirect methods, we favor direct product analysis as the method of choice for determining polymerase processivity. For a definitive determination of processivity, the single requirement is resolution of product lengths. The important limitations in this technique are that the reactions must be allowed to proceed for a time that is long in units of enzyme turnovers compared to $(1 - P)^{-1}$ and reinitiation on previously elongated primers must be prevented (Usually with high 3'-OH:enzyme ratios). Although we have presented data obtained only from synthesis on homopolymer templates, the same analysis can be performed on template-primer combinations containing gaps or nicks in natural DNA molecules of known sequence that have been generated by restriction enzyme digestion (work in progress).

The polymerization of DNA *in vivo* can occur at a variety of sites both in the presence and in the absence of large and small molecule effectors. It is anticipated that processivity measurements along with other mechanistic studies will be required on a variety of templates before a detailed model of function and regulation can be developed.

We have emphasized the importance of measuring processivity as a part of the kinetic characterization of a nucleic acid polymerase. The processivity of each polymerase has doubtless been optimized during evolution so that the resulting balance between dissociation and translocation indicates an important feature of mechanism and regulation. To an extent the processivity of a polymerase is also related to the high fidelity of nucleotide polymerization. In Table III we have separated the kinetic and mechanistic factors that have been identified as contributing to the fidelity of nucleic acid synthesis. The first column shows that the incorporation selectivity ranges between $10^3$ and $10^4$ for most polymerases studied. Considering the finding that app $K_n$'s are considerably larger for misincorporation than for correct incorporation,[13] the selectivity associated with $V/K$ is probably the relevant limiting value to use in comparing incorporation fidelity. Although many enzymes can discriminate between, for example, adenosine and guanosine with selectivities comparable to those shown in Table III, the unexplained fact is that a polymerase *in associa-*

---

[13] F. D. Gillin and N. G. Nossal, *Biochem. Biophys. Res. Commun.* **64**, 457 (1975).

TABLE III
ENZYMIC STEPS RESPONSIBLE FOR SYNTHETIC FIDELITY

| Enzyme | Incorporation selectivity[a] | | Editing selectivity | | Overall fidelity | References[b] |
|---|---|---|---|---|---|---|
| | $V/K$ | $V$ | $3' \rightarrow 5'$ Exonuclease | Nonprocessive synthesis | | |
| *Escherichia coli* DNA polymerase | ? | ? | Yes | Yes | $>10^5$ | c |
| T4 DNA polymerase | $10^3$ | 10 | Yes | ? | $\sim 10^4$ | c, d |
| Calf thymus DNA polymerase | $\sim 10^5$ | ? | No | Yes | $\sim 10^5$ | e |
| AMV reverse transcriptase | $9 \times 10^3$ | $3 \times 10^3$ | No | Yes | $\sim 10^4$ | f, g |
| *Escherichia coli* RNA polymerase | $>2 \times 10^3$ | ? | No | No | $\sim 2 \times 10^3$ | h |

[a] In each case, the value listed here for $V/K$ or $V$ is a ratio of the respective kinetic constant for the correct triphosphate vs. the incorrect triphosphate.
[b] The references cited pertain only to fidelity estimates; processivity estimates are compiled in Table II.
[c] C. F. Springgate and L. A. Loeb, *Proc. Natl. Acad. Sci. U.S.A.* **70**, 245 (1973).
[d] F. D. Gillin and N. G. Nossal, *Biochem. Biophys. Res. Commun.* **64**, 457 (1975).
[e] L. M. S. Chang, *J. Biol. Chem.* **248**, 6983 (1973).
[f] G. Seal and L. A. Loeb, *J. Biol. Chem.* **251**, 978 (1976).
[g] N. Battula, D. K. Dube, and L. A. Loeb, *J. Biol. Chem.* **250**, 8404 (1975).
[h] C. F. Springgate and L. A. Loeb, *J. Mol. Biol.* **97**, 577 (1975).

*tion with a template* can discriminate between all four bases and achieve high-fidelity incorporation for all four nucleotides. Too many interesting ideas have been proposed to explain this fact to allow a review here; suffice it to say that unambiguous experiments demonstrating any of the ideas are lacking.

The special feature found in many nucleic acid polymerases that allows overall fidelity to approach a final value of less than one error in $10^5$ to $10^6$ is second-step editing (i.e., $3'-5'$ hydrolysis of the misincorporated nucleotide). The hydrolytic activity was characterized in early work on purified *E. coli* DNA polymerase I[14] and recognized by Brutlag and Kornberg[15] as playing the crucial editing function. The basic idea was recently popularized with rather more formal notation under the new rubric of "kinetic proofreading" by Hopfield[16,17] who has applied it to describe the fidelity of aminoacylation of tRNA. Other workers[18,19] have reported measurements on the discrete enzymic steps in the aminoacylation that contribute to the second-step fidelity.

In terms of the steady-state mechanism shown here in Fig. 1, the first selectivity is exerted in those steps in which EPn reaches an incorporation central complex. For the second step in fidelity the focus is on EPn + 1. The editing pathway (not shown) would involve those steps in which EPn + 1 is reconverted to EPn via a hydrolytic central complex followed by release of NMP. The quantitative question relating these multiple pathways is: Given a misincorporated nucleotide on the 3'-OH end of a nascent polynucleotide, what is the probability that the EPn + 1 complex will (*a*) react hydrolytically to form EPn; (*b*) translocate to EPn and allow stable incorporation of the error; or (*c*) dissociate to form free enzyme and a 3'-terminated error on the nascent polynucleotide? Bessman and co-workers have quantitatively described the first two possibilities in the T4 DNA polymerase reaction[20,21] and have found phage mutants that produce altered enzyme, with the result that increased hydrolytic activity is correlated with an antimutator phenotype and decreased hydrolytic activity is associated with a mutator phenotype.

Some polymerases do not exhibit an editing function. In these cases, the fidelity of a processive enzyme is limited to incorporation selectivity

[14] I. R. Lehman and C. C. Richardson, *J. Biol. Chem.* **239**, 233 (1964).
[15] D. Brutlag and A. Kornberg, *J. Biol. Chem.* **247**, 241 (1972).
[16] J. J. Hopfield, *Proc. Natl. Acad. Sci. U.S.A.* **71**, 4135 (1974).
[17] T. Yamane and J. J. Hopfield, *Proc. Natl. Acad. Sci. U.S.A.* **74**, 2246 (1977).
[18] F. von der Haar and F. Cramer, *Biochemistry* **15**, 4131 (1976).
[19] A. R. Fersht and M. M. Kaethner, *Biochemistry* **15**, 3342 (1976).
[20] M. J. Bessman, N. Muzyczka, M. F. Goodman, and R. L. Schnar, *J. Mol. Biol.* **88**, 409 (1974).
[21] K.-Y. Lo and M. J. Bessman, *J. Biol. Chem.* **251**, 2475 (1976).

(e.g., RNA polymerase). A nonprocessive enzyme may participate in a high-fidelity DNA synthesis *pathway in vivo* if it dissociates with high probability after a misincorporation event and if other cellular enzymes are available to excise the faulty 3'-OH end. Misincorporation by a processive DNA polymerase ordinarily will be corrected with only a 0.50 probability of retaining the parental sequence.

The theory and methods of product analysis demonstrated in this chapter could obviously be extended to describe the degradative modes of nucleic acid exonucleases. Indeed, the only change required would be to consider a product distribution composed of the several species $R_{m-n}$ instead of $R_{m+n}$. Radioactive labeling at the nondegraded end of the primer being hydrolyzed would allow an analogous quantitation to be performed. An analysis of several DNA exonucleases was recently reported[22] along the lines proposed above, but only the general patterns of the strictly processive vs. distributive extremes were identified. Such work could easily be extended to yield precise estimates of the processivity.

In addition to providing mechanistic and regulatory information on nucleic acid polymerases, a determination of the processivity of a polymerase under a variety of conditions can yield important practical insights. DNA polymerases and reverse transcriptases are used to extend primers in some DNA sequencing methods and in some DNA cloning methods. The processivity in many cases could be chosen by the experimenter in order to optimize such parameters as the specific activity of input primers, length of finished product, etc. In a recent study,[23] it was found that high triphosphate concentrations increased the yield of full length reverse transcripts of the globin mRNA. Although we have not studied the enzyme employed in those experiments (AMV reverse transcriptase), it seems likely that it too must obey the processivity saturation equation demonstrated for *E. coli* DNA polymerase I in this chapter.

### Acknowledgments

We are grateful to Dr. Hans Lehrach for discussions on the Kuhn statistical distributions employed in this study.

NOTE ADDED IN PROOF: Several pertinent papers dealing with polymerase processivity and kinetics have appeared since this manuscript was submitted (March, 1978). The results and conclusions do not appear in Tables II and III; the reader is referred to the following papers.

S. K. Das and R. K. Fujimura, *J. Biol. Chem.* **252**, 8700 (1977).

---

[22] K. R. Thomas and B. M. Olivera, *J. Biol. Chem.* **253**, 424 (1978).
[23] A. Efstratiadis, T. Maniatis, F. C. Kafatos, A. Jeffrey, and J. Vournakis, *Cell* **4**, 367 (1975).

S. K. Das and R. K. Fujimura, *J. Biol. Chem.* **252**, 8708 (1977).
J. D. Engel and P. H. von Hippel, *J. Biol. Chem.* **253**, 935 (1978).
W. B. Helfman, S. S. Hindler, D. H. Shannahiff, and D. W. Smith, *Biochemistry* **17**, 1607 (1978).
S. S. Agarwal, D. K. Dube, and L. A. Loeb, *J. Biol. Chem.* **254**, 101 (1979).
D. J. Galas and E. W. Branscomb, *J. Mol. Biol.* **124**, 653 (1978).
S. K. Das and R. K. Fujimura, *J. Biol. Chem.* **254**, 1227 (1979).
F. Bernardi, M. Saghi, M. Dorizzi, and J. Ninio, *J. Mol. Biol.* **129**, 93 (1979).

## [12] Covalently Interconvertible Enzyme Cascade Systems

*By* P. B. Chock and E. R. Stadtman

Since the discovery of the interconversion between glycogen phosphorylase $b$ and its phosphorylated form, phosphorylase $a$,[1] it has become increasingly apparent that the covalent interconversion of enzymes and proteins plays an important role in cellular regulation.[2,3] By means of the covalent modification and demodification reactions, the enzyme involved is converted between its active and inactive forms. Interestingly, a majority of the known interconvertible enzymes or proteins occupy a key role in metabolism. To date (see Stadtman and Chock[2] and references therein) there are four types of covalent modifications that are known to occur in enzymic systems. They are ($a$) phosphorylation of specific serine residues; ($b$) nucleotidylation of particular tyrosine residues; ($c$) ADP ribosylation of an arginine residue or a yet unidentified site; and ($d$) methylation of a specific aspartate or glutamate residue. The reactions that lead to these covalent modifications are

$$\text{ATP} + \ldots \text{(Ser)} \ldots \longrightarrow \text{ADP} + \ldots \underset{|}{\text{(Ser)}} \ldots$$
$$\quad\quad\quad\quad \underset{|}{\text{OH}} \quad\quad\quad\quad\quad\quad \text{OPO}_3$$

$$\text{NTP} + \ldots \text{(Tyr)} \ldots \longrightarrow \text{PP}_i + \ldots \underset{|}{\text{(Tyr)}} \ldots$$
$$\quad\quad\quad\quad \underset{|}{\text{OH}} \quad\quad\quad\quad\quad\quad \text{NMPO}$$

$$\text{DPN} + \ldots \text{(Arg)} \ldots \longrightarrow \text{nicotinamide} + \ldots \text{(Arg)} \ldots$$
$$\quad\quad\quad\quad \|\text{NH} \quad\quad\quad\quad\quad\quad\quad\quad \|\text{ADPR-N}$$

$$+ \ldots (\quad) \ldots \longrightarrow \text{nicotinamide} + \ldots (\quad) \ldots$$
$$\quad\quad\quad\quad X \quad\quad\quad\quad\quad\quad\quad\quad\quad \text{ADPR-X}$$

$$S\text{-Adenosylmethionine} + \ldots \text{(Glu)} \ldots \longrightarrow S\text{-adenosylhomocysteine} + \ldots \underset{|}{\text{(Glu)}} \ldots$$
$$\quad\quad\quad\quad\quad\quad\quad\quad\quad\quad\quad\quad\quad\quad\quad\quad\quad\quad\quad\quad\quad\quad\quad\quad \text{COOCH}_3$$

---

[1] G. T. Cori and A. A. Green, *J. Biol. Chem.* **151**, 31 (1943).
[2] E. R. Stadtman and P. B. Chock, *Curr. Top. Cell. Regul.* **13**, 53 (1978).
[3] P. Greengard, *Science* **199**, 146 (1978).

In addition to the above reactions, other covalent modifications such as acylation and peptidylylation of proteins are also well established, but their physiological roles have yet to be demonstrated.

In each instance, covalent modification of the so-called "target" enzyme or protein is catalyzed by a "converter" enzyme; i.e., the covalent modification involves the action of one enzyme upon another. Therefore, interconvertible enzyme systems are in effect bidirectional or cyclic cascade systems. These cyclic cascade systems can also function as unidirectional cascades when one of the opposing cascades is incapacitated through the action of allosteric effectors.

We present here a steady-state analysis of cyclic cascade models[4,5] patterned after the phosphorylation–dephosphorylation of glycogen phosphorylase system,[6] mammalian pyruvate dehydrogenase,[7] tyrosine aminotransferase,[8] and the nucleotidylation–denucleotidylation of *Escherichia coli* glutamine synthetase system.[2,9,10] The results reveal that, compared to other regulatory enzymes, interconvertible enzymes can respond to a greater number of allosteric effectors and can integrate the multiple effects into a single output, namely, the specific activity of the target enzyme. This specific activity is determined by the steady-state distribution of active and inactive forms of the interconvertible enzyme, which is established through the dynamic coupling of the forward and the regeneration cascades. Since the steady-state distribution of the active and inactive interconvertible enzyme forms is a function of the kinetic constants of all the cascade enzymes, modulation of the specific activity of the target enzyme is achieved by alteration of the kinetic constants by allosteric or mass action effects. Further, more cyclic cascades exhibit great flexibility with respect to their control patterns, and they possess enormous amplification potential with respect to the concentration of primary stimuli required to obtain a 50% conversion of the target enzyme. In addition, they can generate a sigmoidal response of interconvertible enzymic activity to increasing concentration of an allosteric effector. Moreover, as shown elsewhere,[11] cyclic cascades can function as rate amplifi-

---

[4] E. R. Stadtman and P. B. Chock, *Proc. Natl. Acad. Sci. U.S.A.* **74**, 2761 (1977).
[5] P. B. Chock and E. R. Stadtman, *Proc. Natl. Acad. Sci. U.S.A.* **74**, 2766 (1977).
[6] E. G. Krebs, *Curr. Top. Cell. Regul.* **5**, 99 (1972).
[7] F. Huchs, D. D. Randall, T. E. Roche, M. W. Burgett, J. W. Pelley, and L. J. Reed, *Arch. Biochem. Biophys.* **151**, 328 (1972).
[8] K. L. Lee and J. M. Nickol, *J. Biol. Chem.* **249**, 6024 (1974).
[9] S. P. Adler, J. H. Mangum, G. Magni, and E. R. Stadtman, *in* "Third International Symposium on Metabolic Interconversion of Enzymes" (E. H. Fischer, E. G. Krebs and E. R. Stadtman, eds.), p. 221. Springer-Verlag, Berlin and New York, 1974.
[10] S. P. Adler, D. Purich, and E. R. Stadtman, *J. Biol. Chem.* **250**, 6264 (1975).
[11] E. R. Stadtman and P. B. Chock, *in* "Neuroscience Fourth Study Program" (F. O. Schmitt, ed.), p. 801. MIT Press, Cambridge, Massachusetts (1979).

ers to generate almost explosive increases in catalytic activity of the target enzymes in response to primary stimuli. Finally, we describe a procedure for obtaining the necessary kinetic parameters required to described the cyclic system.

## Model

### Monocyclic Cascade

Figure 1 shows a monocyclic cascade that is achieved by the coupling of two essentially irreversible opposing cascades. The forward cascade is initiated by the binding of an allosteric effector, $e_1$, to the inactive form $E_i$, of a converter enzyme, E (a protein kinase), thereby converting it to its active form, $E_a$. This activated converter enzyme catalyzes the covalent modification (phosphorylation) of the unmodified form, $o$-I, of the interconvertible enzyme, I, thereby converting it to the modified form, $m$-I. (Throughout this manuscript, the prefix, $o$- and $m$- are used to designate the unmodified and modified form of the interconvertible enzyme, respectively.) This cascade is opposed by the regeneration cascade, which is initiated by the binding of an allosteric effector, $e_2$, to the inactive form, $R_i$, of another converter enzyme, R (a phosphoprotein phosphatase), thus converting it to the active form, $R_a$, which catalyzes the conversion of $m$-I back to $o$-I. Dynamic coupling of the forward and regeneration cascades results in the cyclic interconversion of $o$-I and $m$-I and in the conversion of ATP to ADP and $P_i$. When ATP is not a limiting factor, continual cyclic interconversion of $o$-I − $m$-I will occur, and the fraction of interconvertible enzyme in the form of $m$-I will be a function of the steady-state distribution between $m$-I and $o$-I. This steady-state distribution is a complex function of several parameters, including the concentrations of the allosteric effectors, $e_1$ and $e_2$, the dissociation constants, $K_1$, $K_2$ and $K_f$, $K_r$ for the effector–converter enzyme complexes and the converter–interconvertible enzyme complexes, respectively; and the specific activities, $k_f$ and $k_r$, of the converter enzymes, $E_a$ and $R_a$.

*Definition of Terms.* The steady-state analysis of this cascade model (described later) discloses several unique features of interconvertible enzyme systems. To facilitate further discussion of these features, and to illustrate the meaning of various terms used throughout this discussion, a diagrammatic representation is presented in Fig. 2A to show how (depending on the metabolic state) the fractional modification, $[m\text{-I}]/[\text{I}]$, of an interconvertible enzyme, I, can vary as a function of increasing concentrations of the primary effector, $e_1$ (see Fig. 1). A comparison of various curves in Fig. 2A illustrates three important properties of cyclic cascade systems.

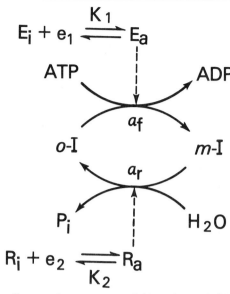

FIG. 1. A monocyclic cascade system. $\alpha_f = k_f/K_f$ and $\alpha_r = k_r/K_r$, where $k_f$ and $k_r$ are specific rate constants for the forward and reverse reactions designated and $K_f$ and $K_r$ are dissociation constants for $o$-I · $E_a$ and $m$-I · $R_a$, respectively; $K_1$ and $K_2$ are dissociation constants for $E_a$ and $R_a$, respectively; $o$-I and $m$-I are covalently unmodified and modified enzyme, respectively. This figure is patterned after Fig. 3 in E. R. Stadtman and P. B. Chock, *Curr. Top. Cell. Regul.* **13**, 53 (1978).

1. Curves b, d, and e show that the maximal amount of the interconvertible enzyme that can be modified with saturating levels of the primary effector can vary.
2. Curves a and c show that the concentration of primary effector that indirectly produces a given fractional modification, $[m\text{-}I]/[I]$, of the interconvertible enzyme can be much lower than the concentration required to produce the same fractional activation $[E_a]/[E]$, of the converter enzyme, E, with which it reacts directly.
3. Curves b and c show that the increment in effector concentration required to produce a given increment in $[m\text{-}I]/[I]$ can vary.

These three properties of cyclic cascades are quantitatively defined as follows: The maximum value of $[m\text{-}I]/[I]$ that can be achieved with saturating concentrations of a primary effector is referred to as the *amplitude* of the modification reaction and is designated $M$ or $([m\text{-}I]/[I])_{max}$. The time-independent relationship between the effector concentrations required to activate E and that required to produce modification of the interconvertible enzyme is referred to as the *signal amplification*, designated SA, and is defined as the ratio

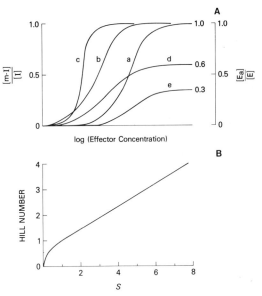

FIG. 2. (A) An illustration of how the fractional modification of an interconvertible enzyme (curves b, c, d, and e) and fractional activation of a converter enzyme (curve a) can vary as a function of a primary effector concentration. The numbers on curves a, d, and e indicate the amplitude. (B) The relationship between the sensitivity index $S$ and the Hill number. This figure is patterned after Fig. 4 in E. R. Stadtman and P. B. Chock, *Curr. Top. Cell. Regul.* **13,** 53 (1978).

$$\frac{\text{Concentration of effector required for}}{\text{Concentration of effector at which } [m\text{-}I]/[I] = 0.5} = \frac{[e]_{0.5E}}{[e]_{0.5I}}$$

For simplicity, in this report it is assumed that the binding is constant for the effector $e_1$ to the converter enzyme E is 1.0, and that the binding exhibits no cooperativity. Therefore, the signal amplification factor, $SA = 1/[e]_{0.5I}$. In this discussion the effector $e_1$, which triggers the covalent modification of the interconvertible enzyme by activating the converter enzyme E (see Fig. 1) is regarded as the primary signal. Amplification of this signal derives from the fact that only a small amount of activated converter enzyme is required to convert a much greater amount of the interconvertible enzyme to its modified form. Of course, the maximum amount of covalent modification is determined by the steady-state distribution between the modified and unmodified form of the interconvertible enzyme, which derived from the cyclic coupling of the modification (forward) and demodification (regeneration) cascades.

It is noteworthy that the signal amplification potential would be infinite if the regeneration cascade is completely incapacitated. This is due to the fact that under the stated condition any amount of activated converter enzyme, no matter how small, can eventually catalyze the complete conversion of the interconvertible enzyme. In addition, it must be pointed out that the signal amplification potential described here is distinctly different from the *catalytic amplification* potential, which is a function of the relative concentrations and the maximal catalytic efficiencies of the converter enzyme and the interconvertible enzyme; i.e., catalytic amplification is equal to the ratio

[I] × maximal catalytic efficiency of [I]/[E]
× maximal catalytic efficiency

of E. In this report, we are concerned more with the problem of signal amplification than with catalytic amplification, even though both types of amplification are properties of interconvertible enzyme cascades.

The relationship between the change in effector concentration $[e_1]$, and the change in fractional modification is referred to as the *sensitivity* and is designated $S$. It is defined as the ratio

$$\frac{8.89 \times \text{(concentration of effector required to attain 50\% maximal amplitude)}}{\text{(concentration of effector required for 90\% maximal amplitude)} - \text{(concentration of effector required for 10\% maximal amplitude)}}$$

$$= \frac{8.89[e_{0.5M}]}{[e_{0.9M}] - [e_{0.1M}]} = S$$

The constant 8.89 in this expression is introduced to yield a reference value of 1.0 for a pure hyperbolic binding isotherm. According to this definition, sensitivity values greater than 1.0 are obtained when $[m\text{-}I]/[I]$ is a sigmoidal function of the effector concentration and values less than 1.0 are obtained for negative cooperativity. The sensitivity index $S$ is therefore similar to the so-called Hill number $n_H$, which is a measure of the steepness of the saturation response to effector concentrations. In fact, the Hill number obtained from the plot of $\log[[m\text{-}I]/([I] - [m\text{-}I])]$ versus $\log[e]$ can also be used as a measure of the sensitivity of cascade systems to effector concentrations. However, the Hill numbers so obtained do not have the same meaning as those assumed in the derivation of the Hill equation. The sigmoidal response observed in the cascade system does not reflect "positive cooperativity" in the binding of effector to multiple binding sites on a given enzyme in the cascade but is due to the fact that the effector influences more than one step in the cascade. Similarly, sensitivity values less than 1.0 are not due to "negative cooperativity" in the effector binding process, but are a consequence of antagonistic effects of an effector acting at more than one step in the cascade. Therefore, it is inappropriate to use the Hill number as a measure of the sensitivity of a cascade system in response to increasing effector concentration. Instead, a method containing no mechanistic implication, such as the sensitivity index, should be used. Figure 2B depicts the relationship between the sensitivity index and the Hill number. When the sensitivity index $S$ is greater than 0.6, it is proportional to the Hill number, whereas the $S$ value decreases much more rapidly with respect to the Hill number when the Hill number is smaller than 0.5.

*Derivation of the Steady-State Equation for the Monocyclic Systems.* For simplicity, in deriving the equation describing the steady-state distribution between $m$-I and $o$-I, it is assumed that ($a$) there is a rapid equilibrium in the formation of the enzyme–enzyme and enzyme–effector complexes; ($b$) the concentration of the enzyme–enzyme complexes are negligible compared to the concentrations of the modified and unmodified or active and inactive enzymes, such that $[E] \simeq [E_i] + [E_a]$, $[I] \simeq [o\text{-}I] + [m\text{-}I]$, $[R] \simeq [R_i] + [R_a]$, where $[E]$, $[I]$, and $[R]$ are total concentrations of E, I, and R, respectively; ($c$) the concentrations of the allosteric effectors, $e_1$ and $e_2$, are maintained at constant levels for any given metabolic state. With these assumptions, the reactions for the interconversion of $o$-I and $m$-I can be written as follows:

$$o\text{-}I + E_a \xrightleftharpoons{K_f} o\text{-}I \cdot E_a \xrightarrow{k_f} E_a + m\text{-}I$$

and

$$m\text{-}I + R_a \xrightleftharpoons{K_r} m\text{-}I \cdot R_a \xrightarrow{k_r} R_a + o\text{-}I$$

It follows that, for any given metabolic condition, a steady state will be established in which the rate of $m$-I formation is equal to the rate of $o$-I regeneration; i.e.,

$$\alpha_f[E_a][o\text{-}I] = \alpha_r[R_a][m\text{-}I]$$

with $\alpha_f = k_f/K_f$ and $\alpha_r = k_r/K_r$, and the concentrations of $E_a$ and $R_a$ are governed by the dissociation constants $K_1$ and $K_2$ and the concentrations of the allosteric effectors $e_1$ and $e_2$, i.e.,

$$[E_a] = \frac{[E][e_1]}{K_1 + [e_1]} \quad \text{and} \quad [R_a] = \frac{[R][e_2]}{K_2 + [e_2]}$$

When $[o\text{-}I]$ is expressed in terms of $[I]$ and $[m\text{-}I]$, the fraction of interconvertible enzyme in the modified form at steady state is given by

$$\frac{[m\text{-}I]}{[I]} = \left[\frac{\alpha_r[R][e_2](K_1 + [e_1])}{\alpha_f[E][e_1](K_2 + [e_2])} + 1\right]^{-1} \quad (1)$$

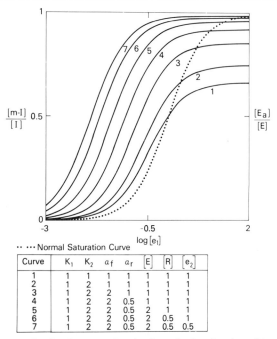

FIG. 3. Computer-simulated curves showing how the fractional modification of an interconvertible enzyme in a monocyclic cascade varies when each cascade parameter, except $K_1$, is altered by a factor of 2 in a successively cumulative manner so as to favor $m$-I formation. This figure is patterned after Fig. 5 in E. R. Stadtman and P. B. Chock, *Curr. Top. Cell. Regul.* **13**, 53 (1978).

Equation (1) was obtained when the role of ATP in the monocyclic cascade was ignored. This is justifiable because for any given metabolic state the intercellular concentration of ATP is maintained at a nearly constant level, which is several orders of magnitude greater than the concentration of the interconvertible enzyme.

The extraordinary capacity of an interconvertible enzyme to respond to allosteric regulation is accomplished by the fact that any one of the 10 parameters ($\alpha_f$ and $\alpha_r$ were ratios of two constants each) in Eq. (1) is susceptible to modulation either directly or indirectly by allosteric interaction with one or more allosteric effectors. Figure 3 illustrates some of the effects on the fractional modification of the interconvertible enzyme as a function of increasing $e_1$ concentration predicted by Eq. (1) when its parameters are changed owing to allosteric interactions. For simplicity, it is assumed for all curves shown in the figure that the dissociation constant $K_1$ for the conversion $E_a \rightleftharpoons e_1 + E_1$ is equal to 1.0, and that the formation of $E_a$ is a normal hyperbolic function of the $e_1$ concentration as shown by the dotted line (i.e., the $[E_a]/[E]$ ratio). Curves 1 through 7 in Fig. 3 show the results obtained when $e_1$ activation of the converter enzyme is linked to the cascade via the catalytic role of $E_a$ in the conversion of $o$-I to $m$-I. This indirect effect of $e_1$ concentration on the fractional modification of the interconvertible enzyme can vary enormously, depending on the magnitude and the number of parameters in the cascade that are varied. For example, curve 1 shows that when all parameters in Eq. (1) are assigned values of 1.0, the concentration of $e_1$ required to obtain 50% modification of the interconvertible enzyme (the $[e_1]_{0.5I}$ value) is equal to the value of $K_1$ (i.e., 1.0); moreover, with saturating levels of $e_1$ only 67% of the interconvertible enzyme can be modified. In other words, under these conditions there is no signal amplification ($1/[e_1]_{0.5I} = 1.0$), and the amplitude $M$ of the modification is only 0.67. However, when each of six other parameters in Eq. (1) is varied by a factor of only 2, in a successively cumulative manner (curves 2 to 7 in Fig. 3) that favors modification of the interconvertible enzyme, there is a progressive increase in both the amplification factor (the $[e_1]_{0.5I}$ value decreases) and the amplitude $M$. A comparison of curves 1 and 7 shows that with only 2-fold changes in all six parameters there is an 80-fold increase in the amplification factor and the amplitude increases from 0.67 to 1.0. In addition, as shown elsewhere[4] that a 2-fold change for the same six parameters in the opposite direction from those depicted in Fig. 3, the values of $M$ decrease progressively from 0.67 to 0.07 for the case when all six parameters are varied simultaneously by a factor of 2 in favor of the unmodified form of the interconvertible enzyme. In essence, variations of these parameters can have two effects on an interconvertible enzyme: (a) they may alter the amplification factor, i.e., the concentration of $e_1$ required to produce 50% conversion of the interconvertible enzyme; (b) they may alter the amplitude of modification, i.e., the fractional activation obtained with saturating levels of $e_1$. The enormous variability in terms of the amplification factor and/or the amplitude predicted by Eq. (1) when its parameters were varied by a factor of 2 is significant because 2-fold changes in the various parameters are well within the range of allosteric effects. Therefore the results described here illustrate the enormous control potential of the cyclic cascade system as compared to other types of enzyme regulation.

The monocyclic cascade system described by Eq. (1) is one of many patterns that are possible, depending upon the nature of the interactions between the allosteric effectors $e_1$ and $e_2$ and the converter enzymes E and R. With the restriction that E must be activated by $e_1$, there are four unique patterns that can occur in the covalent modification systems. Figure 4 depicts these four patterns, and their steady-state expressions are given in the table. The model depicted as case I is the same as that described in Fig. 1. In this case, it is assumed that $e_1$ activates the E-

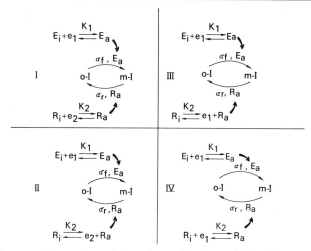

FIG. 4. Four regulatory patterns that can be used in the regulation of monocyclic cascade system. Notations are as described in Fig. 1, except that $K_2$ in II and III is a dissociation constant for $R_i$. This figure is patterned after Fig. 6 in E. R. Stadtman and P. B. Chock, *Curr. Top. Cell. Regul.* **13**, 53 (1978).

converter enzyme and that $e_2$ activates the R-converter enzyme. In contrast to case I, it is assumed in case II that $e_2$ inactivates rather than activates the R enzyme; in case III, it is assumed that one and the same effector, $e_1$, activates the E enzyme but inactivates the R enzyme; and in case IV, it is assumed that $e_1$ activates both the converter enzymes. Experimentally, it has been shown[4,12,13] by Reed and his co-workers that three of the four regulatory patterns described in Fig. 4 are utilized in regulation of the mammalian pyruvate dehydrogenase cascade. Figure 5 illustrates the wide variation in the regulatory patterns elicited by these different types of monocyclic cascades. In this figure the fractional modification of the interconvertible enzyme is expressed as a function of increasing concentration of $e_1$ in each of the four cases, when all parameters but one are held constant. Figure 5A shows that each type of cascade yields a different response pattern to variation in the value of $K_1$, and Fig. 5B shows that a uniquely different pattern of response is obtained for each type of cascade control when the value of $\alpha_f$ is varied. The patterns differ with respect to amplitude, amplification, and sensitivity of the response to

[12] L. J. Reed, F. H. Pettit, T. E. Roche, and P. J. Butterworth, *in* "Protein Phosphorylation in Control Mechanisms" (F. Huijing and E. Y. C. Lee, eds.), p. 83. Academic Press, New York, 1973.
[13] F. H. Pettit, J. W. Pelley, and L. J. Reed, *Biochem. Biophys. Res. Commun.* **65**, 575 (1975).

STEADY-STATE EXPRESSIONS FOR FOUR UNIQUE REGULATORY PATTERNS
OBTAINED FROM THE MONOCYCLIC CASCADE SYSTEM[a]

| Case | $E_i \to E_a$ | | $R_i \to R_a$ | | Steady-state expression for $[m - I]/[I]$ |
|---|---|---|---|---|---|
| | $e_1$ | $e_1$ | $e_1$ | $e_2$ | |
| I | + | 0 | 0 | + | $\left[\dfrac{K_1 + [e_1]}{K_2 + [e_2]} \dfrac{\alpha_r[R][e_2]}{\alpha_f[E][e_1]} + 1\right]^{-1}$ |
| II | + | 0 | 0 | − | $\left[\dfrac{K_1 + [e_1]}{K_2 + [e_2]} \dfrac{\alpha_r[R]K_2}{\alpha_f[E][e_1]} + 1\right]^{-1}$ |
| III | + | 0 | − | 0 | $\left[\dfrac{K_1 + [e_1]}{K_2 + [e_1]} \dfrac{\alpha_r[R]K_2}{\alpha_f[E][e_1]} + 1\right]^{-1}$ |
| IV | + | 0 | + | 0 | $\left[\dfrac{K_1 + [e_1]}{K_2 + [e_1]} \dfrac{\alpha_r[R]}{\alpha_f[E]} + 1\right]^{-1}$ |

[a] +, Activate; −, inactivate; 0, no effect.

the change in $e_1$ concentration. It would be emphasized that the various patterns of response demonstrated by these few examples do not begin to demonstrate the enormous flexibility of the monocyclic cascade systems to allosteric control. With the consideration that more than one positive or negative allosteric effector may react with each enzyme in the cascade, and the fact that these interactions can affect more than one or all of the various parameters in the steady-state equation, and also the fact that in the presence of multiple effectors a given interconvertible enzyme may utilize all four types of regulation, it is apparent that an almost unlimited number of variations in the overall regulatory pattern can be obtained.

In deriving the steady-state equations shown in the table, it was assumed that the concentrations of the $o\text{-}I\cdot E_a$ and $m\text{-}I\cdot R_a$ complexes are negligibly small compared to the concentrations of any other enzyme species in the cascade. With this assumption, the steady-state expression for the fraction of $m\text{-}I$ is independent of the total concentration of I. To determine how this assumption could affect the validity of the conclusion based on the simplified equations in the table, a more general form of the steady-state equation was derived in which no assumption was made for all enzyme concentrations in the cascade. By so doing, a quartic equation with respect to $[m\text{-}I]$ is obtained, and it contains more than 200 terms. The value of $[m\text{-}I]$ was calculated by Newton's method of successive approximation.

Computer-simulated curves obtained with the quartic equation are very similar to those obtained from Eq. (1). In fact, as shown in Fig. 6C, when the conditions assumed in the derivation of Eq. (1) are approximately fulfilled, the saturation curve obtained with the quartic equation is essentially identical to that obtained with Eq. (1). Therefore, it is obvious

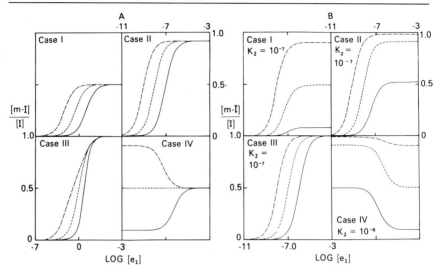

FIG. 5. The effect of various allosteric control patterns, as described in Fig. 4, on the response of the fractional modification of the interconvertible enzyme to variations in the concentration of the primary effector, $e_1$, and the parameters $K_1$ and $\alpha_f$. The curves in cases I, II, III, and IV are computer-simulated curves derived from the corresponding equations in the table, with the assumptions that $[E] = [R]$ and otherwise as follows: (A) $\alpha_f = \alpha_r$ and $K_1 = 10^{-6}$ (solid line), $10^{-7}$ (dotted line), $10^{-8}$ (dashed line); (B) $K_1 = 10^{-7}$, $\alpha_r = 10^5$ and $\alpha_f = 10^4$ (solid line), $10^5$ (dotted line), or $10^6$ (dashed line). This figure is patterned after Fig. 8 in E. R. Stadtman and P. B. Chock, *Curr. Top. Cell. Regul.* **13**, 53 (1978).

that the assumptions made to simplify the derivation of equations in the table do not invalidate their usefulness in analyzing the fundamental characteristics of the monocyclic cascade systems. However, an analysis of the quartic equation demonstrates that the monocyclic cascade systems are in fact even more flexible and can achieve even greater amplification than is predicted by the simplified equations. This is illustrated by the curves in Fig. 6A, showing that, for the conditions stipulated in the legend, the signal amplification factor increases by a factor of 20 when the concentration of I increases from 10 to 1000. This increase in amplification factor is due to the addition of [I] as a variable in the quartic equation. In addition, the greater flexibility with the quartic equation can also be obtained because identical values of $\alpha_f$ and $\alpha_r$ can produce different effects when the absolute values of $k_f$ and $K_f$ or $k_r$ and $K_r$ are varied. In the simplified equations, $k_f$, $K_f$, $k_r$, and $K_r$ appear only as $k_f/K_f(\alpha_f)$ and $k_r/K_r(\alpha_r)$ ratios. Therefore the fractional modification of I calculated with these equations is independent of the absolute values of $k_f$ and $K_f$ or $k_r$ and $K_r$ so long as the $k_f/K_f$ or $k_r/K_r$ ratios are the same. As shown in Fig. 6B, $k_f$, $K_f$, $k_r$, and $K_r$ do not occur in $k_f/K_f$ and $k_r/K_r$ ratios in the

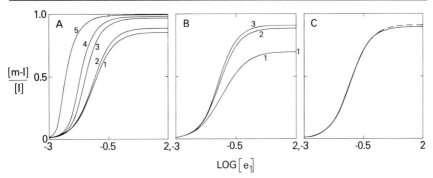

FIG. 6. Computer-simulated curves obtained with the quartic equation. (A) I is varied as follows: [I] = 1.0 (curve 1), 10 (curve 2), 100 (curve 3), 200 (curve 4), and 1000 (curve 5). (B) $K_f$ and $k_f$ are varied such that $\alpha_f$ is constant. Curve 1 is obtained with $K_f = 0.5$, $k_f = 1$; curve 2 with $K_f = 2.5$, $k_f = 5$; and curve 3 with $K_f = 25$, $k_f = 50$. (C) [m-I]/[I] obtained with the quartic equation (solid line) and the simplified equation (dashed line). This figure is patterned after Fig. 9 in E. R. Stadtman and P. B. Chock, *Curr. Top. Cell. Regul.* **13**, 53 (1978).

quartic equation. Hence, both the amplitude and amplification factor vary when $k_f$ and $K_f$ are varied simultaneously such that the $k_f/K_f$ ratio is held constant.

In all the analysis so far it is assumed that the cascade systems are cyclic processes devoid of any net chemical fluxes. However, it is evident from Fig. 1 that each complete cycle in a phosphorylation–dephosphorylation cascade is associated with the net decomposition of ATP to ADP and $P_i$. In deriving the steady-state equation, the role of nucleoside triphosphates, such as ATP and UTP, which serve as substrates in the phosphorylation and nucleotidylylation reactions, has been disregarded because the concentrations of these nucleotides are metabolically maintained at fairly constant levels several order of magnitude greater than the concentrations of the enzymes undergoing covalent modification. Nevertheless, the decomposition of nucleoside triphosphate is required to maintain a particular steady-state level of modified interconvertible enzyme. This nucleotide flux, $J_{NTP}$ (concentration per time unit), is dependent on the concentrations of all enzymes and effectors used in the cyclic cascade and the constants $K_1$, $K_2$, $\alpha_f$, and $\alpha_r$

$$J_{NTP} = \frac{\alpha_f \alpha_r [e_1][e_2][E][R][I]}{\alpha_r[R][e_2](K_1 + [e_1]) + \alpha_f[E][e_1](K_2 + [e_2])} \quad (2)$$

From Eqs. (1) and (2) it is evident that, for a given steady-state level of modified enzyme, the rate of NTP consumption can be varied through the interaction between allosteric effector(s) and the enzyme(s) involved. This NTP flux is by no means wasteful, nor is it a futile process. For example, in the absence of continual ATP consumption, the coupled phosphorylation–dephosphorylation reactions would approach a thermodynamic equilibrium in which the enzymes should be almost completely in the unmodified form. The decomposition of ATP is therefore an essential feature of the cascade mechanism because it provides the free energy needed to maintain the modified and unmodified forms of the various enzymes at metabolite-specified steady-state levels that differ from true thermodynamic equilibrium values. Therefore the consumption of ATP is the price that must be paid to support the elegant cascade-type cellular regulation.

*Bicyclic Cascade Systems*

When the modified form of an interconvertible enzyme in one cycle serves as a converter enzyme to catalyze the covalent modification of an interconvertible enzyme in another cycle, the two cycles become coupled such that the fractional modification of the second interconvertible enzyme is a function of all the parameters in both cycles. Relative to the monocyclic cascade, the bicyclic cascade endows with greater flexibility toward allosteric control; it can achieve a greater amplification of a response to primary stimuli; and under appropriate conditions the interconversion of the interconvertible enzyme in the second cycle can respond with a much higher sensitivity with respect to the change in allosteric effector concentration. To date, two types of bicyclic cascades are known to be involved in the regulation of key enzymes in metabolism. One of these, referred to as opened bicyclic cascade, is exemplified by the cascade involved in the activation of glycogen phosphorylase[6,14]; the other, exemlified by the cascade which regulates the *E. coli* glutamine synthetase[9,10,15], is called closed bicyclic cascade.

*Opened Bicyclic Cascade.* Figure 7A depicts an opened bicyclic cascade. The first cycle in this model is identical to the monocyclic system shown in Fig. 1, except that the role of ATP is not shown here. In the second cycle, the interconvertible enzyme is modified through the catalytic action of the modified form of the interconvertible enzyme in the first cycle ($m$-$I_1$). The allosteric control potential of this cascade can be demonstrated by assuming that the binding of the allosteric effectors, $e_1$, $e_2$, and $e_3$ to the inactive forms of the converter enzymes $E_i$, $R_{1i}$, and $R_{2i}$ is necessary to generate the active forms $E_a$, $R_{1a}$, and $R_{2a}$, respectively.

[14] E. H. Fischer, L. M. G. Heilmeyer, Jr., and R. H. Haschke, *Curr. Top. Cell. Regul.* **4**, 211 (1971).
[15] E. R. Stadtman and A. Ginsburg *in* "The Enzymes (P. Boyer, ed.), 3rd ed., Vol. 10, p. 755. Academic Press, New York 1974.

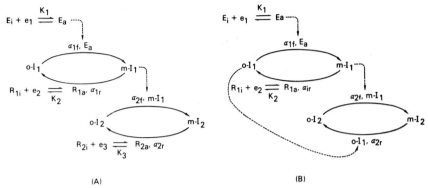

FIG. 7. Schematic representation of (A), an opened bicyclic cascade system, and (B), a closed bicyclic cascade system. The variables $\alpha_{nf}$ and $\alpha_{nr}$ are defined as: $\alpha_{nf} = k_{nf}/K_{nf}$ and $\alpha_{nr} = k_{nr}/K_{nr}$, where $n = 1, 2$; $k_{nf}$ and $k_{nr}$ are specific rate constants for the forward and reverse reaction as designated. $K_1$, $K_2$, $K_3$, $K_{nf}$, and $K_{nr}$ are dissociation constants for the various equilibria as indicated. This figure is patterned after Fig. 11 in E. R. Stadtman and P. B. Chock, *Curr. Top. Cell. Regul.* **13**, 53 (1978).

These active converter enzymes catalyze the following modification reactions.

$$o\text{-}I_1 + E_a \xrightleftharpoons{K_{1f}} o\text{-}I_1 \cdot E_a \xrightarrow{k_{1f}} E_a + m\text{-}I_1$$

$$m\text{-}I_1 + R_{1a} \xrightleftharpoons{K_{1r}} m\text{-}I_1 \cdot R_{1a} \xrightarrow{k_{1r}} R_{1a} + o\text{-}I_1$$

$$o\text{-}I_2 + m\text{-}I_1 \xrightleftharpoons{K_{2f}} o\text{-}I_2 \cdot m\text{-}I_1 \xrightarrow{k_{2f}} m\text{-}I_1 + m\text{-}I_2$$

$$m\text{-}I_2 + R_{2a} \xrightleftharpoons{K_{2r}} m\text{-}I_2 \cdot R_{2a} \xrightarrow{k_{2r}} R_{2a} + o\text{-}I_2$$

in which $K_{1f}$, $K_{2f}$, $K_{1r}$, and $K_{2r}$ are dissociation constants for the enzyme–enzyme complexes and $k_{1f}$, $k_{2f}$, $k_{1r}$, and $k_{2r}$ are the specific rate constants for the reaction designated. With the same assumptions used in deriving Eq. (1), the fractional modification of the interconvertible enzyme in the second cycle is expressed[5] by Eq. (3).

$$\frac{[m\text{-}I_2]}{[I_2]} = \left[ \frac{\alpha_{1r}\alpha_{2r}(K_1 + [e_1])[R_1][R_2][e_2][e_3]}{\alpha_{1f}\alpha_{2f}(K_2 + [e_2])(K_3 + [e_3])[E][I_1][e_1]} + \frac{\alpha_{2r}[R_2][e_3]}{\alpha_{2f}(K_3 + [e_3])[I_1]} + 1 \right]^{-1} \quad (3)$$

where $\alpha_{nf} = k_{nf}/K_{nf}$ and $\alpha_{nr} = k_{nr}/K_{nr}$, with $n =$ the cycle number.

This equation shows that the $[m\text{-}I_2]/[I_2]$ ratio is a function of 18 variables, namely, the concentrations of the enzymes, E, $I_1$, $R_1$, and $R_2$; the concentration of the allosteric effectors, $e_1$, $e_2$, and $e_3$; the specific rate constants, $k_{1f}$, $k_{2f}$, $k_{1r}$, and $k_{2r}$; the dissociation constants, $K_{1f}$, $K_{2f}$, $K_{1r}$,

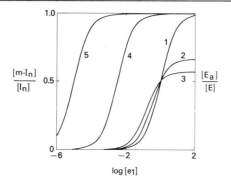

| Curve | $K_1$ | n | $K_{(n+1)}$ $E, I_1$ | $e_{(n+1)}$ $R_n$ | $\alpha_{nf}$ | $\alpha_{nr}$ |
|---|---|---|---|---|---|---|
| 1 | 1 | 0 | | | | |
| 2 | 1 | 1 | 1 | 1 | 1 | 1 |
| 3 | 1 | 2 | 1 | 1 | 1 | 1 |
| 4 | 1 | 1 | 2 | 0.5 | 4 | 0.25 |
| 5 | 1 | 2 | 2 | 0.5 | 4 | 0.25 |

FIG. 8. A comparison of the dependence of the fractional modification of an interconvertible enzyme in a monocyclic and an opened bicyclic cascade system on variations in the concentration of the primary effector, $e_1$, when all the parameters in the cascades (except $K_1 = 1.0$) are varied by a factor of 2 in favor of $m$-I formation. As a point of reference, curve 1 is fractional saturation of $E_a$; curves 2 and 4, and 3 and 5 are obtained with Eq. (1) and Eq. (3), respectively, for the conditions given in the table above. This figure is patterned after Fig. 12 in E. R. Stadtman and P. B. Chock, *Curr. Top. Cell. Regul.* **13**, 53 (1978).

$K_{2r}$, $K_1$, $K_2$, and $K_3$. To illustrate the amplification potential of the bicyclic cascade, the computer-simulated curves generated with Eq. (3) are compared with those generated for a monocyclic system and for a normal binding isotherm (Fig. 8). Curve 1 represents the fractional activation of the first converter enzyme, E, with $K_1 = 1.0$. Curves 2 and 3 depict the fractional modification as a function $e_1$ concentration for the interconvertible enzymes in the first and second cycle when all the parameters in Eq. (3) are set at 1.0 [for the one-cycle system, the parameters $[R_2]$, $[e_3]$, $\alpha_{2r}$, $\alpha_{2f}$, $[I_1]$, and $K_3$ are deleted in the first two terms of Eq. (3)]. Under these conditions, there is no amplification with respect to the fractional modification of the interconvertible enzymes in either cycle; i.e., $[e_1]_{0.5I_n} = 1.0$ for $I_1$ and $I_2$ systems. The amplitude of the fractional modification obtained in the first cycle is 0.67, which is greater than that obtained in the second cycle (0.57). However, the enormous amplification potention of the bicyclic cascade is apparent when $K_1$ is kept at 1.0 and each of the other 8 parameters in the monocycle system and the 16 parameters in the two-cycle system are varied by a factor of 2 so as to favor $m$-$I_1$ and $m$-$I_2$ formation. Under these conditions, curves 4 and 5 were obtained for the fractional modification of $I_1$ and $I_2$, respectively. The value of $[e_1]_{0.5I_n}$ for

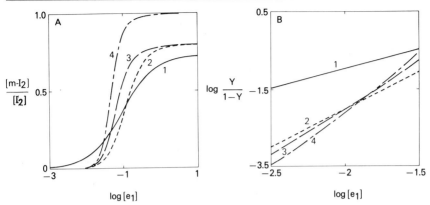

FIG. 9. The dependence of the fractional modification of the target enzyme in an opened bicyclic cascade on the concentration of an allosteric effector affecting up to four different steps in the cascade. (A) Curves 1, 2, 3, and 4 were obtained with Eqs. (4), (5), (6), and (7), respectively. They represent the following conditions: curve 1, $e_1$ activates $E_1$ only while $e_2$ and $e_3$ inactivate $R_{1a}$ and $R_{2a}$, respectively; curve 2, $e_1$ activates both $E_1$ and $m$-I (assuming only the $m$-I$_1$ · $e_1$ complex can catalyze the conversion of $o$-I$_2$ to $m$-I$_2$), and $e_2$ and $e_3$ inactivate $R_{1a}$ and $R_{2a}$, respectively; curve 3, $e_1$ activates both $E_1$ and $m$-I$_1$ and inactivates $R_{1a}$ and $e_3$ inactivates $R_{2a}$; curve 4, $e_1$ activates both $E_1$ and $m$-I$_1$ and inactivates both $R_{1a}$ and $R_{2a}$. (B) Hill plots for the corresponding curves in (A). This figure is patterned after Fig. 13 in E. R. Stadtman and P. B. Chock, *Curr. Top. Cell. Regul.* **13**, 53 (1978).

curves 1, 4, and 5 are 1, $3.12 \times 10^{-3}$, and $9.76 \times 10^{-6}$, respectively. This corresponds to a 320-fold signal amplification in response to $e_1$ stimulation for the monocycle system and a 102,400-fold signal amplification for the two-cycle system. These enormous amplifications are accomplished with only a 2-fold variation in each of the cascade parameters, which is well within the range of allosteric effects. Moreover, the fact that any one, more than one, or all parameters can be varied independently, in response to changes in the concentrations of various effectors, illustrates the enormous flexibility of the bicyclic cascade system in metabolic regulation.

The variations in the regulatory patterns derived from varying the role of effectors in the activation or deactivation of the converter enzymes are more numerous for the bicyclic cascade than those obtained for the monocyclic system described in Fig. 4 because the number of converter enzymes is greater in a bicyclic cascade than in a monocyclic system. It is particularly noteworthy that, when one given effector is assigned to regulate all four converter enzymes in a bicyclic cascade, the cascade system is capable of generating a highly "cooperative type" of response to the change in effector concentration. This property is clearly demonstrated by the curves in Fig. 9A, which depict how the fractional modification of $I_2$, as a function of $e_1$ concentration changes when the role of $e_1$ is enhanced by making it an activator of both forward converter enzymes, E and $m$-I, and inactivator of the regeneration converter enzymes, $R_{1a}$ and $R_{2a}$. The activation of $m$-I, and inactivation of $R_{1a}$ and $R_{2a}$ are governed by the equilibria

$$m\text{-}I_1 + e_1 \underset{}{\overset{K_4}{\rightleftharpoons}} m\text{-}I_1 \cdot e_1$$

$$R_{1a} + e_1 \underset{}{\overset{K_2}{\rightleftharpoons}} R_{1i}$$

$$R_{2a} + e_1 \underset{}{\overset{K_3}{\rightleftharpoons}} R_{2i}$$

Assuming that only the $m$-$I_1 \cdot e_1$ complex is capable of converting $o$-$I_2$ to $m$-$I_2$ and that $[I_1] \simeq [o\text{-}I_1] + [m\text{-}I_1] + [m\text{-}I_1 \cdot e_1]$, the steady-state equations were obtained for each of the following conditions: (i) $e_1$ activates E only, while $e_2$ and $e_3$ inactivate $R_{1a}$ and $R_{2a}$, respectively; (ii) $e_1$ activates both E and $m$-I, (iii) $e_1$ activates both E and $m$-$I_1$ and inactivates $R_{1a}$; and (iv) $e_1$ activates E and $m$-$I_1$ and inactivates both $R_{1a}$ and $R_{2a}$. Equations for each of these four cases, listed in the above order, are as follows:

$$\frac{[m\text{-}I_2]}{[I_2]} = \left[ \frac{\alpha_{1r}\alpha_{2r}(K_1 + [e_1])[R_1][R_2]K_2K_3}{\alpha_{1f}\alpha_{2f}(K_2 + [e_2])[E][I_1][e_1](K_3 + [e_3])} + \frac{\alpha_{2r}[R_2]K_3}{\alpha_{2f}[I_1](K_3 + [e_3])} + 1 \right]^{-1} \quad (4)$$

$$\frac{[m\text{-}I_2]}{[I_2]} = \left[ \frac{\alpha_{1r}\alpha_{2r}(K_1 + [e_1])[R_1][R_2]K_2K_3K_4}{\alpha_{1f}\alpha_{2f}(K_2 + [e_2])(K_3 + [e_3])[E][I_1][e_1]^2} + \frac{\alpha_{2r}(K_4 + [e_1])[R_2]K_3}{\alpha_{2f}[I_1][e_1](K_3 + [e_3])} + 1 \right]^{-1} \quad (5)$$

$$\frac{[m\text{-}I_2]}{[I_2]} = \left[ \frac{\alpha_{1r}\alpha_{2r}[R_1][R_2](K_1 + [e_1])K_2K_3K_4}{\alpha_{1f}\alpha_{2f}(K_2 + [e_1])(K_3 + [e_3])[E][I_1][e_1]^2} + \frac{\alpha_{2r}[R_2](K_4 + [e_1])K_3}{\alpha_{2f}[I_1][e_1](K_3 + [e_3])} + 1 \right]^{-1} \quad (6)$$

$$\frac{[m\text{-}I_2]}{[I_2]} = \left[ \frac{\alpha_{1r}\alpha_{2r}[R_1][e_2](K_1 + [e_1])K_2K_3K_4}{\alpha_{1f}\alpha_{2f}(K_2 + [e_1])(K_3 + [e_1])[E][e_1]^2} + \frac{\alpha_{2r}[R_2](K_4 + [e_1])K_3}{\alpha_{2f}[I_1][e_1](K_3 + [e_1])} + 1 \right]^{-1} \quad (7)$$

Curves 1, 2, 3, and 4 in Fig. 9A were generated with these expressions for conditions (i), (ii), (iii), and (iv) described above, respectively. It is evident that for the conditions given in the figure legend, as the number of converter enzymes regulated by $e_1$ in the bicyclic cascade increased progressively, there is a gradual increase in the steepness of the response of fractional modification of $I_3$ to the $e_1$ concentration. This phenomenon is quantitatively illustrated by the change in the sensitivity index ($S$ value) which was calculated to be 1.0, 3.3, 4.4, and 5.6 for curve 1, 2, 3, and 4, respectively. Expressed in the more conventional (but improper) manner, these $S$ values correspond to Hill numbers of 1.0, 2.0, 2.5, and 3.0, respectively (see Fig. 9B). However, it should be noted from Eqs. 6 and 7, that the $S$ values obtained for cases (iii) and (iv) can be different from these calculated for curves 3 and 4 in Fig. 9A depending upon the values of $K_2$ and $K_3$. Nevertheless, the $S$ values are always greater than 1.0 when $e_1$ is utilized in more than one reaction in the cascade in a manner that favors $m$-$I_2$ formation. In addition, Fig. 9A also shows that by increasing the number of steps being regulated by $e_1$ can cause an increase in both amplitude of the fractional modification of $I_2$ and the amplification factor.

*Closed Bicyclic Cascade.* Figure 7B shows a variation of the bicyclic cascade system which is patterned after the glutamine synthetase cascade. In this bicyclic cascade, both the modified and unmodified forms, $o$-$I_1$ and $m$-$I_1$, of the interconvertible enzyme in the first cycle serve as converter enzymes in the second cycle. Therefore the fractional modification of $I_2$ at steady state is determined by the $[m$-$I_1]/[o$-$I_1]$ ratio, and it is independent of the total $I_1$ concentration. With the same assumptions given before, the steady-state equation for the fractional modification of $I_2$ is given by

$$\frac{[m\text{-}I_2]}{[I_2]} = \left[\frac{\alpha_{1r}\alpha_{2r}[R_1][e_2](K_1 + [e_1])}{\alpha_{1f}\alpha_{2f}[E][e_1](K_2 + [e_2])} + 1\right]^{-1} \quad (8)$$

Equation (8) shows that only 14 parameters are required to determine the fractional modification of $I_2$ for the closed bicyclic cascade. This is comparatively fewer than the 18 parameters needed to describe the opened bicyclic cascade as shown in Eq. (3). Therefore, the closed bicyclic cascade is intrinsically less flexible than the opened bicyclic cascade. Figure 10A illustrates that the fractional modification of $I_2$ in the closed bicyclic cascade increases as the ratio of $[e_1]/[e_2]$ increases, until it reaches a maximum value that is determined by the $\alpha_{1f}\alpha_{2f}/\alpha_{1r}\alpha_{2r}$ ratio. If $e_1$ is required for the activation of both E and $m$-$I_1$ and $e_2$ is required for the activation of both $R_{1i}$ and $o$-$I_1$, the steady-state equation becomes

$$\frac{[m\text{-}I_2]}{[I_2]} = \left[\frac{\alpha_{1r}\alpha_{2r}(K_1 + [e_1])K_3[R_1][e_2]^2}{\alpha_{1f}\alpha_{2f}(K_2 + [e_2])K_4[E][e_1]^2} + 1\right]^{-1} \quad (9)$$

where $K_3$ and $K_4$ are the dissociation constants for $m$-$I_1 \cdot e_1$ and $o$-$I_1 \cdot e_2$ complexes, respectively. Under these conditions, the fractional modification of $I_2$ is a sigmoidal function of the $[e_1]/[e_2]$ ratio (Fig. 10B). In other words, the modification of $I_2$ is more sensitive to the $[e_1]/[e_2]$ ratio.

*Multicyclic Cascade Systems*

Comparison of Eqs. (1), (3), and (8) shows that bicyclic cascades possess much greater allosteric control flexibility and amplification capacity than monocyclic cascades. To gain a better understanding of the rela-

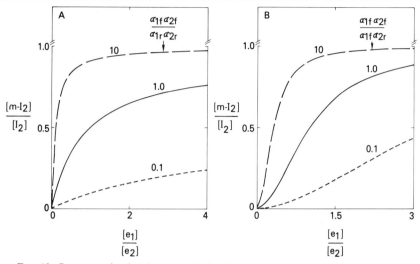

FIG. 10. Computer-simulated curves obtained with Eqs. (8) (A) and (9) (B) to illustrate the level of $[m\text{-}I_2]/[I_2]$ as a function of $[e_1]/[e_2]$ and $\alpha_{1f}\alpha_{2f}/\alpha_{1r}\alpha_{2r}$. The curves are generated with $[R] = [E] = K_1 = K_2 = 1.0$, $K_3 = K_4 = [e_2] = 0.1$. This figure is patterned after Fig. 21 in E. R. Stadtman and P. B. Chock, *Curr. Top. Cell. Regul.* **13**, 53 (1978).

tionship between the number of cycles in a cascade and regulatory characteristics, a steady-state analysis of a multicyclic cascade system containing $n$ cycles as depicted in Fig. 11 was carried out. The cascade is initiated by the binding of an allosteric effector, $e_1$, to the first converter enzyme, $E_i$, and the modified form of the interconvertible enzyme, $m\text{-}I_n$, in one cycle serves as a converter enzyme for the modification of the interconvertible enzyme in the following cycle ($I_{n+1}$). With the same assumptions described above for deriving Eq. (1), the general expression for the fractional modification of $I_n$ at steady state is obtained.[5]

$$\frac{[m\text{-}I_n]}{[I_n]} = \left[ \frac{\alpha_{1r}\alpha_{2r}\alpha_{3r}\cdots\alpha_{nr}[R_1][R_2]\cdots[R_n][e_2][e_3]\cdots[e_{n+1}](K+[e_1])}{\alpha_{1f}\alpha_{2f}\alpha_{3f}\cdots\alpha_{nf}[E][I_1][I_2]\cdots[I_{n-1}][e_1](K_2+[e_2])\cdots(K_{n+1}+[e_{n+1}])} \right.$$

$$+ \frac{\alpha_{2r}\alpha_{3r}\cdots\alpha_{nr}[R_2][R_3]\cdots[R_n][e_3][e_4]\cdots[e_{n+1}]}{\alpha_{2f}\alpha_{3f}\cdots\alpha_{nf}[I_1][I_2]\cdots[I_{n-1}](K_3+[e_3])(K_4+[e_4])\cdots(K_{n+1}+[e_{n+1}])}$$

$$\left. + \cdots + \frac{\alpha_{nr}[R_n][e_{n+1}]}{\alpha_{nf}[I_{n-1}](K_{n+1}+[e_{n+1}])} + 1 \right]^{-1} \quad (10)$$

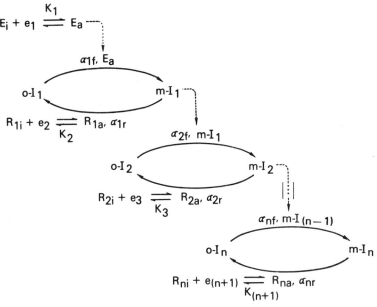

FIG. 11. Schematic representation of a multicyclic cascade system. See Fig. 7 for definition of the symbols. This figure is patterned after Fig. 22 in E. R. Stadtman and P. B. Chock, *Curr. Top. Cell. Regul.* **13**, 53 (1978).

Equation (10) shows that with each additional cycle in the cascade the fractional modification of the last interconvertible enzyme $I_n$, becomes dependent upon eight additional variables, namely, $k_{nf}$, $K_{nf}$, $k_{nr}$, $K_{nr}$, $[e_{n+1}]$, $[R_n]$, $[I_{n-1}]$ and $K_{n+1}$. Furthermore, the fractional modification of $I_n$ is a multiplicative function of all these parameters, any one of which can be varied through changes in the concentration of substrates or allosteric effectors. Therefore each additional cycle in the cascade enhances enormously the amplification potential and allosteric control potential of the system. This is demonstrated by the curves in Fig. 12A. As a point of reference, curve 0 shows the fractional activation (saturation) of E as a function of $e_1$ concentration when $K_1$ is 1.0. This curve yields a $[e_1]_{0.5E}$ value of 1.0. If all parameters in Eq. (10) are assigned values of 1.0, the $[e_1]_{0.5I_n}$ value is also equal to 1.0 for the fractional modification of $I_n$. However, with $K_1$ being held at 1.0 and all other parameters in Eq. (10) that favor the modification reactions are given values of 2, and all parameters that favor the demodification steps are assigned values of 0.5, then the amplification factor is increased by increasing the number of cycles in the cascade. From curves 1, 2, 3, and 4 in Fig. 12A, the amplification factors, $([e_1]_{0.5I})^{-1}$, are calculated to be $3.2 \times 10^2$, $1.02 \times 10^5$, $3.25 \times 10^7$, and $1.05 \times 10^{10}$ for one-, two-, three-, and four-cycle cascades, respectively.

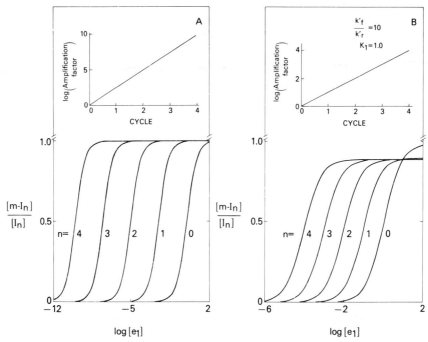

FIG. 12. Computer-simulated curves obtained with Eqs. (10) (A) and (11) (B) to illustrate the dependence of $[m\text{-}I_n]/[I_n]$ on the number of cycles, $n$, as a function of $\log[e_1]$. For (A), $\alpha_{nf} = 4$, $\alpha_{nr} = 0.25$, $[R_n] = [e_{n+1}] = 0.5$, $K_1 = 1$, $K_{n+1} = [E] = [I_{n-1}] = 2$. For (B) $K_1 = 1$, $k'_f/k'_r = 10$. The insets in (A) and (B) depict the linear relationship between $\log([e_1]_{0.5I_n})^{-1}$ and $n$. This figure is patterned after Fig. 23 in E. R. Stadtman and P. B. Chock, *Curr. Top. Cell. Regul.* **13**, 53 (1978).

The inset in Fig. 12A shows that the log of the amplification factor is a linear function of the number of cycles in the cascade.

The exponential property of the signal amplification factor as a function of the number of cycles in the cascade can be better illustrated using the simplified version of Eq. (10) shown below.

$$[m\text{-}I_n]/[I_n] = [(K_1/[e_1] + 1)(k'_r/k'_f)^n + (k'_r/k'_f)^{n-1} + \ldots + k'_r/k'_f + 1]^{-1} \quad (11)$$

Equation (11) is obtained from Eq. (10) by neglecting the role of $e_n$, except $e_1$, and by assuming that the ratios $\alpha_{1r}[R_1]/\alpha_{1f}[E]$ and $\alpha_{nr}[R_{nr}]/\alpha_{nf}[I_{n-1}]$ are the same and equal to $k'_r/k'_f$ for all values of $n$. Figure 12B shows the computer-simulated curves obtained with this simplified equation, assuming $k'_f/k'_r = 10$ and $K_1 = 1.0$. Under these conditions, the amplification factors for one-, two-, three-, and four-cycle cascades are 9.1, 83, 833,

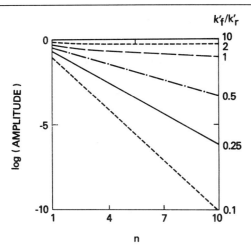

FIG. 13. Computer-generated lines obtained with Eq. (11) to illustrate the linear relationship between the log of the amplitude and $k'_f/k'_r$ as a function of the increasing number of cycles, $n$. This figure is patterned after Fig. 25 in E. R. Stadtman and P. B. Chock, *Curr. Top. Cell. Regul.* **13,** 53 (1978).

and 8333, respectively. The inset in Fig. 12B shows that the log of the amplification factor is proportional to the number of cycles in the cascade.

A more detailed analysis of Eq. (11) reveals several important features of multicyclic cascades. These are the interrelationships between the amplitude of the fractional modification, amplification factor, $k'_f/k'_r$ and the number of cycles in the cascade. Figure 13 shows the relationship between the log of the amplitude and $k'_f/k'_r$ as a function of increasing the number of cycles, $n$. When the ratio $k'_f/k'_r$ is 10 or greater, the amplitude is independent of the number of cycles in the cascade. However, if $k'_f/k'_r$ is less than 10, the log of the amplitude is inversely proportional to the number of cycles in the cascade, and the proportionality constant increases as the ratio of $k'_f/k'_r$ decreases. With respect to the amplification factor, Fig. 14 depicts the existence of a linear relationship between the log of the amplification factor and the number of cycles in the cascade when $k'_f/k'_r$ is equal to or greater than 3.0. The slopes of these linear functions increases as the ratio of $k'_f/k'_r$ increases.

Applications and Examples

Experimentally, the kinetic parameters in Eqs. (1), (3), (8), and (10) and the role of the allosteric effectors involved in a cascade system can be determined either by studying each of the converter enzymes individually with its proper form of interconvertible enzyme as substrate in the pres-

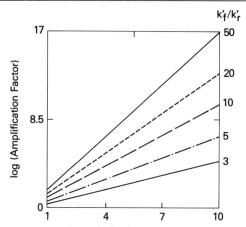

FIG. 14. Computer-generated lines obtained with Eq. (11) to demonstrate the linear relationship between log of amplification factor and $k_f'/k_r'$ as a function of the increasing number of cycles, $n$. This figure is patterned after Fig. 26 in E. R. Stadtman and P. B. Chock, *Curr. Top. Cell. Regul.* **13**, 53 (1978).

ence and in the absence of the allosteric effectors involved, or by investigating the complete cascade system. However, owing to the complexity of the cascade system and the fact that the values of the parameters may vary when they are determined with the whole system, it is essential to utilize the combination of the two approaches mentioned above. Methods for the first approach, in the initial-rate kinetic method, is the primary topic of this volume. Therefore, we will discuss only the second approach here.

As noted earlier, in most cyclic cascade systems, ATP is consumed continually. Therefore, in order to sustain a steady-state level of interconvertible enzyme modification, it is essential to maintain the concentration of ATP at a constant level. Under physiological conditions this is achieved through metabolic regulation. However, for *in vitro* studies the cascade system must be coupled with an ATP regeneration system, e.g., pyruvate kinase and phosphoenolpyruvate. In the presence of the ATP regeneration system and all the enzymes of the cascade, the fractional modification of the interconvertible enzyme will reach a steady-state value. From determinations of this steady-state value as a function of allosteric effector and enzyme concentrations, it is possible to derive the various constants needed for the evaluation of the steady-state equations. For example, in the case of the monocyclic system, Eq. (1) can be rearranged to

$$\frac{[I]}{[m\text{-}I]} - 1 = \frac{\alpha_r[R][e_2]}{\alpha_f[E](K_2 + [e_2])} \left(1 + \frac{K_1}{[e_1]}\right) \quad (12)$$

When $[m\text{-}I]/[I]$ is measured as a function of $1/[e_1]$, a plot of $([I]/[m\text{-}I] - 1)$ versus $1/[e_1]$ will give the value of $-1/K_1$ at the intercept of the abscissa and the value of $\alpha_r[R][e_2]/\{\alpha_f[E](K_2 + [e_2])\}$ at the intercept of the ordinate. With further rearrangement of Eq. (1), the following expression can be obtained:

$$\left(\frac{[I]}{[m\text{-}I]} - 1\right)^{-1} = \frac{\alpha_f[E][e_1]}{\alpha_r[R](K_1 + [e_1])}\left(1 + \frac{K_2}{[e_2]}\right) \quad (13)$$

Equation (13) indicates that a plot of $([I]/[m\text{-}I] - 1)^{-1}$ vs. $1/[e_2]$ will yield the value of $-1/K_2$ at the intercept of the abscissa and $\alpha_f[E][e_1]/\{\alpha_r[R](K_1 + [e_1])\}$ at the intercept of the ordinate. Since $[E]$, $[R]$, $[e_1]$, and $[e_2]$ are known values, the ratio of $\alpha_f/\alpha_r$ can be calculated from the intercept of the ordinate once $K_1$ and $K_2$ are determined. To evaluate $\alpha_f$ and $\alpha_r$, it is necessary to perform the initial-rate study for either the modification cascade or the regeneration cascade, or both.

If $e_2$ is an inhibitor for the converter enzyme, R, the fractional modification of the interconvertible enzyme can be written as

$$\left(\frac{[I]}{[m\text{-}I]} - 1\right)^{-1} = \frac{\alpha_f[E][e_1]}{\alpha_r[R](K_1 + [e_1])}\left(1 + \frac{[e_2]}{K_2}\right) \quad (14)$$

A plot of $([I]/[m\text{-}I] - 1)^{-1}$ vs. $[e_2]$ will give the value of $-K_2$ at the intercept of the abscissa.

A similar approach can be applied to a multicyclic cascade system, except that the number of variables and the complexity of the system will also increase with increase of the number of cycles in the cascade. For an opened bicyclic cascade system, Eq. (3) can be rearranged to

$$\left(\frac{[I_2]}{[m\text{-}I_2]} - 1\right)$$
$$= \frac{\alpha_{2r}[R_2][e_3]}{\alpha_{2f}[I_1](K_3 + [e_3])}\left[\frac{\alpha_{1r}[R_1][e_2]}{\alpha_{1f}[E](K_2 + [e_2])}\left(1 + \frac{K_1}{[e_1]}\right) + 1\right] \quad (15)$$

The intercept at the abscissa from the plot of $([I_2]/[m\text{-}I_2] - 1)$ vs. $1/[e_1]$ will give the values of $-1/K_{obs}$, which is a linear function of $[E]$ or $1/[R_1]$, i.e.,

$$\frac{1}{K_{obs}} = \frac{\alpha_{1f}(K_2 + [e_2])[E]}{\alpha_{1r}[R_1][e_2]K_1} + \frac{1}{K_1}$$

A secondary plot of $1/K_{obs}$ vs. either $[E]$ or $1/[R_1]$ will yield the value of $1/K_1$ at the intercept of the ordinate. Owing to the nature of Eq. (3), it is simpler to proceed from this point to determine $K_3$ with Eq. (16).

$$\left(\frac{[I_2]}{[m\text{-}I_2]} - 1\right)^{-1}$$

$$= \frac{\alpha_{1f}[E](K_2 + [e_2])\alpha_{2f}[I_1]}{\alpha_{2r}[R_2][\alpha_{1r}[R_1][e_2] + \alpha_{1f}[E](K_2 + [e_2])]} \left(1 + \frac{K_3}{[e_3]}\right) \quad (16)$$

The plot of $([I_2]/[m\text{-}I_2] - 1)^{-1}$ vs. $1/[e_3]$ will yield the value of $-1/K_3$ at the intercept of the abscissa, and the intercept at the ordinate, $\beta$, can be expressed as

$$\beta^{-1} = \left(\frac{\alpha_{1r}[R_1](K_1 + [e_1])[e_2]}{\alpha_{1f}(K_2 + [e_2])[E]} + 1\right) \frac{\alpha_{2r}[R_2]}{\alpha_{2f}[I_1]} \quad (17)$$

By varying either $[R_1]$ or $[E]$, or $[R_1]/[E]$ ratio, the value of $\alpha_{2r}/\alpha_{2f}$ can be determined from a plot of $\beta^{-1}$ vs. either $[R_1]$, or $1/[E]$ or $[R_1]/[E]$. Since all the parameters, except $K_2$ and $\alpha_{1r}/\alpha_{1f}$ are known, a plot of $(\beta^{-1} - \alpha_{2r}[R_2]/\alpha_{2f}[I_1])^{-1}$ vs. $1/[e_2]$ will provide values of $K_2$ and $\alpha_{1r}/\alpha_{1f}$. The values of $\alpha_{nr}$ and $\alpha_{nf}$, can be determined only with the initial-rate method.

It should be noted that the above analysis is valid only when the cascade systems studied satisfy the assumptions used to derive Eqs. (1), (3), (8), and (10). When the conditions deviate from the assumptions, a more complex equation should be derived and solved accordingly. A typical example for such a system is the monocyclic cascade of the adenylylation–deadenylylation of glutamine synthetase from *E. coli*.[16] Since the allosteric effectors, glutamine and $\alpha$-ketoglutarate, for this monocyclic system are multifunctional effectors and because three active converter enzyme complexes are involved in both adenylylation and deadenylylation of the interconvertible enzyme, glutamine synthetase, a minimum of 28 constants are needed to describe this monocyclic cascade. Despite the complexity, Rhee *et al.*[16] demonstrated in their study that a steady state is established at a given effector concentration and the fractional modification of glutamine synthetase, as measured by the number of adenylyl groups per mole of enzyme, is a function of the concentration of the allosteric effectors involved as predicted by the cyclic cascade model described here. Furthermore, both the amplification factors and the change in sensitivity index were also observed as a function of the concentration of allosteric effectors.

## Concluding Remarks

The cyclic cascade systems are fundamentally different from the irreversible, unidirectional cascades of proteolytic enzymes of the blood-

[16] S. G. Rhee, R. Park, P. B. Chock, and E. R. Stadtman, *Proc. Natl. Acad. Sci. U.S.A.*, **75**, 3138 (1978).

clotting type[11,17-19] and complement fixation.[20] When triggered by an appropriate alarm signal the unidirectional cascade responds in an explosive manner to produce an avalanche of product needed to meet a specific biological emergency; but having fulfilled its function, the cascades are terminated by a self-destructive process that is initiated by autoregulatory signals. In effect, unidirectional cascades are contingency systems that function as biological ON–OFF switches in response to occasional emergency needs.

On the contrary, the following concepts emerge regarding cyclic cascades as revealed by the simplified model described here.

1. The covalent modification and demodification of an interconvertible enzyme is visualized as a dynamic process by means of which the specific activity of this enzyme is determined through the dynamic coupling of two opposing cascades, thereby establishing a steady-state distribution of active and inactive forms of the interconvertible enzyme. With this concept the activity of an interconvertible enzyme can be varied smoothly and continuously in response to the changes of multiple metabolites, which are allosteric effectors of the cascade enzymes.

2. Since a minimum of three enzymes are involved in each interconvertible enzyme cycle and each of these enzymes can be a separate target for one or more positive or negative metabolite effectors, the interconvertible enzymes can respond to a great number of allosteric effectors and integrate these multiple effects into a single output, namely, the specific activity of the target enzyme in the cascades. Therefore, cyclic cascades possess enormous flexibility in terms of its control patterns.

3. Cyclic cascades are endowed with great amplification potential derived from the fact that the response of the last interconvertible enzyme in a cascade to a primary stimulus is a multiplicative function of various parameters in the cascade. As a consequence of signal amplification, the interconvertible enzymes can respond to effector concentrations well below the dissociation constants of the effector–allosteric enzyme complexes. For example, comparison of curve 7 with the dashed curve in Fig. 3 shows that, if each of the six variables in Eq. (1) undergoes a 2-fold change, the concentration of the effector that causes only 2% activation of the converter

---

[17] E. W. Davie and K. Fujikawa, *Annu. Rev. Biochem.* **44,** 799 (1975).
[18] R. G. Macfarlane, *Proc. R. Soc. London, Ser. B.* **173,** 261 (1969).
[19] E. W. Montroll, *Adv. Chem. Phys.* **26,** 145 (1974).
[20] H. S. Müller-Eberhard, *Annu. Rev. Biochem.* **44,** 697 (1975).

enzyme E can produce 90% activation of the interconvertible enzyme. Furthermore, since the total number of these parameters is directly proportional to the number of cycles in the cascade, a concerted effect of only 2-fold variation in the contributions of each parameter can yield a $3 \times 10^2$-, $10^5$-, $3 \times 10^7$-, and $10^{10}$-fold amplification of a primary stimulus in a one-, two-, three-, and four-cycle cascade, respectively. However, unlike the unidirectional cascades, where high amplification leads to an explosive response to primary stimuli, the amplification capacity of cyclic cascades is susceptible to very fine regulation because each parameter in the cascade can be independently regulated and the system is reversible.

4. Cyclic cascades can generate a sigmoidal response of interconvertible enzyme activity to increasing concentration of an allosteric effector. This property is derived from the fact that there are more steps in the cyclic cascade at which a given effector can interact.

It should be kept in mind that the purpose of developing the simplified model shown here is to reveal more clearly the principle and the usefulness of the complex cyclic cascade in metabolic regulation. The properties of cyclic cascades derived from these simplified models have been confirmed by the results obtained from a much more rigorous treatment of the steady-state equations (Fig. 6) and also from a considerably more complex cycle.[16] The data given in Fig. 6 not only confirm the fundamental characteristics of cyclic cascades revealed by the simplified models, but they show additionally that cyclic cascades are potentially more sensitive to effector stimuli and can achieve an even greater amplification potential than those predicted by the simplified model. The work of Rhee et al.[16] on the monocyclic cascade of glutamine synthetase demonstrates that such a monocycle is quite complex inasmuch as 28 constants are required to describe the system. Despite the complexity, the results confirm the properties predicted by the simplified model.

The capacity of biological systems to take advantage of these elegant regulatory mechanisms depends upon the kinetic constraints, such as the concentrations and turnover numbers of the cascade enzymes and the dissociation constants of the effector–enzyme and enzyme–enzyme complexes. The kinetic analysis[11] of the multicyclic cascade shows that, with reasonable parameters as stated above, a 50% modification of the second and third interconvertible enzymes in a three-cycle cascade can be accomplished in the millisecond time range. With these same kinetic constants, an even faster response can be achieved by proper topographical positioning of the converter and interconvertible enzymes in multienzyme complexes, as occurs with mammalian pyruvate dehydrogenase[7,12] and the aromatic amino acid synthetase complex of *Neurospora*,[21] or by assemblage of the cascade enzymes on a solid support, as occurs in the binding of the glycogen phosphorylase cascade components on glycogen particles.[22] Further kinetic analysis shows[11] that interconvertible enzyme cascade can in fact function as a rate amplifier to generate an almost explosive increase in enzymic activity in response to stimuli.

It is therefore not surprising that in recent years cyclic cascades have been demonstrated to be the basis of regulation of many enzymes that occupy strategic positions in metabolism. Thus, multicyclic cascades are involved in the regulation of glycogen phosphorylase, glycogen synthase, peptide initiation factor eIF-2, and *E. coli* glutamine synthetase; monocyclic cascades are involved in the regulation of adipose tissue triglyceride lipase, pyruvate dehydrogenase, tyrosine aminotransferase, RNA polymerase, phenylalanine hydroxylase, pyruvate kinase, and possibly fatty acid synthetase, phosphofructokinase, and acetyl-CoA carboxylase (for review and references to original literature see Stadtman and Chock[2]).

[21] K. G. Welch and F. H. Gaertner, *Proc. Natl. Acad. Sci. U.S.A.* **72**, 4128 (1975).
[22] F. Meyer, L. M. G. Heilmeyer, Jr., R. H. Haschke, and E. H. Fischer, *J. Biol. Chem.* **245**, 6642 (1970).

## [13] The Use of Alternative Substrates to Study Enzyme-Catalyzed Chemical Modification

### By DONALD J. GRAVES and TODD M. MARTENSEN

Enzyme-catalyzed chemical modification of proteins is an important mechanism for the regulation of their biological activity. The first example of this type of control was documented for the enzyme glycogen phosphorylase.[1,2] Here it was shown that the chemical modification reactions, phosphorylation and dephosphorylation, are coupled to various physiological stimuli, and that the modification reactions have an important consequence on glycogen metabolism.[3] This type of control is now known to be quite common for other enzymes and proteins. Since these enzyme-catalyzed chemical modification reactions are important in regulatory phenomena, a number of approaches have been taken to study them. In this chapter, the use of alternative substrates will be stressed.

[1] C. F. Cori, G. T. Cori, and A. A. Green, *J. Biol. Chem.* **151**, 39 (1943).
[2] E. G. Krebs and E. F. Fischer, *Biochim. Biophys. Acta* **20**, 150 (1956).
[3] T. H. Fischer, A. Pocker, and J. C. Saari, *Essays Biochem.* **6**, 23 (1970).

There are several important ways that alternative substrates can be used.

1. When a protein is a substrate for an enzyme-catalyzed reaction, regulation of the reaction by modifiers can occur by their binding to the protein substrate, the enzyme, or both. To differentiate control by substrate-directed modifiers from enzyme-directed modifiers, an alternative substrate can be used. This could be a peptide containing the amino acid sequence of the modification site, a chemical derivative of the natural substrate or a completely different protein. If modifiers alter the reaction only with the natural substrate, it is assumed that the modifier acts by binding to the substrate. Using both natural and alternative substrates, it is also possible to differentiate various mechanisms of inhibition or activation and determine whether the enzyme as well as the substrate is involved in the regulation.
2. Initial-velocity studies can be done to establish the kinetic mechanism of the reaction. An alternative substrate not affected by various molecular species used in the kinetic studies, e.g., metal ions and nucleotides, would give a system more amenable to analysis. The results could provide a basis for the understanding of the more complex reaction with the natural substrate.
3. Studies could be carried out on the specificity of the reaction. Various protein substrates or peptide fragments of protein substrates might be used. With the numerous and effective methods that are available for chemical synthesis of peptides, the use of various analogs could provide important information regarding specificity. This approach has been useful in determining specificity requirements for protein kinases.[4]
4. Studies could be done on the nature of the interaction of the substrate with the enzyme. With a well-defined small molecular weight alternative substrate, various physical studies, e.g., ultracentrifugation, nuclear magnetic resonance (NMR) spectroscopy, and circular dichroism, could be made. Eventually, the mode of interaction of a peptide with the enzyme might be specified by examination of an enzyme substrate complex by X-ray crystallography.

It is important to establish whether the alternative substrate chosen is a useful model for the study of the enzyme-catalyzed modification reaction. For example, with effectors that are known to act upon the enzyme,

---

[4] B. E. Kemp, D. J. Graves, E. Benjamin, and E. G. Krebs, *J. Biol. Chem.* **252**, 4888 (1977).

but not the substrate, does the enzyme show the same basic catalytic and regulatory features with the two substrates? With phosphorylase $b$ kinase, studies have shown that a tetradecapeptide containing the reactive seryl residue is a useful model for the natural substrate, phosphorylase $b$, in that all the properties seen with phosphorylase $b$ as substrate were also seen with the peptide substrate.[5] This is not true for the alternative substrate, troponin T. Using this substrate, no activation of phosphorylase kinase could be detected by phosphorylation of the enzyme by the cyclic AMP-dependent protein kinase.[6]

Another important aspect is the comparison of the kinetic parameters obtained with the alternative and natural substrates. How good is the alternative substrate? If the natural substrate and model substrate are acted upon by the enzyme at a single site, the comparison is straightforward. For the case where multiple reactions can take place on the natural substrate, but only a single reaction occurs on the alternative substrate, certain limitations need to be considered. For example, if the substrate is a phosphoprotein containing two sites that can be dephosphorylated randomly by a phosphatase, the rate equation for dephosphorylation of one of these sites is

$$\frac{1}{v} = \frac{K_1}{V_1}\frac{1}{S} + \frac{1}{V_1}\left(1 + \frac{K_1}{K_2}\right)$$

where $v$ and $V_1$ are, respectively, the observed velocity and maximal velocity for dephosphorylation of the first site; $K_1$ and $K_2$ are, respectively, the Michaelis constants for the first and second sites. It can be seen from the preceding equation that the intercept contains terms for the two interactions the enzyme can have with its substrate. Because of this, only an apparent maximal velocity and an apparent Michaelis constant can be evaluated; their values will be different from the real values depending on the ratio of $K_1/K_2$. The maximal velocity for the dephosphorylation of the first site would be underestimated if a fraction of the total enzyme added is interacting with the second site, i.e., $K_2 \leq K_1$. Conceivably, a higher $V_m$ value could be obtained with an alternative substrate than with a natural substrate based on these considerations. Likewise, the $K_m$ values for the two substrates might vary on this basis. In this case, the $K_m$ value for dephosphorylation of the first site in the natural substrate could be overestimated.

---

[5] G. M. Carlson, L. B. Tabatabai, and D. J. Graves, *Int. Symp. Metab. Interconversion Enzymes, 4th,* (Arad, Israel), p. 50. Springer-Verlag, Berlin and New York, 1976.
[6] J. T. Stull, C. O. Brostrom, and E. G. Krebs, *J. Biol. Chem.* **247**, 5272 (1972).

Theory

*The Effect of Modifiers on the Kinetics of Enzyme-Catalyzed Chemical Modification: Model A.*

The model illustrated in Scheme 1 involves an enzyme whose single substrate is another enzyme. Both macromolecules are capable of binding effectors with rapid binding equilibria. The model is general in that all

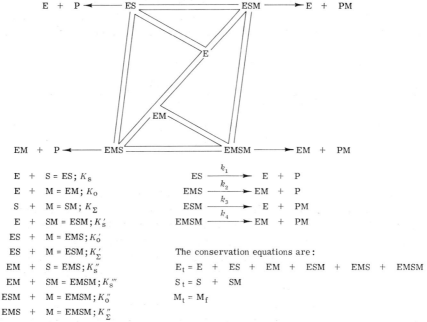

Scheme 1

complexes yield product. The model deals with a single modifier site present on both the enzyme and its substrate.[7]

*A notable aspect of this model is the assumption that the free modifier concentration is equivalent to the total modifier added.* This assumption is valid only for kinetic studies in which the macromolecular substrate is used at much lower concentrations than that of the modifier.[8] A number of authors have considered modifier complexing with substrate.[9-11] The re-

[7] T. M. Martensen and D. J. Graves, *J. Biol. Chem.* **248**, 8333 (1973).
[8] D. J. Cavanaugh, J. Harris, and J. T. Hearon, *J. Am. Chem. Soc.* **77**, 1531 (1955).
[9] J. M. Reiner, "Behavior of Enzyme Systems," p. 154. Van Nostrand-Reinhold, New York, 1969.
[10] J. S. Friendenvold and Maengwyn-Davies, *in* "The Mechanism of Enzyme Action" (W. D. McElroy and B. Glass, eds.), p. 154. John Hopkins Press, Baltimore, Maryland, 1954.
[11] J. L. Webb, "Enzyme and Metabolic Inhibitors," Vol. 1, p. 85. Academic Press, New York, 1963.

sulting complex rate equations were expressed in terms of both total as well as free concentration of substrate and modifier. The identity $M_{total} = M + MS$ and $S_{total} = S + MS$ utilized in the derivations resulted in quadratic expressions containing $M_{total}$ and $S_{total}$ terms, and the resulting equations are of little practical use.

For Scheme 1, an equilibrium approach has been utilized inasmuch as the steady-state method leads to exceedingly complex rate equations containing substrate terms raised to the second power, which could impart nonlinearity to reciprocal plots. An important criterion for the applicability of the equilibrium method is the linearity of reciprocal plots in the presence of modifier.

The general form of the rate equation is

$$\frac{1}{v} = \frac{1}{V_m}\left[\frac{1 + M/K_0' + M/K_\Sigma' + M^2/K_0'K_\Sigma''}{1 + (k_2/k_1)(M/K_0') + (k_3/k_1)(M/K_\Sigma') + (k_4/k_1)(M^2/K_0'K_\Sigma'')}\right]$$
$$+ \frac{K_s}{V_m}\left[\frac{(1 + M/K_\Sigma)(1 + M/K_0)}{1 + (k_2/k_1)(M/K_0') + (k_3/k_1)(M/K_\Sigma') + (k_4/k_1)(M^2/K_0'K_\Sigma'')}\right]\frac{1}{S_t} \quad (1)$$

Equation (1) is general in that either inhibition or activation can be described. Restrictions on the magnitude of the dissociation constants as well as the reactivities of the complexes are specified by the values of the appropriate constants. The dissociation and rate constants are as defined: $V_m = k_1E_0$. $K_0$ represents $K_i$ or $K_a$, depending upon whether a case involves inhibition or activation, respectively. The concentration of product is assumed to be zero.

The specification of identities between the binding of modifier to free substrate, substrate bound to enzyme, or substrate bound to an enzyme–modifier complex also specifies the relationship between other dissociation constants through the following equations.

$$\frac{K_\Sigma}{K_\Sigma'} = \frac{K_s}{K_s'}; \quad \frac{K_\Sigma'}{K_\Sigma''} = \frac{K_0'}{K_0''}; \quad \frac{K_\Sigma}{K_\Sigma''} = \frac{K_s''}{K_s'''}; \quad \frac{K_0}{K_0'} = \frac{K_s}{K_s''}; \quad \frac{K_0}{K_0''} = \frac{K_s'}{K_s'''}$$

A considerable simplification is possible if the modifier influences the enzymic reaction by binding exclusively to the substrate [Eq. (2)] or to the enzyme [Eq. (3)].

$$\frac{1}{v} = \frac{1}{V_m}\left[\frac{1 + M/K_\Sigma'}{1 + (k_3/k_1)(M/K_\Sigma')}\right] + \frac{K_s}{V_m}\left[\frac{1 + M/K_\Sigma}{1 + (k_3/k_1)(M/K_\Sigma')}\right]\frac{1}{S_t} \quad (2)$$

$$\frac{1}{v} = \frac{1}{V_m}\left[\frac{1 + M/K_0'}{1 + (k_2/k_1)(M/K_0')}\right] + \frac{K_s}{V_m}\left[\frac{1 + M/K_0}{1 + (k_2/k_1)(M/K_0')}\right]\frac{1}{S_t} \quad (3)$$

The locus of action of a modifier can easily be distinguished by the use of an alternative substrate. When using the native substrate and an alternative substrate in separate experiments, the modifier should have a similar effect when either substrate is utilized if the modifier affects only the converting enzyme; no effect when the alternative substrate is used if the action of the modifier is substrate directed, and similar effects though differing in magnitude when the modifier affects both substrate and enzyme. Once these three types of action are distinguished, further kinetic studies can be done to elucidate the exact mechanism of action of the modifier.

Substrate-directed effects can be classified into several different categories. Rate equations, conditions, and type of effect observed when the data are plotted in double-reciprocal form are given in Table I. Cases I-A through I-H show that kinetic studies of substrate-directed modifiers can show the same classical patterns obtained from enzyme directed modifiers. If the modifier acts as an inhibitor, and all constants are unequal, reciprocal plots in the presence of modifier would yield mixed inhibition kinetics. Case I-A considers an SI complex that does not bind to the enzyme; inhibition results from the reduction of the effective substrate concentration. Case I-B involves an SI complex that competes with the substrate less efficiently for the binding site of the enzyme but cannot be converted to product. Case I-C deals with an SI complex that competes less efficiently for the binding site of the enzyme, yet whose catalytic constant is unaffected. For case I-D, inhibition results solely from a lower catalytic efficiency of the ESI complex. Case I-E is a mechanism in which binding and catalysis can occur with the SI complex but with a certain identity of binding and rate constants.

A modifier could also serve as a substrate-directed activator, whose observed kinetic effect would be an increase in $V_{max}$ or a decrease in $K_m$, or both. For case I-F, the activator increases catalysis but has no influence upon binding. In case I-G, the SA complex binds more tightly to the enzyme than the free substrate, but the $V_{max}$ is unchanged. Case I-H is the opposite of case I-E of substrate-directed inhibition.

Determination of the mechanism by which an inhibitor or activator functions by binding to both enzyme and substrate is not easily accomplished. However, competitive inhibition kinetics allows the locus of inhibition to be relatively easily determined in a number of cases (Table II). Mechanisms yielding noncompetitive kinetics that regulate enzyme interconversion are more difficult to resolve and are not presented.

Case II-A concerns a modifier that is a partial competitive inhibitor, yet also affects the substrate so that it is not bound to the enzyme. Case II-B is analogous to Case II-A except that the SI complex can bind to the enzyme, but less efficiently than free substrate. Case II-C involves an inhibitor that binds competitively. When inhibitor binds to substrate, the complex is not bound by the enzyme. Case II-D is similar to II-C except that the SI complex binds less effectively than substrate, but the catalytic reaction occurs at the same rate.

# TABLE I
## Substrate-Directed Inhibition and Activation

| Case | Rate equation (inhibition) | Condition | Type of effect |
|---|---|---|---|
| I-A | $\dfrac{1}{v} = \dfrac{1}{V_m} + \dfrac{K_s}{V_m}\left(1 + \dfrac{I}{K'_\Sigma}\right)\dfrac{1}{S_t}$ | $K'_\Sigma \to \infty$<br>$K'_s \to \infty$ | Competitive |
| I-B | $\dfrac{1}{v} = \dfrac{1}{V_m}\left(1 + \dfrac{I}{K'_\Sigma}\right) + \dfrac{K_s}{V_m}\left(1 + \dfrac{I}{K_\Sigma}\right)\dfrac{1}{S_t}$ | $k_3 = 0$<br>$K'_s > K_s$ | Noncompetitive |
| I-C | $\dfrac{1}{v} = \dfrac{1}{V_m} + \dfrac{K_s}{V_m}\left(\dfrac{1 + I/K_\Sigma}{1 + I/K'_\Sigma}\right)\dfrac{1}{S_t}$ | $k_3 = k_1$<br>$K'_\Sigma > K_\Sigma$ | Partial competitive |
| I-D | $\dfrac{1}{v} = \dfrac{1}{V_m}\left[\dfrac{1 + I/K'_\Sigma}{1 + (k_3/k_1)(I/K'_\Sigma)}\right] + \dfrac{K_s}{V_m}\left[\dfrac{1 + I/K_\Sigma}{1 + (k_3/k_1)(I/K'_\Sigma)}\right]\dfrac{1}{S_t}$ | $k_3 < k_1$<br>$K_\Sigma = K'_\Sigma$ | Noncompetitive<br>($1/v$ vs. I-nonlinear) |
| I-E | $\dfrac{1}{v} = \dfrac{1}{V_m}\left[\dfrac{1 + I/K'_\Sigma}{1 + I/K_\Sigma}\right] + \dfrac{K_s}{V_m}\dfrac{1}{S_t}$ | $\dfrac{k_3}{k_1} = \dfrac{K'_\Sigma}{K_\Sigma} < 1$ | Uncompetitive |
| | (activation) | | |
| I-F | $\dfrac{1}{v} = \dfrac{1}{V_m}\left[\dfrac{1 + A/K'_\Sigma}{1 + (k_3/k_1)(A/K'_\Sigma)}\right] + \dfrac{K_s}{V_m}\left[\dfrac{1 + A/K_\Sigma}{1 + (k_3/k_1)(A/K'_\Sigma)}\right]\dfrac{1}{S_t}$ | $k_3 > k_1$<br>$K_\Sigma = K'_\Sigma$ | $V_m$ increase |
| I-G | $\dfrac{1}{v} = \dfrac{1}{V_m} + \dfrac{K_s}{V_m}\left[\dfrac{1 + A/K'_\Sigma}{1 + A/K_\Sigma}\right]\dfrac{1}{S_t}$ | $k_3 = k_1$<br>$K'_\Sigma < K_\Sigma$ | $K_m$ decrease |
| I-H | $\dfrac{1}{v} = \dfrac{1}{V_m}\left[\dfrac{1 + A/K'_\Sigma}{1 + A/K_\Sigma}\right] + \dfrac{K_s}{V_m}\left(\dfrac{1}{S_t}\right)$ | $\dfrac{k_3}{k_1} = \dfrac{K'_\Sigma}{K_\Sigma} > 1$ | $V_m$ increase<br>$K_m$ increase |

TABLE II
Substrate- and Enzyme-Directed Inhibition

| Case | Rate equation competitive cases | Condition | Effect[a] NS | Effect[a] AS |
|---|---|---|---|---|
| II-A | $\dfrac{1}{v} = \dfrac{1}{V_m} + \dfrac{K_s}{V_m}\left(\dfrac{1 + I/K_1}{1 + I/K_1'}\right)(1 + I/K_\Sigma)\dfrac{1}{S_t}$ | $k_1 = k_2$ | C | C |
|  |  | $K_1 < K_1'$ | NL | NL |
|  |  | $K_\Sigma' \to \infty$ | C | C |
| II-B | $\dfrac{1}{v} = \dfrac{1}{V_m} + \dfrac{K_s}{V_m}\left[\dfrac{(1 + I/K_1)(1 + I/K_\Sigma)}{1 + I/K_1' + I/K_\Sigma' + I^2/K_1'K_\Sigma''}\right]\dfrac{1}{S_t}$ | $k_1 = k_2 = k_3 = k_4$ | | |
|  |  | $K_1 < K_1'$ | NL | NL |
|  |  | $K_\Sigma < K_\Sigma'$ and/or $K_\Sigma''$ | | |
| II-C | $\dfrac{1}{v} = \dfrac{1}{V_m} + \dfrac{K_s}{V_m}(1 + I/K_1)(1 + I/K_\Sigma)\dfrac{1}{S_t}$ | $k_2, k_3,$ and $k_4 = 0$ | C | C |
|  |  | $K_1' \to \infty$ | NL | L |
|  |  | $K_\Sigma' \to \infty$ | | |
| II-D | $\dfrac{1}{v} = \dfrac{1}{V_m} + \dfrac{K_s}{V_m}\left[\left(\dfrac{1 + I/K_\Sigma}{1 + I/K_\Sigma'}\right)(1 + I/K_1)\right]\dfrac{1}{S_t}$ | $k_2$ and $k_4 = 0$ | C | C |
|  |  | $k_1 = k_3$ | NL | L |
|  |  | $K_1' \to \infty$ | | |
|  |  | $K_\Sigma < K_\Sigma'$ | | |

[a] NS, native substrate; AS, alternative substrate; C, competitive; NL, nonlinear; L, linear.

Tables I and II summarize the kinetic characteristics of the inhibition and activation mechanisms previously considered. For these cases it may be seen that the use of both reciprocal and Dixon plots allows the mechanism of an inhibitor to be determined with the exception of cases II-A from II-B and II-C from II-D. Cases II-A and II-B could be differentiated from one another by the characteristics of the inhibition profiles obtained from the Dixon plots. It can be shown that for cases II-A

$$\lim_{I \to \infty} \frac{1}{v} = \frac{1}{V_m} + \frac{K_s}{S_t V_m} \frac{K_I' I}{K_I K_\Sigma}$$

and for II-B

$$\lim_{I \to \infty} \frac{1}{v} = \frac{1}{V_m} + \frac{K_s}{S_t V_m} \frac{K_I' K_\Sigma''}{K_I K_\Sigma}$$

Thus, for case II-A, $1/v$ increases indefinitely as I increases, while for case II-B, $1/v$ reaches a constant value at high I. Case II-C can be differentiated from case II-D by conducting rate experiments at $S_t$ and I at a constant ratio $(I = S_t)$. For case II-C it can be shown that under these conditions, for a $1/v$ vs. I plot

$$\lim_{I \to \infty} \frac{1}{v} = \frac{\alpha K_s}{V_m} \frac{1}{K_I K_\Sigma}$$

while for Case II-D:

$$\lim_{I \to \infty} \frac{1}{v} = \frac{\alpha K_s}{V_m} \frac{K_\Sigma'}{K_I K_\Sigma}$$

Thus, for case II-C, the $1/v$ vs. I plot is parabolic; and for case II-D, the plot profile is sigmoidal.

## The Effect of Modifiers on the Kinetics of Enzyme-Catalyzed Chemical Modification: Model B

The model illustrated in Scheme 2 is for a two-substrate system. It involves an interconverting enzyme that catalyzes the modification of a macromolecular substrate, A, which has a binding site for ligand, M, with the second substrate, B, which does not bind ligands. The model assumes that the binding of all species shown is rapid and reversible. The model shows two rate-determining steps involving the interconversion of substrates to products. As shown in Scheme 1 for a one-substrate system, free and total modifier concentration are assumed to be equivalent.

Equation (4) is the general form of the rate equation.

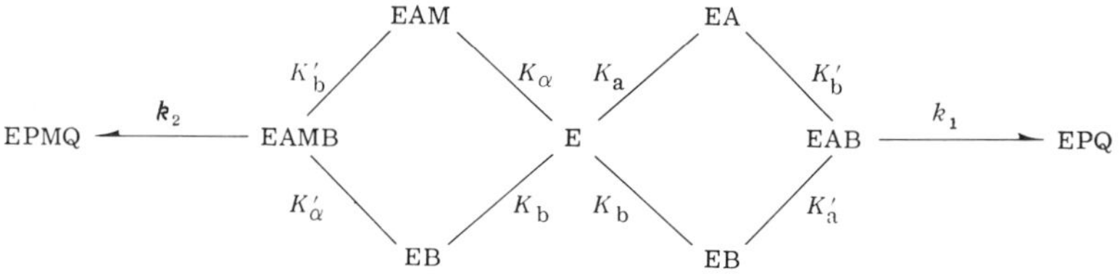

```
                    EAM              EA
              K'_b  /       K_α  K_a       K'_b
    k_2          /        \  |  /        \          k_1
EPMQ ←——— EAMB            E              EAB ———→ EPQ
              K'_α  \       K_b  K_b      K'_a
                    EB               EB
```

E + A = EA; $K_a$

E + B = EB; $K_b$

A + M = AM; $K_\Sigma$

EA + B = EAB; $K'_b$

EB + A = EAB; $K'_a$

E + AM = EAM; $K_\alpha$

EB + AM = EAMB; $K'_\alpha$

EA + M = EAM; $K'_\Sigma$

EAB + M = EAMB; $K''_\Sigma$

EAB $\xrightarrow{k_1}$ EPQ

EAMB $\xrightarrow{k_2}$ EPMQ

The conservation equations are:

$E_0$ = E + EA + EB + EAB + EAM + EAMB

$A_0$ = A + AM

$M_0 = M_f$

Scheme 2

$$\frac{V_m}{v} = \frac{1 + M/K''_\Sigma}{1+\phi} + \frac{K'_a}{A_0}\left(\frac{1 + M/K_\Sigma}{1+\phi}\right) + \frac{K'_b}{B}\left(\frac{1 + M/K'_\Sigma}{1+\phi}\right) \\ + \frac{K_a K'_b}{A_0 B}\left(\frac{1 + M/K_\Sigma}{1+\phi}\right) \quad (4)$$

where $\phi = (k_2/k_1)(M/K''_\Sigma)$ and $V_m = k_1 E_0$. Three identities by which the binding constants of modifier to A, EA, and EAB are related to other binding constants are

$$\frac{K_\Sigma}{K'_\Sigma} = \frac{K_a}{K_\alpha}; \quad \frac{K'_\Sigma}{K''_\Sigma} = \frac{K'_b}{K'''_b}; \quad \frac{K_\Sigma}{K''_\Sigma} = \frac{K'_a}{K'_\alpha}; \quad \frac{K_\alpha}{K'_\alpha} = \frac{K_b}{K'''_b}; \quad \frac{K_a}{K'_a} = \frac{K_b}{K'_b}$$

Two cases of substrate-mediated effects are analyzed. Mechanism I: A substrate modifier, M, binds to A to give a complex, AM, which does not bind to the enzyme. Thus, E + AM ≠ EAM, and EB + AM ≠ EAMB ($K_\alpha$ and $K'_\alpha \to \infty$). It follows from the identities above that $K'_\Sigma$ and $K''_\Sigma \to \infty$. The rate equation for this case is

$$\frac{V_m}{v} = 1 + \frac{K'_b}{B} + \left(K'_a + \frac{K_a K'_b}{B}\right)\left(1 + \frac{M}{K_\Sigma}\frac{1}{A_0}\right)$$

Double-reciprocal plots with inhibitor present give competitive kinetics vs. A and noncompetitive kinetics vs. B. This mechanism can be distinguished from a substrate-mediated inhibitor that prevents binding of B to the EAM complex (EAM + B ≠ EAMB), or the turnover of EAMB ($k_2 = 0$). In these cases competitive inhibition will not be seen with respect to A. Mechanism I seems to describe the inhibitory effect of glucose 6-phosphate on the phosphorylase kinase reaction.[12] Inhibition can be explained by the lack of binding of the phosphorylase $b$–glucose 6-phosphate complex to phosphorylase kinase. Mechanism II: A modifier, M, binds to the substrate to increase the binding affinity of substrate A ($K_\alpha = K'_\alpha < K_a = K'_a$). The rate equation for this case is

$$\frac{V_m}{v} = 1 + \frac{K'_b}{B} + \left(K'_a + \frac{K_a K'_b}{B}\right)\left(\frac{1 + M/K_\Sigma}{1 + M/K''_\Sigma}\right)\frac{1}{A_0}$$

Reciprocal plots of $1/v$ vs. A in the presence of M will have a common ordinate intercept. Plots of $1/v$ vs. B will not. This mechanism can be distinguished from modifier binding to A which increases the binding of B to the EAM complex, or the turnover of the EAMB complex ($k_2 > k_1$). In these cases a common ordinate intercept will not be seen in plots with respect to A.

Practical Aspects

Probably the most important consideration in comparing rates of reaction with natural and alternative substrates is whether one has satisfactorily measured the initial velocities of the reactions. Use of radiolabeled substrates, e.g., $^{32}$P-phosphorylated proteins or peptides greatly facilitates the analyses. We would not recommend following the reaction by indirect assays, e.g., loss of phosphorylase $a$ activity upon dephosphorylation. It is less sensitive and might give misleading information. In some cases the loss of phosphate from phosphorylase $a$ does not parallel the loss of enzymic activity.[13] A direct assay for product formation is best, and there are numerous procedures described for following phosphorylation[14] and dephosphorylation reactions.[15]

Perhaps it is an obvious point, but the reaction conditions ought to be the same for comparing rates with the two substrates, e.g., pH, buffer, and temperature. The two substrates used should be as similar as possible. We recommend the use of a peptide fragment of the natural substrate as an alternative substrate. The two substrates should be used at their $K_m$ values when modifiers are being evaluated. If one substrate were more saturating than another, differences in percent activation or inhibition could be seen. A wrong interpretation of these data could be made easily. With saturating substrate, effects of modifiers on binding could be

---

[12] J.-I. Tu and D. J. Graves, *Biochem. Biophys. Res. Commun.* **53**, 59 (1973).
[13] S. S. Hurd, D. Teller, and E. H. Fischer, *Biochem. Biophys. Res. Commun.* **24**, 79 (1966).
[14] E. M. Reimann, D. A. Walsh, and E. G. Krebs, *J. Biol. Chem.* **246**, 1986 (1971).
[15] T. M. Martensen, J. E. Brotherton, and D. J. Graves, *J. Biol. Chem.* **248**, 8323 (1973).

missed. In order to evaluate the precise mode of action, we suggest that the kinetic constants be determined. It is important to remember that the modifier should be in excess for our analyses to apply.

## Examples

The type of analysis we suggest has been applied to the study of the action of modifiers on the phosphorylase phosphatase reaction.[15,16] Illustrations of reciprocal plots expected for some cases of modifier action (Tables I and II) that apply to phosphorylase phosphatase are shown in Fig. 1.

The effect of a modifier that forms an inactive complex, SI, which is

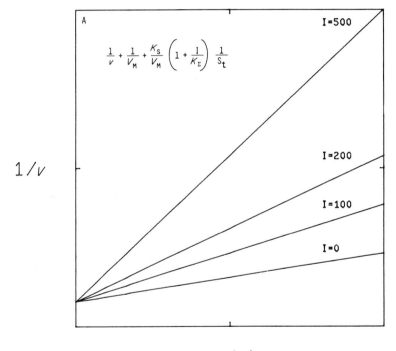

FIG. 1. Theoretical plots for the effect of modifiers on enzyme-catalyzed chemical modification. (A) Mechanism I-A, $K_s = 1$ $\mu M$, $K_\Sigma = 100$ $\mu M$, $V_m = 1$; (B) mechanism I-C, $K_s = 1$ $\mu M$, $K_\Sigma = 100$ $\mu M$, $K'_\Sigma = 250$ $\mu M$, $V_m = 1$; (C) mechanism II-C, $K_s = 1$ $\mu M$, $K_\Sigma = 100$ $\mu M$, $K_I = 20$ $\mu M$, $V_m = 1$; (D) mechanism I-F, $K_s = 1$ $\mu M$, $K_\Sigma = 100$ $\mu M$, $V_m = 1$, $k_3/k_1 = 3.5$; (E) mechanism I-G, $K_s = 1$ $\mu M$, $K_\Sigma = 100$ $\mu M$, $K'_\Sigma = 30$ $\mu M$, $V_m = 1$.

[16] T. M. Martensen, J. E. Brotherton, and D. J. Graves, *J. Biol. Chem.* **248**, 8329 (1973).

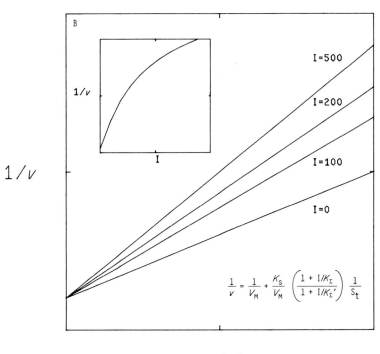
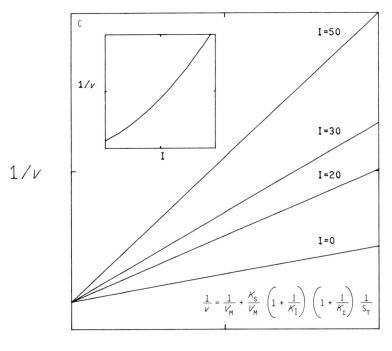

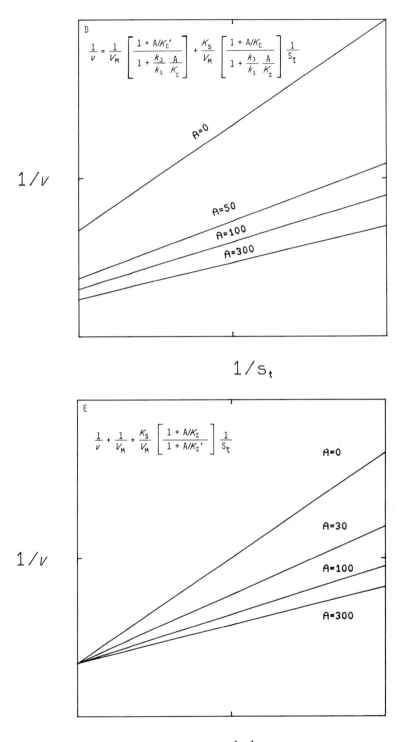

not recognized as a substrate (mechanism I-A) is shown in Fig. 1A. This mechanism was used to explain the inhibition of the phosphorylase phosphatase reaction by AMP.[15] Figure 1B shows that a secondary plot is useful to distinguish whether an inhibitory substrate modifier complex can bind to enzyme. The curvilinear nature of the plot shown in the inset is characteristic for an SI complex that binds less efficiently to enzyme yet is turned over at the same rate (mechanism I-C). The inhibition of the phosphorylase phosphatase reaction by glucose 1-phosphate showed these characteristics. Figure 1C illustrates the effect of a modifier that binds to enzyme as a competitive inhibitor, and also binds to substrate, rendering it unable to be bound by the enzyme (Mechanism II-C). The nonlinear plot seen in the inset illustrates this case of inhibition at two loci. The inhibition by $Mg^{2+}$ of the phosphorylase phosphatase reaction showed this pattern.[15] This pattern could also be shown by Case II-D. Figure 1D describes the kinetic behavior of a modifier that activates the reaction by enhancing the turnover of an ESA complex, Mechanism 1-F. A common intersection point on the abscissa is characteristic of this mechanism. The activation of the phosphorylase phosphatase reaction by glucose 6-phosphate can be described by this mechanism.[16] Figure 1-E shows the results expected for an activator that increases the ability of an SM complex to bind to interconverting enzyme. This corresponds to mechanism I-G. A common ordinate intercept is observed. Glycogen activation of the phosphorylase phosphatase reaction followed this pattern.[16]

Limitations

The equilibrium approach, not the steady state, has been used for the derivation of our rate equations. Thus, if the mechanism is steady state, our system does not apply. An important criterion of the applicability of the equilibrium method is the linearity of reciprocal plots in the presence of modifier. Also, we have pointed out that the modifier would be in excess of the substrate, so that the assumption $M_{total} = M_{free}$ is valid. This will be true for many cases, but unfortunately there are some examples where modifier binds so tightly to substrate that this assumption will not hold. Therefore, modifier action in these cases cannot be analyzed by our equations.

It is conceivable that a mechanism could exist where a modifier binds to the enzyme yet affects only the activity with the natural substrate. If the investigator did not know that the modifier bound to the enzyme, he might conclude that the action of the modifier was substrate directed since no effect was seen with the alternative substrate. In the analyses, it was assumed that alternative substrates did not bind modifiers. Any appreciable binding of modifiers to the alternative substrates would make the distinction between substrate-directed and enzyme-directed effects less clear.

## Concluding Remarks

The use of alternative substrates in conjunction with natural substrates in the study of enzymic covalent modification reactions has important value. Alternative substrates can be used for various purposes as we have indicated, but we have stressed one aspect, their use in determining the site and mechanism of action of modifiers. The determination of the site of action of a modifier is an important goal in terms of understanding enzyme regulation. The next phase, and a more complex part, is the analysis of the action of these modifiers. Kinetics is just one approach that can be taken to understand the molecular basis of modifier action.

## [14] Enzyme Kinetics of Lipolysis

### By R. VERGER

A large number of cellular enzymes are located in membranes, and the growing interest in the mechanisms controlling their activities has incited a number of biochemists to study in detail certain aspects of heteregeneous catalysis. Because of the difficulties encountered in the purification of these membrane proteins, our knowledge of their kinetic behavior is still very scanty. This explains why most experimental approaches used so far deal with model systems based on the combination either of (a) an immobilized enzyme + soluble substrate or (b) a soluble enzyme + insoluble substrate.

The immobilized enzyme system and its kinetic implications have been discussed in reviews by McLaren and Packer[1] and Goldstein,[2] as well as in chapter [9] of this volume. For more general aspects of the behavior of proteins at interfaces (kinetics and mechanism of adsorption, conformation of protein molecules at interfaces, the equilibrium aspects of adsorption and reactions at interfaces) the reader is referred to the reviews of Macritchie.[2a,2b]

---

[1] A. D. McLaren and L. Packer, *Adv. Enzymol.* **33**, 245 (1970).
[2] L. Goldstein, this series, Vol. 44, p. 397.
[2a] F. Macritchie, *in* "Interfacial Synthesis" (F. Millich and C. E. Carraher, eds.), Vol. 1, p. 103. Marcel Dekker, New York, 1977.
[2b] F. Macritchie, *in* "Advances in Protein Chemistry" (C. B. Anfinsen, J. T. Edsall, and F. M. Richards, eds.), Vol. 32, p. 283. Academic Press, New York, 1978.

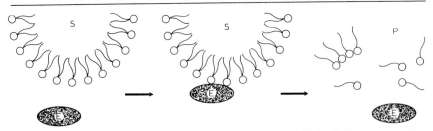

FIG. 1. Schematic picture of lipolysis. From R. Verger and G. H. de Haas, *Annu. Rev. Biophys. Bioeng.* **5**, 77 (1976).

In this review, we limit ourselves to studies based on approach (*b*). It will come as no surprise that most of the work reported so far deals with lipolysis. The naturally occurring (phospho)lipids are important building stones of the biological membranes. They are water insoluble and spontaneously form molecular aggregates, such as monomolecular films, bilayers, emulsions, liposomes or micelles. In addition, a number of soluble enzymes that play an important role in such biological events as digestion, lipid transport, and lipid metabolism are known and have been isolated. Recently, phospholipases were used in the determination of asymmetric phospholipid distribution in membranes.[3] The fact that it is the substrate that forms molecular aggregates makes lipolysis a very attractive system for studies of interfacial enzyme kinetics.

The highly schematic picture of lipolysis illustrated by Fig. 1, in which a soluble enzyme, E, transforms an insoluble substrate, S, into soluble products, P, raises several questions. How does the enzyme interact with the interface? Is this interaction reversible? Does the contact modify the architecture of the lipid–water interface and/or of the enzyme? Is the catalytic mechanism of the enzymic reaction altered?

*Activation of Lipolytic Enzymes by Interfaces*

One of the most characteristic and intriguing features of lipolytic enzymes is their activation by interfaces. This phenomenon was recognized very early by Holwerda *et al.*[4] and Schønheyder and Volqvartz,[5] who, using tricaproin as substrate for pancreatic lipase, showed clearly that the rate of breakdown of a molecular solution of this glyceride is very slow,

---

[3] B. Roelofsen and R. F. A. Zwaal, *in* "Methods in Membrane Biology" (E. D. Korn, ed.), Vol. 7, p. 147. Plenum, New York, 1976.
[4] K. Holwerda, P. E. Verkade, and A. H. A. de Willingen, *Recl. Trav. Chim. Pays-Bas* **55**, 43 (1936).
[5] F. Schønheyder and K. Volqvartz, *Acta Physiol. Scand.* **9**, 57 (1945).

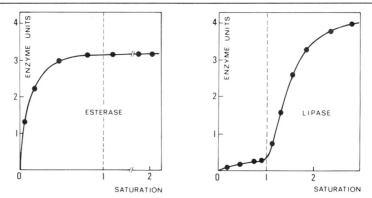

FIG. 2. Hydrolysis of triacetin by horse liver esterase and porcine pancreatic lipase. Reaction rates are given as a function of substrate concentration, which is expressed in multiples of saturation. From L. Sarda and P. Desnuelle, *Biochim. Biophys. Acta* **30**, 513 (1958).

but that, once the substrate solubility is exceeded, the enzymic activity increases dramatically.

In 1958, when highly purified preparations of the enzyme became available, the problem was reinvestigated by Sarda and Desnuelle[6] in a more quantitative way. They clearly demonstrated a fundamental difference between an ordinary well known esterase and pancreatic lipase based upon ability or inability to be activated by interfaces. This is illustrated in Fig. 2. In contrast to the esterase, which shows a normal Michaelis–Menten activity dependence on substrate concentration, the lipase displays almost no activity with the same substrate when it is in the monomeric state. However, when the solubility limit of triacetin is exceeded, there is a sharp increase in enzyme activity with the same substrate in the emulsified state. Apparently, the esterase is active only on molecularly dispersed substrates. The lipases appear to constitute a special class of esterases capable of hydrolyzing plurimolecular aggregates at a high rate.

This peculiar behavior of pancreatic lipase turned out not to be limited to emulsions. Entressangles and Desnuelle[7] showed that, in isotropic solutions of short-chain triglycerides containing NaCl, small micellar aggregates were formed on which lipase displayed a maximal velocity similar to that measured with the same substrate in the emulsified state. More recently, Brockman and co-workers reported a 1000-fold increase in the rate of hydrolysis of tripropionin by pancreatic lipase in the presence of siliconized glass beads.[8] The interpretation given by these investigators is

---

[6] L. Sarda and P. Desnuelle, *Biochim. Biophys. Acta* **30**, 513 (1958).
[7] B. Entressangles and P. Desnuelle, *Biochim. Biophys. Acta* **159**, 285 (1968).
[8] H. L. Brockman, J. H. Law, and F. J. Kézdy, *J. Biol. Chem.* **248**, 4965 (1973).

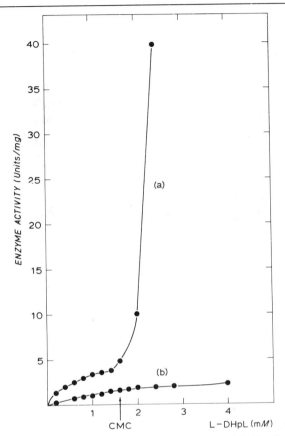

FIG. 3. Hydrolysis of 1,2-diheptanoylglycerol-3-*sn*-phosphorylcholine (L-DHpL) by porcine pancreatic phospholipase $A_2$ (curve a) and its zymogen (curve b). Reaction rates are given as a function of substrate concentration. From W. A. Pieterson, J. C. Vidal, J. J. Volwerk, and G. H. de Haas, *Biochemistry* **13**, 1455 (1974). Copyright by the American Chemical Society.

that a spherical glyceride monolayer forms around the beads to give rise to an interface similar to those in micelles or emulsions.

Pieterson *et al.*[9] and Misiorowski and Wells[10] have shown that other lipolytic enzymes are also strongly activated by substrate aggregation. The latter group found an increase of about $10^4$-fold in enzyme activity in an ethyl ether medium in which small water droplets containing phospholipase $A_2$ and $Ca^{2+}$ are coated by a monolayer of dioctanoyllecithin.

Figure 3 shows how the activity of porcine pancreatic phospholipase $A_2$ depends on the substrate concentration (curve a). In this case, a

[9] W. A. Pieterson, J. C. Vidal, J. J. Volwerk, and G. H. de Haas, *Biochemistry* **13**, 1455 (1974).
[10] R. L. Misiorowski and M. A. Wells, *Biochemistry* **13**, 4921 (1974).

water-soluble short-chain lecithin, 1,2-diheptanoyl-glycero-3-$sn$-phosphorylcholine, was used as substrate. Below the critical micelle concentration (CMC), where only monomeric lecithin molecules are present, phospholipase A activity remains very low and seems to follow normal Michaelis–Menten kinetics. When the monomers aggregate above the CMC, there occurs a strong increase in lipolytic activity indicating that the micellar aggregates are a much better substrate for the enzyme than the molecules dispersed in water.

The naturally occurring zymogen of pancreatic phospholipase $A_2$ has about the same activity as the enzyme on monomeric short-chain lecithins (curve b). However, above the CMC the formation of substrate aggregates does not accelerate the hydrolysis. There must be a special fit between the respective geometries of the enzyme active site and aggregated substrate inducing a very large activation effect.

As one might expect, not all interfaces are equivalent in this effect. From bulk studies with pancreatic phospholipase $A_2$, it has been shown that homologous lecithins with acyl chain lengths varying only from 6 to 10 carbon atoms are hydrolyzed at very different rates.[11] For example, under similar conditions of ionic strength the enzyme hydrolyzes the dioctanoyl derivative with a specific activity of 6000 $\mu$mol min$^{-1}$ mg$^{-1}$, whereas the didecanoyllecithin is not hydrolyzed at all. In contrast, it was found by the monolayer technique that lecithins with acyl chain lengths varying from 8 to 12 carbon atoms are hydrolyzed at about the same rate.[12] This difference in response to chain length can be assumed to be related, not to the different techniques (bulk and monolayer) used, but to a strong dependence of enzyme activity on the "quality" of the interfacial structure of the substrate. A direct confirmation of this view is provided by the fact that equimolar mixtures of 3-$sn$-dioctanoyllecithin + 1-$sn$-didecanoyllecithin and of 1-$sn$-dioctanoyllecithin × 3-$sn$-didecanoyllecithin are hydrolyzed in bulk at the same rate.[13] In the former mixture only the dioctanoyllecithin is hydrolyzed; in the latter, only the didecanoyl derivative, because of the stereospecificity of the enzyme. These results confirm those obtained in the monolayer system. Once the "quality" of the interface is equalized, the enzyme acts with the same velocity.

Adamich and Dennis[14,15] recently reported that the specificity of phospholipase $A_2$ in mixed Triton–phospholipid micelles could be reversed when more than one lipid was present. This interesting reversal of appar-

[11] G. H. de Haas, P. P. M. Bonsen, W. A. Pieterson, and L. L. M. Van Deenen, *Biochim. Biophys. Acta* **239**, 252 (1971).
[12] G. Zografi, R. Verger, and G. H. de Haas, *Chem. Phys. Lipids* **7**, 185 (1971).
[13] R. Verger, M. C. E. Mieras, and G. H. de Haas, *J. Biol. Chem.* **248**, 4023 (1973).
[14] M. Adamich and E. A. Dennis, *Biochem. Biophys. Res. Commun.* **80**, 424 (1978).
[15] M. Adamich and E. A. Dennis, *J. Biol. Chem.* **253**, 5121 (1978).

ent substrate specificity was explained as either a lipid–lipid interaction leading to a conformational change of the substrate at the lipid–water interface or as a result of the "dual phospholipid" model of Roberts et al.[16] One can imagine that each lipid substrate could have a variable affinity and/or susceptibility for the enzyme at the interface. Thus, the enzyme can bind and catalyze each class of lipid molecules *very differently* as a function of the overall quality of the interface and independently of variable associations with the interface. This may be compared to the effects of solvent properties on enzyme activities in conventional aqueous enzymology. As discussed later in more detail in the section on examples, the influence of bile salts on pancreatic lipase activity constitutes another example of the influence exerted by the quality of the interface on the kinetic properties of lipolytic enzymes.

As shown in the following sections, the definition and measurement of the concentration of insoluble substrates constitute a first obstacle in interfacial enzyme kinetic studies. In enzyme reactions that follow Michaelis–Menten kinetics, in which the enzyme and the substrate are water soluble, substrate concentation is always expressed in moles per liter. With insoluble or partly soluble substrates that form a lipid–water interface, the enzymic reaction takes place at the interface and is therefore two-dimensional in character. The term "supersubstrate" has been proposed by Brockerhoff and Jensen[17] to describe the matrix in which a substrate molecule is located.

Interfaces occur in nature or can be prepared experimentally as emulsions, micelles or liposomes, and monolayers. The practical aspects of this review, therefore, are classified according to the physical state of the different substrates used.

## Theory

### Kinetic Model for Lipolysis of Insoluble Lipids

In order to explain the pathways of lipolysis, several investigators have proposed a reversible enzyme adsorption to, or penetration into, the interface.[13,18,19] Moreover, the first step was supposed to precede the formation of the enzyme substrate complex and was found to be rate limiting under certain circumstances. Figure 4 illustrates a simple model proposed

---

[16] M. F. Roberts, R. A. Deems, and E. A. Dennis, *Proc. Natl. Acad. Sci. U.S.A.* **74**, 1950 (1977).
[17] H. Brockerhoff and R. G. Jensen, "Lipolytic Enzymes," p. 17. Academic Press, New York, 1974.
[18] H. Brockerhoff, *Biochim. Biophys. Acta* **212**, 92 (1970).
[19] I. R. Miller and J. M. Ruysschaert, *J. Colloid Interface Sci.* **35**, 340 (1971).

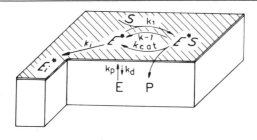

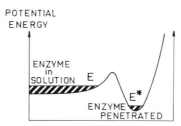

FIG. 4. Proposed model for the action of a soluble enzyme at an interface. From R. Verger, M. C. E. Mieras, and G. H. de Haas, *J. Biol. Chem.* **248,** 4023 (1973) modified by J. Rietsch, F. Pattus, P. Desnuelle, and R. Verger, *J. Biol. Chem.* **252,** 4313 (1977).

by Verger et al.[13] and modified by Rietsch et al.[20] to describe kinetically the action of a lipolytic enzyme at an interface.

The model consists basically of two successive equilibria. The first describes the penetration[21] of a water-soluble enzyme into an interface ($E \rightleftarrows E^*$). This is followed by a second equilibrium, in which one molecule of penetrated enzyme binds a single-substrate molecule giving the complex $E^*S$. This is the equivalent in two dimensions of the classical Michaelis–Menten equilibrium. Once the complex $E^*S$ is formed, the catalytic steps take place, regenerating the enzyme in the form $E^*$ along with liberation of the products. The only case considered is one in which all the products of the reaction are soluble in the water phase, diffuse rapidly way, and induce no change with time in the physicochemical properties of the interface.

[20] J. Rietsch, F. Pattus, P. Desnuelle, and R. Verger, *J. Biol. Chem.* **252,** 4313 (1977).
[21] Penetration or adsorption will be used interchangeably. Both refer to a rate-limiting step preceding the formation of the enzyme–substrate complex.

## Interfacial Enzyme Inactivation

Some years later, an irreversible step converting the penetrated enzyme E* into an inactive form $E_i^*$ and competing with the formation of the productive complex E*S was added to the scheme. This progressive inactivation of lipolytic enzymes at the interface has been reported to occur in lipid monolayers at low surface pressure[19,20,22] as well as at a hexadecane–water interface,[23] siliconized glass beads surface,[24] and triglyceride–water interface.[23,25] As a consequence of these consecutive penetration and inactivation steps, the lipolytic enzyme kinetics are expected to be qualitatively controlled by an adsorption flux responsible for an initial lag period and an inactivation flux tending to decrease the reaction rate. The kinetics are strictly linear either when both fluxes equilibrate temporarily or when the penetration step is fast and the inactivation rate is negligible.

## Reversibility of the Adsorption Step

Dervichian et al.[26] have interpreted their kinetic data obtained by the monolayer technique as an indication of irreversible binding of the enzyme at the interface. This was confirmed at low surface pressure by transfer experiments using radioactively labeled enzyme.[20] It can be postulated that irreversible or quasi irreversible adsorption would provide a kinetic advantage for the enzyme, the substrate concentration around the adsorbed enzyme being close to that existing in a pure lipid phase. However, at high surface pressure, the reversibility of the adsorption was checked by Pattus et al.[27]

Using pure snake venom phospholipase $A_2$, Wells[28] investigated enzyme kinetics on a micellar lipid system. His finding that freely moving dibutyryllecithin monomers competively inhibit the hydrolysis of the very large dioctanoyllecithin micelles strongly supports the idea that the enzyme is reversibly absorbed by the micellar aggregates. In what follows, the kinetic treatments will be made, as in the usual case, under the as-

---

[22] H. Cohen, B. W. Shen, W. R. Snyder, J. H. Law, and F. J. Kézdy, *J. Colloid Interface Sci.* **56**, 240 (1976).
[23] H. Brockerhoff, *J. Biol. Chem.* **246**, 5828 (1971).
[24] W. E. Momsen and H. L. Brockman, *J. Biol. Chem.* **251**, 378 (1976).
[25] A. Vandermeers, M. C. Vandermeers-Piret, J. Rathé, and J. Christophe, *Biochim. Biophys. Acta* **370**, 257 (1974).
[26] D. G. Dervichian, C. Préhu, and J. P. Barque, *C. R. Acad. Sci. Ser. D* **276**, 839 (1973).
[27] F. Pattus, A. J. Slotboom, and G. H. de Haas, *Biochemistry* **13**, 2691 (1979).
[28] M. A. Wells, *Biochemistry* **13**, 2248 (1974).

sumption that the formation of the enzyme–substrate complex is reversible.

The penetration of the enzyme into the interface is assumed to confer on the enzyme a new conformation (E*) having much more efficient catalytic properties than the soluble enzyme (E). The penetration step is thus considered to be distinct from any more general and unspecific adsorption that might precede it. Support for this conformational change comes from ultraviolet and fluorescence spectroscopy measurements showing that some chromophoric residues of pancreatic phospholipase $A_2$, in sharp contrast to its zymogen, give different signals in the presence or in the absence of a lipid–water interface.[29]

The model presented in Fig. 4 is valid for enzymes other than lipolytic enzymes. Theoretically, it can be applied to describe the action of any soluble enzyme on an insoluble substrate. Furthermore, for the kinetic treatment of the model, the shape of the interface does not impose any restrictions. It may be spherical or ellipsoidal (micelles, emulsions, liposomes) or planar (monolayers). According to the proposed model and assuming for the sake of simplicity that the enzyme inactivation is negligible, the general formula that describes the release of the products (P) of the reaction as a function of time ($t$) is

$$P = (C/B)\{t + \tau_1[\exp(-t/\tau_1) - 1] + \tau_2[\exp(-t/\tau_2) - 1]\} \quad (1)$$

in which the factors $B$ and $C$ are known functions of the individual rate constants.[13]

As will be shown later, $\tau_1$ and $\tau_2$ are the induction times of the equilibria $E \rightleftarrows E^*$ and $E^* \rightleftarrows E^*S$, respectively. The general equation (1) is valid during the steady state and the presteady state for soluble enzymes acting at interfaces according to the above model.[13]

*Steady-State Conditions.* When time tends to infinity (steady-state conditions), the general equation (1) reduces to

$$P = (C/B)[t - (\tau_1 + \tau_2)] \quad (2)$$

Then, the expression of the enzymic velocity is

$$v_b = v_m \frac{I}{V} = \frac{k_{cat}E_0 \cdot S}{S + K_m^*} \frac{S(I/V)}{S(I/V) + (k_d/k_p) \cdot (K_m^*S/\,S + K_m^*)} \quad (3)$$

See Verger *et al.*[13] for the definition of these parameters. This formula describes the steady state in bulk and monolayer conditions.

BULK CONDITIONS. From Eq. (3), one can immediately derive the ex-

---

[29] M. C. E. Van Dam-Mieras, A. J. Slotboom, W. A. Pieterson, and G. H. de Haas, *Biochemistry* **14**, 5387 (1975).

pression of $V_{max}$ and $K_m$ obtained generally in bulk conditions by measuring the enzymic velocity as a function of the substrate concentration (proportional to $I/V$).

$$V_{max} = \frac{k_{cat}E_0 \cdot S}{S + K_m^*} \quad (4)$$

$$K_m = \frac{k_d}{k_p} \times \frac{K_m^* S}{S + K_m^*} \quad (5)$$

$V_{max}$ and $K_m$ are complex functions of several constants, including $S$ defined as substrate concentration and expressed as the number of molecules per unit surface. However, under bulk conditions, it is possible by increasing the amount of substrate (area of interface per unit volume) to reach a situation in which all the enzyme is in the penetrated form ($E^*$ + $E^*S$). This has been generally interpreted as giving the maximal velocity $V_{max}$.[17,30] It should be realized that, in bulk conditions, the effective substrate concentration (molecules per unit surface) cannot be changed by adding more lipid. This means that an unknown fraction of $E^*$ is converted to $E^*S$ and, therefore, the $V_{max}$ and $K_m$ values measured under bulk conditions are only apparent values. For a better evaluation of the influence of varying the substrate concentration (molecules per unit surface), the monolayer technique must be used.

MONOLAYER CONDITIONS. A major difference between the monolayer and the bulk system lies in the fact that their ratios of interface area to volume ($I/V$) are of different orders of magnitude. In the monolayer systems, this ratio is usually about 1 cm$^{-1}$, depending upon the depth of the trough, whereas in the bulk system it can be as high as $10^5$ cm$^{-1}$, depending upon the amount of lipid used. As a consequence, bulk conditions allow the adsorption of nearly all the enzyme at the interface, whereas with a monolayer only 1 enzyme molecule out of 2500 may be at the interface.[13,31] This corresponds to the surface covered by a ball in a football field. When ($I/V$) is very small, as in the monolayer system, we find that

$$E \gg (E^* + E^*S)I/V$$

and therefore Eq. (3) of the steady state reduces to

$$v_m = (k_{cat}E_0 \cdot S)/[(k_d/k_p)K_m^*] \quad (6)$$

This is analogous to the formula derived by Miller and Ruysschaert using the monolayer technique.[19]

---

[30] M. Dixon and E. C. Webb, "Enzymes," 2nd ed., p. 92. Academic Press, New York, 1964.
[31] E. A. Dennis, *Arch. Biochem. Biophys.* **158**, 485 (1973).

When one looks at Eq. (6), it is evident that the enzymic velocity measured by the monolayer technique must be linearly dependent upon the enzyme concentration in the subphase. In fact, this is what is generally found experimentally. It can also be seen that theoretically the velocity should depend linearly upon the substrate concentration, $S$. It has been reported that, in a limited range of substrate concentrations, $V_m$ and $S$ have a quasi-linear relationship.[8,19,32]

*Presteady-State Conditions.* The general equation (1) possesses an asymptote. When the time tends to infinity, the equation of this asymptote is represented by Eq. (2)

$$P = (C/B)t - C/B(\tau_1 + \tau_2)$$

where, as is usual, the intercept of this asymptote with the time axis is designated induction time, or $\tau$. In the general case, the induction time ($\tau$) is the sum of two independent induction times $\tau_1$ and $\tau_2$ reflecting, respectively the penetration-desorption and the interfacial Michaelis–Menten equilibrium.

Under certain experimental conditions, it is likely that one of the two induction times is much less than the other, and thus the interpretation of the kinetic analysis during the presteady state is simpler.[13]

The zero-order trough, consisting of two compartments connected by a narrow surface canal (see Fig. 12), gives, as expected, linear kinetics after injection of the venom enzyme under the monolayer.[33] However, as shown in Fig. 5a (left panel), this is not the case when pancreatic phospholipase A is injected under a film of dinonanoyl lecithin.[34] In fact, one observes that the velocity given by the slope of the recorded curve increases with time and seems to approach an asymptotic limit indicated by the dashed line; the intercept between the asymptote and the time axis is the induction time, $\tau$. This behavior is in sharp contrast with the kinetics shown in Fig. 5a, right panel, which is obtained by injection of pure phospholipase $A_2$ from snake or bee venom under the same lecithin film.

What is the reason for the lag period observed for the pancreatic enzyme? One can immediately rule out the possibility that it is due to slow mixing of the injected enzyme below the surface film. It has been shown

[32] R. Verger, J. Rietsch, M. C. E. Van Dam-Mieras, and G. H. de Haas, *J. Biol. Chem.* **251**, 3128 (1976).
[33] R. Verger and G. H. de Haas, *Chem. Phys. Lipids* **10**, 127 (1973).
[34] The bulk hydrolysis of dinonanoyllecithin by pancreatic phospholipase A is also characterized by a long lag period. A quantitative approach in this case, however, is more complicated because of the presence of the products of the reaction (fatty acids), which reach rather high concentrations. The combination of an accelerating effect at the early stages of the reaction with the inhibitory effects later on produces S-shaped kinetics that cannot be analyzed accurately.

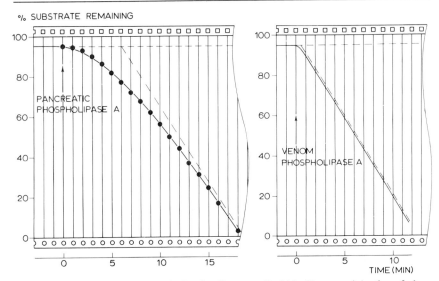

FIG. 5a. Kinetics of the hydrolysis of a dinonanoyllecithin film upon injection of phospholipase $A_2$ from different sources. The continuous curves are the tracings from the barostat recorder. The points as well as the dashed lines are computed values. From R. Verger, et al. M. C. E. Mieras, and G. H. de Haas, *J. Biol. Chem.* **248**, 4023 (1973).

that complete mixing is attained in less than 30 sec. A slow diffusion of enzyme from the bulk to the surface through an unstirred layer can be excluded as well. Langmuir and Schaeffer[35] have shown that, upon stirring of the subphase, the thickness of the unstirred layer is of the order of 0.01 mm. Using the diffusion constant[36] of pancreatic phospholipase $A_2$, $1.35 \times 10^{-6}$ cm² sec⁻¹, one can estimate that under these conditions the lag period would be of the order of seconds. Second, the venom phospholipases, which have been reported[37,38] to possess diffusion constants similar to that of the pancreatic enzyme, do not show a lag period. Finally, the hydrolysis by pancreatic phospholipase $A_2$ of other substrates, such as phosphatidylglycerol and its lysyl derivative, is not characterized by a lag period.[13]

To explain the unusually long induction time observed during the hydrolysis of lecithin films by pancreatic enzyme, one may assume that the enzyme penetrates slowly into the monolayer. In other words, taking into

[35] I. Langmuir and V. F. Schaeffer, *J. Am. Chem. Soc.* **59**, 2400 (1937).
[36] G. H. de Haas, N. M. Postema, W. Nieuwenhuizen, and L. L. M. Van Deenen, *Biochim. Biophys. Acta* **159**, 103 (1968).
[37] M. A. Wells and D. J. Hanahan, *Biochemistry* **8**, 414 (1969).
[38] R. A. Shipolini, G. L. Callewaert, R. C. Cottrell, S. Doonan, C. A. Vernon and B. E. C. Banks, *Eur. J. Biochem.* **20**, 459 (1971).

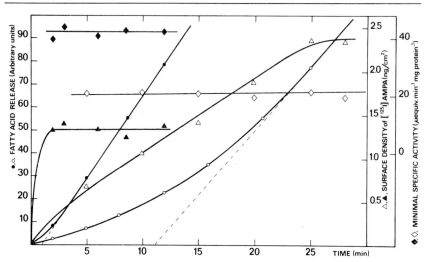

FIG. 5b. Adsorption kinetics of [$^{125}$I]AMPA (E-amidinated phospholipase $A_2$) on didecanoyllecithin monolayers at 10 dynes/cm, pH 6.0 (▲) and pH 8.5 (△). Progress of the reaction at pH 6.0 (●) and pH 8.5 (○). Minimal specific activity at pH 6.0 (◆) and pH 8.5 (◇). From F. Pattus, A. J. Slotboom, and G. H. de Haas, *Biochemistry* **13**, 2691 (1979). Copyright by the American Chemical Society.

account the initial model, the establishment of the equilibrium E ⇌ E* could be the rate-limiting step in the overall enzymic process. Once the enzyme has penetrated, the formation of the interfacial Michaelis complex and the liberation of products starts immediately. As an alternative explanation, one can also imagine that the formation of this complex (E* + S ⇌ E*S) or an enzyme conformational change could be rate-limiting. In order to discriminate between these two possibilities, Pattus *et al.*[27] measured simultaneously as a function of time the enzyme activity and the amount of radioactive enzyme in excess at the interface as shown in Fig. 5b. Under conditions of short (1.4 min) or long (11 min) induction times, the amount of radioactive enzyme present in or close to the monolayer increased continuously up to saturation value. This increase was found to be closely related to the increase in enzyme activity during the same period of time. As a consequence, the specific activity of the enzyme remained constant during all the time of the kinetic experiment. These data demonstrate clearly that the induction times observed during the hydrolysis of lipid monolayers by lipolytic enzymes are related to a slow enzyme penetration into the interface.

The concept of a penetration step was confirmed by studying the influence of surface pressure on the lag period using two different lipolytic enzymes, lipase and phospholipase $A_2$ of pig pancreas acting on

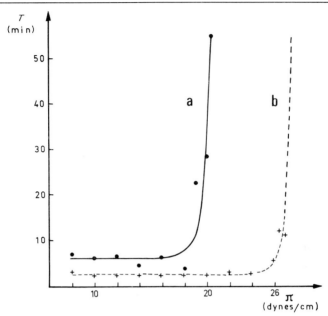

FIG. 6. Variations with surface pressure of the induction times. Hydrolysis of monolayers of didodecanoylphosphatidylethanolamine by pancreatic lipase (curve a) and pancreatic phospholipase $A_2$ (curve b). From R. Verger, J. Rietsch, M. C. E. Van Dam-Mieras, and G. H. de Haaas, *J. Biol. Chem.* **251**, 3128 (1976). Copyright by the American Chemical Society.

1,2-didodecanoyl-*sn*-glycero-3-phosphorylethanolamine monolayers. As shown in Fig. 6, the two enzymes behave differently. For each enzyme, there is a characteristic critical packing density of the substrate molecules above which the enzyme cannot penetrate into the film.[32] Recently, Rothen *et al.* (personal communication) measured the influence of surface pressure on the induction time of four different phospholipases $A_2$. The authors could unambiguously classify these enzymes according to their penetration power. This type of information can be of practical importance in choosing a lipolytic enzyme to degrade the lipid moiety of biological membranes.[3,39]

### Kinetic Models Applicable to Partly Soluble Amphiphilic Lipids

At low concentration in water, some amphiphilic lipids are freely soluble as monomers. However, at the critical micellar concentration (CMC),

---

[39] L. L. M. Van Deenen, R. A. Demel, W. S. M. Geurts van Kessel, H. H. Kamp, B. Roelofsen, A. J. Verkleij, K. W. A. Wirtz, and R. F. A. Zwaal, in "The Structural Basis of Membrane Function" (Y. Hatefi and L. Djavadi-Ohaniance, eds.), p. 21, Academic Press, New York, 1976.

the limit of molecular solubility is reached and aggregation of the molecules into particles starts. Such micellar solutions are optically clear, and the lipid molecules in the micelle are in rapid equilibrium with the molecules in free solution. In principle, the same or separate active sites on a lipolytic enzyme can be involved during the hydrolysis of monomers and micelles of soluble amphiphilic lipids. Wells[28] has proposed a two-substrate model in which the same site of an enzyme interacts and catalyzes either the reaction of a soluble amphiphilic lipid as a monomer or present in a micelle. The concentration of the first substrate is the monomer concentration in true molecular solution. The concentration of the second substrate is the concentration of the same lipid molecules, which are incorporated into a micellar structure (substrate in the aggregated form). Rate equations were derived based on the kinetic theory of Michaelis and Menten assuming independent kinetic parameters for these two physical forms of the same substrate molecule.

On the basis of this two-substrate model, Gatt and Bartfai[40] described rate equations and simulation curves for enzymic reactions in an attempt to interpret the numerous "irregular" $v$ vs. S curves observed in lipid enzymology when using a partially soluble lipid as substrate. Thirteen simulated $v$ vs. S curves were obtained, most of which were not hyperbolic. Furthermore, these authors stated that such kinetic curves may be obtained and interpreted solely on the basis of the nature of the physical dispersion of the lipid substrate, disregarding any possible effect of the latter on the structure of the enzyme itself. An alternative, though probably less likely, model proposed by the same authors, proposes an enzyme with two separate sites, one for monomers, the second for micelles, each obeying Michaelis–Menten kinetics. However, this latter assumption is not supported by the data obtained by Wells[28] using snake venom phospholipase $A_2$. Dibutyryllecithin monomers behave as a competitive inhibitor in the hydrolysis of dioctanoyllecithin micelles with a $K_i = 40$ m$M$, a value identical with the $K_m$ for the hydrolysis of dibutyryllecithin. These results clearly indicate that a lipid molecule in a micellar structure and a monomer can compete for the same site on the enzyme. Gatt and Bartfai,[41] on the basis of the Michaelis–Menten model and Langmuir adsorption isotherm, analyzed theoretically several other factors that might be responsible for the "irregular" kinetic curves frequently observed in lipid enzymology: the enzyme is partly adsorbed, via noncatalytic sites, to lipid aggregates. The substrate is adsorbed into an added protein, such as albumin. The latter case is further divided into those cases where addition of albumin decreases, or alternatively increases, the reaction rates.

---

[40] S. Gatt and T. Bartfai, *Biochim. Biophys. Acta* **488,** 1 (1977).
[41] S. Gatt and T. Bartfai, *Biochim. Biophys. Acta* **488,** 13 (1977).

Although all rate equations were based on the Michaelis–Menten kinetic theory, most of the simulated $v$ vs. S curves were not hyperbolic and some of the $v$ vs. E curves were not linear.

## Surface Dilution Model Applicable to Mixed Micelles

Deems et al.[42] proposed a detailed surface-dilution kinetic scheme for the action of lipolytic enzymes on mixed micelles of phospholipid and surfactant (Triton)

$$E + A \underset{k_{-1}}{\overset{k_1}{\rightleftharpoons}} EA$$

$$EA + B \underset{k_{-2}}{\overset{k_2}{\rightleftharpoons}} EAB \underset{k_{-3}}{\overset{k_3}{\rightleftharpoons}} EA + Q$$

where E is the enzyme, A is the mixed micelle, and B is the phospholipid substrate in the mixed micelle. This scheme takes into account quantitatively the involvement of the lipid–water interface in the action of this enzyme toward substrate in macromolecular complexes. This requires the enzyme to undergo at least two binding steps before catalysis occurs; during the first step the enzyme must be sequestered at the surface. The interface is thus considered to be the first substrate in a bisubstrate enzyme reaction. The concentration of this first substrate is expressed in units of area per unit volume. Once the enzyme is bound to the surface, it then specifically binds a phospholipid molecule in its active site. This model is conceptually similar to the one described previously for lipolysis of insoluble lipids.[13] Nevertheless, the mixed micellar system is interesting because in principle the substrate concentration in the interface can be varied easily by changing the molar ratio of detergent to phospholipid. This system allows the activity of the enzyme to be followed by standard kinetic techniques. Using phospholipase $A_2$, phospholipase C, and a highly purified membrane-bound phosphatidylserine decarboxylase, Dennis et al.[42,43] were able to separate and to give a numerical value to the association constant of each enzyme with the lipid–water interface and to the binding constant of the phospholipid molecule as part of the interface. These calculations, however, were based on the simplifying assumption that the "quality of the interface" does not vary with changes in composition of the mixed micelle. Moreover, it is not clear whether the detergent molecules really behave as neutral dilutors without any affinity for the enzyme or should be considered as competitive inhibitors for the binding of substrate monomers by the enzyme.

[42] R. A. Deems, B. R. Eaton, and E. A. Dennis, J. Biol. Chem. **250**, 9013 (1975).
[43] T. G. Warner and E. A. Dennis, J. Biol. Chem. **250**, 8004 (1975).

## Practical Aspects

### Emulsified Substrates

Using as substrate a long-chain triglyceride emulsion, Benzonana and Desnuelle[44] demonstrated that lipase adsorbs to the emulsified particles and that this adsorption follows a Langmuir isotherm (Fig. 7, ●—●). In addition, they measured the lipase activity as a function of the quantity of emulsion (Fig. 7, ○—○). The correlation between adsorption and enzyme activity was expressed by the following scheme:

$$E_{(solution)} + S_{(emulsion)} \underset{k_2}{\overset{k_1}{\rightleftharpoons}} (ES)_{(emulsion)} \xrightarrow{k_3} products$$

The formal resemblance of this scheme to the well-known Michaelis–Menten model still tempts many authors to determine the kinetic constants $K_m$ and $V_{max}$ for enzymes acting on insoluble substrates. Benzonana and Desnuelle convincingly showed that the quantity of interface as determined by the size of the emulsion droplets strongly influences the value of $K_m$. They compared the rates of lipolysis of coarse and fine emulsions of a substrate and found (Fig. 8, left panel) that the Michaelis constants differ. If the substrate concentrations are expressed as area/volume rather than as weight/volume, the Lineweaver–Burk plots for different emulsions of the same substrate coincide (Fig. 8, right panel), and a single $K_m$ value is obtained independent from the degree of dispersion of the substrate.

The significance of the determination of an interfacial $K_m$ having the dimensions of an area/volume instead of weight/volume has been questioned by Mattson et al.[45] and by Brockerhoff.[18] Brockerhoff argued that such a $K_m$ might be the dissociation constant of the enzyme–interface complex ($k_3$ very small) and be independent of the chemical nature of the substrate, and he predicted that quite similar values of $K_m$ might be found for quite dissimilar combinations of lipid–water interface and protein. In other words, lipid–water interfaces would behave as unspecific surfaces and adsorb many proteins with the same affinity. Mattson et al.[45] suggested that the orientation of the ester molecule at the oil–water interface and the specificity of the enzyme for its substrate are the main factors determining the rate of hydrolysis. Savary[46] also shared this point of view. However, it seems evident that this controversy is only apparent, because the two groups were not investigating the same step of the lipolytic reac-

---

[44] G. Benzonana and P. Desnuelle, *Biochim. Biophys. Acta* **105**, 121 (1965).
[45] F. H. Mattson, R. A. Volpenhein, and L. Benjamin, *J. Biol. Chem.* **245**, 5335, (1970).
[46] P. Savary, *Biochim. Biophys. Acta* **248**, 149 (1971).

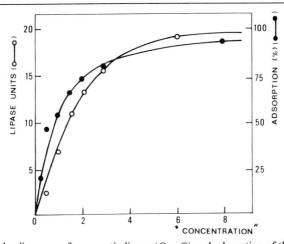

FIG. 7. Michaelis curve of pancreatic lipase (○—○) and adsorption of the enzyme by a long-chain triglyceride emulsion (●—●). The "concentration" of the emulsions is expressed in arbitrary units. From G. Benzonana and P. Desnuelle, *Biochim. Biophys. Acta* **105**, 121 (1965).

tion. Mattson and Volpenhein[47] and Brockerhoff,[18] using saturating interfacial substrate concentrations where all enzyme molecules are adsorbed to the interface, studied the affinity of the enzyme for its specific substrate in the interface. Therefore, they are probably justified in suggesting that the orientation of the substrate molecules at the oil–water interface is a highly important factor in lipolysis. Benzonana and Desnuelle,[44] on the other hand, studied the adsorption of the water-soluble enzyme to a lipid–water interface, a step that has to precede the interfacial reaction, and they defined an affinity constant between the enzyme and the interface. This definition emphasizes the influence of the quantity of emulsion surface on the enzyme velocity. Furthermore, the determination of such an interfacial $K_m$ value requires that the quality of the lipid–water interface remain constant during the experiment because evidence has since become available that the amount of lipolytic enzyme adsorbed varies with the quality of different lipid–water interfaces (see next sections).

A more serious objection to determining interfacial $K_m$ values in area/volume as proposed by Benzonana and Desnuelle is one of a practical nature. First, it is not easy to determine accurately the surface area of emulsion droplets. Second, when detergents are needed to stabilize the emulsion, the calculations of $K_m$ in area/volume are questionable because the interfacial area occupied by the detergent is unknown. The earlier ap-

[47] F. H. Mattson and R. A. Volpenhein, *J. Lipid Res.* **10**, 271 (1969).

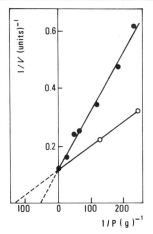 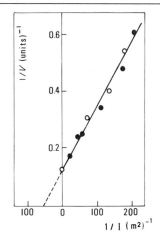

FIG. 8. Lineweaver–Burk plot of the lipolysis of two oil emulsions, coarse (●—●) and fine (○—○). From G. Benzonana and P. Desnuelle, *Biochim. Biophys. Acta* **105**, 121 (1965).

proach used by Sarda and Desnuelle[6] expressed $K_m$ as molar interfacial concentration (number of moles of substrate in the interface per unit volume). However, it also leads to difficulties in defining substrate concentration in the presence of detergents. This is one of the main reasons why investigators in the field of interfacial enzyme kinetics have looked for lipid–water interfaces that can be characterized in physicochemical terms better than is possible with emulsions. One system is the micellar solution.

Micellar Substrates

The preceding section dealt with emulsified lipid substrates, i.e., compounds in which the hydrophobic properties dominate their hydrophilic character. These compounds tend to minimize contact with water and favor the formation of rather large emulsion droplets (Fig. 9, I). Usually, trace amounts of detergents, which concentrate at the oil–water interface, are needed to stabilize the emulsion droplet. Upon reinforcement of the hydrophilic part by charged groups, as in a phospholipid for instance, the compound is still insoluble in water because of hydrophobic interactions between the acyl chains. However, water becomes soluble in the lipid polar headgroups, and this results in a swollen structure usually consisting of alternating lipid bilayers and layers of water (Fig. 9, IIa, IIb). Upon a further shift of the hydrophobic–hydrophilic balance, i.e., by acyl chain shortening or by reduction of the number of long acyl chains in the molecule, compounds are obtained that at low concentration are freely

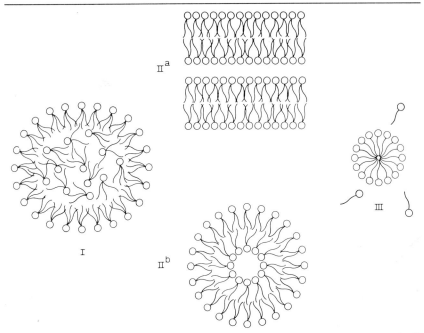

FIG. 9. Schematic picture of an oil emulsion (I), a lipid bilayer (IIa, IIb), and a micelle (III). From R. Verger and G. H. de Haas, *Annu. Rev. Biophys. Bioeng.* **5,** 77 (1976).

soluble in water. However, at the CMC, the limit of molecular solubility is reached and aggregation of the molecules into spherical or rod-shaped particles, the so-called micelles, starts (Fig. 9, III). Such micellar solutions are optically clear, and the lipid molecules in the micelle are in rapid equilibrium with the molecules in free solution. Examples of micellar lipids are soaps and detergents, e.g., sodium dodecyl sulfate, bile salts, Triton X-100, lysolecithin.

It might be assumed that the kinetics of lipolysis measured on lipid micelles (type III) would be much easier to interpret than that studied with emulsified substrates (type I). In the micelle every lipid molecule is present at the interface and able, in principle, to interact with the enzyme molecule. However, the natural substrates of phospholipases are long-chain phospholipids that preferentially form aqueous dispersions of type II[a]; the enzyme kinetics cannot easily be studied in such a system.[48] It has been known for a long time that the closely packed bilayer structure of phospholipids is a very poor substrate for most phospholipases, probably because the enzymes have difficulty penetrating into the interface and

---

[48] S. Gatt, A. Herzl, and Y. Barenholz, *FEBS Lett.* **30,** 281 (1973).

reaching their substrate molecules. This has been demonstrated in a convincing way for pancreatic phospholipase $A_2$ by Op den Kamp et al.[49,50]

Quantitative rate measurements of enzymic hydrolysis of natural long-chain phospholipids have been possible only after transformation of the type II dispersion into micellar systems. Two techniques have been suggested to accomplish this. The first of these is the use of organic solvents, i.e., the well known moist ether system introduced by Hanahan in 1952.[51] In this solvent, long-chain phospholipids form inverted micelles that contain the enzyme in their aqueous interior. Many venom phospholipases are strongly activated in this system, as reported by Wells and colleagues.[10,52,53] They demonstrated the existence of various micellar species depending on the amount of ether present, and they showed that venom phospholipase $A_2$ can distinguish between the various micellar structures. So far, our lack of knowledge of the detailed physicochemical properties of this interesting ether–water system has limited its use mainly to preparative purposes.

The second method of lipid dispersion involves the addition of detergents that transform the lipid bilayer structure into optically clear solutions of mixed micelles. Mainly two detergents have been used extensively: sodium deoxycholate and the neutral Triton X-100. The advantage of detergents is evident: the mixed micelle is rapidly attacked by most lipolytic enzymes, and linear kinetics are often obtained because of the removal of the lipolytic products from the lipid–water interface.[54,55] In addition, the lowering of the apparent p$K$ of long-chain fatty acids in the presence of certain detergents makes continuous pH-state titrations possible. Therefore, various convenient assay procedures used for the purification of lipolytic enzymes are based on such systems.[56,57,60]

---

[49] J. A. F. Op den Kamp, J. de Gier and L. L. M. Van Deenen, *Biochim. Biophys. Acta* **345**, 253 (1974).
[50] J. A. F. Op den Kamp, M. T. Kauerz, and L. L. M. Van Deenen, *Biochim. Biophys. Acta* **406**, 169 (1975).
[51] D. J. Hanahan, *J. Biol. Chem.* **195**, 199 (1952).
[52] P. H. Poom and M. A. Wells, *Biochemistry* **13**, 4928 (1974).
[53] M. A. Wells, *Biochemistry* **13**, 4937 (1974).
[54] F. H. Mattson and R. A. Volpenhein, *J. Am. Oil Chem. Soc.* **43**, 286 (1966).
[55] G. Benzonana and P. Desnuelle, *Biochim. Biophys. Acta* **164**, 47 (1968).
[56] P. Desnuelle, M. J. Constantin, and J. Baldy, *Bull. Soc. Chim. Biol.* **37**, 285 (1955).
[57] W. Nieuwenhuizen, H. Kunze, and G. H. de Haas, this series, Vol. 32, p. 147.
[58] S. Gatt and Y. Barenholz, *Annu. Rev. Biochem.* **42**, 61 (1973).
[59] S. Gatt, Y. Barenholz, I. Borkovski-Kubiler, and Z. Leibovitz-Ben Gershon, *in* "Sphingolipids, Sphigolipodases and Allied Disorders" (B. W. Volk and S. M. Aronson, eds.), p. 237. Plenum, New York, 1972.
[60] E. A. Dennis, *J. Lipid Res.* **14**, 152 (1973).

Dennis reported kinetic studies on a highly purified venom phospholipase $A_2$ using as substrate long-chain lecithins in the presence of Triton X-100.[31,42,60] The neutral character of Triton X-100 undoubtedly constitutes a definite advantage over the anionic bile salt, because the presence of the latter complicates studies on pH effects or metal ion involvement.[17,54,55] With liposomal dispersions of pure egg or synthetic dipalmitoyllecithin, no enzyme action was found. However, upon increasing the amount of Triton, Dennis observed for both phospholipids a rise in enzymic activity as more and more bilayer phase was converted into mixed micelles of Triton + lecithin. At a molar ratio of Triton/lecithin of 2/1, all phospholipid molecules were reported to be in mixed micelles, and maximal enzyme activity was observed. Because mixed micelles, in contrast with liposomes, are hydrolyzed by the enzyme, it is not surprising that the dependence of enzyme velocity on Triton concentration initially follows a saturation curve. More interesting is the decrease in enzyme activity that is observed at molar ratios of Triton/lecithin higher than 2/1. The continuous decrease of enzyme activity with increasing Triton concentration was explained by the decreasing amount of phospholipid per unit surface area ("surface dilution" of substrate).[31] It has never been rigorously proved, however, that Triton X-100 is an inert spacer. Sundler et al.[61] recently purified a phospholipase C that specifically hydrolyzes phosphatidylinositol. With phospholipid as the substrate in mixed micelles with Triton X-100, they showed that the rate of catalysis was highly dependent on the detergent/phospholipid ratio, as previously reported by Dennis[60] for other phospholipases. The decreased catalytic rates observed at high Triton/phospholipid ratios, however, were not attributed by the authors to substrate dilution on the micellar surface, since substrate dilution with phosphatidylcholine, either in mixed micelles (at a constant Triton/phospholipid ratio) or in bilayer vesicles, had no effect on catalysis.[61] Depending on the enzyme and lipid used, however, more or less enzyme present in the interface could be bound to the substrate or to the dilutor, as a function of respective interfacial affinity constants. If the interfacial enzyme–phosphatidylinositol complex is more easily dissociated by Triton molecules than by phosphatidylcholine, this would explain the apparently contradictory results of Sundler et al.[61]

As discussed by Verger and de Haas,[62] some lipids or detergent in a mixed interface could behave as weak or strong competitive inhibitors or as pure inert substrate dilutors. When enzyme activity on mixed micellar substrates is studied, it must be borne in mind that not only substrate den-

---

[61] R. Sundler, A. W. Alberts, and P. R. Vagelos, *J. Biol. Chem.* **253**, 4175 (1978).
[62] R. Verger and G. H. de Haas, *Annu. Rev. Biophys. Bioeng.* **5**, 77 (1976).

sity or substrate dilution is affected, but the "interfacial quality" of the surface is also changed. Dennis assumed that further addition of Triton did not induce changes in the physical state of the micelles and only increases the total number of the mixed Triton–lecithin micelles, "all generally similar in size and shape to pure Triton micelles."[60] It should be emphasized that the presence of different amounts of Triton X-100 has a pronounced influence on the size of the particles, as shown by Dennis using agarose filtration.[63] An alternative interpretation of the decrease in enzyme activity with increasing Triton X-100 concentration was given by Kaplan and Teng[64] for triglyceride lipase. These authors postulated a competition between pure Triton X-100 micelles and mixed lipid–Triton micelles for the binding site of the enzyme. Experimental support for the coexistence of pure detergent micelles and mixed phospholipid–detergent micelles has been provided by Dervichian[65] for a bile salt phosphatidylcholine system and by Yedgar et al.[66] for a sphingomyelin–Triton X-100 system.[58] Olive and Dervichian[67] showed with the lecithin and deoxycholate system that at different lipid/detergent ratios several micellar forms may exist.

To avoid these problems, several investigators have proposed the use of substrates that, in the absence of detergents, form micelles of type III. For obvious reasons, isotropic solutions of triglycerides or phospholipids in the absence of detergents can be obtained only by shortening of the fatty acyl chain. Entressangles and Desnuelle[7] in 1968 reported that the short-chain triglycerides triacetin and tripropionin give molecularly dispersed solutions in aqueous medium of low ionic strength. The absence of molecular aggregates in these solutions was checked by light-scattering. As could be expected, pancreatic lipase hydrolyzed the monomeric substrate molecules very slowly. In the presence of 0.1 $M$ NaCl, however, a considerable enhancement of lipase activity was observed and the presence of aggregates in the isotropic solutions was shown by dye solubilization. Salt induces the formation of lipid aggregates that are rapidly hydrolyzed by lipase. Notwithstanding the fact that these micellar aggregates are much smaller than emulsified particles of the same substrate obtained upon saturation of the solution, it could be shown by extrapolation that lipase has the same $V_{max}$ value at infinite concentration of the emulsion and of the micellar aggregate.

In kinetic studies of phospholipases, short-chain synthetic phospho-

[63] E. A. Dennis, *Arch. Biochem. Biophys.* **165**, 764 (1974).
[64] A. Kaplan and M. H. Teng, *J. Lipid Res.* **12**, 324 (1971).
[65] D. G. Dervichian, *Adv. Chem. Ser.* **84**, 78 (1968).
[66] S. Yedgar, Y. Barenholz, and V. G. Cooper, *Biochim. Biophys. Acta* **363**, 98 (1974).
[67] J. Olive and D. G. Dervichian, *Bull. Soc. Chim. Biol.* **50**, 1409 (1968).

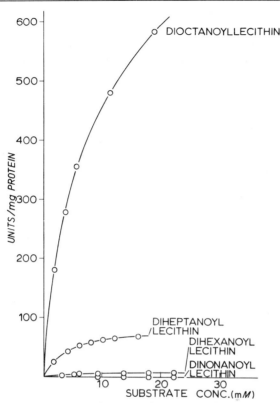

FIG. 10. Comparison of Michaelis curves of the hydrolysis of dinonanoyl-, dioctanoyl-, diheptanoyl-, and dihexanoyllecithin by pancreatic phospholipase $A_2$. From G. H. de Haas, P. P. M. Bonsen, W. A. Pieterson, and L. L. M. Van Deenen, *Biochim. Biophys. Acta* **239**, 252 (1971).

lipids have also proved to be very valuable. De Haas et al.[11] reported that lecithins containing two similar fatty acid chains of 6, 7, or 8 carbon atoms form micellar solutions in water that are effectively degraded by the pancreatic phospholipase. The kinetic parameters have been determined and the role of the essential cofactor $Ca^{2+}$ was investigated. Kinetic studies using isotropic micellar systems seem to be very attractive. $V_{max}$ and $K_m$ values can be easily obtained and expressed in the usual units (see Fig. 10). The large differences in $V_{max}$ value found for pancreatic phospholipase $A_2$ acting on lecithins containing fatty acid of different chain lengths cannot be explained by a different chemical reactivity of the susceptible ester bond. Even minor differences in the lecithin structure cause a change in the architecture of the lipid–water interface. Such changes strongly influence the distribution coefficient of the enzyme between bulk and lipid–water interface. The nature of the changing interfacial parameter is still unknown, but it is evident that the "microscopic quality" of the various lipid–water interfaces is different. Therefore, $K_m$ and $k_{cat}$ values, determined in isotropic micellar systems, are also apparent quantities.

Slotboom et al.[68] synthetized two enantiomeric 2-sn-phosphatidylcholines containing hexanoyl and dodecanoyl acyl chains. Mixed micelles and mixed monolayers, containing variable proportions of the two stereoisomers, are characterized by identical physicochemical properties. The use of the substrate analog n-tetradecylphosphorylcholine as a micellization agent and the description of a kinetic treatment under steady-state conditions for the action of a soluble enzyme on an interface consisting of two substrates ($S_1$ and $S_2$) and a competitive inhibitor (i), alllowed Slotboom et al.[68] to determine the interfacial kinetic parameters $k_{cat}$ and $K_m^*$. Dodecanoic acid is released from the most susceptible isomer about 13 times more rapidly than hexanoic acid from the stereoisomer in spite of the higher $K_m^*$ of the former.

Since lipolytic enzymes act *in vivo* on aggregated substrates, it is clear that the action of these enzymes on their proper substrates should be thoroughly studied under conditions where lipid–water interfaces are present. Because many experimental difficulties are encountered with emulsions and micelles, several investigators have preferred to use the monolayer technique.

Monolayers of Substrates

Spreading of a lipid at the air–water interface as a monomolecular film has been intensively studied over 60 years. For classical references, see the books of Adam,[69] Davis and Rideal,[70] and Gaines.[71] There are at least three major reasons for using lipid monolayers as substrate for lipolytic enzymes.

1. The technique is highly sensitive and very little lipid is needed to obtain kinetics. This argument can often be decisive when one uses synthetic or rare lipids, although the advantage of high sensitivity may partly be offset by the requirement for pure substances and for extreme cleanliness during the experimentation.
2. During the course of the reaction, one can monitor one or several

[68] A. J. Slotboom, R. Verger, H. M. Verheij, P. H. M. Baartmans, L. L. M. van Deenen, and G. H. de Haas, *Chem. Phys. Lipids* **17**, 128 (1976).
[69] N. K. Adam, "The Physics and Chemistry of Surfaces." Dover, New York, 1968.
[70] J. T. Davis and E. Rideal. "Interfacial Phenomena." Academic Press, New York, 1963.
[71] G. L. Gaines, "Insoluble Monolayers at Liquid–Gas Interfaces." Wiley (Interscience), New York, 1966.

physicochemical parameters characteristic of the monolayer film: surface pressure, potential, radioactivity, etc. These variables often give unique information on the evolution of the reaction.
3. Probably the most important reason fundamentally is the possibility of varying the "quality of interface" that is determined by the orientation of the molecules, molecular and charge density, water structure, fluidity, etc. One further advantage of the monolayer technique as compared to bulk methods is the possibility of transferring the film from one aqueous subphase to another.[72-74]

By contrast, a shortcoming of the monolayer technique is the adsorption or denaturation of many enzymes at lipid–water interfaces.[1,19,23,75] This phenomenon is related to the interfacial free energy: surface films at high interfacial energy (low surface pressure, $\pi$) are highly denaturing, whereas films at low interfacial energy (high $\pi$) are much less denaturing. Different solid–water interfaces used as containers for dilute enzyme solutions can cause adsorption and denaturation of protein. Hydrophobic materials having high interfacial free energies, e.g., Teflon, can adsorb large quantities of proteins.[76,77] Fortunately, this phenomenon is slow.[20]

A new field of investigation was opened in 1935 when Hughes[78] used the monolayer technique for the first time to study enzymic reactions. He observed that the rate of the phospholipase A-catalyzed hydrolysis of a lecithin film measured by the decrease of surface potential was considerably reduced when the number of lecithin molecules per square centimeter was increased. To explain this fact, Hughes suggested that "the lecithinase embodies also a spacing of two active groups in the distended lecithin molecule for the maximum probability of reaction." Since this early work, several laboratories have used the monolayer technique to follow lipolytic activities with triglycerides and phospholipids as substrate. These studies can be tentatively divided roughly into four groups. Either long-chain lipids were used and their surface density (number of molecules per square centimeter) was not controlled; or short-chain lipids were applied, again without control of surface density; or short-chain lipids were used at constant surface density. Recently, a method was developed (in our laboratory) to study the hydrolysis of long-chain lipids

---

[72] J. C. Skou, *Biochim. Biophys. Acta* **31**, 1 (1959).
[73] T. Yamashita and H. B. Bull, *J. Colloid Interface Sci.* **27**, 19 (1968).
[74] P. Fromherz, *Biochim. Biophys. Acta* **225**, 382 (1971).
[75] L. K. James and L. G. Augustein, *Adv. Enzymol.* **28**, 1 (1966).
[76] S. R. Turner, M. Litt, and W. S. Lynn, *J. Colloid Interface Sci.* **48**, 100 (1974).
[77] J. L. Brash and D. J. Lyman. "The Chemistry of Biosurfaces." Dekker, New York, 1971.
[78] A. Hughes, *Biochem. J.* **29**, 437 (1935).

with control of surface density. As we shall see, it is not surprising that this classification is roughly chronological.

## Group 1. Long-Chain Lipid Monolayers without Control of Surface Density

Bangham and Dawson[79] and Dawson[80] investigated several types of phospholipases that catalyze the hydrolysis of acyl ester bonds in glycerophosphatides. More recently, Quinn and Barenholz[81] compared the activity of phosphatidylinositol phosphodiesterase against substrate in dispersions and as monolayers at the air–water interface. All these authors followed the enzymic hydrolysis of monolayers of $^{32}$P-labeled phospholipids by measuring the loss of surface radioactivity that occurs as the water-soluble $^{32}$P-labeled reaction products leave the surface and are no longer detected by the counter. Dawson *et al.* noticed that at surface pressures above 30 dyn cm$^{-1}$ enzymic hydrolysis of the lecithin film did not occur unless a small amount of an anionic amphipathic substance was introduced in the film. This result led the authors[82] to the conclusion that the sign of the zeta potential on the substrate surface is critical for the initiation of enzyme activity.

Following the original technique described by Hughes in 1935, Colacicco and Rapport[83] and Shah and Schulman[84] measured the action of snake venom phospholipase A on lecithin monolayers. Both groups spread the lipid film on an enzyme solution and followed the fall in surface potential. Assuming that the reaction products remain in the film, the fall in surface potential will be proportional to the number of lecithin molecules transformed to lysolecithin and fatty acid, and it can be used to express the activity of the venom. Depending on the degree of unsaturation of the lecithin used, the optimal activity of the venom phospholipase was encountered at surface pressures between 12 and 25 dynes cm$^{-1}$, values that are different from the 29–33 dynes cm$^{-1}$ found by Dawson[80] using $^{32}$P-labeled lecithin. However, for the surface-potential technique to be valid, it is essential that the products of the reaction remain in the film, whereas for correct surface radioactivity measurements using $^{32}$P-labeled lecithin it is imperative that the lysolecithin molecules rapidly leave the film. With natural long-chain phospholipids, it is clear that these two ex-

[79] A. D. Bangham and R. M. C. Dawson, *Biochem. J.* **75,** 133 (1960).
[80] R. M. C. Dawson, *Biochem. J.* **98,** 53c (1966).
[81] P. J. Quinn and Y. Barenholz, *Biochem. J.* **149,** 199 (1975).
[82] R. M. C. Dawson, this series, Vol. 14, p. 633.
[83] G. Colacicco and M. M. Rapport, *J. Lipid Res.* **7,** 258 (1966).
[84] D. O. Shah and J. H. Schulman, *J. Colloid Interface Sci.* **25,** 107 (1967).

treme situations are never encountered. In our opinion, the reliability of both techniques is doubtful. If at low $\pi$, the products of the reaction remain in the film, what is their influence on the rate of enzymic reaction? If at high $\pi$ the products leave the surface, is this desorption process quantitative and not rate limiting? The validity of both techniques to follow enzymic hydrolysis of phospholipid monolayers has been discussed also by Colacicco.[85]

Another example showing the limitations of monolayer kinetics obtained with long-chain phospholipids are studies of the action of phospholipase D on surface films of [$^{14}$C]choline-labeled lecithin.[86] The hydrolytic reaction was followed by the surface radioactivity technique. Upon cleavage of the polar headgroup, phosphatidic acid was formed and remained in the film. The presence of negatively charged phospholipid appeared to have a profound influence on the enzymic reaction. Especially at high surface pressure, the authors observed an increase of the hydrolysis rate with time. The lag period in the kinetic runs was attributed to an accumulation of autocatalytic amount of phosphatidic acid that initiates a fast reaction. Although such a mechanism might explain the observed results, other interpretations are possible. Especially, it cannot be precluded that a facilitated penetration of enzyme molecules into a gradually more negatively charged surface film would give rise to increasing amounts of enzyme in the monolayer.

An interesting attempt to follow more quantitatively the phospholipase C hydrolysis of [$^3$H]choline-labeled lecithin by direct surface counting has been reported by Miller and Ruysschaert.[19] They showed that the diffusion-controlled escape rate of radioactive phosphorylcholine from the film was much higher than the rate of enzymic cleavage. Therefore, the latter step is rate limiting and controls the decrease in surface radioactivity. Assuming a Michaelis–Menten type of enzymic reaction, preceded by a reversible adsorption step of the enzyme to the surface

$$E_{bulk} \xrightleftharpoons{K_{ads}} E_{ads} + S \xrightarrow{k_{cat}} E + P$$

they derive the following equation for the velocity of the enzyme reaction in the interface:

$$v = (k_{cat}/K'_m)e_0 \cdot s$$

where $K'_m = K_m/K_{ads}$, $e_0$ is the enzyme concentration in a layer adjacent to the surface, and $s$ is the substrate concentration expressed in molecules/cm$^2$.

---

[85] G. Colacicco, *Nature (London)* **233**, 202 (1971).
[86] R. H. Quarles and R. M. C. Dawson, *Biochem. J.* **113**, 697 (1969).

This equation requires that the enzymic velocity be a linear function of the substrate concentration in the interface. The authors showed experimentally that, in a limited range of substrate concentrations, $v$ is indeed linearly dependent on the surface concentration. Deviations from linearity were found at lower surface concentration, and the velocity was slowing down with time; this behavior was explained by irreversible adsorption leading to inactivation of the enzyme at the interface. Also at higher surface concentrations, much lower enzyme activities were found than are required by the above equation, and this deviation was interpreted to result from steric hindrance in the substrate–enzyme interaction. Unfortunately, the use of an impure enzyme preparation and a racemic substrate did not allow more quantitative conclusions to be drawn.

*Group 2. Short-Chain Lipid Monolayers without Control of Surface Density*

In order to overcome the above-described technical difficulties inherent to the use of long-chain lipids, Olive[87] and Dervichian[88] were the first to use synthetic short-chain lipids, which upon enzymic hydrolysis yield readily soluble products. Since then, several groups have used short-chain lipids as substrates in monolayer studies of lipolytic enzymes.

Garner and Smith[89] studied the action of pancreatic lipase on a series of octanoyl glycerides and analogous octanoate esters spread as monolayers at the air–water interface. With the exception of trioctanoin, all substrates yielded soluble reaction products and the authors followed the course of the reaction at constant area by the fall in surface pressure with time. The rate of hydrolysis was determined near the collapse pressure, where the decrease in film pressure was found to be linear with respect to time.

Interestingly, the hydrolysis rate of a variety of substrates measured by the monolayer technique[89] correlated with the corresponding rates measured by Derbesy and Naudet[90] using the classical bulk titrimetric procedure. A similar monolayer study has been reported by Lagocki *et al.*,[91] who also investigated the kinetics of lipase hydrolysis of monolayers of trioctanoin and dioctanoin. They reported that the rate of the reaction

---

[87] J. Olive, "Contributation à l'enzymologie superficielle des lipides," Ph.D. thesis, Univ. of Paris, 1969.
[88] D. G. Dervichian, *Biochimie* **53**, 25 (1971).
[89] C. W. Garner and L. C. Smith, *Biochem. Biophys. Res. Commun.* **39**, 672 (1970).
[90] M. Derbesy and M. Naudet, *Rev. Fr. Corps Gras* **4**, 225 (1972).
[91] J. W. Lagocki, J. H. Law, and F. J. Kézdy, *J. Biol. Chem.* **248**, 580 (1973).

was independent of the surface pressure of the substrate. Although these authors showed that during the hydrolysis of trioctanoin the formation of insoluble 1,2-diglyceride did not greatly affect the hydrolysis of the triglyceride in the mixed film, we want to emphasize that in general the production of insoluble reaction products should be avoided. In our discussion on long-chain monolayers, we have seen how the presence of insoluble reaction products can influence the rate of lipolysis; if such products are at the same time substrates for the enzyme in a consecutive, albeit slower, reaction, interpretation of the kinetic results becomes more complicated.[91] A second point of discussion concerns the technique used by the above groups, i.e., studying the rate of hydrolysis at constant surface area. This method is simpler than working at constant surface pressure. However, the former technique can be used only in those cases where the rate constant of hydrolysis is independent of the surface pressure. Although such a situation may be found in a certain limited surface pressure range, it is definitely not a general phenomenon; this has been demonstrated in several laboratories.

Brockman et al.[8] studied the enhancement of hydrolysis of tripropionin by pancreatic lipase in the presence of siliconized glass beads. This is a very interesting interfacial system that can be considered as a spherical monolayer of tripropionin at a liquid–solid interface. The binding of the enzyme to the surface was shown to be reversible and diffusion controlled. In addition, the hydrolytic reaction on the surface appeared to be first order with respect to the amount of adsorbed enzyme and first order with respect to the concentration of tripropionin at the solid–liquid interface. This latter result is in agreement with the findings of Miller and Ruysschaert[19] (see above). Brockman et al.[8] reported an enhancement of the velocity on the surface of the siliconized glass beads of 3 orders of magnitude as compared to the homogeneous reaction, and they ascribed this activating effect to the increased local concentration of the substrate at the interface. It should be remarked, however, that although the authors measured directly the fraction of enzyme adsorbed on the interface this determination was done in the absence of tripropionin. From their results, it can be calculated that the ratio of tripropionin molecules present as monomer in solution to tripropionin molecules adsorbed to the interface is about 300. Taking into account that the $K_m$ of lipase for tripropionin monomers is on the order of 10 m$M$,[92] we see that the large amount of substrate in the bulk phase could act as a "competitive inhibitor."[28]

---

[92] B. Entressangles and P. Desnuelle, *Biochim. Biophys. Acta* **341**, 437 (1974).

*Group 3. Short-Chain Lipid Monolayers with Controlled Surface Density*

Dervichian,[88] using short-chain lipids spread at the air–water interface, described for the first time a technique to follow the rate of enzyme reactions at constant surface pressure as shown schematically in Fig. 11. The kinetics of the surface reaction were followed by continuously recording the decrease in area of the film with time. Because most lipolytic reactions taking place at the air–water interface have been shown to be surface pressure dependent, this technique seems to be optimally suited to study interfacial kinetics. Moreover, use of a fully automatized barostat as proposed by Verger and de Haas[33] makes the technique rather simple and accurate. It must be remarked, however, that the technique is limited to those cases where the substrate forms a stable surface film at the air–water interface and where all enzymic reaction products are freely water soluble and diffuse immediately away into the bulk phase. Therefore, until now only acylesterases have been studied in some detail with this method. Because of the requirement of highly purified lipolytic enzymes, it is not surprising that most reports dealt with the lipolytic enzymes, lipase, and phospholipase $A_2$. These conditions of solubility explain why surface studies on lipase action have been generally limited to the short-chain triglycerides or to diglycerides containing 8–10 carbon atoms in the acyl chain. In the case of phospholipase $A_2$, so far only phospholipids containing fatty acids composed of 8–12 carbon atoms have been studied.

To bring enzyme and substrate together, two techniques are extensively used; namely, spreading of the lipid film on a homogeneous enzyme solution or injection of the enzyme under a preformed stabilized surface monolayer. The injection technique requires efficient stirring of the subphase in order to obtain a homogeneous bulk solution rapidly. In the spreading technique, stirring is not needed. This might be considered as advantageous for the stability of the film; however, in both cases the possibly rate-limiting step of product desorption from the interface can be overcome by efficient stirring.[93] A possible drawback of the spreading technique is the adsorption and/or denaturation that can occur with some enzymes at interfaces during the time required to clean the surface. Consequently, there is a possibility of trapping of the enzyme in the interface during the spreading of the film that would lead in general to higher reaction rates than those observed in the injection technique.[93] Several types of troughs have been used to study enzyme kinetics. The most generally used as a simple rectangular one giving nonlinear kinetics. To obtain rate constants, a semilogarithmic transformation of the data is required. This

---

[93] G. Zografi, R. Verger, and G. H. de Haas, *Chem. Phys. Lipids* **7**, 185 (1971).

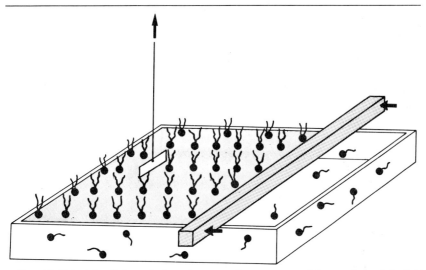

FIG. 11. Principle of the technique for measuring enzyme activity on short-chain lipid monolayers with controlled surface density.

drawback was overcome by a new trough design consisting of two compartments connected by a narrow surface canal.[33] As shown on the right side of Fig. 12, the recorded kinetics obtained with this trough is linear in contrast to the nonlinear plots obtained with the usual one-compartment trough.

Brockman et al.[94] investigated the rate of lipase hydrolysis of trioctanoin and 1,2-dioctanoin films also at constant surface pressure. Their method is based upon measurement of the amount of substrate that must be added to a monolayer to maintain constant surface pressure during the course of the enzymic reaction. Although the idea of an "open" system, in which substrate molecules disappearing from the monolayer are continuously replaced, is attractive, the presence of organic solvents at the surface might influence the surface parameters of the film. Especially in the case of phospholipids, where less volatile solvents must be used to spread the film, such effects cannot be excluded. Moreover, surface heterogeneity might influence the kinetics at higher surface pressures where the films are no longer in an ideal liquid-expanded state. On the other hand, the usefulness of this technique was demonstrated clearly in the lipase-catalyzed hydrolysis of a surface film of trioctanoin. The experiments not only confirmed the surface pressure dependence of the enzymic reaction, but also showed the complexity of the kinetics in lipolytic reac-

[94] H. L. Brockman, F. J. Kézdy, and J. H. Law, *J. Lipid Res.* **16**, 67 (1975).

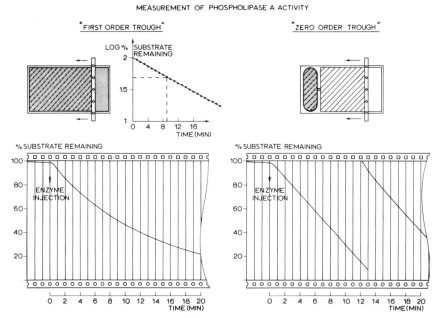

FIG. 12. Measurement of phospholipase A activity. Comparison between the recorded kinetic plots obtained with the "first-order trough" (left side) and with the "zero-order trough" (right side). From R. Verger and G. H. de Haas, *Chem. Phys. Lipids* **10,** 127 (1973).

tions where the first hydrolysis product is a substrate in a consecutive lipolytic reaction. Kinetic equations were derived for a two-step reaction yielding an insoluble intermediate and soluble final products.

*Influence of Surface Pressure on Enzyme Velocity.* In agreement with the results of studies on long-chain phospholipids,[78,83,84,86] most investigators using soluble short-chain lipids also report an optimum in the velocity-$\pi$ profile.[93,95,96] The exact value of this $\pi$ optimum varies considerably depending on the particular enzyme–substrate combination used, and differences in the experimental technique, such as spreading or injection, may also cause variations in the optimal pressure. Several qualitative interpretations have been given to explain this phenomenon. Agreeing with the original interpretation given by Hughes,[78] later workers tried to explain the existence of an optimum in the velocity-$\pi$ profile by emphasizing the importance of molecular orientation of the molecules in the monolayer for an optimal fit between the active site of the enzyme and the substrate molecule. Shah and Schulman[84] stated:

[95] J. Olive and D. G. Dervichian, *Biochimie* **53,** 207 (1971).
[96] S. Esposito, M. Sémériva, and P. Desnuelle, *Biochim. Biophys. Acta* **302,** 293 (1973).

The compression of a monolayer results in an increase of surface concentration of molecules and simultaneously in a decrease of the intermolecular spacing in the monolayer. The former increases the rate of hydrolysis by increasing the frequency of collision between enzyme and substrate molecules, whereas the latter decreases the rate by preventing the penetration of the enzyme molecules in the monolayer. These counterbalancing factors, which directly influence the rate of hydrolysis, determine the optimum surface pressure for hydrolysis of lecithin monolayers.

Another interpretation of the $\pi$-optimum profile was given by Esposito et al.[96] These authors supposed that lipase upon adsorption at the interface acquires a functional conformation for intermediary values of the free interfacial energy (and consequently of the film pressure $\pi$). Lower and higher values would lead to inactive forms due to denaturation or insufficient conformational change of the enzyme molecule. Cohen et al.,[22] working at variable surface pressure, reported the existence of two enzyme classes with respect to the effect of surface pressure on the rate of hydrolysis of substrate monolayers. On one hand, they found two enzymes that show continuously increasing activity with increasing surface pressure, and two that have extensive regions of pressure independence. They further showed that the pressure-independent enzymes were quite resistant to surface denaturation, while the pressure-dependent enzymes undergo rapid surface denaturation at air–water interfaces. On the basis of this correlation, they attributed the pressure dependency to the denaturation of the enzyme at surface domains that are not covered by substrate molecules. Finally, Cohen et al.[22] stated that "surface denaturation of enzymes is sufficient to explain the phenomenon of surface pressure dependency." However, such a conclusion is weakened by a limitation inherent to the monolayer technique: the amount of enzyme that is really involved in catalysis is unknown. If, in a velocity-$\pi$ profile, the amount of enzyme adsorbed to the monolayer was a function of $\pi$, then the plotted velocities would be apparent values and the $\pi$ optimum would have no evident meaning.

Verger et al.[32] and Pattus et al.[27] have reinvestigated this problem using radio-labeled pancreatic phospholipase $A_2$. They measured simultaneously the enzyme activity and the amount of radioactive enzyme in excess at the interface in the presence of a lecithin monolayer. This enzyme was found to be resistant to interfacial inactivation.[20] Figure 13 summarizes the main results obtained at different surface pressures of the film. The amount of radioactive enzyme present in or close to the interface decreases linearly with increasing surface pressure.

A criticism of this experiment is that the surface radioactivity measured does not represent only those enzyme molecules directly involved in the catalysis, but also an unknown amount of protein present close to the monolayer in a concentration higher than the bulk value, but not in-

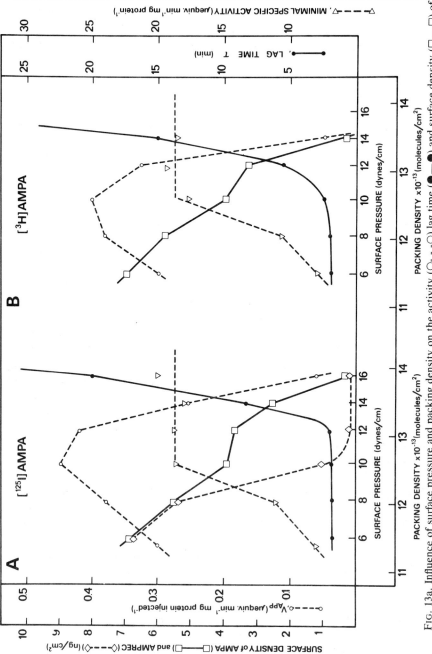

FIG. 13a. Influence of surface pressure and packing density on the activity (○---○) of [125I]AMPA (B) acting on C$_{10}$PC (didecanoyllecithin) monolayer. From F. Pattus, A. J. Slotboom, and G. H. de Haas, *Biochemistry* **13**, 2691 (1979). Copyright by the American Chemical Society.

volved in enzymic hydrolysis of the film. As shown in Fig. 13a, Pattus *et al.*[27] measured the amount of the radioactively labeled zymogen of pancreatic phospholipase $A_2$ adsorbed to lecithin monolayers at different surface pressures. Above 10 dyne/cm, a negligible quantity of zymogen was adsorbed to the film whereas below 10 dyn/cm, comparable amounts of zymogen and active phospholipase $A_2$ were found in excess at the interface. This adsorption at low surface pressure could be considered as nonspecific. The ratio of observed enzyme activity and the amount of protein as determined by surface radioactivity gives a minimal specific activity of the enzyme. The minimal specific activity increases continuously with $\pi$ without reaching an optimum value in the range of surface pressures investigated. This result is qualitatively in agreement with the findings of Miller and Ruysschaert[19] and Brockman *et al.*[8], who reported a linear dependence of enzyme velocity on surface substrate density.

*Lag Periods in Lipolysis.* A rather common observation reported by many authors working on kinetics of lipolytic enzymes is the presence of lag periods in the hydrolysis of both emulsions or micelles and monolayers.[8,13,86,97] Brockman *et al.*,[8] studying lipase hydrolysis of tripropionin adsorbed to siliconized glass beads, described a short lag period in the range of 15 sec, and they found a linear relationship between the lag time and enzyme concentration. They suggested that this short presteady state is due to diffusion-controlled enzyme adsorption. Usually, however, such lag phases are much longer. The origin of the slow hydrolysis phase has not been extensively studied. Very often the interpretation is given that the products of the enzyme reaction, such as lysolecithin or fatty acid in the case of phospholipase A or phosphatidic acid in the case of phospholipase D, "activate" the enzyme by improving the substrate dispersion or promote the enzyme substrate interaction through charge effects.

Because such lag periods, in general, prevent measurements of initial velocity, one usually attempts to suppress them, e.g., by the addition of surface-active agents such as bile salts or Triton. It seems evident, however, that a study of such presteady state phases might yield valuable information on the initial stages of the interaction between lipolytic enzymes and lipid–water interfaces. Such studies should be preferentially done on monolayers of short-chain lipids where the perturbing influence of increasing amounts of reaction products can be excluded.

*Mixed-Lipid Monolayers.* The first attempt to study the hydrolysis of mixed films at constant surface pressure was made by Zografi *et al.*[93] using a first-order trough with L- and D-dioctanoyllecithin mixtures. The authors concluded that inhibition of the phospholipase by inhibitors present in mixed films cannot be studied by the monolayer technique.

[97] A. F. Rosenthal and M. Pousada, *Biochim. Biophys. Acta* **164**, 226 (1968).

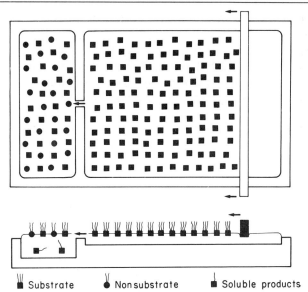

FIG. 13b. Principle of the method for the study of enzymic lipolysis of mixed monomolecular films. From G. Piéroni and R. Verger, *J. Biol. Chem.* (in press).

A new application of the zero-order trough was proposed recently by Piéroni and Verger[98] for studying the hydrolysis of mixed monomolecular films at constant surface density and constant lipid composition as schematically shown in Fig. 13b. It was found that the amount of bound pancreatic lipase decreases linearly with increasing proportions of lecithin in mixed triglyceride–lecithin films kept at constant pressure. This suggests that penetration of mixed films by pancreatic lipase shows a specificity for the triglyceride; i.e., enzyme penetration into an interface is affected by "interfacial quality." When a pure triglyceride film was progressively diluted with lecithin, the minimal specific activity of pancreatic lipase exhibited a bell-shaped curve: a mixed film containing only 20% trioctanoylglycerol was hydrolyzed at the same rate as a monolayer of pure triglyceride.

Using the same technique as described in Fig. 13b, Barnenholz and Verger (unpublished experiments) have studied the lecithin hydrolysis in lecithin–sphingomyelin mixed monomolecular films. The authors observed that the activity of the phospholipase $A_2$ from *Vipera berus* venom was a function of the lecithin concentration. Sphingomyelin acted as a substrate diluter not changing the activity–surface pressure profile. Simi-

---

[98] G. Piéroni and R. Verger, *J. Biol. Chem.* (in press).

lar results were obtained with the pancreatic phospholipase $A_2$ at low surface pressure (10 dyne/cm). In contrast, at higher pressures the mixed monolayers show phase separation and the pancreatic phospholipase activity was 10- to 16-fold increased over the expected value predicted by substrate dilution alone. This increase was explained mainly by a larger amount of enzyme that penetrated into the interface. It seems that the packing defects occurring in membranes through phase transition and phase separation give rise to change in interaction with proteins including lipolytic enzymes.[49,50]

*Group 4. Long-Chain Lipid with Control of Surface Density*

Recently, Scow et al.[99] used triolein films as substrate for lipoprotein lipase, and Rothen et al. (personal communication) studied the hydrolysis of long-chain phospholipid monolayers by different phospholipases $A_2$. In both cases, a large excess of serum albumin was present in the water subphase in order to solubilize the products. This step was found to be not rate limiting. The authors could obtain, with natural long-chain lipids, linear kinetics very similar to that described for short-chain lipids using a zero-order trough.[33] Lairon et al.[100] reported a new technique that allows one to follow the lipolysis of monomolecular films in the presence of bile salts using a "zero-order" trough. The effects of bile salts, the bile lipoprotein complex, and colipase on pancreatic lipase hydrolysis of 1,2-dilaurin films were studied at different surface pressures. Albumin and bile salts are tensioactive and prevent measurements below 22–23 dynes/cm and 31 dynes/cm, respectively, when the barostat technique is used. Nevertheless, the upper surface pressure range is considered to be closer to the lipid packing existing in natural membranes, chylomicrons, etc., making this method valuable for study of quantitative lipolytic enzyme kinetics with natural lipids.

Examples and Applications: Kinetic Aspects of Lipolysis
  by the Lipase-Colipase System

This section is not intended to give an exhaustive view of the lipase–colipase system. For the general aspects of this subject, see a recent review of Sémériva and Desnuelle.[101] Here, we will restrict ourselves to the discussion of few illustrative kinetic problems.

---

[99] R. O. Scow, P. Desnuelle, and R. Verger, *J. Biol. Chem.* **254**, 6456 (1979).
[100] D. Lairon, M. Charbonnier-Augiére, G. Nalbone, J. Léonardi, J. C. Hauton, G. Pieroni, F. Ferrato, and R. Verger, *Biochim. Biophys. Acta* (submitted).
[101] M. Sémériva and P. Desnuelle, *Adv. Enzymol. Relat. Areas Mol. Biol.* **48**, 319 (1979).

The hydrolysis by pancreatic lipase of emulsified substrates or soluble substrates adsorbed at hydrophobic interfaces is inhibited by bile salts below the critical micellar concentration.[102,103] A point of great interest is that this effect is fully reversed[102-105] by a small cofactor (MW about 10,000) synthesized by the pancreas, carried together with lipase to the duodenum by pancreatic juice[106] and designated colipase.[107]

A kinetic illustration of the influence of detergents on the interfacial lipase activity is provided by simultaneous measurements of enzyme velocity and induction time as a function of detergent concentration and shown in Fig. 14A.[108] The induction time increases in parallel to a decrease in lipase activity until the latter cannot be measured. It is clear that the establishment of a steady state characterized by a constant enzyme velocity in the presence of detergents is not an instantaneous process and requires several minutes. By analogy with monolayer studies of the presteady state conditions (see the preceding section on the kinetic model for lipolysis of insoluble lipids), it is likely that the induction times observed also during lipolysis under bulk conditions are related to a slow penetration of the enzyme between the lipid molecules organized at the interface. The existence of such unusually long induction times seems a rather general feature of the interfacial enzyme kinetics of lipolysis. As originally noticed by Vandermeers et al.,[109] the addition of colipase shortens the lag period at neutral and alkaline pH in the presence of bile salts. This fact is further illustrated from the work of Borgström[108] as shown in Fig. 14B. Increase in colipase concentration results in a decrease of the lag phase until it is almost completely abolished, with a concomitant increase of the enzymic velocity measured under steady-state conditions.

Several mechanisms have been proposed to account for the inhibitory effect of bile salts upon lipase activity; they can be divided into two groups. A first hypothesis is that bile salts as well as other tensioactive compounds accumulate on the substrate surface and thus prevent lipase

---

[102] M. F. Maylié, M. Charles, M. Astier, and P. Desnuelle, *Biochem. Biophys. Res. Commun.* **52**, 291 (1973).
[103] B. Borgström and G. Erlanson, *Eur. J. Biochem.* **37**, 60 (1973).
[104] B. Borgström and C. Erlanson, *Biochim. Biophys. Acta* **242**, 509 (1971).
[105] R. G. H. Morgan and N. E. Hoffman, *Biochim. Biophys. Acta* **248**, 143 (1971).
[106] R. G. H. Morgan, J. Barrowman, and B. Borgström, *Biochim. Biophys. Acta* **175**, 65 (1969).
[107] M. F. Maylié, M. Charles, C. Gache, and P. Desnuelle, *Biochim. Biophys. Acta* **229**, 286 (1971).
[108] B. Borgström, *Biochim. Biophys. Acta* **488**, 381 (1977).
[109] A. Vandermeers, M. C. Vandermeers-Piret, J. Rathé, and J. Christophe, *FEBS Lett.* **49**, 334 (1975).

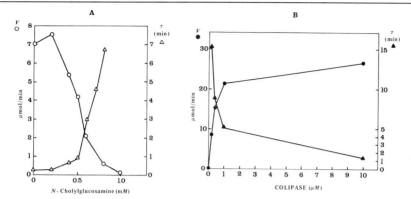

FIG. 14. The effect of $N$-cholylglucosamine concentrations on lag time ($\triangle$—$\triangle$) and steady-state velocity ($\bigcirc$—$\bigcirc$) during the pancreatic lipase hydrolysis of tributyrylglycerol (A). Variations of the same parameters as a function of colipase concentration at 4 m$M$ $N$-cholylcholamine (B). From B. Borgström, *Biochim. Biophys. Acta* **488**, 381 (1977).

adsorption.[103,106,110,111] This interpretation is based on the fact that when increasing the substrate surface area, either by increasing the amount of emulsion added or its degree of dispersion, a higher bile salt concentration is needed for inhibition.[103] However, the undetermined increase in the emulsion area can at the same time increase the amount of adsorbed bile salts and thus reduce its concentration in solution. The fact that a higher bile salt concentration is needed for inhibition is not necessarily the result of the building of a detergent monolayer on the substrate surface. Another hypothesis on the origin of the inhibition by bile salts has been proposed by Momsen and Brockman.[112] In the tripropionin glass beads system, the S-shaped plot of lipase activity vs. taurodeoxycholate concentration suggested to the authors a cooperative formation of a lipase(bile salt)$_4$ complex which has a negligible affinity for the bile salt-covered surface. Lipase inhibition has been reported to be nearly complete well below the classical bile salts CMC and to take place in the bile salt concentration range where occurred the formation of "premicelles," which are tetramers.[113] This hypothesis is in agreement with the recent findings of Lairon *et al.*[114] showing by hydrophobic affinity chromatography that the presence of a bile salts coat on the hydrophobic gels does not significantly

[110] B. Borgström, *J. Lipid Res.* **16**, 411 (1975).
[111] C. Chapus, H. Sari, M. Sémériva, and P. Desnuelle, *FEBS Lett.* **58**, 155 (1975).
[112] W. E. Momsen and H. L. Brockman, *J. Biol. Chem.* **251**, 384 (1976).
[113] D. M. Small, *in* "The Bile Acids" (P. Nair and D. Kritchevsky, eds.), p. 249. Plenum, New York, 1971.
[114] D. Lairon, G. Nalbone, H. Lafont, J. Leonardi, N. Domingo, J. Hauton, and R. Verger, *Biochemistry* **17**, 205 (1978).

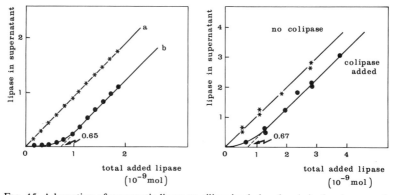

FIG. 15. Adsorption of pancreatic lipase to siliconized glass beads in the absence of substrate. Left panel; No added bile salts. Curves a and b, assays without and with beads, respectively. Right panel: Enzyme adsorption in the presence of 0.85 m$M$ taurodeoxycholate without (★—★) or with (●—●) added colipase. From C. Chapus, H. Sari, M. Sémériva, and P. Desnuelle, *FEBS Lett.* **58**, 155 (1975).

hinder lipase binding. However, the exact interpretation of the origin of bile salt inhibition has to await more direct experimental data of the simultaneous lipase, colipase, and bile salts interfacial adsorptions.[119]

The binding of lipase to siliconized glass beads,[111] to the same beads coated with the substrate tripripionin,[24] and to tributyrin emulsions[110,115] was independently shown in several laboratories to be prevented by bile salts and fully restored by saturating amounts of colipase. The parallel variations of activity and binding strongly suggest that both bile salts and colipase control lipase activity through their opposite effect on the interfacial fixation of the enzyme. The interfacial adsorption behavior of lipase on siliconized glass beads is illustrated in Fig. 15. A very important point revealed by the curves of Fig. 15 is that the adsorption of lipase to siliconized glass beads is completely abolished by a taurodexycholate concentration not exceeding 0.8 m$M$ and that it is fully restored by the addition of 3 moles of colipase per mole of lipase. It has been suggested that the rate of adsorption of lipase is controlled by diffusion through the unstirred layer at the substrate interphase.[8] According to this hypothesis, it would be difficult to imagine that the presence of detergent and colipase could affect adversely the diffusion rate of the lipase molecule. Furthermore, this diffusion process should be kinetically controlled by the hydrodynamic parameters of the diffusing molecule and the viscosity of the medium, and it was calculated to be a rapid process under turbulent conditions.[13] In

---

[115] A. Vandermeers, M. C. Vandermeers-Piret, J. Rathé, and J. Christophe, *Biochem. Biophys. Res. Commun.* **69**, 790 (1976).

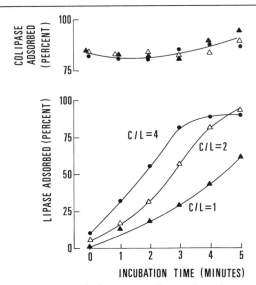

FIG. 16. Time course of pancreatic lipase and colipase adsorption to tributyrylglycerol in the presence of 4 mM taurodeoxycholate. Incubation media containing lipase and colipase in a molar ratio (C/L) of 1 (▲—▲), 2 (△—△), and 4 (●—●). From A. Vandermeers, M. C. Vandermeers-Piret, J. Rathé, and J. Christophe, *Biochem. Biophys. Res. Commun.* **69**, 790 (1976).

contrast, it is easily understandable that a modification by detergents of the interfacial quality will reduce the adsorption rate of lipase and that the presence of colipase, which behaves as an interfacial anchor for lipase, will increase it. This latter point is clearly illustrated in Fig. 16.

One can see that when studied as a function of time, colipase adsorption on the interface did not parallel that of lipase, and was in fact a more rapid event. The percentage of adsorbed colipase was constant with time whereas the rate of lipase adsorption was dependent upon the amount of colipase already bound to the interface. This finding strongly suggests that each molecule of bound colipase acts as a binding site for lipase on the tributyrin surface in the presence of bile salts. It is clear that the rate of lipase adsorption is the limiting step of lipolysis at low colipase concentrations. It is not a surprise that under such conditions the kinetics parameters analogous to "$K_m$" and "$V_m$" measured independently by Maylié *et al.*[102] and Vandermeers *et al.*[115] are very difficult to interpret in quantitative terms. Since colipase promotes lipase binding on tributyrin emulsion in the presence of bile salts,[110,115] as well as on bile salt micelles,[102,116] it is

[116] M. Charles, H. Sari, B. Entressangles, and P. Desnuelle, *Biochem. Biophys. Res. Commun.* **65**, 740 (1975).

to be expected that taurodeoxycholate, at concentrations higher than its critical micellar concentration, acts as a competitive inhibitor on lipolysis.[111,115] Kinetics experiments performed by Vandermeers et al.[115] support this assumption. However, this interpretation was recently challenged by Borgström[108] on the basis of experiments showing that colipase desorbed from the substrate interface at increasing nonionic detergent concentrations well before the appearance of micelles in the medium. At this point, it must be stressed that the mechanisms of interfacial desorption of colipase and lipase by detergents are unknown, and there is no evidence at the present time that neutral and ionic detergents behave similarly.

Brockerhoff[23] reported that pancreatic lipase was rapidly and irreversibly inactivated at hexadecane–water or triglyceride–water interfaces. Bile salts or albumin could prevent this loss of activity. Several other groups have confirmed this observation and have measured the rate constant for lipase inactivation in the presence or the absence of colipase.[24,25,117,118] If lipase activity is controlled by a penetration or adsorption flux responsible for an initial lag period and an inactivation flux tending to decrease the reaction rate, the obtained kinetics are expected to be S-shaped.[20] It was shown by different groups that one of the colipase effects is to reduce the interfacial inactivation rate of lipase.[24,25,117,118] It is probable that interfacial inactivation of lipase and colipase protection are rather nonspecific phenomena. For instance air–water, hydrocarbon–water, or silicone–water are highly inactivating interfaces. Other proteins, such as albumin, also can reduce this inactivation rate, resulting in an increase of lipase activity. However, the kinetically most important role of colipase is probably to reverse the inhibitory effect of amphipathic compounds such as bile salts on the lipase-catalyzed hydrolysis of glycerides.

The monomolecular film system is ideally simple and was used with negatively charged didodecanoylphosphatidylglycerol monolayers. In this system, the phosphatidylglycerol films were reported to be optimally digested at 18 dyne/cm by lipase with, as for phospholipase $A_2$, an absolute requirement for calcium.[32,117] In the presence of EDTA, no enzyme activity and no surface excess of radioactively labeled lipase could be detected. A point of interest using these films was to find that colipase stimulated lipase activity. This stimulation was pressure dependent. Relatively weak between 11 and 20 dyne/cm, it sharply increased beyond this upper limit. More light was shed on the origin of this surface pressure depen-

---

[117] R. Verger, J. Rietsch, and P. Desnuelle, J. Biol. Chem. 252, 4319 (1977).
[118] P. Canioni, R. Julien, J. Rathelot and L. Sarda, Lipids **12,** 393 (1977).

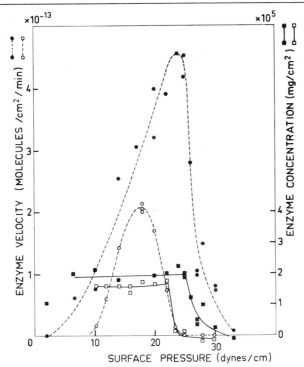

FIG. 17. Pressure dependence of the colipase effect on the hydrolysis of phosphatidylglycerol films by pancreatic lipase. Surface radioactivity excess (□—□) and lipase activity (○---○) vs. film pressures in the absence of colipase. Same variations (■—■) and (●---●) in the presence of colipase. From R. Verger, J. Rietsch, and P. Desnuelle, *J. Biol. Chem.* **252**, 4319 (1977).

dency in the high-pressure range when the adsorption of lipase to the film, with or without colipase, was evaluated. Figure 17 shows that this parameter remains practically constant up to a critical pressure of 22 dyn/cm, above which it sharply decreases. In other words, within a very narrow surface pressure range of 1 dyne/cm, the properties of the interface are modified in such a way as to prevent lipase penetration into the phosphatidylglycerol film. This finding is to be compared to the kinetic observations that the lag period, characteristic of enzyme-catalyzed film digestions, increases abruptly beyond a critical pressure.[32] Both phenomena are probably related to poor enzyme penetration at high surface pressure.

A further point of interest revealed by Fig. 17 is that the above-defined critical pressure was shifted from 22 to 30 dyne/cm by the presence of colipase in the subphase. This fact provides a clear illustration of the favorable effect of colipase on the penetration of lipase into a charged interface. In fact, there is a surface pressure interval between 23 and 30 dyne/cm where the presence of colipase is absolutely required for lipase to be enzymically active on phosphatidylglycerol monolayers. More recently Piéroni and Verger[98] reported that pancreatic lipase can bind to a film of pure lecithin at 12 dyne/cm only in the presence of colipase. It should be pointed out, however, that in the case of phosphatidylcholine films no enzymic activity could be detected, even though lipase can penetrate the lecithin film in the presence of colipase. This lipid binding in the absence of detectable hydrolysis can be viewed as a physiological advantage, since Lairon et al.[119] reported that colipase is required for the formation of the lipase–bile lipoprotein complex (containing lecithin) and that bile lipids are needed to direct the adsorption of this lipolytic entity onto the emulsified substrate. In the context of this model, it can be seen that bile lipids must be resistant to lipolysis.

Concluding Remarks: Hypotheses to Explain the "Activation by Interfaces"

The jump in activity that occurs upon binding of a lipolytic enzyme to certain interfaces can be attributed either to the aggregated state of the substrate or to a new property of the enzyme.

*Local Concentration of Substrate Molecules*

This simple and apparently logical explanation for the interfacial "activation" has been clearly expressed by Brockman et al.[8] The existence at the interface of an ordered array of lipid molecules in a high local concentration could create a situation similar to the concentration of lipid molecules in a pure bulk-lipid phase. This should allow more frequent productive collisions with a soluble enzyme molecule than could occur if monomeric lipid molecules tumble free in solution. It is evident that an increase in substrate concentration would lead to a proportional increase in velocity. However, it seems difficult to explain reported rate accelerations of $10^3-10^4$ only by a concentration increase. Moreover, if this explanation were correct, it would be difficult to understand why the esterases behave so differently from the lipases (cf. Fig. 2) and especially why two structurally related proteins like pancreatic phospholipase A and its zymogen behave so differently toward phospholipids in micellar form (cf. Fig. 3). Finally, if one accepts the two-dimensional Michaelis–Menten model of lipolysis shown in Fig. 4, then a second objection against the hypothesis

[119] D. Lairon, G. Nalbone, H. Lafont, J. Leonardi, N. Domingo, J. C. Hauton, and R. Verger, *Biochemistry* **17**, 5263 (1978).

would result, i.e. high local substrate concentrations would be the main cause of the activating effect by interfaces. According to Michaelis and Menten, in a monomeric solution of substrate the maximal velocity is reached when all enzyme in solution is in the form ES. This would also be the upper limit if the enzyme is present at an interface. If the complexes ES (bulk) and E*S (interface) were identical, it is hard to imagine accelerating factors of 1000 and more due to local concentration of substrate.

*Lowering of the Energy of Activation upon Substrate Aggregation*

In general, enzyme catalysis in monomeric substrate solution seems difficult to explain by favorable entropy factors because "immobilization" of two freely moving molecules on the enzyme active site lowers their translation energy and therefore diminishes the entropy. This would result in an increase of $\Delta G$ and a lowering of the velocity. This unfavorable effect might be compensated for, however, by substrate aggregation. Although substrate aggregation by itself lowers the entropy of the system (size of the aggregate, loss of rotational and translational energy), it can be argued that the orientation of the substrate molecules in the interface is much more favorable for interaction with the enzyme than in the case of free monomers tumbling in solution. Such orientation effects, causing a fall in activation energy, have been proposed by many authors.[18,28,47,84]

Using $^1$H NMR, Roberts *et al.*[120,121] have shown the existence of a large chemical shift difference between the $\alpha$-methylene groups of the two fatty acyl chains of phospholipids in Triton X-100–phospholipid mixed micelles. The authors suggested that phospholipid molecules adopt a unique conformation in all micellar environments, be they pure phospholipid micelles or mixed micelles. In this conformation, the *sn*-1-$\alpha$-methylene protons have indistinguishable chemical shifts and are in a more hydrophobic (shielded) environment than the strongly differentiated protons of the sn-2-$\alpha$-methylene group. However, differential shielding of the two protons on the *sn*-2 carbon is not a prerequisite for enzymic hydrolysis by phospholipase $A_2$.

*Hydration State of the Substrate*

It has been suggested[122] that lipid molecules organized at interfaces would be poorly hydrated, in contrast to the heavily hydrated monomeric molecules moving freely in aqueous solution. Assuming that lipolytic en-

---

[120] M. F. Roberts and E. A. Dennis, *J. Am. Chem. Soc.* **99**, 6142 (1977).
[121] M. F. Roberts, A. A. Bothner-By, and E. A. Dennis, *Biochemistry* **17**, 935 (1978).
[122] H. Brockerhoff, *Biochim. Biophys. Acta* 159, 296 (1968).

zymes are characterized by weak nucleophilic properties of their active site, one finds shielding of the ester bonds by water molecules to impede monomer hydrolysis. However, upon aggregation and dehydration of the substrate, effective hydrolysis would be possible. Some support for this hypothesis might be derived also from a paper by Entressangles and Desnuelle,[92] who showed that monomeric lipid molecules can be hydrolyzed by lipase at a relatively high rate if low concentrations of certain organic solvents are present. These authors suggested that the rate acceleration observed might be caused by a change in water structure around the monomeric lipid molecules. The most extensive studies on the role of water in lipolysis have been reported by Wells and colleagues.[10,52,53] It has been known already for many years that phospholipases from various snake venoms are highly active toward lecithin dissolved in moist ether. In such systems, the lecithin substrate forms reversed micelles. Misiorowski and Wells[10] showed that, upon addition of increasing amounts of water to lecithin dissolved in dry ether, different micellar species are formed. These species differ in degree of hydration and also show dramatic differences in their susceptibility to phospholipase A attack. Although we feel that the hydration state of the substrate could be an important factor in the lipolysis of aggregates, the systems presently available to investigate this attractive hypothesis are too complicated to allow definite conclusions.

*Area per Substrate Molecule or Surface Pressure*

In agreement with the original hypothesis of Hughes[78] and Shah and Schulman,[84] de Haas et al.[11] suggested that under bulk conditions also the area per molecule of substrate in the micelle (or the charge density) might be one of the most important factors determining the rate of lipolysis. Soon afterward, however, the monolayer work of Zografi et al.,[12] in which the same synthetic substrates were used, showed that three different short-chain lecithins present in a monolayer at an air–water interface were hydrolyzed with a maximal velocity at 8 dyne cm$^{-1}$ surface pressure. These maximal velocities were found to be very similar. However, as discussed previously, the amount of enzyme adsorbed to the monolayer was a function of $\pi$ and consequently the velocity–$\pi$ optimum had no direct meaning. Recently, Pattus et al.[27] reinvestigated this problem using $C_8$ to $C_{12}$ short-chain lecithins spread as monolayers at the air–water interface. With all the lecithins used, the profiles of induction time upon lipid surface density were found to be superimposable. In all cases, the same lipid packing (corresponding to a fixed area of 75 Å$^2$ per molecule but to variable surface pressures) was critical for pancreatic phospholipase $A_2$ penetration. Above this critical lipid packing, the induction time increased abruptly and no enzymic activity could be detected.

As an alternative possibility, it was suggested by Esposito et al.[96] that the surface pressure $\pi$ of the film, which determines the free interfacial energy, might govern the functional conformation of the adsorbed enzyme molecule. From the early work of Shah and Schulman,[84] as well as from results of Verger et al.[32] and a recent study of Pattus et al.,[27] it is clear that this hypothesis cannot be correct either. As discussed in the section Monolayers of Substrates, the observed maxima in velocity-$\pi$ profiles disappear when they are related to the interfacial excess of enzyme.

## Lipid-Induced Enzyme Aggregation

Recently, Roberts et al.[16] proposed a dual phospholipid model for phospholipase $A_2$ catalysis of interfacial phospholipid present in mixed micelles. As shown in Fig. 18, the enzyme must first bind $Ca^{2+}$ before it can bind phospholipid. Once $Ca^{2+}$ is bound, the enzyme binds one phospholipid molecule at the interface. This binding causes a conformational change in the enzyme that leads to dimerization. A second phospholipid is then bound by the dimer at a functional active site and catalysis occurs. These two phospholipid molecules have quite distinct roles. One sequesters the enzyme to the interface; the other is needed for subsequent catalysis. According to this scheme, lipid substrate is essential for enzyme aggregation and it is the resulting asymmetric dimer unit that is the active form of the enzyme. Two types of experiments support this hypothesis. First, phospholipid induces the aggregation of enzyme, as shown by the dimethyl suberimidate cross-linking experiments.[16] Aggregation occurs in the presence of a large excess of mixed micelles so that the increase in cross-linking should not merely reflect a concentration effect on the enzyme by the mixed micelles. Second, chemical modification experiments with p-bromophenacyl bromide suggest that an asymmetric dimer exists.[123] Only 0.5 mol of histidine was modified per mole of enzyme, yet the derivative was inactive. Furthermore, the same authors using gel chromatography found that under no conditions could binding of the enzyme to the pure Triton micelle be observed. In contrast, mixed micelles containing Triton and either phospholipid or fatty acid or lysophospholid could bind the enzyme. One would expect lateral diffusion of phospholipid in the mixed micelle to be quite rapid. This may be the primary means for bringing the second phospholipid in contact with the functional catalytic subunit.

Finally, Roberts et al.[16] concluded that "although the surfactant

---

[123] M. F. Roberts, R. A. Deems, T. C. Mincey, and E. A. Dennis, *J. Biol. Chem.* **252**, 2405 (1977).

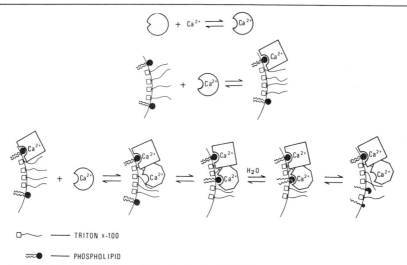

FIG. 18. Schematic diagram of the "dual-phospholipid" model for the action of phospholipase $A_2$ toward phospholipids contained in mixed micelles. Enzyme first binds $Ca^{2+}$ and undergoes a conformational change that allows it to bind to phospholipid in the mixed micelle. The presence of interfacial phospholipid causes the enzyme to form an asymmetric dimer. One subunit of this dimer is responsible for binding to the interface via phospholipid, and the other hydrolyzes an accessible phospholipid. Lateral diffusion of phospholipid in the mixed micelle may be involved before the catalytic subunit binds phospholipid. Alternatively, two enzyme molecules bound to interfacial phospholipid may associate to form the asymmetric dimer directly. Once catalysis occurs, the products may diffuse away from the enzyme and either be retained in the mixed micelle or released into the solution. From M. F. Roberts, R. A. Deems, and E. A. Dennis, *Proc. Natl. Acad. Sci. U.S.A.* **74,** 1950 (1977).

Triton X-100 acts as a pure surface dilutor in relation to the enzyme, it may in the process of solubilizing phospholipids also serve to alter phospholipid conformation. This could make the phospholipid more susceptible to phospholipase $A_2$ binding." In other words, one can expect the interfacial quality of a lipid substrate to be affected by the presence of variable amounts of surfactant in mixed micelles. This means that the dissociation constant for the enzyme–phospholipid and/or the enzyme–mixed micelle complex could change at variable phospholipid–Triton molar ratios.

*Orientation and Conformational Change of the Enzyme at Interfaces*

Several years ago, Desnuelle *et al.*,[124] in an attempt to explain the activating effect of substrate aggregation on lipolysis, suggested that a conformational change might occur in the enzyme molecule upon interaction

---

[124] P. Desnuelle, L. Sarda, and G. Ailhaud, *Biochim. Biophys. Acta* **37,** 570 (1960).

with certain lipid–water interfaces. A simple and direct demonstration of the validity of this hypothesis would be the observation of an enhancement of the hydrolysis rate of lipid monomers in the presence of lipid–water interfaces. Wells,[28] however, showed that the presence of micellar L-dioctanoyllecithin did not accelerate the hydrolysis of molecularly dispersed L-dibutyryllecithin. This observation does not preclude an essential conformational change of the enzyme upon interaction with the micelle: in the above hypothesis the active site of the enzyme is thought to be located at the interface, and it seems obvious that the hydrolysis of monomers will be accelerated only if these "monomers" are incorporated into the interface.

Roholt and Schlamovitz[125] reported an increase in the rate of phospholipase A hydrolysis of dihexanoyllecithin at a concentration below the CMC upon addition of long-chain lysolecithin to the reaction medium. Bonsen et al.[126] showed that the same substrate dihexanoyllecithin is hydrolyzed much more rapidly even if the added micellar phospholipid is a competitive inhibitor for the enzyme, such as the optical antipode D-dioctanoyllecithin. Both examples might be explained in the same way: when the monomers are incorporated into a lipid–water interface, they can be hydrolyzed at rates comparable with pure micellar substrates. More recently, Chapus et al.[127] showed that the deacylation of isolated [$^3$H]acetyl lipase was accelerated by at least a hundred times in the presence of siliconized glass beads. This suggests that a part of the activation of lipase at interfaces may be due to a conformational change resulting from adsorption.

It seems logical that a lipolytic enzyme upon approaching a lipid–water interface will orient itself in such a way that its active site comes into contact with a single substrate molecule in the interface. The idea of the presence of an additional site on the enzyme being topographically and functionally distinct from the classical active site and responsible for the reversible attachment of the protein to the interface has independently been postulated by two groups.[13,128] Because the exact structural description of this site is not known, it is not surprising that various names have been proposed to characterize this region. Brockerhoff[17] suggested the name "Supersubstrate binding site," whereas Verger et al.[13] described

---

[125] O. A. Roholt and M. Schlamowitz, *Arch. Biochem. Biophys.* **94**, 364 (1961).
[126] P. P. M. Bonsen, G. H. de Haas, W. A. Pieterson, and L. L. M. Van Deenen, *Biochim. Biophys. Acta* **270**, 364 (1972).
[127] C. Chapus, M. Sémériva, C. Bovier-Lapierre, and P. Desnuelle, *Biochemistry* **15**, 4980 (1976).
[128] H. Brockerhoff, *Chem. Phys. Lipids* **10**, 215 (1973).

this area as a "penetration site." Pieterson[129] favored the expression "anchoring site" and Desnuelle (personal communication) proposed the more general term "interface recognition site" (IRS). Direct experimental evidence for the existence of such an "interface recognition site" in porcine pancreatic phospholipase $A_2$ was obtained by Pieterson et al.,[130] who found that the naturally occurring zymogen of phospholipase A binds and hydrolyzes monomeric substrate molecules with an efficiency comparable to that of the active enzyme. However, in contrast to the enzyme, the zymogen has no affinity for organized lipid–water interfaces. That the IRS and the active site are located in topographically distinct regions on the enzyme surface was shown by chemical modification techniques. Specific blocking of the $\alpha$-$NH_2$ group in the N-terminal amino acid of phospholipase $A_2$ switched off its affinity for lipid–water interfaces without seriously affecting monomer binding and hydrolysis.[29,131] In addition, upon chemical modification of the active site residue His,[48] which leads to an enzymatically inactive phospholipase $A_2$, the affinity constant of the modified protein for lipid–water interfaces remained unchanged.[130]

Pancreatic lipase can also be chemically modified by reaction of three histidine residues with diethyl pyrocarbonate or by reaction of five carboxyl groups with carbodiimide (5N-lipase) or by esterification of one serine residue with diethyl $p$-nitrophenyl phosphate (DP-lipase).[132] In the three cases, the activity on emulsified substrates is abolished. The modification of histidine residues results also in a loss of activity on dissolved substrates, suggesting that the essential histidine is at (or close to) the active site. The ability of lipase to be adsorbed on siliconized glass beads is not impaired in this reaction. By contrast, 5N-lipase is still able to hydrolyze dissolved monomeric substrates and to absorb onto siliconized glass beads. Therefore, the essential carboxyl group is assumed to play an important role in interfacial activation. Finally, since DP-lipase is still fully active on dissolved $p$-nitrophenyl acetate, the serine residue is more likely implicated in the recognition and the binding to interfaces, as confirmed by the inability of DP-lipase to be adsorbed onto siliconized glass beads.

Indirect evidence for the occurrence of a conformational change in pancreatic phospholipase $A_2$ upon interaction with certain isotropic micellar systems has been obtained recently by both ultraviolet-difference and fluorescence spectroscopy.[29] In agreement with the earlier hy-

---

[129] W. A. Pieterson, Mechanism of action of phospholipase $A_2$ and its zymogen on short-chain lecithins, Ph.D. thesis, Utrecht Univ., the Netherlands, 1973.
[130] W. A. Pieterson, J. C. Vidal, J. J. Volwerk, and G. H. de Haas, Biochemistry 13, 1455 (1974).
[131] A. J. Slotboom and G. H. de Haas, Biochemistry 14, 5394 (1975).
[132] C. Chapus and M. Sémériva, Biochemistry 15, 4988 (1976).

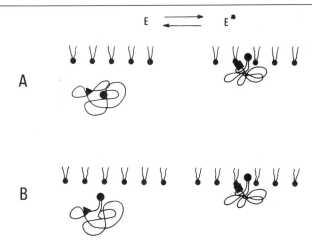

FIG. 19. Speculative representations of the penetration of a lipolytic enzyme in a lipid–water interface. From R. Verger and G. H. de Haas, *Annu. Rev. Biophys. Bioeng:* **5**, 77 (1976).

pothesis[130] that the N-terminal part of the phospholipase $A_2$ molecule might play an important function in the recognition process of lipid–water interfaces, it was found that the single Trp residue located at position 3 of the chain was heavily perturbed upon interactions with micelles. This interaction appeared to be specific for organized lipid structure and dependent on pH and $Ca^{2+}$ concentration.[13,29] [Gly¹]-phospholipase $A_2$ behaves like the native enzyme on lecithin monolayers. DL-Ala⁻¹ and [D-Ala¹]phospholipases $A_2$, although they are active in this system, showed a weaker lipid penetration capacity.[133] Moreover, the specific tryptic cleavage of the Arg-Ser between location 6 and 7 of the phospholipase $A_2$ chain was shown to be inhibited in the presence of $Ca^{2+}$ ions and micellar lipid interface.[29] This observation seems to locate the IRS in a region embracing at least the apolar N-terminal sequence of 6 amino acid residues: Ala, Leu, Trp, Gln, Phe, Arg.

It is conceivable that the recognition site (●) is already present in the soluble enzyme molecules (as illustrated in Fig. 19B), or else is induced during the interaction of the enzyme with the interface ("induced fit") (as illustrated in Fig. 19A). In either case the catalytic groups (▲) of the active site might be brought into proper alignment for optimal action during this process (■). With either model, the "quality" of the interface would act as a powerful regulator of catalytic activity because these enzymes are

[133] F. Pattus, A. J. Slotboom, and G. H. de Haas, *Biochemistry* **13**, 2703 (1979).

held at the interface by a specific site. The same reasoning can be applied to the control mechanism of membrane-bound enzymes. Also, in this case the fine architecture of the membrane, as determined by its metabolic activity, could modulate the interaction between enzyme and membrane and hence control the required catalytic power.

## Acknowledgments

The fruitful discussions with Professor P. Desnuelle (Marseilles University) and his constructive suggestions during the preparation of the manuscript are highly appreciated. Professor G. H. de Haas (Utrecht University, The Netherlands) kindly gave permission to reproduce original documents from the work cited in footnote 62. Thanks are due to Dr. R. O. Scow (NIH Bethesda, Maryland), to Dr. F. Pattus (Marseilles University), and to Dr. C. Rothen (Bern, Switzerland) for personal communications prior to publication. I am grateful to Mme M. Th. Nicolas for typing the manuscript.

# Author Index

Numbers in parentheses are reference numbers and indicate that an author's work is referred to although the name is not cited in the text.

## A

Adam, N. K., 364
Adamich, M., 344
Adler, S. P., 298, 310(9, 10)
Agarwal, S. S., 297
Ailhaud, G., 388
Ainslie, G. R., Jr., 8(4), 10, 140, 143, 164(8), 193, 194(5), 196, 207, 208(5), 211, 214(5), 217(5), 222(5)
Alberts, A. W., 361
Albery, W. J., 110
Allen, J. D., 251, 255, 257, 258, 260, 263, 264, 265, 266, 267, 268, 269, 270, 271, 273(3), 274(6), 30)
Allison, R. D., 7, 9(36), 11, 13(17), 22(17), 42, 43(63), 45(63)
Anderson, C., 225
Anderson, J. W., 9(27), 11
Anderson, W. F., 180
Andrews, J. J., 58
Antonini, E., 146
Aoshima, H., 260, 270, 271(32)
Ashby, B., 219
Assaf, Y., 275
Astier, M., 378
Atkins, G. L., 145
Atkinson, D. E., 162
Augustein, L. G., 365

## B

Baartmans, P. H. M., 364
Badger, M. R., 58
Badway, J. A., 219
Bailey, J. M., 274
Bais, R., 219
Baker, H. A., 9(36), 11
Baldwin, R. L., 224, 226
Baldy, J., 360
Balinsky, D., 8(5), 10
Balinsky, J. B., 9(28), 11
Ballou, D. P., 55, 56
Bambara, R. A., 291, 292

Bangham, A. D., 366
Banks, B. E. C., 351
Barden, R. E., 219
Bardsley, W. G., 141, 181, 182
Barenholz, Y., 359, 360, 362, 366, 390(59)
Barque, J. P., 347
Barrowman, J., 378, 379(106)
Bar-Tana, J., 48, 50(3), 52(3), 71
Bartfai, T., 354
Bates, D. J., 200, 204, 216(17)
Battula, N., 282, 294
Baughn, R. L., 95
Bearer, C. F., 191, 203, 219, 222
Bender, M. L., 109
Benedict, C. R., 96
Benjamin, E., 326
Benjamin, L., 356
Bennett, W., 225
Benyon, J. H., 73
Ben-Zeev, O., 71
Benzonana, G., 356, 357, 358, 360, 361(55)
Bernardi, F., 297
Bessman, M. J., 295
Betz, G., 8(10, 11), 10
Bevington, P. R., 156, 157, 158(34), 203, 205(23)
Bigeleisen, J., 83, 85, 94, 100(2), 103(2)
Biltonen, R. L., 148
Birnbaumer, L., 219
Black, M. K., 34
Blackburn, M. N., 189
Blangy, D., 167, 172(54)
Blattner, P., 110
Bloch, K., 219
Blout, E. R., 249
Boches, F. S., 10(50), 11
Bock, J. L., 83
Bock, P. E., 195, 220
Börning, H., 167
Boezi, J. A., 279
Bommerson, J. C., 91, 94(21), 110
Bonsen, P. P. M., 344, 363, 386(11), 389
Borgström, B., 378, 379, 380(110), 382

# AUTHOR INDEX

Borkenhägen, L. F., 9(44), 11
Borkovski-Kubiler, I., 360, 390(59)
Borowitz, J. L., 94
Bothner-By, A. A., 385
Botts, J., 7
Bovier-Lapierre, C., 389
Boyer, P. D., 3, 5, 7, 8(1, 3, 12), 9(24, 32, 33, 34), 10, 11, 13, 17, 20, 26, 27, 29, 33, 34, 38, 41, 46, 60, 62(2), 66(2), 68(2), 71(2), 74, 76, 77, 78, 80(21), 81(2), 82
Brandts, J. F., 223
Branscomb, E. W., 297
Brash, J. L., 365
Brdiger, W. A., 9(24, 25, 40, 41), 11, 38
Brennan, M., 223
Brewer, C. F., 224
Britton, H. G., 7, 52, 53(6)
Brockerhoff, H., 345, 347, 349(17), 356, 357, 361(17), 365(23), 382, 385, 389
Brockman, H. L., 342, 347, 350(8), 369, 371, 375, 379, 380(8), 382(24), 384
Brostrom, C. O., 327
Brothers, C., 258, 261, 263(19), 264(19)
Brotherton, J. E., 335, 336, 339(15, 16)
Brown, G., 230, 246, 247(20)
Brown, R. D., III, 224
Brunori, M., 179, 181(86)
Brutlag, D., 295
Bryan, D. M., 60, 62(2), 66(2), 68(2), 71(2), 81(2)
Buc, H., 167, 172(54)
Buc, J., 142, 143(14), 193, 207(7), 210(6), 211(6), 215, 216(32), 217, 219, 221(6, 7, 32, 34)
Buchholz, K., 244
Bull, H. B., 365
Bunting, P. S., 227, 236, 237, 238, 239, 242(1), 246(1), 247
Burgett, M. W., 298, 324(7)
Burk, D., 142
Butterworth, P. J., 306, 324(12)
Byrne, W. L., 3, 9(42, 43), 11, 36

## C

Cahn, R. D., 41
Callewaert, G. L., 351
Canioni, P., 382
Caplan, S. R., 246
Caputo, A., 146
Carlson, C. A., 219

Carlson, G. M., 327
Carroll, R. B., 139
Carvalho, M. G. C., 78
Cavanaugh, D. J., 328
Cedar, H., 9(26), 11
Chaio, Yu-Bin, 219
Chamberlin, M. J., 282, 291
Chang, L. M. S., 291, 294
Changeux, J.-P., 140, 142, 165, 167, 171
Chao, J., 9(38), 11
Chapus, C., 379, 380, 382(111), 389, 390(132), 391
Charbonnier-Augière, M., 377
Charles, M., 378, 381
Chay, T. R., 174
Chipman, D. M., 249, 260, 272(3), 275
Chock, P. B., 297, 298, 300, 301, 304, 305(4), 306, 308, 309, 311, 312, 313, 316, 317, 318, 319, 320, 322, 323, 324(11, 16), 325
Choi, T., 291, 292
Chow, Y., 283, 291
Christeller, J. T., 96
Christophe, J., 347, 378, 380, 381, 382(25, 115)
Cleland, W. W., 3, 7, 8(4, 5, 9, 14, 18, 19), 9(22), 10, 11, 23, 30, 32, 33, 34, 35, 39, 50, 52, 84, 104, 105, 106, 107, 109, 113(2), 120, 121, 123, 278
Cohen, G. N., 191, 226
Cohen, H., 347, 373
Cohen, P. P., 65
Cohn, M., 64, 65, 74, 75(19), 83
Colacicco, G., 366, 367, 372(83)
Cole, F. X., 9(30, 31), 11
Cole, K. W., 219
Collins, C. J., 86
Colman, R. F., 219
Colombo, G., 221
Colosimo, A., 179, 181(86)
Colowick, S. P., 36, 53
Compton, J. G., 189
Constantin, M. J., 360
Conway, A., 170, 171(55), 174(55), 190(55)
Cook, G. H., 219
Cook, P. F., 109
Cook, R. A., 190
Cooper, V. G., 362
Cori, C. F., 325
Cori, G. T., 297, 325
Cornish-Bowden, A., 157, 158, 173, 174, 182, 215, 219, 221(33)

## AUTHOR INDEX

Cottrell, R. C., 351
Craig, H., 93
Cramer, F., 295
Cross, D. G., 42, 162
Crusberg, T. C., 128
Czerlinski, G. H., 165, 189(50)

### D

Dahlquist, F. W., 148, 155, 156, 173(32), 184, 190(100)
Dahms, A. S., 76
Daka, N. J., 241
Dalziel, K., 142, 171
Daniels, F., 83
Dann, L. G., 52, 53(6)
Darnall, D. W., 139
Darvey, I. G., 7, 12(20)
Das, S. K., 296, 297
Datta, A., 67
Davidson, N., 282
Davie, E. W., 323
Davis, J. S., 9(28), 11
Davis, J. T., 364
Dawson, R. M. C., 366, 367, 372(86), 375(86)
Deems, R. A., 345, 355, 361(42), 387, 388
Degani, C., 76
de Gier, J., 360, 377(49)
de Haas, G. H., 341, 342, 343, 344, 345(13), 346, 347, 348, 349(13), 350, 351, 352, 353, 355(13), 359, 360, 361, 363, 364, 370, 371(33), 372(93), 373, 374, 375, 377(33) 380(13), 382(33), 383(32), 386, 387(32), 389, 390, 391
Dellweg, H., 256
Dembo, M., 189
de Meis, L., 78
Demel, R. A., 353
Dennis, A. W., 8(5), 10
Dennis, E. A., 344, 345, 349, 355, 360, 361, 362, 385, 387, 388
Derbesy, M., 368
DeRiel, J. K., 175
Dervichian, D. G., 347, 362, 368, 370, 372
Desnuelle, P., 341, 342, 356, 357, 358, 360, 361(55), 362, 369, 372, 373(96), 377, 378, 379, 380, 381, 382, 383, 387(96), 388, 389
Deutsch, J., 71
de Willingen, A. H. A., 341

di Franco, A., 219
Dinovo, E. C., 20
Dittmer, J. C., 256
Dixon, M., 349
Domingo, N., 379, 384
Doonan, S., 351
Dorizzi, M., 297
Dorozhko, A. I., 214, 215, 217(27)
Dostrovsky, I., 64, 65(7)
Doudoroff, M., 9(37), 11
Drysdale, G., 64, 65(8)
Dube, D. K., 282, 291, 294, 297
Dubois, M., 256
Dunn, G. E., 83, 100(6)
Dunnill, P., 236
Durham, A. C. H., 139
Dygert, S., 258

### E

Eargle, D. H., Jr., 71
Eaton, B. R., 355, 361(42)
Eby, D., 190
Edelhoch, H., 180, 181
Edelstein, S. J., 165, 174(51)
Efstratiadis, A., 296
Eisenberg, H., 219
Elson, E. L., 226
Engasser, J. M., 227, 247(3)
Engel, J. D., 297
Engel, P. C., 142, 181
Engers, H. D., 9(40, 41), 11
Engström, L., 35
Entressangles, B., 342, 362, 369, 381
Epstein, S., 64
Erlanson, C., 378
Erlanson, G., 378, 379(103)
Esposito, S., 372, 373, 387

### F

Farrar, Y. J. K., 10(47), 11
Ferdinand, W., 181
Ferrato, F., 377
Fersht, A. R., 226, 295
Filmer, D., 140, 147(6), 165(6), 168, 169(6), 170(6), 171
Fincham, J. R. S., 219
Fischer, E. F., 325
Fischer, E. H., 310, 325, 335
Fischer, T. H., 325

# AUTHOR INDEX

Fisher, H. F., 42, 162
Fisher, J. R. R., 162, 163(44)
Fitzgerald, P., 109
Fletcher, S. J., 110
Fletterick, R. J., 156, 157(35), 158(35), 159(35), 162(35), 180
Florida, D., 258
Fodge, D. W., 52, 53(5)
Fouchier, F., 188, 214
Freer, S. T., 225
French, D., 255, 256, 274
French, T. F., 220
Frey, P. A., 10(49), 11
Frieden, C., 42, 45, 161, 163, 175, 176, 178, 192, 193, 194(3), 195, 200, 201(3), 204, 207(3), 216(17), 217(3), 219, 220
Friendenvold, J. S., 328
Fromherz, P., 365
Fromm, H. J., 3, 7, 8(12, 13), 9(32), 10, 11, 12, 29(7), 30, 38, 39, 40, 41, 43, 45, 162, 176
Frost, A. A., 19
Fry, A., 83, 100(5)
Fujikawa, K., 323
Fujimori, H., 273
Fujimura, R. K., 296, 297

## G

Gache, C., 378
Gaertner, F. H., 219, 325
Gaines, G. L., 364
Galas, D. J., 297
Gallego, E., 189, 190(105), 191(105)
Garay, R. P., 174
Garces, E., 8(14), 10, 34, 35
Garel, J. R., 220, 225(36)
Garland, C. W., 202
Garner, C. W., 368
Garrett, C., 128, 134(9)
Gasser, F. J., 10(46), 11
Gates, R. E., 162
Gatt, S., 354, 359, 360, 362(58), 390(59)
Gefter, M. L., 291
Geneste, P., 109
Gennis, L. S., 190
Geurts van Kessel, W. S. M., 353
Gibson, Q. H., 55, 56(10)
Gilbert, H. R., 195
Gilles, K. A., 256
Gillin, F. D., 293, 294

Ginodman, L. M., 10(45), 11
Ginsburg, A., 310
Girotti, A. W., 221
Glaser, L., 197, 219, 223(13)
Gold, A. M., 9(39), 11
Goldbeter, A., 175
Goldman, R., 246
Goldstein, L., 227, 228, 230, 246, 247(2), 340
Goldthwait, D. A., 139
Goodall, D., 189, 190(105), 191(105)
Goodman, M. F., 295
Gorenstein, D. G., 84, 86(15)
Graves, D. J., 9(33, 38), 11, 326, 327, 328, 334, 335, 336, 339(15, 16)
Green, A. A., 297, 325
Green, M., 111
Greengard, P., 297
Greenspan, M. D., 219
Grimshaw, C. E., 106, 123(3)
Gulbinsky, J. S., 8(18), 11
Gurney, R. W., 252

## H

Haber, J. E., 171
Hackney, D. D., 74, 77, 78, 82
Hagerman, P. J., 224
Halvorson, H. R., 223
Hamilton, J. K., 256
Hammes, G. G., 43, 165, 187(49), 188, 189, 190, 219, 220
Hanahan, D. J., 351, 360
Hanes, C. S., 40
Hansen, J. N., 20
Harington, J. S., 9(28), 11
Harris, B. G., 52, 53(5)
Harris, J., 328
Hartshorn, S. R., 109
Haschke, R. H., 310, 325
Hass, L. F., 3, 9(42), 11, 36
Hassid, W. Z., 9(37), 11
Hatano, H., 260, 270, 271(32)
Hatfield, G. W., 201, 207(18), 219
Hathaway, J. A., 162
Hauton, J. C., 377, 379, 384
Hayatsu, H., 132, 133(13), 135(13)
Hearon, J. T., 328
Heilmeyer, L. M. G., Jr., 310, 325
Helfman, W. B., 297
Helmreich, E., 197, 219, 223(13)
Henis, Y. I., 155

Herlihy, J. M., 110
Hersh, L. B., 10(52), 11
Herzfeld, J., 174, 190(66), 191(66)
Herzl, A., 359
Hess, G. P., 276
Hill, A. V., 144, 145(16)
Hill, C. M., 141
Hill, D. E., 219, 220
Hinberg, I., 232
Hindler, S. S., 297
Hiromi, K., 255, 258, 259, 260, 261, 269, 270, 271(32), 273, 274(31), 275, 276
Ho, C., 174
Hoberman, H. D., 7, 8(2), 10
Hofer, H. W., 9(23), 11
Hoffman, N. E., 378
Hogg, J. L., 109
Holler, E., 276
Holwerda, K., 341
Hopfield, J. J., 174, 295
Horn, A., 167
Horovitz, M., 7
Horvath, C., 227, 247(3)
Hosotani, T., 275
Howlett, G. J., 189
Hu, A., 74
Huchs, F., 298, 324(7)
Huestis, W. H., 173
Hughes, A., 365, 372, 386
Hunston, D. L., 155
Hurd, S. S., 335
Hutton, R. L., 74, 77, 82
Hyman, R. W., 282

I

Ingham, K. C., 180, 181
Isbell, A. F., 279
Isogai, Y., 146
Iwasa, S., 260, 270, 271
Iwatsubo, M., 219, 226

J

Jackson, W. J. H., 178, 179(84), 180(84), 181(84)
Jacob, F., 140
Jallon, J. M., 219
James, E., 32
James, L. K., 365
Janin, J., 189, 226

Janson, C. A., 39
Janson, C. J., 76
Jarabak, R., 219
Jeffrey, A., 287, 296
Jencks, W. P., 10(52), 11
Jensen, D. E., 141
Jensen, R. G., 345, 349(17), 361(17), 389(17)
John, M., 256
Johnson, G. F., 9(38), 11
Johnson, R. M., 9(39), 11
Jones, J. M., 109
Josephs, R., 219
Jovin, T. M., 279, 291
Julien, R., 382

K

Kaethner, M. M., 295
Kafatos, F. C., 296
Kagan, Z. S., 214, 215, 217(27)
Kallen, R. G., 10(50), 11
Kamp, H. H., 353
Kaplan, A., 362
Kaplan, N. O., 35, 41
Karlin, A., 246
Karplus, M., 171
Kasvinsky, P. J., 156, 157, 158, 159(35), 162(35)
Katchalski, E., 246
Kato, M., 260, 261
Katzy, B., 248
Kauerz, M. T., 360, 377(50)
Kawai, M., 275
Kedem, O., 246
Keech, B., 219
Kelly, R. C., 141
Kemp, B. E., 326
Kemp, R. G., 221
Kennedy, E. P., 9(44), 11
Kenyon, G. L., 71
Kézdy, F. J., 342, 347, 350(8), 368, 369(8), 371, 373(22), 375(8), 380(8), 384(8)
Khandker, R., 193, 219, 220(8)
Kim, K.-H., 219
Kirkpatrick, D. S., 10(48), 11
Kirsch, J. F., 95
Kirschner, K., 189, 190(105), 191(105)
Kirshner, K., 214, 219
Kirtley, M. E., 147, 169(25), 170(25), 176, 190
Kisliuk, R. L., 128

Klein, F. S., 64, 65(7), 94
Klotz, I. M., 139, 145, 155, 156, 173(33)
Kluetz, M. D., 95
Knowles, J. R., 110
Kobayashi, T., 233, 240
Koenig, S. H., 224
Kornberg, A., 295
Korus, R., 232
Koshland, D. E., Jr., 140, 143, 147, 157, 158, 165, 168, 169(6, 25), 170, 171, 173, 174, 176, 182, 183(47),184, 189, 190, 261
Kosow, D. P., 37, 38
Kraus, H. M. J., 219
Kraut, J., 225
Krebs, E. G., 298, 310(6), 325, 326, 327, 335
Krishnaswamy, P. R., 47
Kuchel, P. W., 177
Kunze, H., 360
Kurganov, B. I., 148, 175(28), 214, 215, 217(27)

## L

Labouesse, B., 220, 225(36)
Lad, P. M., 219, 220
Lafont, H., 379, 384
Lagocki, J. W., 368, 369(91)
Laidler, K. J., 227, 230, 232, 233, 236, 237, 238, 239, 240, 241, 242, 243, 244, 245, 246(1, 15), 247, 248
Laing, W. A., 96
Lairon, D., 377, 379, 384
Lamaty, G., 109
Lamb, C. J., 177
Landsperger, W. J., 52, 53(5)
Langerman, N. R., 139
Langmuir, I., 351
Law, J. H., 342, 347, 350(8), 368, 369(8, 91), 371, 373(22), 375(8), 380(8), 384(8)
Lazdunski, M., 183, 184
Leary, R., 128
Lee, K. L., 298
Lee, L. F., 279
Lee, M. W., 94
Lehman, I. R., 291, 295
Leibovitz-Ben Gershon, Z., 360
Leinbach, S. S., 279
Leonardi, J., 377, 379, 384
Levine, L., 41
Levitzki, A., 155, 165, 171(47), 174(47), 180, 183(47), 184, 190(99)

Lewis, C. A., Jr., 110
Li, L. H., 258, 263(19), 264(19)
Licho, V., 71
Lietzke, M. H., 86
Lillford, P. J., 190
Lilly, M. D., 236
Limburg, J. A., 87, 123
Lineweaver, H., 142
Litt, M., 365
Litwin, S., 48, 50(3), 52(3)
Lo, K.-Y., 295
Loeb, L. A., 282, 291, 294, 297
Long, J. W., 54, 184, 190(100)
Lorimer, G. H., 58
Loudon, G. M., 189
Lowenstein, J. M., 8(16), 10, 195, 219
Lueck, J. D., 38, 39
Lutsenko, N. G., 10(45), 11
Lyman, D. J., 365
Lynn, W. S., 365

## M

McClure, W. R., 198, 279, 283, 291
MacColl, A., 83, 85, 100(4), 103(4)
McConnell, H. M., 173
McDonald, G. G., 83
McDonald, R., 225
Macfarlane, R. G., 323
McGhee, J. D., 141
McHenry, C. S., 128, 129(10), 130(11)
McLaren, A. D., 340, 365(1)
Madsen, N. B., 9(40, 41), 11, 156, 157(35), 158(35), 159(35), 162(35)
Maengwyn-Davies, 328
Magni, G., 298, 310(9)
Malhotra, O. P., 183, 184(95), 190(95)
Mandel, F., 94
Mangum, J. H., 298, 310(9)
Maniatis, T., 278, 296
Marcus, H. B., 10(50), 11
Marlier, J. F., 95, 96(28)
Marshall, M., 65
Martensen, T. M., 328, 335, 336, 339(15, 16)
Mata-Segreda, J. F., 110
Matsuda, A., 127
Matsuno, R., 269, 273, 274(31)
Mattson, F. H., 356, 357, 360, 361(54), 385(47)
Maylié, M. F., 378
Mehler, A. H., 10(51), 11, 20

## AUTHOR INDEX

Meister, A., 34, 47, 58
Melander, L., 83, 85(3), 100(3), 103(3), 128(12), 129
Mellman, W. J., 10(50), 11
Meloche, H. P., 109
Metzenberg, R. L., 65
Metzger, B. E., 197, 219, 223(13)
Meunier, J.-C., 142, 143(14), 193, 207(7), 210(6), 211(6), 215, 216(32), 217, 219, 221(6, 7, 32, 34)
Meyer, F., 325
Midelfort, C. F., 34, 65, 71(10), 75(10)
Mieras, M. C. E., 344, 345(13), 346, 347(13), 349(13), 351, 355(13), 375(13), 380(13), 389(13), 391(13)
Mildvan, A. S., 282
Millen, W. A., 9(24), 11, 38
Miller, I. R., 345, 347(19), 349, 350(19), 365(19), 367, 375
Miller, W. G., 65
Mills, R. C., 3, 9(32), 11
Mincey, T. C., 387
Minton, A. P., 146
Misiorowski, R. L., 342, 360(10), 386
Mitton, C. G., 84
Mizushima, M., 255, 269(6)
Moffet, F. J., 9(25), 11
Monod, J., 140, 142, 165, 167, 171, 172(54)
Monti, C. T., 109
Montroll, E. W., 323
Mook, W. G., 91, 94(21), 110
Momsen, W. E., 347, 379, 382(24)
Morales, M. F., 7
Morales, N. F., 248
Morgan, R. G. H., 378, 379(106)
Morita, Y., 260, 261
Morrison, J. F., 7(10), 8(17, 19), 11, 23(10), 32, 33, 161, 162(38)
Mourad, N., 34
Mouttet, C., 175, 186(78), 188, 214
Müller-Eberhard, H. S., 323
Muzyczka, N., 295
Myerson, A. L., 83

### N

Nagao, A., 275
Nakajima, T., 275
Nalbone, G., 377, 379, 384
Nari, J., 175, 186(78), 188, 214
Narinesingh, D., 239, 241

Naudet, M., 368
Navarro, A., 219
Neet, K. E., 139, 140, 143(8), 164(8), 191, 193, 194(5), 196, 197, 199, 201(14), 203(14), 207, 208(5), 211(5), 214(5), 216(5, 14, 16), 217(5), 219, 222
Neff, N. F., 291
Negishi, K. K., 132, 133(13), 135(13)
Nemethy, G., 140, 147(6), 165(6), 168, 169(6), 170(6), 171
Neuhaus, F. C., 9(43), 11
Ngo, T. T., 232, 236, 238, 239, 240, 241(13), 242, 243, 244, 245, 246(15), 247
Nichol, L. W., 178, 179(84), 180, 181(84, 91)
Nickol, J. M., 298
Niekamp, C. W., 191
Nieuwenhuizen, W., 351, 360
Ninio, J., 297
Nitta, Y., 255, 258, 259(22), 260, 261(22), 269, 275
Noat, G., 30, 32
Noble, R. W., 174
Norman, A. W., 34
Norris, T. H., 19
Nossal, N. G., 293, 294
Numata, C., 258, 259(22), 260, 261(22)

### O

O'Connell, E. L., 10(51), 11, 20, 48, 50(3), 52(3), 56, 57(12)
O'Driscoll, K. F., 232
Ogata, R. T., 173
Ogawa, S., 174
Ogez, J. R., 180
Ohnishi, M., 269, 273, 274(31), 276
O'Leary, M. H., 83, 92, 95, 96(28), 98(8, 9, 10), 99(8), 102(9), 110, 123
Olive, J., 362, 368, 372
Olivera, B. M., 296
O'Neil, J. R., 64
Ono, S., 255, 258, 259(22), 260, 261(22), 269(8), 275
Op den Kamp, J. A. F., 360, 366(49, 50)
Ouellet, L., 248

### P

Packer, L., 340, 365(1)
Palmer, G. A., 55, 56(11)
Pamiljans, V., 47

Papas, T. S., 9(29), 11
Park, R., 322
Parks, R. E., Jr., 34
Pattus, F., 346, 347, 352, 365(20), 373, 374, 375, 382(20), 386, 391
Paulus, H., 175
Pazur, J. H., 255, 256(5)
Pearson, R. G., 19
Pelley, J. W., 298, 306, 324(7)
Penefsky, H. S., 67
Perriard, E. R., 128, 130(11)
Peterkofsky, A., 9(29), 11
Peters, B. A., 199, 216(16)
Pettigrew, D. W., 175, 176
Pettit, F. H., 306, 324(12)
Piéroni, G., 376, 377, 384
Pieterson, W. A., 342, 343, 344, 363, 386(11), 389, 390
Plowman, K. M., 10(47), 11, 21
Pocker, A., 325
Pogolotti, A. L., Jr., 125, 128, 134(9)
Pohl, S. L., 219
Poom, P. H., 360, 386(52)
Postema, N. M., 351
Pousada, M., 375
Powers, S. G., 58
Préhu, C., 347
Pulkownik, A., 260, 271
Pullman, M. E., 67
Purich, D. L., 7, 8(13), 9(36), 10, 11, 13(17), 22(17), 30, 34, 41, 42, 43, 45, 162, 176

## Q

Quarles, R. H., 367, 372(86), 375(86)
Quinn, P. J., 366

## R

Rabin, B. R., 192
Racker, R., 67
Raftery, M. A., 173
Ramaiah, A., 221
Randall, D. D., 298, 324(7)
Rao, G. V. K., 258, 263(19), 264(19)
Rapport, M. M., 366, 372(83)
Rathé, J., 347, 378, 380, 381, 382(25, 115)
Rathelot, J., 382
Raushel, F. M., 9(22), 11, 120

Raushel, R. W., 52
Ray, W. J., Jr., 10(48), 11, 54, 201, 207(18), 219
Rebers, P. A., 256
Reed, L. J., 298, 306, 324(7, 12)
Regan, D. L., 236
Reibach, P. H., 96
Reid, T. W., 76
Reimann, E. M., 335
Reiner, J. M., 161, 328
Reisler, E., 219
Reno, J. M., 279
Rhee, S. G., 322, 324
Rhoads, D. G., 8(16), 10
Rhodes, G., 282
Ricard, J., 30, 32, 142, 143(14), 175, 186(78), 188, 193, 207(7), 210, 211(6), 214, 215, 216(32), 217, 219, 221(32, 34), 222(6, 7)
Richardson, C. C., 295
Rideal, E., 364
Rietsch, J., 346, 347(20), 350, 353, 365(20), 373(20, 32), 382, 383
Rife, J. E., 104, 106, 107, 123
Risley, J. M., 65
Roberts, D. V., 177
Roberts, M. F., 345, 385, 387, 388
Robertus, J. D., 225
Robyt, J. F., 255, 274
Roche, T. E., 298, 306, 324(7, 12)
Rodbell, M., 219
Roelofsen, B., 341
Roholt, O. A., 389
Roque, P., 109
Roscelli, G. A., 10(48), 11
Rose, G., 71
Rose, I. A., 3, 10(51), 11, 20, 34, 37, 38, 48, 50, 52(3), 56, 57(12), 65, 71(10), 75(10), 109
Rosen, G., 74, 82
Rosenthal, A. F., 375
Rossi-Fanelli, A., 146
Rubery, P. H., 177
Rubin, M. M., 167, 171(53)
Rubinow, S. I., 108
Rudolph, F. B., 30, 40
Rübsamen, H., 193, 219, 220(8)
Rugh, W., 244
Rupley, J. A., 276
Russell, P. J., Jr., 8(15), 10
Ruysschaert, J. M., 345, 347(19), 349, 350(19), 365(19), 367, 375

# AUTHOR INDEX

## S

Saari, J. C., 325
Saghi, M., 297
Sakoda, M., 273
Sander, E. G., 132
Santi, D. V., 35, 125, 126, 127, 128, 129(10), 130(11), 133, 134(9)
Sarda, L., 341, 342, 358, 382, 388
Sari, H., 379, 380, 381, 382(111)
Saroff, H. A., 146, 174, 180, 181
Savary, P., 356
Sawyer, C. B., 95
Scatchard, G., 148
Schachman, H. K., 189
Schaeffer, V. F., 351
Schartz, J. H., 9(26), 11
Schimerlik, M. I., 23, 104, 106, 107, 123
Schimmel, P. R., 9(30, 31), 11, 43, 188, 189(102)
Schindler, M., 275
Schlamowitz, M., 389
Schlesinger, P. A., 174, 190(66), 191(66)
Schlessinger, J., 180
Schmitt, J. A., 83
Schnaar, R. L., 295
Schønheyder, F., 341
Schowen, R. L., 84, 109, 110
Schramm, V., 161, 162(38)
Schulman, J. H., 366, 372, 385(84), 386, 387
Schuster, I., 189, 190(105), 191(105)
Seal, G., 294
Segal, I. H., 162, 165(43), 171
Segel, I. H., 21, 180
Selegny, E., 230, 246, 247(20)
Sémériva, M., 372, 373(96), 377, 379, 380, 382(111), 387(96), 389, 390(132), 391
Seydoux, F. J., 174, 183, 184(95), 190(95)
Shah, D. O., 366, 372, 385(84), 386, 387
Shannahiff, D. H., 297
Sharon, N., 249, 260, 272(3), 275
Shaw, J.-F., 9(21), 11, 25
Shen, B. W., 347, 373(22)
Shepherd, J. B., 9(28), 11
Sherman, L. A., 291
Shibata, S., 276
Shill, J. P., 140, 143(8), 164(8), 193, 194(5), 196(5), 197, 199(14), 201(14), 203(14), 207, 208(5), 211(5), 214(5), 216(5, 14), 217(5), 219, 222(5)

Shiner, V. J., 109
Shipolini, R. A., 351
Shoemaker, D. P., 202
Shulman, R. G., 174
Silman, H. I., 246
Silverstein, E., 3, 7(3, 7), 8(1, 3, 6, 7, 8, 12), 9(20), 10, 11, 24, 25, 27, 29, 39
Simcox, P. D., 26
Simon, W. A., 9(23), 11
Simplicio, J., 190
Skarstedt, M. T., 9(20), 11, 39
Skou, J. C., 365
Slater, J. P., 282
Sleep, J. A., 74, 76, 77, 80(21), 82
Slotboom, A. J., 347, 348, 352, 364, 373(27), 374, 375(27), 386(27), 390, 391
Small, D. M., 379
Smith, D. W., 297
Smith, E., 8(17), 11
Smith, E. C., 162
Smith, F., 256
Smith, G. D., 177
Smith, L. C., 368
Smith, W. G., 9(21), 11, 25
Snyder, W. R., 347, 373(22)
Søcow, R., 377
Sommer, H., 128, 129(10)
Spradlin, J., 258, 261, 263(19), 264(19)
Springgate, C. F., 294
Stadtman, E. R., 297, 298, 300, 301, 304, 305(4), 306, 308, 309, 310, 311, 312, 313, 316, 317, 318, 319, 320, 322, 323, 324(11, 16), 325
Stanley, H. E., 174
Staverman, W. H., 91, 94(21), 110
Steitz, T. A., 180, 225
Stempel, K. E., 60
Stenkamp, R., 225
Stern, M. J., 84
Stewart, T. A., 162, 163(44)
Stokes, B. O., 5
Storer, A. C., 215, 219, 221(33)
Stoute, V., 109
Stryer, L., 59
Stull, J. T., 327
Sturgill, T., 148
Sturtevant, J. M., 191
Su, S., 8(15), 10
Suetsugu, N., 275
Suganuma, T., 269, 274(31)
Sulebele, G., 8(6, 7, 8), 10, 24

Sund, H., 28
Sundaram, P. V., 230
Sundler, R., 361
Switzer, R. L., 26
Sygusch, J., 156, 157(35), 158(35), 159(35), 162(35)
Szabo, A., 171

## T

Tabatabai, L. B., 327
Tamir, I., 282
Tate, P. W., 221
Taube, H., 111
Taylor, P., 8(10), 10
Tedesco, T. A., 10(50), 11
Teipel, J., 143, 182
Tejwani, G. A., 221
Teller, D., 335
Teng, M. H., 362
Terenna, B., 249
Theorell, H., 28
Thoma, J. A., 251, 255, 256, 257, 258, 260, 261, 262, 263, 264, 265, 266, 267, 268, 269, 270, 271, 272, 273(30), 274
Thomas, D., 230, 246, 247(20)
Thomas, K. R., 296
Thompson, C. J., 145, 156, 174
Thompson, R. C., 249
Thompson, V. W., 8(9), 10
Thomson, J. F., 110
Thornton, E. R., 111
Todhunter, J. A., 7, 9(36), 11, 13(17), 22(17), 34, 42, 43(63), 45(63)
Torgerson, E., 272
Tornheim, K., 195
Travaglini, E. C., 282
Trénel, G., 256
Trentham, D. R., 83
Troughton, J. H., 96
Truffa-Bachi, P., 191
Tseng, J. K., 9(39), 11
Tsong, T. Y., 226
Tu, J.-I., 334, 335(12)
Turner, S. R., 365
Tweedale, A., 230

## U

Uhr, M. L., 8(9), 10
Umbarger, H. E., 201, 207(18), 219

Urberg, M., 95
Uyemura, D., 291, 292

## V

Vagelos, P. R., 361
Vance, D., 219
Van Dam-Mieras, M. C. E., 348, 350, 353, 373(32), 383(32), 387(32), 390(29), 391(29)
Van Deenen, L. L. M., 344, 351, 353, 360, 363, 364, 377(49, 50), 386(11), 389
Vandermeers, A., 347, 380, 381, 382
Vandermeers-Piret, M. C., 347, 378, 380, 381, 382(25, 115)
van de Sande, H., 287
Van Etten, R. L., 65
Velick, S. F., 191
Verger, R., 341, 344, 345(13), 346, 347(20), 348, 349(13), 350, 351, 353, 355(13), 359, 361, 364, 365(20), 370, 371(33) 372, 373, 375, 376, 377, 379, 380(13), 382, 383, 384, 386(12), 387, 389, 391
Verheij, H. M., 364
Verkade, P. E., 341
Verkleij, A. J., 353
Vernon, C. A., 351
Veron, M., 191
Vidal, J. C., 342, 343, 390
Viratelle, O. M., 174
Vogel, P. C., 84
Volpenhein, R. A., 356, 357, 360, 361(54), 385(47)
Volqvartz, K., 341
Volwerk, J. J., 342, 343, 390
von der Haar, F., 295
von Hippel, P. H., 141, 297
Vournakis, J., 296

## W

Waight, R. D., 141
Wålinder, O., 35
Walker, G. J., 255, 260, 271
Walsh, D. A., 335
Walz, F. G., Jr., 249
Warner, T. G., 355
Watari, H., 146
Wataya, Y., 126, 127, 128, 132(13), 133, 135(13)
Webb, E. C., 349

Webb, J. L., 328
Webb, M. R., 83
Weber, G., 174
Webster, R. W., Jr., 35
Wedding, R. T., 34
Wedler, F. C., 5, 7, 9(34, 35), 10(46), 11, 13, 17, 26, 33, 34
Welch, G. R., 219
Welch, K. G., 325
Wells, B. D., 162, 163(44)
Wells, M. A., 256, 342, 347, 351, 354, 360, 369(28), 385(28), 386, 389
Westhead, E. W., 219
Westly, J., 219
Whitehead, E. P., 140, 164(7), 165(7), 174 212, 213
Wilkinson, G. N., 290
Willi, A. V., 83, 100(7)
Wilson, I. B., 76
Winnick, M. A., 109
Wint, S., 110
Winzor, D. J., 178, 179(84), 180, 181(84, 91)
Wirtz, K. W. A., 353
Witzel, H., 192, 193, 219, 220(8)
Wolfenden, R., 110
Wolff, J., 219
Wolfsberg, M., 83, 85, 100(2), 103(2)
Womack, F. C., 53
Wong, J. T.-F., 40
Wong, L.-J., 10(49), 11
Wood, H. G., 219

Wootton, J. C., 219
Wu, C.-W., 165, 187(49), 189(49), 190(49)
Wu, J. W., 10(50), 11
Wright, H. T., 225
Wyman, J., 140, 142, 146, 165, 167(48), 171, 174(22), 179, 181(86)

X

Xuong, N. H., 225

Y

Yagil, G., 7, 8(2), 10
Yakovlev, V. A., 214, 215, 217(27)
Yamane, T., 295
Yamashita, T., 365
Yap, W. T., 174
Yapp, C. J., 110
Yedgar, S., 362
Young, A. P., 95
Young, O. A., 9(27), 11

Z

Zatman, L. J., 36
Zervos, C., 202
Zetterqvist, Ö., 35
Zografi, G., 344, 370, 372(93), 375, 386
Zwaal, R. F. A., 341, 353
Zwilling, E., 41

## Subject Index

### A

Abortive complex, *see also* Shifted binding, nonproductive binding
  depolymerase, 273, 275–276
  isotope exchange, and, 32–33, 39–45
  isotope trapping, and, 58
Acceleration factor, polysaccharide depolymerase, 264, 271
Acetaldehyde
  deuterated form, 110
  hydrate, 110
  tris (hydroxymethyl) amino methane interaction, 22
Acetate kinase, exchange properties, 9, 39
Acetoacetate decarboxylase, isotope effect, 94–95
Acetone, deuterium isotope effect, 109
Acetylcholineesterase, immobilized
  activation energy, 243, 245
  enzyme concentration effect, 237, 238
  immobilization on nylon tubing, 240, 242
  Michaelis constant, flow rate effect, 238, 240
  temperature effect, 243–244
  Thiele function, 232, 236
Acetyl-CoA carboxylase
  cascade control, 325
  cooperativity, 218
  hysteresis, 218
Acetylene, isotope effect, 108–109
$N$-Acetylglucosamine, lysozyme interaction, 275
$N$-Acetyltryptophan nitrophenyl ester, chymotrypsin cooperativity, 220
Acid phosphatase, 76
Acid, carboxylic, isotope effect, 108–109
Actin, myosin interaction, 82
Activated ping pong mechanism, 9, 39
Activation, *see also* Cooperativity, burst phenomenon
  aspartate transcarboxylase, 177
  cascade control system, 297–325
  chemical modification, enzyme-catalyzed, 326–335
  concerted, 167–168
  cooperative, 140
  energetics, 225–226, 242–245, 385
  fractional, converter enzyme action, 301
  glass bead, 369, 389
  hysteresis, and, 196
  interface, lipase enzymes, 341–345, 384–392
  isotope trapping, and, 58
  micelle, 360
  phosphofructokinase, 221
  phospholipase, 367, 375
  phosphorylase phosphatase, 339
  protomer–oligomer equilibrium, 181
  pyruvate kinase, 218
Acylation, enzyme, 298
Acylesterase, 370
Adair model, *see also* Cooperativity
  constants, 154, 156, 159
  equation, 169
Adenosine diphosphate ribosylation, enzymatic, 297
Adenosine 5'-monophosphate, phosphorylase phosphatase inhibitor, 339
Adenosine triphosphatase
  calcium dependent, reaction pathway, 76
  mitochondrial F-1, 82
  myosin, 76, 82, 83, 247–248
  sodium, potassium-activated, 76
Adenosine 5'-triphosphate
  aspartate transcarbamylase, binding, 189
  $\gamma$-oxygen-18 labeled, 67
  regeneration system, 320
Adenylate cyclase, *see* Adenyl cyclase
Adenylate kinase
  enzyme contamination, 24
  isotope exchange properties, 8
Adenyl cyclase, 218
Adsorption effect, lipolysis, 345–353, 365, 375
Aggregation, enzyme, 387–388
Alanylproline, 225–226
Albumin, lipid interaction, 354, 377, 382
Alcohol dehydrogenase
  equilibrium perturbation, 120, 123
  half-site reactivity, 184
  isotope exchange, 8, 28–29

# SUBJECT INDEX

## A

Alcohol, equilibrium isotope effect, 108, 109, 111
Aldehyde hydrate, equilibrium isotope effect, 108, 109, 110
Aldolase
  isotope exchange, 10, 20
  reaction quenching, 57
Alkaline phosphatase
  flip–flop mechanism, 185
  negative cooperativity, 185
  phosphate–water exchange, 83
  phosphorus-31 magnetic resonance, 83
  reaction pathway, 76
Alkylation, half-site reactivity, 183
Allosterism, *see also* Site–site interaction
  cascade control systems, 298–322
  cooperativity, 139, 161–162
  isotope exchange probes, 24
  isotope trapping method, 58
Alternative substrate, 325–340
Amide hydrolysis, isotope effect, 95, 103
Amines, equilibrium isotope effect, 108, 110
Aminex A-27, nucleotide purification, 128, 134
Amino acid, equilibrium isotope effect, 108
Ammonia, equilibrium isotope effect, 108
Amphiphilic lipids, 353–355, 366
Amplification, regulatory enzyme, 295–322
Amplitude, cascade control systems, 295–322
α-Amylase
  binding energy, 260, 267
  bond cleavage frequency, 257, 269, 270
  computer model, 265–266
  cyclic dextrin, 256
  nonclassical kinetics, 269
  subsite mapping, 269, 270, 272–273, 261–269
  substrate, multimolecular reaction, 273–274
  transglycosylase activity, 269
β-Amylase
  repetitive attack, 274, 276
  subsite mapping, 258–261
γ-Amylase, *see* Glucoamylase
Anchoring site, 390
Anilino-naphthalene-sulfonate, gonadotropin binding, 181
Anomer, depolymerase binding, 275
Apparent cooperativity

bisubstrate mechanism, 162–163
concerted symmetrical model, 167–168
Arginine decarboxylase, isotope effect, 100–102
Arginine kinase, isotope exchange, 8
Arginyl residue, 297
Arginyl-tRNA synthetase, isotope exchange, 9
Aromatic amino acid synthetase, cascade control, 325
Arrhenius plot, immobilized enzyme, 244
L-Asparaginase, immobilized enzyme kinetics, 237–239
Asparagine synthetase, isotope exchange, 9
Aspartate kinase, isotope exchange, 9, 25
Aspartate residue, methylation, 297
Aspartate transcarbamylase
  activation, 177
  concerted model, 173
  isotope exchange, 10
  negative cooperativity, 189
  rapid relaxation measurement, 188–189
  sequential model, 173
Aspartic β-semialdehyde dehydrogenase, hysteresis, 215
Aspartokinase-homoserine dehydrogenase complex
  activation energetics, 225–226
  cooperativity, 191–192, 218, 222–223
  feedback inhibition, 191–192
  hysteresis, 196, 218, 222–223
ATP, *see* Adenosine 5'-triphosphate

## B

Barostat, 370
Barrier, subsite, 273–273
Bicarbonate, isotope effect study, 108, 120
Bicyclic cascade system, 310–315
Bilayer, substrate–water, 340, 359–361
Bile salt
  glass bead, interaction, 380–381
  lag period, suppression, 375
  lipase activity, effect, 345, 377, 378, 382
  micelle, 359
  surface pressure, effect, 377
  tensioactivity, 377
  zero-order trough, 377
Binary complex desorption, 50–52
Binding, *see* Cooperativity

# SUBJECT INDEX

$N,O$-Bis(trimethylsilyl)acetamide, 71–72
Bisubstrate mechanism, see also specific enzyme
  immobilized system, 241–242
  isotope exchange method, 3–46
  kinetic cooperativity, 162–164
Blood-clotting cascade, 322–323
Bond cleavage frequency
  definition, 254
  determination, 256–257
  multimolecular substrate reaction, 274
Bridge oxygen, 75
5-Bromo-2'-deoxyuridine, 126, 128, 132–133
5-Bromo-2'-deoxyuridine monophosphate, 126–128, 132–135
$p$-Bromophenacyl bromide, phospholipase modification, 387
Buffer effect, 198–199
Bulk conditions, lipolysis, 348–349
Burst phenomenon, see also Hysteresis
  coupled enzymes, 197, 198
  hexokinase, 216, 221

## C

Calcium ion
  adenosine triphophatase, 76
  lipase, interaction, 382
  phospholipase $A_2$, interaction, 363, 382, 387, 391
Carbamyl-phosphate synthetase, isotope trapping study, 58
Carbodiimide, lipase modification, 390
Carbonate, equilibrium isotope effect, 108, 110
Carbon-13 isotope effect, 86–95,
  arginine decarboxylase, 100–102
  carbon-14, relative effect, 84
  decarboxylation, 100
  equilibrium constant change, 108, 110
  methanol boiling point, 94
Carbon dioxide
  carbonic anhydrase, 120
  carbon isotope effect, 86–95
  contamination, 93, 102
  distillation, 91
  equilibrium isotope effect, 109
  measurement, 92
  nitrogen removal, 91
  ribulose-1,5-bisphosphate activation, 58
  solubility, 93–94
  water removal, 91
Carbon monoxide, isotope effect, 95
Carboxylation reaction, isotope effect, 96
Cascade control system, 297–325
Catalytic amplification, 302
Catalytic effect, 174–176
$N$-Cholylglucosamine, lipase interaction, 379
Chorismate synthase, hysteresis, 218
Chylomicron, 377
Chymotrypsin
  activation energetics, 226
  burst, 199
  conformational change, 224–225
  hysteresis, 217, 220
  negative cooperativity, 220
  nitrogen isotope effect, 95
  zymogen conformational change, 225
Citrate, hexokinase interaction, 216
  phosphofructokinase activation, 220–221
Citrate cleavage enzyme, isotope effect, 10
Classification, hysteretic effect
  cooperativity
    monomeric, 221
    oligomeric, 222
  incidental to function, 217–220
  physiological damping mechanism, 220–221
CMC, see Critical micelle concentration
Coenzyme A transferase, isotope exchange, 10
Colipase
  adsorption, 381–382
  bile salt interaction, 380
  inhibition, 378
  lipase interaction, 377, 378, 384
  role, 382
  systhesis, 378
Competitive inhibition
  detergent, 361–362
  lipase, 369, 382
  micelle, and, 364
  phospholipase $A_2$, 347
  subsite model, 252
  substrate-directed, 331
Complement fixation cascade, 323
Complex cooperativity, 140, 141, 181–185
Computer program
  equilibrium perturbation, 115–117
  hysteresis, 203–205

# SUBJECT INDEX

subsite mapping, 264–265
Concanavalin A, conformational change, 223–226
Concerted cooperativity
  assumptions, 165–166
  catalytic effect, 174–175
  comparison with other model, 171–174
  constant, 154
  equilibrium expression, 166–167
  exclusive binding model, 166, 172
  hysteresis, 214–215
  modifier action, 167
  negative, 172–175
  nonexclusive binding model, 166, 172
Condensation, depolymerase subsite mapping, 273–274
Conformation change
  activation energetics, 225–226
  calcium binding, 387
  enzyme, interfacial, 388–392
  hexokinase burst, 199
  hysteresis, 194
  isotope trapping, 58
  lipase, 344, 348
  lipid, 385, 388
  phospholipase $A_2$, 348, 387–388
  ribonuclease, 220
  specificity, reversal, 344
  structural basis, 223–225
  surface pressure, 387
Contamination
  carbon dioxide, 91–93
  isotope effect study, 91–95
  isotopically enriched compound, 94
  substrate, 25–26
Cooperativity, *see also* Concerted cooperativity, Sequential cooperativity, Hysteresis
  apparent, 162–163
  catalytic effect, 174–176
  comparative aspects, 171–174
  concerted-symmetrical model, 165–168
  constants, 153–159
  data analysis, 150–151
  definition, 139, 159–160
  equilibrium model, 141, 151–153, 159–164, 185–189
  estimation of, 142–153
  Γ coefficient, 142–143, 149, 150, 151
  heterotropic interaction, 140–141

Hill model, 144–147, 150–151
homotropic interaction, 140
hysteretic systems, 163–164, 192–226
induced-fit model, 185–189
isotope exchange, 4, 17
kinetic aspects, 141, 159–164
Kurganov method, 148
lipase–colipase system, 379
negative, 140, 181–183
nomenclature, 159–160
positive, 139, 140
protomer–oligomer equilibrium, 178–181
rate equation, 151–153
$R_s$, 147, 149, 150
$R_v$, 143, 144, 149, 150, 151
$S_{0.5}$, 146–147, 150, 152
Scatchard plot, 145, 147, 148, 150
sequential model, 168–171
site–site interactions, 164–178
Sturgill-Biltonen method, 148
Cornish-Bowden-Koshland fitting protocol, 158
Coupling enzymes, hysteresis, 197–198
Covalent interconversion, 297–340
Cratic free energy, 251–252
Creatine kinase, mechanism, 7, 23, 32–33, 39
Critical micelle concentration, 343–354
Critical packing density, lipolysis, 353, 386
Cross-linking, phospholipase $A_2$, 387
Curvature, reciprocal plot, 142–144
Cyclic AMP, positive cooperative, 180
2′,3′-Cyclic cytosine monophosphate, ribonuclease, 220
Cycling-time, perturbation method, 290, 292–293
Cyclodextrin
  glucotransferase, 255
  hexaamylose, 255
  preparation, 255–256
Cyclohexanol, equilibrium isotope effect, 108, 120, 122
Cyclohexanone, equilibrium isotope effect, 109, 120, 122
Cysteine, thymidylate synthetase mechanism, 132–133
Cysteine, formation, thymidylate synthetase, 132–135
Cytosine-5′-triphosphate, aspartate transcarbamylase inhibitor, 189

## D

Damping mechanism, hysteresis, 194–196, 217–221
Data processing, subsite model, 256–258, 264–269
Deadend branch, hysteretic effect, 206–208
Decarboxylation, isotope effect, 84–85, 87–93, 99–103, 123
Dehalogenation, thymidylate synthetase mechanism, 132–135
Dehydrogenase, see specific dehydrogenase
  equilibrium isotope effect, 106–107
  isotope exchange, 4, 8, 40–41
Denaturation, surface pressure, 365, 373
Density, substrate, 361–362, 366–377
2'-Deoxycytidylate hydroxymethyltransferase, mechanism, 127
Deoxyribonucleic acid binding protein, indefinite association, 139
2'-Deoxythymidine 5'-monophosphate purification, 128
  thymidylate synthetase interaction, 125
2'-Deoxyuridylate hydroxymethylase, mechanism, 127
2'-Deoxyuridylate hydroxymethyl transferase, mechanism, 127
$S$-[5-(2'-Deoxyuridyl)] cysteine formation, 133
Depolymerase
  isotope exchange, 24
  polysaccharide subsite mapping, 248–277
Desorption rates, isotope trapping method, 47–59
Detergent, see specific detergent
  colipase effect, 382
  competitive inhibition, 355, 361
  emulsion, 357–358
  fatty acid pK shift, 360
  glass bead adsorption 380–381
  micelle, 359–360
Deuterium equilibrium isotope effect, 106–110
Dextranase
  *Streptococcus mutans* enzyme, 260, 271
  subsite mapping, 271
  substrate, 255

Dibutyryllecithin, phospholipase $A_2$ inhibition, 347, 354
1-sn-Didecanoyllecithin, phospholipase $A_2$ substrate, 344
3-sn-Didecanoyllecithin, phospholipase $A_2$ substrate, 344
1,2-Didecanoyl-sn-glycero-3-phosphorylethanolamine, 352–353
Didodecanoylphosphatidylglycerol monolayer, lipase action, 382
Diethyl pyrocarbonate, lipase modification, 390
Diffusion controlled reaction
  immobilized enzyme, 227–237, 239–242
  lipolysis, 369, 375, 380
1,2-Diheptanoylglycerol-3-sn-phosphorylcholine, phospholipase $A_2$ substrate, 343–344
Diheptanoyllecithin, phospholipase $A_2$ substrate, 363
Dihexanoyllecithin, phospholipase $A_2$ substrate, 363, 389
5,6-Dihydrocytosine derivative, substitution reaction, 127
5,6-Dihydropyrimidine intermediate, thymidylate synthetase mechanism, 125–128, 132–135
Dihydroxyacetone phosphate, deuterium isotope effect, 108
Diisopropylethylamine, preparation, 72
Dilution, lipolysis substrate effect, 361–362
Dimethyl suberimidate, phospholipase modification, 387
Dinonanoyllecithin, phospholipase $A_2$ substrate, 343, 354, 363, 375, 389
Dipalmitoyllecithin, phospholipase $A_2$ substrate, 361
2,3-Diphosphoglycerate, hemoglobin regulation, 140
Diphosphopyridine nucleotide, see also Nicotinamide adenine dinucleotide, oxidized, isotope effect, 108–109
reduced
  equilibrium isotope effect, 108–109
  instability, 119
Dissociation constant, subsite model, 253, 276
Distributive synthesis, 279, 288
Dixon plot, 333

# SUBJECT INDEX

DNA-dependent RNA polymerase, *see* RNA polymerase
DNA polymerase I
  editing mechanism, 295
    change, 288, 292
    product distribution, 289
  fidelity, 273, 294
  general mechanism, 279
  kinetic pathway, 282
  nucleotide binding, 282
  processivity, 288–292
  selectivity, 293, 294
DNA polymerase II
  fidelity, 294
  pathway, 295
  processivity, 291
  selectivity, 294
Double-label method, isotope effect, 96–98
Double-reciprocal plot
  constant, 153
  cooperativity, 142–144, 150–152
  curvature, 142–143
  immobilized enzyme, 150–155, 236–238
  inflection, 183
DPN, *see* Diphosphopyridine nucleotide
Dual phospholipid model, lipase action, 345, 388

## E

Editing, nucleic acid polymerase, 294–296
Effector, *see also* Modifier, Activation, Inhibition
  cascade control system, 299–322
  concerted cooperativity, 168
  sequential cooperativity, 170
Elongation reaction, *see* Processivity
Emulsion
  lipase substrate, 340, 341, 348, 356–358
  stabilization, 358
Enolase, equilibrium isotope effect, 110
Enthalpy, activation, 225–226
Entropy, activation, 225–226, 385
Equilibrium isotope effect, 129–132
Equilibrium isotope exchange, *see* Isotope exchange
Equilibrium cooperativity model, 164–185
Equilibrium perturbation method,
  example, 122–125
  experimental protocol, 117–120

  interpretation, 120–122
  pH effect, 124
  theory, 104–117
Error, isotope effect, 86, 91–95
Esterase, interface activation, 341–342
Ester hydrolysis, isotope effect, 95–98, 103
17$\beta$-Estradiol dehydrogenase, isotope exchange, 8
Ethane, equilibrium isotope effect, 108
Ethanol, equilibrium isotope effect, 110
Ether, micelle inversion, and, 360
Exchange method, *see* Isotope exchange
Exodextranase substrate, 255
Exponential decay, 201–204

## F

Fatty acid synthetase, cascade control, 325
Fatty acyl-CoA synthetase, isotope exchange, 9
Feedback inhibition, *see* Cooperativity, Hysteresis
Fidelity, nucleic acid polymerases, 293–295
Flip–flop enzyme mechanism, 184, 185
Flow rate, immobilized enzyme, 233–236, 238–240
Fluorescence, phospholipase $A_2$, 390–392
5-Fluoro-2'-deoxyuridine monophosphate, thymidylate synthetase mechanism, 125–131
Fluorodinitrobenzene, half-site reactivity, 183
Flux, oxygen-18, 80–81
Fractional activation, cascade converter enzyme, 301
Fractional modification, cascade system, 299–301, 304–315
Fructokinase
  isotope exchange, 9
  isotope trapping, 52
Fructose bisphosphate, pyruvate kinase activation, 218
Fructose kinase, *see* Fructokinase
Fumarase
  equilibrium isotope effect, 110
  equilibrium perturbation, 112
  isotope exchange, 20
  mechanism, 124–125

# SUBJECT INDEX

β-(2-Furyl)acryloyl phosphate, half-site reactivity, 183

## G

Γ, immobilized enzyme parameter, 142–143, 149–151, 234
Galactokinase, isotope exchange, 8
Galactose-1-phosphate uridylyltransferase, isotope exchange, 10
β-Galactosidase, immobilized, 236, 237, 239–241
Gene-32 protein, bacteriophage, 141
Glass bead immobilization, 342, 347, 369, 375, 379, 381, 389–390
Glucoamylase
 anomeric specificity, 275
 binding energetics, 260, 261
 subsite mapping, 258–261
Glucokinase
 cooperativity, 218, 221
 hysteresis, 215, 218, 221
Glucose 6-phosphatase, isotope exchange, 9, 36–37
Glucose 1-phosphate, phosphorylase phosphatase inhibitor, 339
Glucose 6-phosphate, phosphorylase phosphatase inhibitor, 335
Glucose 6-phosphate dehydrogenase, as nonequilibrium exchange tool, 37–38
Glutamate dehydrogenase
 alanine deamination, 8
 equilibrium perturbation, 123
 Hill plot, 182
 hysteresis, 218
 isotope exchange, 8
Glutamine synthetase
 cascade control, 298, 310, 315, 322, 324
 isotope exchange, 5, 7, 9, 23, 33–34, 35
 isotope trapping, 47–48
γ-Glutamylcysteine synthetase, isotope exchange, 9
γ-Glutamyl phosphate, glutamine synthetase mechanism, 34, 47–48
Glyceraldehyde 3-phosphate dehydrogenase
 conformational change, 174, 190
 cooperativity, 170, 174, 190–191, 219
 half-site reactivity, 183
 hysteresis, 219
 negative cooperativity, 170, 174
 preparation of [γ-$^{18}$O]ATP, 67
 sequential cooperativity, 170, 174
Glycerophosphatide, hydrolysis, 366
Glycogen phosphorylase phosphatase, activation, 339
Glycogen particle, cascade control mechanism, 325
Glycogen phosphorylase
 allosteric control, 162
 alternative substrate, 327
 bicyclic cascade, 310, 325
 conformational change, 223
 debranching enzyme effect, 24
 end-labeled substrate, 255
 hysteresis, 194, 197
 isotope exchange, 9, 24
 kinetic constant determination, 215
Glycogen synthase, see Multicyclic cascade control
Glycolysis, oscillation, 195, 217–221
Gonadotropin, fluorescence titration, 181
Graphical analysis, hysteresis, 200–203
Guanosine 5'-triphosphate
 glutamate dehydrogenase regulation, 218
 hexokinase hysteresis, 216
Guggenheim method, exponential decay, 202–203

## H

Half-site reactivity, 183–185
Heavy atom isotope effect, see Isotope effect
Hemerythrin, oligomerization, 139
Hemiacetal, isotope effect, 108
Hemiketal, isotope effect, 108
Hemoglobin
 cooperativity model, 173–174
 2,3-diphosphoglycerate effect, 140
 Hill coefficient, 146
 positive cooperativity, 139
Heterotropic cooperativity, 140
Hexokinase
 conformational change, 225–226
 cooperativity, 218, 221
 half-site reactivity, 184
 hysteresis, 197, 215–216, 218, 221–222
 isomerization, 197
 isotope exchange, 4, 7, 8, 25, 29–32, 37–38
 mnemonic mechanism, 221

negative cooperativity, 216, 221–222
nucleotide site, 180
Hill cooperativity model
　bicyclic cascade control, 313
　coefficient, 144–147, 149
　concerted symmetrical model, 166–167
　curvilinear plot, 182
　interaction energy, 144–147
　limitation, 151
　oligomerization effect, 181
Homotropic cooperativity, 140
20β-Hydroxysteroid dehydrogenase, isotope exchange, 8
Hysteresis, see also Cooperativity
　activation energetics, 225–226
　analysis, 196–199
　artifact, 196–200
　classification, 217–220
　graphical analysis, 200–203
　isomerization, 194
　isotope trapping, 58
　model, 205–213
　oligomerization, 214–217
　oxygen-18 exchange, 82
　physiological role, 195–196
　statistical method, 203–205
　structural basis, 223–225

## I

Immobilized enzyme kinetics
　analysis, 236–242
　pH effect, 246–247
　principles, 228–236
　temperature effects, 242
Inactivation
　burst phenomenon, 198
　cyclic cascade control, 306, 307
　glass surface effect, 198
　interface, 347, 368
　lipase, 382
Incorporation selectivity, nucleic acid polymerase, 294
Incubation, see Preincubation
Induced-fit model, 185–189
Induction time, lipolysis, 348, 350, 378–379, 386–387
Inhibition
　alternative substrate, 326, 335
　concerted model, 167–168
　cooperative, 140, 162, 177

hysteretic effect, 195–196, 222–223
isotope exchange, 32, 34–36
product, nucleic acid polymerase, 280
protomer–oligomer equilibria, 181
substrate, 141
Inorganic pyrophosphatase
　partition coefficients, 82
　$^{18}$O-phosphate, 61, 65
　reaction pathway, 76
Interconversion, see Cascade control
Interface activation, see also Lipolysis
　adsorption effect, 345–346
　bile salt interaction, 345
　chain length effect, 344
　charge density, 386
　energy, 385
　enzyme aggregation, 387–388
　enzyme inactivation, 347, 368, 372–375
　free energy of adsorption, 365
　hydration state, 385–386
　insoluble lipid effect, 345–353
　lipolytic enzymes, 341–345, 384–392
　Michaelis constant, 356–358
　orientation effect, 357, 385
　penetration, insoluble lipid, 345–346
　size effect, 356–358, 362
　substrate concentration, 384–385
　surface dilution model, 355
　surface pressure, 386–387
Intermediate
　isotope exchange, 25–26, 34, 36–38
　isotope trapping, 47–59
　thymidylate synthetase, 125–127, 132–135
Intrinsic isotope effect, 99–100
Inverted micelle, 360, 386
Ionic strength
　isotope exchange, 27, 32
　micelle formation, 362
Irreversible enzyme, isotope exchange, 36–38
Isoamylase, 255
Isocitrate, equilibrium isotope effect, 108
Isocitrate dehydrogenase
　allosterism, 162
　equilibrium perturbation, 123
　isotope exchange, 8
Isoleucine tRNA synthetase, isotope exchange, 9
Isomerism
　activation energetics, 225–226

hysteresis, 194, 197
Isoordered mechanism, nucleic acid polymerase, 278–279
Isotope chase, see Isotope trapping
Isotope effect, see specific enzyme, see specific method
Isotope exchange, see also Oxygen-18 methods
　abortive complex interaction, 39–45
　enzyme stability, 23–24
　equilibrium
　　away from, 36–38
　　equations, 7–18
　　experimental protocol, 19–27, 31
　　initial rate method, 32–33
　　ionic strength effect, 27, 32
　　kinetic parameter, 17, 18, 25, 31–32
　　pH effect, 23
　　profile, 13, 17
　　quench, 25
　　rapid equilibrium model, 6–7, 17–18
　　side reaction effect, 24
　　substrate inhibition, 34–36
　　substrate–product ratio, 22–23
　　substrate stability, 24
　　theory, 7–18, 30, 32, 33, 44
　　types, 4–5
　example, 8–11
　substrate synergism, 38–39
Isotope partition, see Isotope trapping
Isotope trapping method, 47–59,
　complex formation, 53
　data analysis, 56–57
　extension, 57–59
　kinetic equation, 49–52
　limitation, 57–59
　mixing procedure, 53–56
　reaction condition, 56–57
　termination, 56–57

## K

$K_\psi$, 152, 153
KB cell type, processivity, 291
α-Ketoglutarate, cascade control effect, 322
Kinases, isotope exchange, 8–9
Kinetic cooperativity, see Cooperativity, see also Hysteresis
K system, allosterism, 167–168
Kuhn distribution law, 282
Kurganov cooperativity method, 148

## L

Lactate, equilibrium isotope effect, 108
Lactate dehydrogenase, isotope exchange, 8, 19–20
Lag, see Hysteresis, see also Lipolysis
Langmuir isotherm, 354, 356
Least squares method, cooperativity, 156–159
Lecithin, see specific compound
Lecithinase, see Phospholipase $A_2$
Lichrosorb $C_{18}$, nucleotide purification, 128
Ligand exclusion model, 42, 162
Lipase action, see Lipolysis
Lipid micelle, 355–365
Lipolysis, see Interface activation, see also Lipase, Phospholipase
　see specific lipid
Lipoprotein lipase, surface density, 377
Liposome, 340, 345, 348, 361
Lysolecithin
　micelle, 359
　product effect, 375
Lysozyme
　$N$-acetylglucosamine interaction, 275
　subsite binding energy, 260, 275

## M

Macroscopic parameters, subsite binding, 252–254
Malate, equilibrium isotope effect, 108–111, 122–124
Malate dehydrogenase
　equilibrium perturbation, 123
　half-site reactivity, 184
　isotope exchange, 8, 23, 25–26
Malic enzyme
　equilibrium perturbation, 104–105, 122–123
　isotope trapping, 51–52
Maltodextrin phosphorylase, isotope exchange, 9
Maltooligosaccharide, preparation, 255
Maltose, glucoamylase interaction, 259–260
Mapping, see Subsite mapping
Marquart algorithm, 157–158, 205
Mass-transfer coefficient, 234
Mesotartrate, malic enzyme inhibitor, 52
Methanol, isotope effect, 94, 108–110
Methotrexate, thymidylate synthetase, 128

Methylamine, equilibrium isotope effect, 110
5, 10-Methylenetetrahydrofolate, 126–127, 130–131
Micelle
  conformation, 385
  critical concentration, 359
  inverted, 360, 386
  ionic strength effect, 362
  lag time effect, 375
  preparation, 360, 361
  substrate, 340, 342, 347, 353–355, 358–364
  surface dilution model, 355, 361–362
  transformation, 360–364
Mnemonic enzyme system, 193, 210–212, 221, see also Hysteresis
Modifier effect
  cooperativity, 167–168
  covalent enzyme interconversion, 325–340
  hysteretic effect, 196
  protomer–oligomer equilibrium, 181
  substrate, 161–162
Monocyclic cascade control, 299–310
Monolayer
  advantage, 364–365
  enzyme inactivation, 347
  limitation, 365, 367
  lipolysis, 349–350, 366–368, 370–372, 375–377
  substrate, 340, 342–343, 364–377
Multicyclic cascade control, 315–325
Multisite cooperativity, 160–162
Myosin
  actin interaction, 82
  ATP hydrolysis, 82
  phosphorus-31 magnetic resonance, 83
  reaction pathway, 76

## N

$n_H$, see Hill coefficient
NADase, see NAD nucleosidase
NAD nucleosidase, isotope exchange, 36
Negative cooperativity, see also Cooperativity
  chymotrypsin, 220
  concerted model, 172–173
  definition, 140
  double-reciprocal plot, 142–144
  half-site reactivity, 183–185
  hexokinase, 216, 221–222
  Hill plot, 144–147, 150–153
  hysteresis, 164, 195, 196, 205, 206, 210, 212, 214
  maximal velocity determination, 145
  mixed, 181–183
  product effect, 177
  protomer–oligomer equilibrium, 180
  Scatchard plot, 147–148
  sensitivity index, 302
  sequential model, 170, 175
Nick translation, polymerase processivity, 292
Nitrogen-15 isotope effect, 108, 110
Noncompetitive inhibition, 331, 334
Nonhyperbolic behavior, see Cooperativity, see also Hysteresis
Nonprocessive synthesis, nucleic acid polymerase, 294
Nonproductive binding
  α-amylase, 269
  depolymerase, 292, 299
Northrop isotope effect method, 121–122
Nucleic acid polymerase
  fidelity, 293–295
  mechanism, 278–282
  processivity, 277–297
Nucleoside diphosphatase, complex dissociation, 161–162
Nucleoside diphosphate kinase
  contaminating exchange reaction, 24
  isotope exchange, 8, 34–36, 39
Nucleotide regenerating system, 320
Nucleotidylation, 297–298

## O

Oligomerization
  hysteretic effect, 214–217
  isotope exchange, 46
Oligosaccharide
  bond cleavage frequency, 256–257
  degradation, 271
  Michaelis parameter, 257–278
  preparation, 255–256
  radiolabel, 255–258
Ordered bi bi mechanism
  equilibrium perturbation, 105–115,
  hysteresis, 208–210
  isotope exchange, 5–7, 17–18, 22
    equation, 14
    limit, 12–13

rate profile, 18
  Wedler-Boyer protocol, 13
  Theorell-Chance mechanism, 15, 18
Ordered bi uni mechanism
  isotope exchange, 13, 14, 18, 36–37
Ordered mechanism, *see also* Specific mechanism
  hysteresis, 208–210
  isotope exchange, 5–7
  rapid equilibrium, 42–45
Organic solvent, micellization, 360, 386
Orientation effect, lipolysis, 357, 385, 388–392
Oscillatory kinetics, hysteresis, 195, 217, 220–221
Oxaloacetate, isotope exchange, 108–110
Oxygen-18 methods
  application, 75–83
  bridge position, 75
  distribution equation, 72–75
  flux calculation, 80–81
  glutamine synthetase, 5
  isotope effect, 95, 109–111
  methanol, boiling point, 94
  partition coefficient, 76–77
  pattern, 81–82
  phosphate, 68–72
  phosphate–water exchange, 77–79
  phosphorus-31 magnetic resonance, 74
  preparation, 61–62, 65–67, 71–72
  spillover correction, 73
Oxygenation, hemoglobin, 139–140, 146, 164–165

## P

Papain, nitrogen isotope effect, 95
Partial competitive inhibition, 331
Partition analysis, 50
Partition coefficient, 76–77, 99–100, 231
Partition factor, 99–100, 102
Penetration power, lipolysis, 345–348, 352, 390
Pepsin, isotope exchange, 10
Peptide initiation factor, cascade control, 325
Peptidylation, 298
Perturbation, *see* Equilibrium perturbation
pH effect
  bile salt micelle, 361
  hysteresis, 198–199, 216

immobilized enzyme, 246–247
  isotope effect, 100
  isotope exchange, 23
  phospholipase $A_2$, conformation, 391
Phenylalanine hydroxylase, cascade control, 325
Phosphatase, cascade control, 299
Phosphate, *see also* Oxygen-18 method
  molybdate complex, 69–71
  negative cooperativity, 185
Phosphatidic acid, product effect, 367, 375
Phosphatidyl choline, phospholipase C, 361
Phosphatidyl glycerol, phospholipase $A_2$, 351
Phosphatidylinositol
  phosphodiesterase, 366
  phospholipase C, 361
Phosphatidylserine decarboxylase, mixed micelle, 355
Phosphoenol pyruvate
  ATP regeneration, 320
  carboxykinase, 218
  carboxylase, 96
  equilibrium isotope effect, 108
Phosphoenzyme intermediate, 34–35, 76
Phosphofructokinase
  activation, 221
  burst analysis, 197
  cascade control, 325
  equilibrium shift, 220
  hysteresis, 195
  inactivation, 221
  physiological damping, 217, 220–221
Phosphoglucomutase, isotope exchange, 10
2-Phosphoglycerate, equilibrium isotope effect, 108
Phosphoglycerate kinase, [$\gamma$-$^{18}$O]ATP preparation, 67
Phospholipase $A_2$
  adsorption, 347–348
  aggregation, 387–388
  anchoring site, 390
  bilayer, 355–360
  chemical modification, 387, 390
  competitive inhibition, 347, 354
  conformational change, 348, 387–389
  cross-linking, 387
  dibutyryllecithin monomers, inhibition, 347, 354
  diffusion constant, 351
  dimerization, 387

dimethylsuberimidate modification, 387
dual-phospholipid model, 388
ether, contamination, 360, 386
interface activation, 343–344
interface recognition site, 390
lag period, product effect, 375
micelle interaction, 355, 360, 361, 363–367
penetration, 390
pressure effect, 352–353, 366–367, 373–374
presteady state, 350–353
single-site model, 354
supersubstrate binding site, 389
tryptic cleavage, 391
ultraviolet difference spectra, 390–392
Phospholipase C
  mixed micelle, 355, 361
  monolayer, 367
  specificity, 361
  Triton X-100, 361
Phospholipase D
  lag period, 367, 375
  monolayer action, 367
  product effect, 367, 375
Phosphoprotein phosphatase
  alternative substrate method, 327
  cascade control, 299
Phosphoribosylpyrophosphate synthetase, isotope exchange
Phosphorus-31 magnetic resonance, see Nuclear magnetic resonance
Phosphorylase, see Glycogen phosphorylase
Phosphorylase phosphatase
  activation, 339
  adenosine monophosphate, inhibition, 339
  alternative substrate effect, 336–339
  glucose 1-phosphate, inhibition, 339
  magnesium ion effect, 339
Phosphorylation
  alternative substrate effect, 327
  glycogen phosphorylase, 297, 298
  monocyclic cascade model, 299–300
  pyruvate dehydrogenase, 298
  tyrosine aminotransferase, 298
Phosphoserine phosphatase, isotope exchange, 9
Physiological damping, see Damping
Ping pong mechanism
  activated, 39

isotope exchange behavior, 16, 18, 21, 25–26, 39
substrate synergism, 38–39
Poisson distribution, nucleic acid polymerase, 283, 288
Positive cooperativity, see Cooperativity
Preincubation effect, enzyme activity, 198–199
Premicelle, formation, 379
Pressure, surface effect on lipolysis, 347–373
Presteady state condition, lipolysis, 350–353, 375
Primary kinetic isotope effect, see Isotope effect
Processivity
  change, 288, 292
  cycling-time method, 291, 292
  example, 291–292
  product distribution, 277–297
  strict, 279
  theory, 278–287
Product
  activation, 317, 375
  accumulation, hysteresis, 197–198
  concentration, 238–240
  cooperativity, 176–178
  inhibition, 195–196, 280
  length distribution, depolymerase, 271, 282–284, 286, 288–290
Progress curve
  burst, 20, 203, 215
  hysteresis, 197, 200–204
Proofreading, DNA polymerase I, 295
Propane, equilibrium isotope effect, 108
Protein kinase
  alternative substrate effect, 327
  cyclic AMP-dependent, 180, 327
  monocyclic cascade control, 299
  specificity, 326
Pullanase, substrate, 255
Pyrimidine methylase, mechanism, 127
Pyrophosphate
  inhibition, nucleic acid polymerase, 280
  oxygen-18 labeling, 66
Pyruvate, equilibrium isotope effect, 108–110
Pyruvate carboxylase, hysteresis, 218
Pyruvate dehydrogenase, cascade control, 298, 306, 324–325
Pyruvate kinase
  activation, 218

ATP regeneration, 320
cascade control, 325
cooperativity, 218
hysteresis, 218
isotope trapping, 52

## R

Random mechanism
concerted binding model, 175–176
isotope exchange, 6–7, 9–18, 42–45
kinetic cooperativity, 162–164
negative cooperativity, 162–163
rapid equilibrium model, 6–7, 42–45
Rapid equilibrium mechanism
chemical modification, enzyme catalyzed, 328
cooperativity, 159, 161, 163, 210–211
cyclic cascade model, 303
hysteresis, 201, 213
micelle lipolysis, 354, 359
protomer–oligomer interconversion, 178
sequential model, 170
Rapid relaxation technique, conformational change, 185–189
Rate determining step
aspartokinase, 226
conformational transition, 226
isotope exchange, 4, 7, 17, 32
lipolysis, 345, 352, 367
phospholipase C, monolayer, 367
secondary isotope effect, 126, 128, 132, 135
Reacting enzyme sedimentation, 215, 218
Repetitive attack, depolymerase, 274
Reverse transcriptase
fidelity, 294
processivity, 291, 296
selectivity, 294
Rhodanese
cooperativity, 218
hysteresis, 218
Ribitol dehydrogenase, abortive complex, 40
Ribonuclease
conformational change, 220, 223, 225–226
cooperativity, 217, 220
hysteresis, 217, 220
proline isomerization, 223
Ribose phosphate pyrophosphokinase, *see*
Phosphoribosylpyrophosphate synthetase
Ribosylation, *see* ADP ribosylation
Ribulose 1,5-bisphosphate carboxylase, 58, 95
RNA polymerase
cascade control, 325
DNA-dependent system, 281
editing mechanism, 296–297
*Escherichia coli* system, 291
rate expression, 282
$R_s$, definition, 147, 149–150
$R_v$, definition, 143–144, 149–151

## S

$S_{0.5}$, definition, 146–147, 150–155
Salt effect, isotope exchange, 32–33
Scatchard plot, 147–148, 151, 155
constant, 154–155
Hill coefficient, 148–149
maximal velocity, 145
Schiff's base, isotope effect, 102
Secondary isotope effect
dehalogenation, 132–135
equilibrium perturbation, 122
method, 128–129, 130–134
rate determining step, 126, 128, 132
thymidylate synthetase, 125–135
Selectivity, nucleic acid polymerase, 293–295
Sensitivity, cyclic cascade
bicyclic system, 314–315
definition, 302–303
Hill number, relation, 301–303
plot, 301
simulation, 306–308
Sequential model, cooperativity model, 168–171
Serine residue
chemical modification, 340
phosphorylation, 297
Serum albumin, protective effect, 198
Shifted binding, depolymerases, 273
Shikimate dehydrogenase, isotope exchange, 8
Short-chain fatty acyl-CoA synthetase, isotope exchange, 9
Side reactions
depolymerase, 273–274
isotope exchange, 24

# SUBJECT INDEX

Signal amplification, cascade control, 300–318
Site–site interaction, see Cooperativity
Sodium-Potassium adenosine triphosphatase, mechanism, 76
Solid-matrix enzymes, 228–245
Sorbitol dehydrogenase, use in equilibrium perturbation, 120
Statistical fitting
 burst data, 203–205
 cooperativity data, 156–159
 Cornish-Bowden protocol, 158
 hysteresis, 203–205
 Marquardt algorithm, 157–158
Strict processivity, 279
Stugill-Biltonen method, cooperativity, 148
Subsite mapping
 assumption, 275–276
 binding energetics, 251–252, 275
 complex formation, 275–276
 data collection, 256–258
 dextranases, 271
 equations, 249–254
 examples, 258–273
 features, 272–273
 limitations, 273–276
 Michaelis constant, 276–277
 oligosaccharide degradation, 271
 polysaccharide depolymerases, 248–277
 practical aspect, 254–256
Substrate
 activation, 330–331
 aggregation effect, 341–345, 346, 384–392
 alternative, 325–340, 342–343
 analogs, cooperativity, 176–178
 density, 361–362
 dilution effect, 361
 emulsion, 340, 342
 inhibition, 34–36, 141–142, 163, 175, 206
 isotope exchange, 24, 26
 subsite binding, 255–256
 synergism quotient, 38–39
Succinyl-CoA synthetase
 isotope exchange, 9
 substrate synergism, 38
Sucrose phosphorylase, isotope exchange, 9
Surface dilution model, lipolysis, 355, 361
Surfactant
 competitive inhibition, 355
 conformational effect, 387–388
 mixed micelle formation, 355
Synergism, substrate, 38–39
Synthetases, isotope exchange, 9

## T

Taurodeoxycholate, lipase effect, 379–382
Teflon, protein adsorption, 365
Template challenge method, 291–293
Tensioactive compounds, 377–378
Ternary complex desorption, isotope trapping, 52
$N$-Tetradecylphosphorylcholine, micellization, 364
Theorell-Chance mechanism, isotope exchange, 15, 18
Thiele function, 231–232, 236
Threonine deaminase, hysteresis, 219
Thymidine kinase, labeled nucleotide preparation, 128
Thymidylate synthetase
 cystine formation, 132–135
 dehalogenation activity, 132–135
 intermediate, 127, 129
 mechanism, 126–129, 132, 133
 preparation, 128
Transition probability function, 77
Transition state, isotope effect, 98–100
Triton X-100, lipolysis, 359–362, 385
Trough, controlled surface density, 350–377
Trypsin, immobilized, 244
Tyrosine nucleotidylation, 297

## U

Ultraviolet-difference spectra, phospholipase, 390–392
Uncompetitive inhibition, 331
Unidirectional cascade reaction, 322–323
Urease, pH optimum shift, 247

## V

Velocity
 subsite model, definition, 252–253
 lipolysis, surface pressure effect, 372–375
$v_{n\,max}$, 152–153
$V_{system}$, cooperativity, 168

## W

Water, *see* Oxygen-18 method
Wedler-Boyer protocol, isotope exchange, 13, 26–27, 33–34

## Z

Zeta potential, monolayer, 366